AF677983

Zinc in Health and Disease Conditions

Zinc in Health and Disease Conditions

Guest Editor

Xiang-Ping Chu

MDPI

Basel • Beijing • Wuhan • Barcelona • Belgrade • Novi Sad • Cluj • Manchester

Guest Editor
Xiang-Ping Chu
Department of Biomedical
Sciences
University of
Missouri-Kansas City
Kansas City
USA

Editorial Office
MDPI AG
Grosspeteranlage 5
4052 Basel, Switzerland

This is a reprint of the Special Issue, published open access by the journal *Biomolecules* (ISSN 2218-273X), freely accessible at: https://www.mdpi.com/journal/biomolecules/special_issues/Disease_Conditions.

For citation purposes, cite each article independently as indicated on the article page online and as indicated below:

Lastname, A.A.; Lastname, B.B. Article Title. *Journal Name* **Year**, *Volume Number*, Page Range.

ISBN 978-3-7258-4143-1 (Hbk)
ISBN 978-3-7258-4144-8 (PDF)
https://doi.org/10.3390/books978-3-7258-4144-8

Contents

About the Editor . . . vii

Preface . . . ix

Ken Ichiro Tanaka, Sachie Shiota, Okina Sakakibara, Mikako Shimoda, Ayaka Takafuji, Misaki Takabatake, et al.
Exacerbation of Elastase-Induced Emphysema via Increased Oxidative Stress in Metallothionein-Knockout Mice
Reprinted from: *Biomolecules* **2022**, *12*, 583, https://doi.org/10.3390/biom12040583 . . . 1

Ewelina A. Dziedzic, Jakub S. Gąsior, Agnieszka Tuzimek, Justyna Paleczny, Mirosław Kwaśny, Marek Dąbrowski and Piotr Jankowski
No Association of Hair Zinc Concentration with Coronary Artery Disease Severity and No Relation with Acute Coronary Syndromes
Reprinted from: *Biomolecules* **2022**, *12*, 862, https://doi.org/10.3390/biom12070862 . . . 14

Shunfeng Liu, Nan Wang, Yaqian Long, Zhuan Wu and Shouhong Zhou
Zinc Homeostasis: An Emerging Therapeutic Target for Neuroinflammation Related Diseases
Reprinted from: *Biomolecules* **2023**, *13*, 416, https://doi.org/10.3390/biom13030416 . . . 26

Amber W. Sun, Michelle H. Wu, Madhumathi Vijayalingam, Michael J. Wacker and Xiang-Ping Chu
The Role of Zinc in Modulating Acid-Sensing Ion Channel Function
Reprinted from: *Biomolecules* **2023**, *13*, 229, https://doi.org/10.3390/biom13020229 . . . 41

Zhe Li, Yang Liu, Ruixue Wei, V. Wee Yong and Mengzhou Xue
The Important Role of Zinc in Neurological Diseases
Reprinted from: *Biomolecules* **2023**, *13*, 28, https://doi.org/10.3390/biom13010028 . . . 65

Martina Maywald and Lothar Rink
Zinc in Human Health and Infectious Diseases
Reprinted from: *Biomolecules* **2022**, *12*, 1748, https://doi.org/10.3390/biom12121748 . . . 80

Mellisa Renteria, Ofek Belkin, Justin Aickareth, David Jang, Majd Hawwar and Jun Zhang
Zinc's Association with the CmPn/CmP Signaling Network in Breast Cancer Tumorigenesis
Reprinted from: *Biomolecules* **2022**, *12*, 1672, https://doi.org/10.3390/biom12111672 . . . 109

Tong Shen, Qing Zhao, Yumin Luo and Tao Wang
Investigating the Role of Zinc in Atherosclerosis: A Review
Reprinted from: *Biomolecules* **2022**, *12*, 1358, https://doi.org/10.3390/biom12101358 . . . 122

Ruize Sun, Jue Wang, Juan Feng and Bin Cao
Zinc in Cognitive Impairment and Aging
Reprinted from: *Biomolecules* **2022**, *12*, 1000, https://doi.org/10.3390/biom12071000 . . . 147

Yan Wan and Bingkun Zhang
The Impact of Zinc and Zinc Homeostasis on the Intestinal Mucosal Barrier and Intestinal Diseases
Reprinted from: *Biomolecules* **2022**, *12*, 900, https://doi.org/10.3390/biom12070900 . . . 178

Hui-Liang Zhang, Xiao-Chuan Wang and Rong Liu
Zinc in Regulating Protein Kinases and Phosphatases in Neurodegenerative Diseases
Reprinted from: *Biomolecules* **2022**, *12*, 785, https://doi.org/10.3390/biom12060785 . . . 193

About the Editor

Xiang-Ping Chu

Dr. Chu is a distinguished neuroscientist known for his research on ion channels, membrane receptors, and their roles in neurological and psychological disorders. He earned his MD from Jiangsu University in 1989 and a PhD from Fudan University in 1999, both in the People's Republic of China. In 2000, he began postdoctoral training at the Legacy Research Institute in Portland, Oregon, where he received an American Heart Association (AHA) fellowship in 2002 and rose to Assistant Scientist. In 2008, Dr. Chu joined the University of Missouri–Kansas City (UMKC) School of Medicine as a tenure-track Assistant Professor. He established an advanced electrophysiology lab and was promoted to Associate Professor with tenure in 2014. His NIH- and AHA-funded research focuses on cellular and molecular mechanisms underlying stroke, addiction, and other brain disorders. Dr. Chu has published over 100 peer-reviewed papers and five book chapters in leading journals such as *Cell*, *The Lancet*, *Neuron*, *Nature Neuroscience*, and *PNAS*. His H-index is 38. As a dedicated educator, he teaches medical neuroscience and physiology for medical students and serves on the doctoral faculty. He received UMKC's Provost's Award for Excellence in Teaching (2025) and the Betty M. Drees Excellence in Mentoring Award (2024). He reviews over 80 journals and serves on editorial boards for *Frontiers in Neuroscience*, *Biomolecules*, and *Scientific Reports*. He also reviews grants for major funding agencies including the AHA, the National Natural Science Foundation of China, the UK's Medical Research Council, and others. Dr. Chu's work continues to impact on the understanding and treatment of neurological diseases.

Preface

Zinc, the second most abundant trace element in the human body, is indispensable for a wide range of essential biological functions. Far beyond serving as a cofactor or structural component of enzymes and proteins, zinc acts as a pivotal signaling molecule that regulates cellular activities and systemic processes. Its role spans across diverse domains, including immune response, neural function, redox biology, and cellular signaling. As a cofactor for over 300 enzymes and a key player in transcription regulation through zinc-finger proteins, zinc maintains cellular homeostasis and plays critical roles in maintaining overall health.

The growing body of research has shed light on the importance of zinc homeostasis, the tight regulation of zinc levels across tissues and cellular compartments, in both maintaining physiological balance and preventing disease. Disruptions in the zinc balance—whether through deficiency, overload, or impaired regulation—have been linked to a variety of diseases, including neurodegenerative disorders, immune dysfunctions, cancer, and metabolic diseases.

One of the most compelling areas of zinc research lies in its involvement in the nervous system, where it modulates neuronal excitability and synaptic plasticity. Its dual role in both protecting and potentially contributing to disease underscores its complexity: while zinc can have a protective function in neurodegenerative diseases, excess zinc can, in certain contexts, exacerbate neuronal damage, such as in the case of seizures or ischemia. This duality is also observed in other systems such as the immune response, where zinc's deficiency impairs immune function, while excess zinc can suppress immune activity.

Zinc's involvement in oxidative stress and inflammation has made it a key focus in the study of chronic diseases like diabetes, cardiovascular disease, and cancer. Through its antioxidant properties, zinc protects cells from oxidative damage, yet its dysregulation may contribute to disease progression, as seen in some cancers, where altered zinc transport mechanisms promote tumorigenesis.

The motivation behind this Special Issue is to bring together emerging perspectives on zinc's complex biological roles, with a focus on its physiological significance and its involvement in disease mechanisms. By exploring zinc from both basic science and clinical perspectives, the aim is to offer new insights into how zinc functions as a double-edged sword in both health and disease. This Special Issue will span contributions from cutting-edge mechanistic studies, clinical research, and translational applications, helping to bridge the gap between laboratory research and clinical practice.

The Special Issue is intended for a diverse audience, including researchers, clinicians, and students who are interested in the molecular mechanisms of zinc and its implications for human health. We hope that the articles herein will not only advance the understanding of zinc's dual roles but also foster discussions on potential therapeutic applications. Zinc-related therapies are already being explored in clinical settings, and ongoing research is likely to unveil new avenues for intervention in disease prevention and treatment.

We would like to express our sincere gratitude to the authors who have contributed their expertise to this Special Issue. Their work, alongside the support of colleagues and reviewers, has been instrumental in bringing this Special Issue to fruition. We also acknowledge the researchers and funding organizations whose ongoing work continues to expand our understanding of zinc's significance in both health and disease.

In conclusion, we hope that this Special Issue will serve as a valuable resource for advancing research and clinical practice, helping to unlock the full therapeutic potential of zinc in a variety of medical conditions. Its complex nature, as both a nutrient and signaling molecule, makes it an essential target for continued exploration and innovation in biomedical sciences.

Xiang-Ping Chu
Guest Editor

biomolecules

MDPI

Article

Exacerbation of Elastase-Induced Emphysema via Increased Oxidative Stress in Metallothionein-Knockout Mice

Ken Ichiro Tanaka [1,*], Sachie Shiota [1], Okina Sakakibara [1], Mikako Shimoda [1], Ayaka Takafuji [1], Misaki Takabatake [1], Yoshito Kadota [2], Takashige Kawakami [2], Shinya Suzuki [2] and Masahiro Kawahara [1]

[1] Laboratory of Bio-Analytical Chemistry, Research Institute of Pharmaceutical Sciences, Faculty of Pharmacy, Musashino University, 1-1-20 Shinmachi, Nishitokyo 202-8585, Japan; s1743142@stu.musashino-u.ac.jp (S.S.); s1743037@stu.musashino-u.ac.jp (O.S.); s1343142@stu.musashino-u.ac.jp (M.S.); s1243088@stu.musashino-u.ac.jp (A.T.); s1443123@stu.musashino-u.ac.jp (M.T.); makawa@musashino-u.ac.jp (M.K.)

[2] Faculty of Pharmaceutical Sciences, Tokushima Bunri University, Tokushima 770-8514, Japan; ykadota@ph.bunri-u.ac.jp (Y.K.); tkawakami@ph.bunri-u.ac.jp (T.K.); suzukis@ph.bunri-u.ac.jp (S.S.)

* Correspondence: k-tana@musashino-u.ac.jp

Abstract: Although the pathogenesis of chronic obstructive pulmonary disease (COPD) is not yet fully understood, recent studies suggest that the disruption of the intracellular balance of oxidative (such as reactive oxygen species (ROS)) and antioxidant molecules plays an important role in COPD development and progression. Metallothionein is an endogenous metal-binding protein with reported ROS scavenging activity. Although there have been many publications on the protective effects of metallothionein in the kidney and liver, its role in COPD models such as elastase- or cigarette smoke (CS)-induced lung injury is unknown. Thus, in the present study, we analyzed the elastase-induced lung injury model using metallothionein-knockout (MT-KO; MT-1 and -2 gene deletion) mice. The expression of MT-1 and MT-2 in the lungs of MT-KO mice was markedly lower compared with that in the lungs of wildtype (WT) mice. Porcine pancreatic elastase (PPE)-induced lung injury (alveolar enlargement and respiratory impairment) was significantly exacerbated in MT-KO mice compared with WT mice. Additionally, PPE-induced increases in the number of inflammatory cells, inflammatory cytokines, and cell death in lung tissue were significantly more pronounced in MT-KO mice compared with WT mice. Finally, using an in vivo imaging system, we also found that PPE-induced ROS production in the lungs was enhanced in MT-KO mice compared with WT mice. These results suggest that metallothionein may act as an inhibitor against elastase-induced lung injury by suppressing ROS production. These results suggest that metallothionein protein, or compounds that can induce metallothionein, could be useful in the treatment of COPD.

Keywords: COPD; oxidative stress; metallothionein; ROS; in vivo imaging system; elastase

1. Introduction

Chronic obstructive pulmonary disease (COPD) is a serious health problem in which progressive and irreversible airflow limitations are observed, and it is caused by an abnormal inflammatory response to cigarette smoke (CS) and air pollution. Long-acting β_2-agonists (LABAs) and muscarinic receptor antagonists (LAMAs) are used clinically for bronchodilation in the treatment of COPD [1,2]. Although steroid therapy is commonly used to treat inflammation associated with COPD, there are some cases in which steroid therapy does not sufficiently suppress COPD disease progression because of resistance to steroid therapy in part of the inflammatory processes [1,3]. Therefore, it is important to discover new proteins that inhibit the onset and exacerbation of COPD and to develop new types of anti-inflammatory compounds for treating COPD.

While the pathogenesis of COPD is not yet fully understood, recent studies suggest that the cellular redox state, determined by the balance between oxidative molecules such as reactive oxygen species (ROS) and antioxidant molecules, plays an important role in this process [4]. In fact, it has been confirmed that oxidative stress is higher in the bronchoalveolar lavage fluid (BALF), sputum, and exhaled breath condensate of COPD patients compared with healthy subjects [5,6]. Moreover, in addition to the fact that CS contains high concentrations of ROS, leukocytes, particularly neutrophils, activated by CS produce excess ROS, such as superoxide anion, via nicotinamide adenine dinucleotide phosphate (NADPH) oxidase [7,8]. Thus, antioxidant molecules have attracted attention as candidates for COPD treatment, and their effects have been confirmed in COPD animal models induced by elastase or CS [9,10].

Metallothioneins are intracellular cysteine-rich proteins that maintain metal homeostasis. They were first isolated from horse kidney as cadmium-binding proteins in 1957. To date, four metallothionein isoforms (MT-1, -2, -3, and -4) have been identified in mammals. Among these metallothionein isoforms, MT-1 and MT-2 in particular have been found to be expressed in most organs [11]. Metallothionein can be induced not only by heavy metals, such as cadmium, but also by less biologically toxic metals, such as zinc (Zn) [12]. In addition to detoxifying heavy metals such as cadmium by binding to them, metallothionein also exerts antioxidant effects by binding ROS to their own thiol groups, thus exerting protective effects in disease models in various organs [13]. Many studies have shown a protective role for metallothionein in the kidney and liver; for example, studies using metallothionein-knockout (MT-KO) mice have shown that metallothioneins have protective effects against hypoxia-induced renal injury and drug-induced liver injury [14–16]. In contrast, there are few reports describing metallothionein protection in the lungs. Specifically, previous studies using MT-KO mice showed that metallothionein inhibits acute lung injury caused by paraquat or nickel, but similar protection has not yet been investigated in a COPD model [17,18]. We speculate that metallothionein proteins, or compounds that can induce metallothioneins, may be useful in the treatment of COPD.

Polaprezinc, which is used clinically as an anti-ulcer drug, is a chelate compound consisting of carnosine and Zn [19]. We previously conducted experiments with the hope that polaprezinc would exert a cytoprotective effect by inducing metallothionein proteins. As a result, we found that polaprezinc clearly inhibits lung epithelial cell death induced by cadmium, a toxic metal contained in CS, by suppressing ROS production via metallothionein induction [20]. However, no studies to analyze the efficacy of metallothionein in the COPD models of porcine pancreatic elastase (PPE)-induced or CS-induced lung injury have been conducted using genetically modified mice. Here, we analyzed PPE-induced lung injury in MT-KO mice and their WT littermates. Furthermore, we also analyzed the PPE-induced ROS production in both types of mice using an in vivo imaging system.

2. Materials and Methods

2.1. Chemicals and Animals

Diff-Quik staining solution was purchased from Sysmex (Kobe, Japan). Luminal-based chemiluminescent probe (L-012) and isoflurane were obtained from Fujifilm Wako Pure Chemical Corporation (Tokyo, Japan). Novo-Heparin (5000 units), suitable for injection, was purchased from Mochida Pharmaceutical (Tokyo, Japan). The RNeasy® kit was obtained from Qiagen (Hilden, Germany), PrimeScript™ II 1st strand cDNA Synthesis kit was obtained from Takara Bio (Shiga, Japan), and THUNDERBIRD® NEXT SYBR qPCR Mix was obtained from Toyobo (Osaka, Japan). MT-KO mice (129/Sv-MT1MT2^{tm1Bri}) developed by Masters et al. [21] and 129/SvCPJ (WT) mice were purchased from Jackson Laboratory (Bar Harbor, ME, USA). To maintain the same genetic background, only mice of the 129/SvCPJ strain were used for mating. Mice were housed in a controlled room (temperature, 22 °C; humidity, 55% ± 5%) under a 12-h light/dark cycle, fed a controlled diet (MF, Oriental Yeast Co., Ltd., Tokyo, Japan), and given ad libitum water. Mice aged 8–10 weeks were used in the experiments. The sex of the mice did not correlate with the extent of PPE-induced lung injury (Supplementary Figure S1). Since the number of mice born is not large, both male and female mice were used in this study. The experiments and procedures described here were carried out in accordance with the Guide for the Care and Use of Laboratory Animals as adopted and promulgated by the National Institutes of Health (Bethesda, MD, USA) and were approved by the Animal Care Committee of Musashino University.

2.2. Preparation of BALF

BALF was collected by cannulating the trachea and lavaging the lung twice with 1 mL of sterile saline containing 50 units/mL heparin. Approximately 1.8 mL of BALF was routinely recovered from each mouse, and the total cell number in the BALF was counted using a hemocytometer. After centrifugation with a Cytospin4® (Thermo Fisher Scientific, Waltham, MA, USA), the cells were stained with Diff-Quik reagents, and the ratio of neutrophils or macrophages to the total cell number was determined. The amount of protein present in the BALF was evaluated by using the Bradford method (Takara Bio).

2.3. Measurement of ROS by In Vivo Imaging Analysis

In vivo imaging of ROS in mice was performed as previously described [22,23]. We used the FUSION chemiluminescence imaging system (Vilber Lourmat, Collégien, France). The ROS-sensing chemiluminescent probe, L-012, in sterile saline (75 mg/kg) was intraperitoneally administered to mice 24 h after the administration of PPE. The mice were euthanized 15 min after the L-012 administration, and their lungs were promptly dissected and imaged (5-min exposure). All data were analyzed using the FUSION chemiluminescence imaging system software.

2.4. Histological and Immunohistochemical Analyses

Lung tissue samples were fixed in 10% formalin neutral buffer solution for 24 h at a pressure of 25 cm H_2O and then embedded in paraffin before being cut into 4-μm-thick sections. For histological examination, sections were stained first with Mayer's hematoxylin and then with 1% eosin alcohol solution (H&E staining). Samples were mounted with malinol and inspected with the aid of an Olympus BX51 microscope (Tokyo, Japan).

To determine the mean linear intercept (MLI; an indicator of airspace enlargement), 20 lines (800 μm) were drawn randomly on the image of an H&E-stained section, and the intersection points with the alveolar walls were counted. For example, if there are 10 intersection points with the alveolar walls on an 800-μm line, the MLI = 80 μm. Twenty MLI points were measured for each tissue section, and the mean values are shown in figure The morphometric analysis at the light microscopic level was conducted by an investigator blinded to the study protocol.

For the immunohistochemical analysis of 8-OHdG, sections were first incubated with proteinase K (20 μg/mL) for 15 min at 37 °C, then with 0.3% hydrogen peroxide for 5 min, and finally with 2.5% goat serum for 10 min. They were then incubated with an antibody against 8-OHdG (1:200 dilution) for 12 h, followed by an incubation with a peroxidase-labeled polymer conjugated to goat anti-mouse immunoglobulin for 1 h. After 3,3′-diaminobenzidine was applied to the sections, they were incubated with Mayer's hematoxylin. Finally, the sections were mounted in malinol and examined using a microscope. Image J software was used to calculate the 8-OHdG-positive area as a percentage of the total.

For terminal deoxynucleotidyl transferase biotin-dUTP nick end labeling (TUNEL) assays, sections were first incubated with proteinase K (20 μg/mL) for 15 min at 37 °C, then with terminal deoxynucleotidyl transferase and biotin 14-ATP for 1 h at 37 °C, and finally with Alexa Fluor 488 conjugated to streptavidin and 4′,6-diamidino-2-phenylindole (DAPI; 5 μg/mL) for 2 h. The sections then were mounted in Vectashield and examined using a fluorescence microscope. Image J software was used to count the number of TUNEL-positive cells.

2.5. Real-Time Reverse Transcription (RT) Polymerase Chain Reaction (PCR) Analysis

Total RNA was extracted from lung tissue using an RNeasy kit in accordance with the manufacturer's protocol. Samples were reverse-transcribed using the PrimeScript™ II 1st strand cDNA Synthesis kit. The synthesized cDNA was used in real-time PCR experiments with THUNDERBIRD® Next SYBR qPCR Mix and analyzed with a Bio-Rad (Hercules, CA, USA) CFX96™ real-time system and CFX Manager™ software. Specificity was confirmed by an electrophoretic analysis of the reaction products and the inclusion of template- or reverse transcriptase-free controls. To normalize the amount of total RNA present in each reaction, glyceraldehyde-3-phosphate dehydrogenase (Gapdh) cDNA was used as an internal standard. Primers were designed using the Primer-BLAST website (https://www.ncbi.nlm.nih.gov/tools/primer-blast/, accessed on 25 January 2022). Primer sequences will be provided upon request.

2.6. Treatment of Mice with PPE

After anesthetization with isoflurane, mice were intratracheally administered PPE (10 U/kg) in sterile saline using a P200 micropipette. During PPE administration, the mouse's nostrils were blocked so that the solutions were inhaled from the mouth into the respiratory tract as the animal breathed. The control group received sterile saline intratracheally.

2.7. Analyses of Lung Mechanics and Respiratory Functions

The measurement of lung mechanics was performed with a computer-controlled small-animal ventilator (FlexiVent, SCIREQ, Montreal, QC, Canada) [24,25]. After mice were anesthetized with a triple anesthetic combination (0.75 mg/kg medetomidine, 4 mg/kg midazolam, and 5 mg/kg butorphanol), a tracheotomy was performed, and an 8-mm section of metallic tube was inserted into the trachea. Mice were then mechanically ventilated at a rate of 150 breaths/min, using a tidal volume of 8.7 mL/kg and a positive end-expiratory pressure of 2–3 cm H_2O. Total lung capacity (TLC), total respiratory system elastance, tissue elastance and tissue damping were determined using FlexiVent software (version 5.3; SCIREQ, Montreal, QC, Canada).

Determination of the forced expiratory volume in 0.05 s (FEV0.05)/forced vital capacity (FVC) or forced expiratory volume in 0.1 s (FEV0.1)/FVC ratio was performed with the same computer-controlled small-animal ventilator connected to a negative pressure reservoir (SCIREQ) [24,25]. Mice were tracheotomized and ventilated as described above. The lungs were inflated to a pressure of 30 cm H_2O over 1 s and held at this pressure. After 0.2 s, the pinch valve (connected to the ventilator) was closed, and after 0.3 s, the shutter valve (connected to the negative pressure reservoir) was opened for exposure of the lungs to the negative pressure. The negative pressure was held for 1.5 s to ensure complete expiration. The FEV/FVC was determined using FlexiVent software.

2.8. Statistical Analysis

All values are expressed as the mean $\pm$ S.E.M. A two-way ANOVA followed by the Tukey test or the Student's *t*-test for unpaired results was used to evaluate differences between three or more groups or between two groups, respectively. Differences were considered to be significant for values of $p < 0.05$.

3. Results

3.1. Expression of Metallothionein in Lung Tissues

Lung tissues were harvested from wildtype (WT) mice and MT-KO mice, and the expression of Mt-1 and Mt-2 in the harvested lung tissues was examined by real-time RT-PCR. As shown in Figure 1, the expression of Mt-1 and Mt-2 in the excised lung tissue was significantly lower in the MT-KO mice. Thus, we next examined the effect of metallothionein on elastase-induced lung injury in these two types of mice.

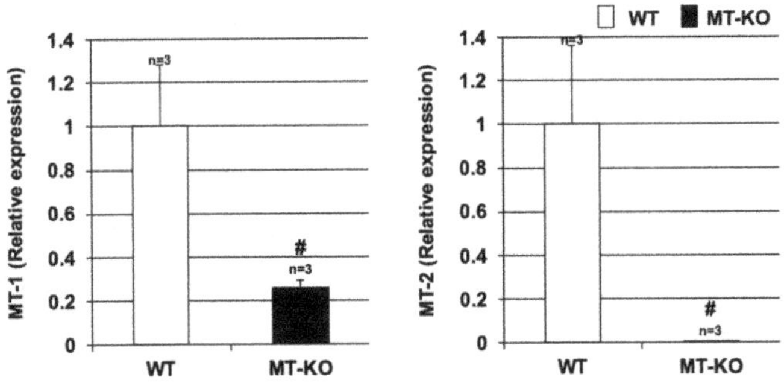

Figure 1. Expression of metallothionein in lung tissues. Lung tissues were harvested from wildtype (WT) or MT-1 and -2 gene knockout (MT-KO) mice, and total RNA was extracted from these tissues. The resulting samples were subjected to real-time RT-PCR performed with a specific primer set for each gene. Values were normalized to Gapdh and are expressed relative to the WT. Values are the mean ± S.E.M.; # $p < 0.05$ (#, vs. WT). The experiments that produced the data shown in this figure were performed at least twice.

3.2. Effect of Metallothionein on PPE-Induced Lung Injury

We first examined the effect of metallothionein in the PPE-induced lung injury model, which is one of the animal models of COPD. It is known that PPE administration to mice induces alveolar enlargement and respiratory impairment as evidenced by decreased lung elastance levels and FEV/FVC ratios [10]. Alveolar enlargement was observed in both WT and MT-KO mice after an intratracheal administration of PPE, but this enlargement was more pronounced in the MT-KO mice. We used H&E-stained images to measure the MLIs and found that the MLI was significantly higher in MT-KO mice than in WT mice after PPE administration. In contrast, there was no difference in the MLIs of these mice in the absence of PPE (Figure 2A,B). These results suggest that metallothionein may have a protective effect against alveolar enlargement in COPD.

The FEV_1/FVC ratio, measured using a spirometer, is the most important indicator for the diagnosis and examination of COPD [2]. In COPD animal models, in addition to decreased FEV/FVC ratios, the levels of elastance, which is an index of lung stiffness, also decrease [10]. As shown in Figure 2C, PPE treatment decreased the levels of total elastance (i.e., elastance from the whole lung, including the bronchi, bronchioles, and alveoli) and tissue elastance (elastance from the alveoli) in both types of mice. Moreover, the TLC, and tissue damping were measured as additional indices, and the results show a significantly accelerated PPE-dependent increase in the TLC and a decrease in tissue damping in MT-KO mice, along with worsened respiratory function in MT-KO mice (Supplementary Figure S2). In addition, the 0.05-s and 0.1-s rates ($FEV_{0.05}/FVC$ and $FEV_{0.1}/FVC$ ratio, respectively), which are equivalent to the FEV_1/FVC ratio in humans, were also decreased in both types of mice. The decrease in all indices was more pronounced and statistically significant in the MT-KO mice than in the WT mice. Taken together, the results shown in Figure 2 suggest that metallothionein may have an inhibitory effect on the development of COPD.

3.3. Effect of Metallothionein on the PPE-Induced Pulmonary Inflammatory Response and Pulmonary Cell Death

We next compared the increases in the numbers of inflammatory cells in BALF at 3 days after exposure to PPE in WT and MT-KO mice (Figure 3). Under these experimental conditions, the number of total inflammatory cells in BALF increased in a PPE-dependent manner in both WT and MT-KO mice, but the number of total inflammatory cells after PPE treatment was significantly higher in MT-KO mice than in WT mice. In addition, there was no difference in the number of macrophages after PPE administration in either mouse type, but the number of neutrophils involved in acute inflammation was significantly higher in the PPE-exposed MT-KO mice compared with the WT mice. The protein concentration in

BALF is an indicator of lung injury and edema. As shown in Figure 3B, treatment with PPE increased the BALF protein concentration in both types of mice, but this increase was more pronounced in the MT-KO mice.

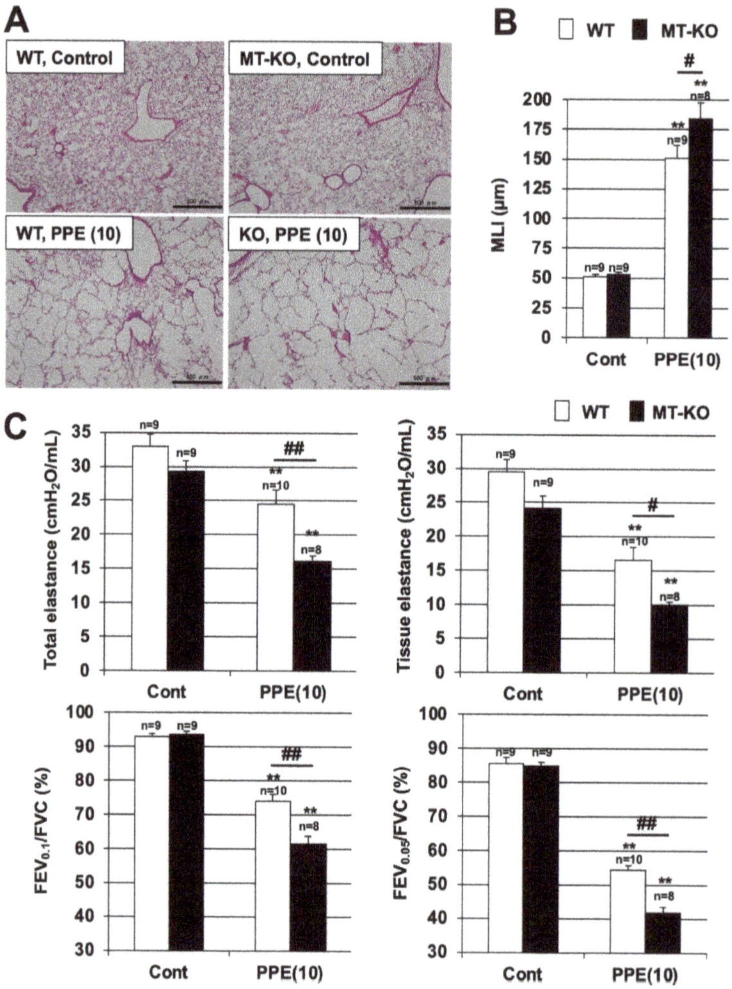

Figure 2. Effect of metallothionein on PPE-induced pulmonary damage. (**A–C**) Wildtype (WT) and MT-1 and -2 gene knockout (MT-KO) mice were treated with sterile saline (Cont) or PPE (10 U/kg) only once, on day 0. Sections of pulmonary tissue were prepared on day 14 and subjected to histopathological examination (H&E staining) (scale bar, 500 μm) (**A**). Airspace size was estimated by determining the MLI as described in the Materials and Methods (**B**). Total elastance, tissue elastance, $FEV_{0.05}/FVC$, and $FEV_{0.1}/FVC$ values were determined on day 14 as described in the Materials and Methods (**C**). Values are the mean ± S.E.M. # $p < 0.05$; ** or ## $p < 0.01$ (**, vs. Control; # or ##, vs. WT). The experiments that produced the data shown in this figure were performed at least twice.

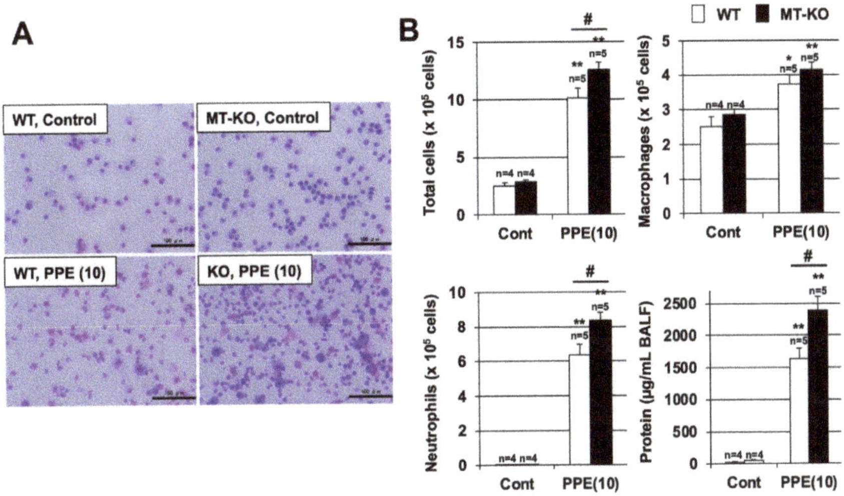

Figure 3. Effect of metallothionein on PPE-induced acute lung injury. (**A**,**B**) Wildtype (WT) and MT-1 and -2 gene knockout (MT-KO) mice were treated with sterile saline (Cont) or PPE (10 U/kg) only once, on day 0. BALF samples were collected 72 h after the intratracheal administration of saline or PPE. (**A**) BALF cells were deposited onto slides by using a Cytospin® 4 cytocentrifuge then stained with Diff-Quik reagents and visualized via light microscopy (scale bar, 100 μm). (**B**) The numbers of total cells, macrophages, and neutrophils were determined. The amount of protein present in the BALF was determined by the Bradford method. Values are the mean ± S.E.M. * or # $p < 0.05$; ** $p < 0.01$ (* or **, vs. Control; #, vs. WT). The experiments that produced the data shown in this figure were performed at least twice.

In PPE-induced lung injury models and COPD patients, it has been reported that inflammatory cytokine production is induced in the lungs, which is involved in the promotion of inflammatory responses in the lungs. Thus, we monitored the expression of pro-inflammatory cytokines by using real-time RT-PCR. As shown in Figure 4, the mRNA expressions of tumor necrosis factor-α (TNF-α), interleukin (IL)-1β, IL-6, and macrophage inflammatory protein 2 (MIP2) were induced by PPE treatment. The levels of TNF-α, IL-6, and MIP2 mRNA expression after PPE treatment were higher in MT-KO mice than in WT mice. In contrast, the levels of IL-1β mRNA expression after PPE treatment in both types of mice were similar. Together, the data shown in Figures 3 and 4 suggest that metallothionein may have an inhibitory effect on PPE-dependent inflammatory responses.

Pulmonary cell apoptosis also plays an important role in the pathogenesis of COPD and PPE-induced pulmonary emphysema. Thus, we analyzed cell death in the lungs of MT-KO and WT mice after PPE administration using the TUNEL assay. While few TUNEL-positive cells (i.e., those that succumbed to cell death) were observed in the PPE-untreated groups of mice, the number of TUNEL-positive cells after PPE administration trended higher in WT mice and increased significantly in MT-KO mice (Figure 5). These results suggest that metallothionein may also have an inhibitory effect on PPE-dependent pulmonary cell death. Therefore, we speculate that metallothionein suppresses PPE-induced pulmonary emphysema through its anti-inflammatory and cell death-inhibitory effects.

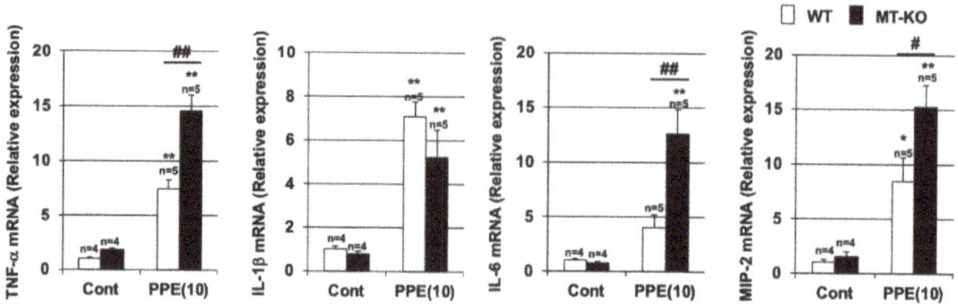

Figure 4. Effect of metallothionein on PPE-induced pulmonary inflammatory response. Wildtype (WT) and MT-1 and -2 gene knockout (MT-KO) mice were treated with sterile saline (Cont) or PPE (10 U/kg) only once, on day 0. Total RNA was extracted from the lungs 72 h after PPE administration and subjected to real-time RT-PCR using a specific primer set for each gene. Values were normalized to Gapdh and are expressed relative to the Control. Values are the mean ± S.E.M. * $p < 0.05$; ** or ## $p < 0.01$ (* or **, vs. Control; # or ##, vs. WT). The experiments that produced the data shown in this figure were performed at least twice.

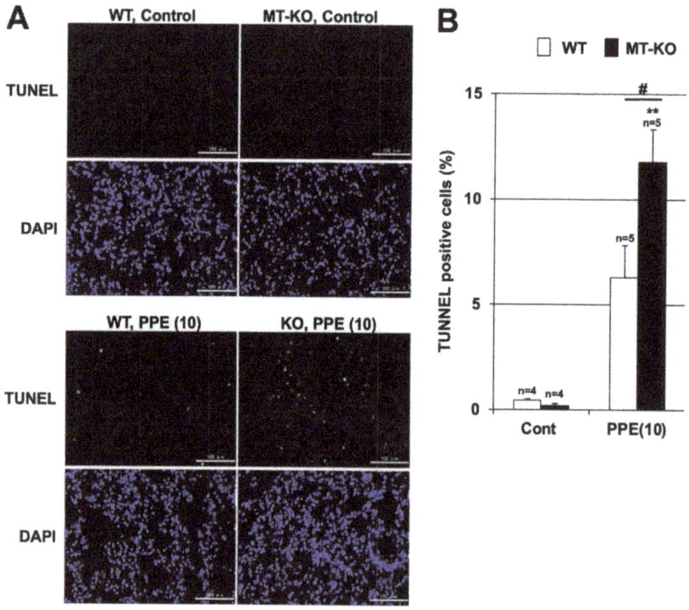

Figure 5. Effect of metallothionein on PPE-induced pulmonary cell death. (**A**,**B**) Wildtype (WT) and MT-1 and -2 gene knockout (MT-KO) mice were treated with sterile saline (Cont) or PPE (10 U/kg) only once, on day 0. Sections of pulmonary tissue were prepared 72 h after PPE administration. (**A**) Sections were subjected to terminal deoxynucleotidyl transferase biotin-dUTP nick end labeling (TUNEL) assay and 4,6-diamidino-2-phenylindole (DAPI) staining (scale bar, 100 μm). (**B**) The numbers of TUNEL-positive cells were counted. Values are the mean ± S.E.M. # $p < 0.05$; ** $p < 0.01$ (**, vs. Control; #, vs. WT). The experiments that produced the data shown in this figure were performed at least twice.

3.4. Effect of Metallothionein on PPE-Induced ROS Production

ROS are involved in the onset and progression of various lung injuries, such as acute respiratory distress syndrome (ARDS) and COPD [26,27]. Because metallothioneins have ROS scavenging activity [13], we used the FUSION in vivo imaging system to monitor

ROS levels in our model. Locations with high levels of ROS are indicated by color (highest to lowest levels shown in red, yellow, green, and blue, respectively) in Figure 6. The intratracheal administration of PPE increased ROS production in the lungs of both types of mice (Figure 6). The lungs of MT-KO mice produced more ROS in response to PPE administration compared with those of WT mice. In contrast, there was almost no ROS production in the control-treated WT or MT-KO mice. A quantitative analysis using the standard software provided with FUSION showed that PPE-dependent ROS production was significantly enhanced in the MT-KO mice as compared with WT mice (Figure 6B).

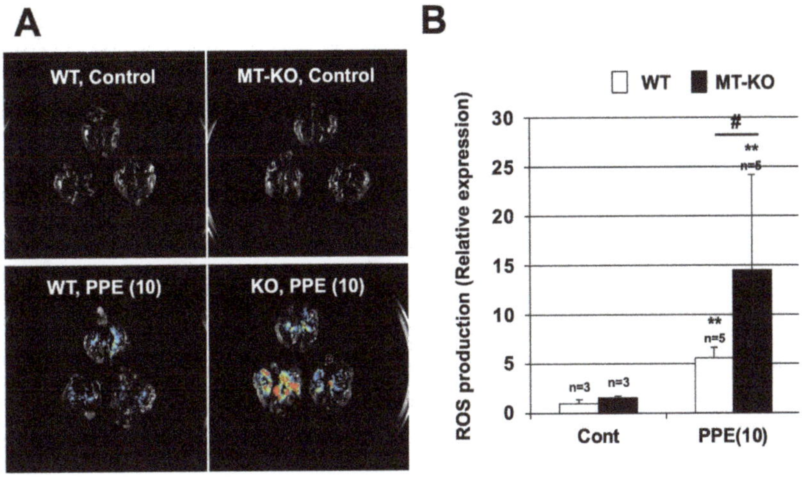

Figure 6. Effect of metallothionein on PPE-induced ROS production. (**A**,**B**) Wildtype (WT) and MT-1 and -2 gene knockout (MT-KO) mice were treated with sterile saline (Cont) or PPE (10 U/kg) only once, on day 0. Luminescent probe (L-012, 75 mg/kg) was administered 24 h after the PPE administration. Isolated lungs were imaged using a FUSION chemiluminescence imaging system (**A**). The summed pixel intensity of the ROS signal was determined using standard software for FUSION (**B**). Values are the mean ± S.E.M. # $p < 0.05$; ** $p < 0.01$ (**, vs. Control; #, vs. WT). The experiments that produced the data shown in this figure were performed at least twice.

To identify oxidative stress in another way, specific antibodies against 8-OHdG (a marker of oxidative damage to DNA) were used. As shown in Figure 7, there was a marked increase in the 8-OHdG-positive area, i.e., an increase in the level of oxidative stress, in the lung tissue of MT-KO mice treated with PPE. In contrast, only a slight increase in the 8-OHdG-positive area was observed in the lung tissue of WT mice treated with PPE.

Finally, to investigate the involvement of oxidative stress in the progression of PPE-induced pulmonary emphysema, we analyzed the ROS production 7 days after PPE administration by using an in vivo imaging system. As shown in Supplementary Figure S3, both types of mice produced less ROS at 7 days after PPE administration than they did at 24 h after PPE administration. However, the level of ROS production at 7 days after PPE administration in the lung tissue of MT-KO mice was significantly higher than that of WT mice. These results suggest that ROS production is enhanced at both the onset and progression stages of PPE-induced pulmonary emphysema in MT-KO mice, resulting in exacerbation of PPE-induced pulmonary emphysema in MT-KO mice relative to that in WT mice.

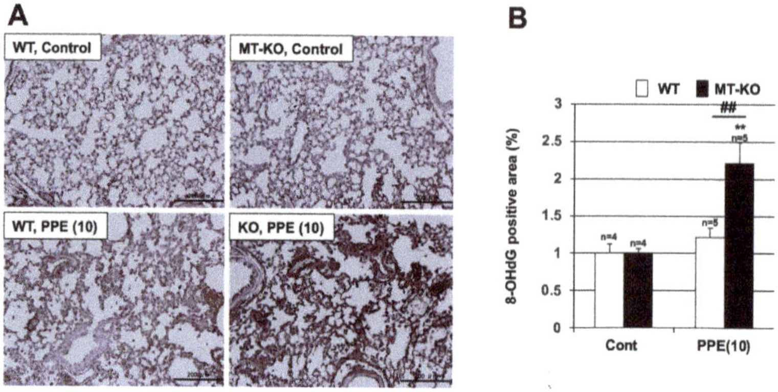

Figure 7. Effect of metallothionein on PPE-induced oxidative stress. (**A**,**B**) Wildtype (WT) and MT-1 and -2 gene knockout (MT-KO) mice were treated with sterile saline (Cont) or PPE (10 U/kg) only once, on day 0. (**A**) Sections of pulmonary tissue were prepared 72 h after PPE administration and subjected to 8-hydroxydeoxyguanosine (8-OHdG) staining (scale bar = 200 μm). (**B**) The 8-OHdG-positive area was measured. Values are the mean ± S.E.M. ** or ## $p < 0.01$ (**, vs. Control; ##, vs. WT). The experiments that produced the data shown in this figure were performed at least twice.

4. Discussion

In the present study, we found that PPE-dependent lung injury was exacerbated in MT-KO mice compared with their WT littermates. By using an in vivo imaging system, we observed that PPE-dependent ROS production was also significantly enhanced in MT-KO mice. These results suggest that metallothionein may act as an inhibitor against elastase-induced lung injury by suppressing ROS production, which is one of the causes of COPD (Figure 8). Previous analyses using metallothionein-deficient mice have shown that metallothionein is effective against various diseases, including some respiratory diseases. In contrast, there have been few studies on the efficacy of metallothionein against COPD, so it is notable that the present study using MT-KO mice provides evidence regarding the efficacy of metallothionein against COPD.

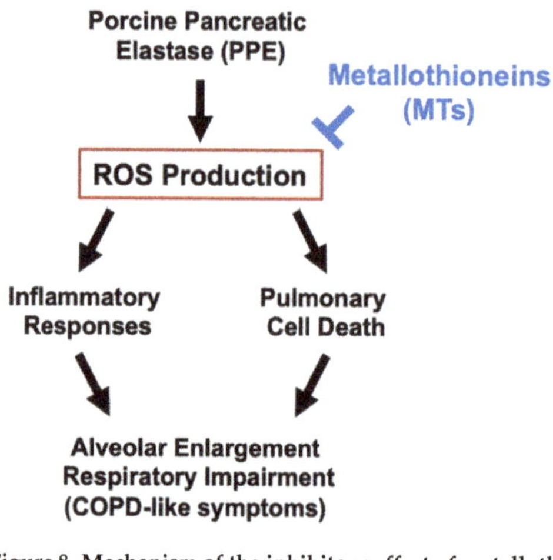

Figure 8. Mechanism of the inhibitory effect of metallothioneins on PPE-induced lung injury.

As described above, metallothionein is known to inhibit the ROS production caused by various stresses [13]. We therefore focused our analysis on the antioxidant effect of metallothionein as a protective mechanism against lung injury, and we demonstrated for the first time, using an in vivo imaging system, that PPE-dependent ROS production is enhanced in MT-KO mice. This work is the first to demonstrate visually that metallothionein inhibits PPE-dependent oxidative stress. Notably, ROS are not only a short-term exacerbating factor of COPD, e.g., following exposure to stress such as CS, but also a cause of chronic COPD exacerbation through their involvement in chronic inflammation and fibrosis [28]. In this study, ROS production was significantly increased in MT-KO mice at not only 24 h but also 7 days after PPE exposure. This increase in ROS levels at 7 days after PPE exposure may be related to chronic inflammation and fibrosis; the details of this mechanism should be investigated in future work. Such research will lead to a better understanding of the molecular mechanisms by which metallothionein inhibits lung injury.

One particularly interesting finding of this study is that the amounts of total elastance and tissue elastance tended to be lower in MT-KO mice than in WT mice, not only after PPE exposure but also in the untreated state. It has been reported that MT-KO mice have a shorter lifespan because of progressive aging [29,30]. Our findings, in addition to those of previous reports, suggest that the induced lower lung function of MT-KO mice may be one reason for their shorter lifespan relative to WT mice. Moreover, the results of this study also suggest the possibility of using metallothionein as a predictor of COPD onset or exacerbation. To the best of our knowledge, no studies have yet analyzed the correlation between the metallothionein amount in the blood, BALF, or lung tissue and COPD susceptibility. If such an analysis can be conducted, it may be possible not only to apply metallothionein inducers as therapeutic agents but also to use metallothionein as a predictor of the onset or exacerbation of COPD.

In our previous analysis, we found that polaprezinc and zinc acetate, which are approved zinc-containing drugs, suppressed the alveolar epithelial cell injury caused by cadmium, a metal found in cigarette smoke, suggesting that metallothionein inducers such as polaprezinc may be effective against COPD [20]. Considering the results of our previous studies together with the results of the present study, it is conceivable that metallothionein-inducing compounds may have efficacy as COPD therapeutic agents. In particular, our results showing that metallothionein inhibits the injury-induced decrease in FEV/FVC ratio, which is a diagnostic index of COPD in clinical practice, are very significant. However, several issues remain to be addressed prior to the clinical application of metallothionein inducers as a COPD treatment. Specifically, it is essential to evaluate this question using a lung injury model for long-term CS exposure. Future work will need to examine a lung injury model with long-term CS exposure using MT-KO mice and their WT littermates and to analyze the efficacy of compounds that induce metallothionein, such as polaprezinc, on the lung injury model induced by long-term CS exposure.

5. Conclusions

In this study, we found that both PPE-dependent lung injuries were exacerbated in MT-KO mice. Furthermore, we observed that PPE-dependent ROS production was also significantly enhanced in MT-KO mice. These results suggest that metallothionein may act as an inhibitor against elastase-induced lung injury by suppressing ROS production. In conclusion, metallothionein protein, or compounds that can induce metallothionein, may be useful in the treatment of COPD.

Supplementary Materials: The following supporting information can be downloaded at: https://www.mdpi.com/article/10.3390/biom12040583/s1, Figure S1: Effect of metallothionein on PPE-induced pulmonary damage; Figure S2: Effect of metallothionein on PPE-induced respiratory dysfunction; Figure S3: Effect of metallothionein on ROS production after 7 days of PPE administration.

Author Contributions: Study conception and design, K.I.T. and M.K.; Acquisition, analysis and/or interpretation of data, K.I.T., S.S. (Sachie Shiota), O.S., M.S., A.T. and M.T.; Provision of experimental materials, Y.K., T.K. and S.S. (Shinya Suzuki); Final approval and overall responsibility for the published work; K.I.T. All authors have read and agreed to the published version of the manuscript.

Funding: This study was supported in part by Smoking Research Foundation.

Institutional Review Board Statement: The experiments and procedures described here were carried out in accordance with the Guide for the Care and Use of Laboratory Animals as adopted and promulgated by the National Institutes of Health and were approved by the Animal Care Committee of Musashino University.

Informed Consent Statement: Not applicable.

Data Availability Statement: The data that support the findings of our study are available from the corresponding author upon reasonable request.

Conflicts of Interest: The authors declare no conflict of interest.

References

1. Barnes, P.J.; Stockley, R.A. COPD: Current therapeutic interventions and future approaches. *Eur. Respir. J.* **2005**, *25*, 1084–1106. [CrossRef] [PubMed]
2. Vogelmeier, C.F.; Criner, G.J.; Martinez, F.J.; Anzueto, A.; Barnes, P.J.; Bourbeau, J.; Celli, B.R.; Chen, R.; Decramer, M.; Fabbri, L.M.; et al. Global Strategy for the Diagnosis, Management, and Prevention of Chronic Obstructive Lung Disease 2017 Report. GOLD Executive Summary. *Am. J. Respir. Crit. Care Med.* **2017**, *195*, 557–582. [CrossRef] [PubMed]
3. Alsaeedi, A.; Sin, D.D.; McAlister, F.A. The effects of inhaled corticosteroids in chronic obstructive pulmonary disease: A systematic review of randomized placebo-controlled trials. *Am. J. Med.* **2002**, *113*, 59–65. [CrossRef]
4. Mak, J.C. Pathogenesis of COPD. Part II. Oxidative-antioxidative imbalance. *Int. J. Tuberc. Lung Dis.* **2008**, *12*, 368–374. [PubMed]
5. Antus, B.; Harnasi, G.; Drozdovszky, O.; Barta, I. Monitoring oxidative stress during chronic obstructive pulmonary disease exacerbations using malondialdehyde. *Respirology* **2014**, *19*, 74–79. [CrossRef] [PubMed]
6. Drost, E.M.; Skwarski, K.M.; Sauleda, J.; Soler, N.; Roca, J.; Agusti, A.; MacNee, W. Oxidative stress and airway inflammation in severe exacerbations of COPD. *Thorax* **2005**, *60*, 293–300. [CrossRef] [PubMed]
7. Drummond, G.R.; Selemidis, S.; Griendling, K.K.; Sobey, C.G. Combating oxidative stress in vascular disease: NADPH oxidases as therapeutic targets. *Nat. Rev. Drug. Discov.* **2011**, *10*, 453–471. [CrossRef]
8. Boukhenouna, S.; Wilson, M.A.; Bahmed, K.; Kosmider, B. Reactive Oxygen Species in Chronic Obstructive Pulmonary Disease. *Oxid. Med. Cell. Longev.* **2018**, *2018*, 5730395. [CrossRef]
9. Tanaka, K.-I.; Tanaka, Y.; Miyazaki, Y.; Namba, T.; Sato, K.; Aoshiba, K.; Azuma, A.; Mizushima, T. Therapeutic Effect of Lecithinized Superoxide Dismutase on Pulmonary Emphysema. *J. Pharmacol. Exp. Ther.* **2011**, *338*, 810–818. [CrossRef]
10. Tanaka, K.; Ishihara, T.; Sugizaki, T.; Kobayashi, D.; Yamashita, Y.; Tahara, K.; Yamakawa, N.; Iijima, K.; Mogushi, K.; Tanaka, H.; et al. Mepenzolate bromide displays beneficial effects in a mouse model of chronic obstructive pulmonary disease. *Nat. Commun.* **2013**, *4*, 2686. [CrossRef]
11. Davis, S.R.; Cousins, R.J. Metallothionein expression in animals: A physiological perspective on function. *J. Nutr.* **2000**, *130*, 1085–1088. [CrossRef] [PubMed]
12. Schanz, M.; Schaaf, L.; Dippon, J.; Biegger, D.; Fritz, P.; Alscher, M.D.; Kimmel, M. Renal effects of metallothionein induction by zinc in vitro and in vivo. *BMC Nephrol.* **2017**, *18*, 91. [CrossRef] [PubMed]
13. Ruttkay-Nedecky, B.; Nejdl, L.; Gumulec, J.; Zitka, O.; Masarik, M.; Eckschlager, T.; Stiborova, M.; Adam, V.; Kizek, R. The role of metallothionein in oxidative stress. *Int. J. Mol. Sci.* **2013**, *14*, 6044–6066. [CrossRef] [PubMed]
14. Lian, Y.; Zhao, J.; Wang, Y.M.; Zhao, J.; Peng, S.Q. Metallothionein protects against isoniazid-induced liver injury through the inhibition of CYP2E1-dependent oxidative and nitrosative impairment in mice. *Food Chem. Toxicol.* **2017**, *102*, 32–38. [CrossRef] [PubMed]
15. Wu, H.; Zhou, S.; Kong, L.; Chen, J.; Feng, W.; Cai, J.; Miao, L.; Tan, Y. Metallothionein deletion exacerbates intermittent hypoxia-induced renal injury in mice. *Toxicol. Lett.* **2015**, *232*, 340–348. [CrossRef]
16. Lv, Y.; Zhang, B.; Xing, G.; Wang, F.; Hu, Z. Protective effect of naringenin against acetaminophen-induced acute liver injury in metallothionein (MT)-null mice. *Food Funct.* **2013**, *4*, 297–302. [CrossRef]
17. Ren, M.; Wang, Y.M.; Zhao, J.; Zhao, J.; Zhao, Z.M.; Zhang, T.F.; He, J.; Ren, S.P.; Peng, S.Q. Metallothioneins attenuate paraquat-induced acute lung injury in mice through the mechanisms of anti-oxidation and anti-apoptosis. *Food Chem. Toxicol.* **2014**, *73*, 140–147. [CrossRef] [PubMed]
18. Wesselkamper, S.C.; McDowell, S.A.; Medvedovic, M.; Dalton, T.P.; Deshmukh, H.S.; Sartor, M.A.; Case, L.M.; Henning, L.N.; Borchers, M.T.; Tomlinson, C.R.; et al. The role of metallothionein in the pathogenesis of acute lung injury. *Am. J. Respir. Cell Mol. Biol.* **2006**, *34*, 73–82. [CrossRef]
19. Fuji, Y.; Matsura, T.; Kai, M.; Kawasaki, H.; Yamada, K. Protection by polaprezinc, an anti-ulcer drug, against indomethacin-induced apoptosis in rat gastric mucosal cells. *Jpn. J. Pharmacol.* **2000**, *84*, 63–70. [CrossRef]
20. Kimura, K.; Nakano, Y.; Sugizaki, T.; Shimoda, M.; Kobayashi, N.; Kawahara, M.; Tanaka, K.I. Protective effect of polaprezinc on cadmium-induced injury of lung epithelium. *Metallomics* **2019**, *11*, 1310–1320. [CrossRef]
21. Masters, B.A.; Kelly, E.J.; Quaife, C.J.; Brinster, R.L.; Palmiter, R.D. Targeted disruption of metallothionein I and II genes increases sensitivity to cadmium. *Proc. Natl. Acad. Sci. USA* **1994**, *91*, 584–588. [CrossRef] [PubMed]

22. Tanaka, K.I.; Tamura, F.; Sugizaki, T.; Kawahara, M.; Kuba, K.; Imai, Y.; Mizushima, T. Evaluation of Lecithinized Superoxide Dismutase for the Prevention of Acute Respiratory Distress Syndrome in Animal Models. *Am. J. Respir. Cell Mol. Biol.* **2017**, *56*, 179–190. [CrossRef] [PubMed]
23. Tanaka, K.I.; Kubota, M.; Shimoda, M.; Hayase, T.; Miyaguchi, M.; Kobayashi, N.; Ikeda, M.; Ishima, Y.; Kawahara, M. Thioredoxin-albumin fusion protein prevents urban aerosol-induced lung injury via suppressing oxidative stress-related neutrophil extracellular trap formation. *Environ. Pollut.* **2021**, *268*, 115787. [CrossRef] [PubMed]
24. Yamashita, Y.; Tanaka, K.I.; Yamakawa, N.; Asano, T.; Kanda, Y.; Takafuji, A.; Kawahara, M.; Takenaga, M.; Fukunishi, Y.; Mizushima, T. Chemical modification-mediated optimisation of bronchodilatory activity of mepenzolate, a muscarinic receptor antagonist with anti-inflammatory activity. *Bioorg. Med. Chem.* **2019**, *27*, 3339–3346. [CrossRef] [PubMed]
25. Sugizaki, T.; Tanaka, K.I.; Asano, T.; Kobayashi, D.; Hino, Y.; Takafuji, A.; Shimoda, M.; Mogushi, K.; Kawahara, M.; Mizushima, T. Idebenone has preventative and therapeutic effects on pulmonary fibrosis via preferential suppression of fibroblast activity. *Cell. Death. Discov.* **2019**, *5*, 146. [CrossRef]
26. Sarma, J.V.; Ward, P.A. Oxidants and redox signaling in acute lung injury. *Compr. Physiol.* **2011**, *1*, 1365–1381. [CrossRef]
27. Tasaka, S.; Amaya, F.; Hashimoto, S.; Ishizaka, A. Roles of oxidants and redox signaling in the pathogenesis of acute respiratory distress syndrome. *Antioxid. Redox. Sign.* **2008**, *10*, 739–753. [CrossRef]
28. Checa, J.; Aran, J.M. Airway Redox Homeostasis and Inflammation Gone Awry: From Molecular Pathogenesis to Emerging Therapeutics in Respiratory Pathology. *Int. J. Mol. Sci.* **2020**, *21*, 9317. [CrossRef]
29. Kadota, Y.; Aki, Y.; Toriuchi, Y.; Mizuno, Y.; Kawakami, T.; Sato, M.; Suzuki, S. Deficiency of metallothionein-1 and -2 genes shortens the lifespan of the 129/Sv mouse strain. *Exp. Gerontol.* **2015**, *66*, 21–24. [CrossRef]
30. Swindell, W.R. Metallothionein and the biology of aging. *Ageing Res. Rev.* **2011**, *10*, 132–145. [CrossRef]

biomolecules

MDPI

Article

No Association of Hair Zinc Concentration with Coronary Artery Disease Severity and No Relation with Acute Coronary Syndromes

Ewelina A. Dziedzic [1,2,*], Jakub S. Gąsior [3], Agnieszka Tuzimek [2], Justyna Paleczny [4], Mirosław Kwaśny [5], Marek Dąbrowski [6] and Piotr Jankowski [2,7]

1 Medical Faculty, Lazarski University in Warsaw, 02-662 Warsaw, Poland
2 Department of Internal Medicine and Geriatric Cardiology, Centre of Postgraduate Medical Education, 01-813 Warsaw, Poland; agnieszkatuzimek@gmail.com (A.T.); piotr.jankowski@interia.pl (P.J.)
3 Department of Pediatric Cardiology and General Pediatrics, Medical University of Warsaw, 02-091 Warsaw, Poland; jgasior@wum.edu.pl
4 Department of Pharmaceutical Microbiology and Parasitology, Faculty of Pharmacy, Wroclaw Medical University, 50-556 Wroclaw, Poland; justyna.paleczny@student.umed.wroc.pl
5 Institute of Optoelectronics, Military University of Technology, 00-908 Warsaw, Poland; miroslaw.kwasny@wat.edu.pl
6 Department of Cardiology, Bielanski Hospital, 01-809 Warsaw, Poland; mardab@vp.pl
7 Department of Epidemiology and Health Promotion, School of Public Health, Center of Postgraduate Medical Education, 01-826 Warszawa, Poland
* Correspondence: ewelinadziedzic82@gmail.com; Tel.: +48-792-207-779

Abstract: Cardiovascular diseases (CVDs) are the leading cause of death worldwide. Although zinc (Zn) was reported to have antioxidant, anti-inflammatory and protective properties in CVDs, its association with coronary artery disease (CAD) is still unclear. As methods commonly used to assess Zn levels in blood and urine do not show the full picture of the microelement supply, in this study, the nutritional status of Zn in patients with angiographically confirmed CAD was assessed using inductively coupled plasma optical emission spectrometry. We found no association between Zn and the severity of CAD evaluated with the Coronary Artery Surgery Study Score (p = 0.67). There were no statistically significant differences in Zn levels between patients with acute coronary syndrome and those with stable CAD (p = 0.937). A statically significant negative correlation was observed between Zn content and serum triglyceride concentration ($p < 0.05$). Patients with type 2 diabetes mellitus were found to have a significantly lower hair Zn content compared to non-diabetic individuals ($p < 0.01$). The role of Zn in the pathogenesis of CAD and its complications need further well-designed research as the moderation and supplementation of Zn dietary intake could be a simple intervention to reduce the CVDs risk.

Keywords: zinc; coronary artery disease; myocardial infarction; acute coronary syndrome

1. Introduction

Despite great efforts to advance prophylaxis and treatment, cardiovascular diseases (CVDs) are still the leading cause of death in the world. In 2019, CVDs were responsible for 17.9 million deaths worldwide, which is 85% of deaths from myocardial infarctions or strokes. These numbers are growing, as experts predict that in 2030 CVDs-related deaths could rise to 24 million annually [1]. Despite Zinc (Zn) being one of the key microelements of the human body [2], its role in CVDs pathogenesis has not yet been firmly established. The available data generally show unfavourable levels of this metal in cardiology patients as low concentrations were found particularly in patients with CAD [3], heart failure [4–6] and atrial fibrillation [7]. Moreover, patients with left ventricular hypertrophy and atrial fibrillation had an inverse correlation of Zn concentration with heart muscle thickness [8].

However, data on the relationship between Zn concentration and coronary artery disease (CAD) are limited [9–14] and are mainly based on serum Zn concentration. Analytical methods commonly used to assess Zn levels in blood and urine do not show the full picture of microelement supply, as Zn content in serum is easily influenced by the time of the day, the type of meal eaten before obtaining the sample, the serum protein concentration, as well as the natural homeostasis mechanisms controlling the Zn concentration utilizing tissue storage [15]. Hair sample analysis has several advantages over serum samples. The concentration of Zn in a hair sample is about 100 times higher than in serum, and it is not as labile as serum concentration, which makes it perfect for long-term nutrition assessment [16]. In addition, the hair sample better reflects the recent excessive exposure to metals, as cations are quickly transferred from blood to tissue storage [17,18]. Hair samples are considered a good retrospective marker of microelement nutrition in the previous 6–8 weeks. As the concentration of Zn in hair is reported to reflect its content in other tissues [19–21], we decided to explore this approach.

This research aims to determine whether the Zn content in hair samples measured by inductively coupled plasma optical emission spectrometry (ICP-OES) correlates with the progression of CAD and acute coronary syndrome (ACS). In ICP-OES, elements are excited using heat from an argon plasma. During de-excitation, the atoms emit light with a spectrum consisting of lines specific to a particular element, allowing for the determination of the elemental composition of the sample [22]. This method has found an application in physiological samples due to its remarkable sensitivity and versatility. As obtaining a hair sample is simpler and less invasive than phlebotomy, this method could become an appropriate screening tool for patients at risk of CAD.

2. Materials and Methods

2.1. Study Population

This study is based on data obtained from 133 patients (37 women and 96 men) who underwent coronary angiography to assess the extent of CAD between 2013 and 2017 in the Department of Cardiology of Bielanski Hospital, Warsaw, Poland and agreed to participate in the study in writing. The analysis included patients whose hair was not dyed or permanently waved in segments of at least 3 cm, measuring from the scalp. Patients with an active neoplastic disease, significantly increased inflammatory markers, chronic kidney disease above stage III, taking medications or dietary supplements containing zinc, or using shampoos with an increased content of the bio-element were excluded from the study. Patients with a history of previous MI who were treated with coronary angioplasty were included in this study; however, those with thrombosis or restenosis were excluded. All patients lived in Warsaw and they had no history of occupational exposure to chemical elements. The study was carried out according to the principles of the Declaration of Helsinki and was approved by the bioethics committee of the Medical University of Warsaw.

2.2. Coronary Angiography

Coronary angiography is the default method for assessing stenosis in CAD [23]. The examination was performed by radial or femoral artery access. The severity of CAD was classified by three independent cardiologists using the Coronary Artery Surgery Study Score (CASSS). This scale reflects the stenosis of one, two or three arteries through a sum of points (0–3), which were assigned as follows: 1 point for stenosis of the main coronary artery (right coronary artery, circumflex branch, or anterior descending branch) exceeding 70% and 2 points for stenosis of the left main coronary artery greater than 50%. Inconclusive findings between moderate or severe stenosis were decided using fractional flow reserve measurement. The diagnosis of acute coronary syndrome (ACS) was based on criteria from the European Society of Cardiology guidelines, including the increased concentration of markers of myocardial injury with the coexistence of at least one of the following: symptoms of stenocardia, changes in the ECG suggestive of ischemia, results of imaging tests showing myocardial necrosis or coronary artery thrombus identified in coronary angiography [24].

2.3. Laboratory Tests

Non-permed and non-dyed hair samples, weighing between 0.2 and 0.3 g, were obtained from a few separate scalp sites at the back of the head, close to the skin. In preparation for inductively coupled plasma optical emission spectrometry (ICP-OES) the samples were washed using non-ionic detergent (Triton X-100, Sigma Aldrich Sp. z.o.o., Poznań, Poland) water solution (1:100) in an ultrasonic bath for 5 min, then rinsed sequentially with high-purity water, acetone and water and then dried to constant mass. The dry samples of hair, 0.15 g each, were dissolved in 4 mL of 65% nitric acid (Merck, Darmastadt, Germany) and 1 mL of 30% hydrogen peroxide (Merck) in a closed polypropylene tube (8 mL), then incubated in 80 °C for 30 min in a microwave station. After cooling to room temperature, the samples were diluted to a final volume of 10 mL with Milli-Q water and then analysed using an ICP-OES spectrometer (iCAP7400, Thermo Scientific, Waltham, MA, USA). The concentration of zinc in the solution, and then the total content in the hair samples, were calculated according to the previously determined standard curve.

2.4. Statistical Analysis

The Shapiro–Wilk test was used to assess the distribution of data. The chi-square statistic was used to identify associations between dichotomous and categorical data. The Mann–Whitney test was used to compare the values between two groups of patients. Kruskal–Wallis analysis by rank was used to determine the dependence between more than two groups. The R Spearman correlation test was used to evaluate the relationship between the variables. A two-sided p-value < 0.05 was regarded as statistically significant. Statistical analysis and figures were performed and created with Statistica 13 (StatSoft Inc., Tulsa, OK, USA). GraphPad Prism 5 (GraphPad Software Inc., San Diego, CA, USA, 2005) was used to create figures.

3. Results

3.1. Population Characteristics

The median age of the study population was 65 years (range: 37–95). The median BMI value was 27.7 kg/m^2 (range: 16.9–54.1). A total of 39 (29.3%) participants had a normal body weight, 53 (39.9%) were overweight and 41 (30.8%) patients were classified as obese. A history of type 2 diabetes mellitus (t2DM) or diagnosis during the current hospitalization was found in 42 (31.6%) patients and pre-diabetes in 7 (5.3%) patients. On the basis of the lipid profile (total cholesterol—TC, LDL and HDL cholesterol, triglycerides—TG), hyperlipidaemia was assessed in 123 patients and diagnosed in over half of them despite statin treatment, i.e., in 55 (41.4%). Hypertension was present in 114 (85.7%) patients. Acute coronary syndrome (ACS) as the cause of hospitalization was diagnosed in 67 (50.4%) patients, while stable CAD was the cause in 66 (49.6%) patients. A history of myocardial infarction (MI) was noticed in 40 (30.1%) patients. Active smoking during the study was declared by 40 (30.1%) patients, and 17 (12.8%) patients had smoked in the past. Insignificant changes in the coronary arteries (CASSS 0) were found in only 22 (16.5%) patients. One-vessel coronary disease (CASSS 1) was found in 34 (25.6%) patients, two-vessel (CASSS 2) in 46 (34.6%) and three-vessel (CASSS 3) in 31 (23.3%) patients. The median Zn concentration was 166 parts per million (ppm) (range: 25–495).

3.2. Association between Zn Level and Severity of CAD

Table 1 presents results of measurements for the study group according to CASSS level. A significant difference in sex distribution was observed between CASSS groups. There was also a significant difference in distribution of patients with history of previous MI and cause of hospitalization.

Table 1. Association between selected parameters, including Zn level and CAD stages.

	CASSS 0	CASSS 1	CASSS 2	CASSS 3	*p*-Value
N	22	34	46	31	-
Sex (♀/♂)	10/12	12/22	6/40	9/22	0.025
Age (years)	66 (54–85)	66 (48–89)	63 (37–84)	68 (52–95)	0.455
BMI (kg/m^2)	28.5 (21.1–46.3)	27.7 (16.9–40.9)	27.9 (22.3–54.1)	26.8 (18.7–45.3)	0.476
Cause of hospitalization (ACS/stable CAD)	5/17	22/12	22/24	18/13	0.016
t2DM (no/yes/pre-diabetes) **	17/4/1	21/13/0	29/13/4	17/12/2	—
t2DM (no/yes)	17/4	21/13	29/13	17/12	0.355
Hyperlipidaemia (no/yes) *	10/8	17/16	24/20	17/11	0.911
HDL (mg/dL) *	46.6 (28.6–66.5)	47.3 (25.7–97.4)	47.1 (21.3–77.5)	44.5 (18.6–81.3)	0.292
LDL (mg/dL) *	106.9 (26.6–171.7)	100.4 (26.2–175.3)	88.7 (26.9–244.3)	76.3 (31.8–228.3)	0.418
TG (mg/dL) *	124.8 (47.8–281.0)	104.2 (60.6–357.6)	111.6 (42.6–367.8)	110.3 (43.4–332.4)	0.545
TC (mg/dL) *	178.2 (73.3–255.9)	166.9 (101.7–256.7)	159.9 (94.1–334.1)	154.5 (70.0–310.8)	0.316
Hypertension (no/yes)	5/17	6/28	4/42	4/27	0.421
History of MI (no/yes)	22/0	26/8	27/19	18/13	0.002
Smoking (no/yes/ex-smokers) **	16/5/1	22/10/2	19/17/10	19/8/4	—
Smoking (no/yes)	16/5	22/10	19/17	19/8	0.254
Zn (ppm)	161 (70–214)	169 (74–226)	169 (81–495)	166 (25–403)	0.670

*—assessed in 123 patients; **—three subgroups of patients due to the low number of patients in the selected subgroup; data for three and two groups are presented separately for statistical purposes.

3.3. *Difference in Zn Level between Patients with Stable CAD and Patients with ACS*

Significant differences were observed between patients with ACS and stable CAD in LDL and triglyceride (TG) levels (Table 2).

Table 2. Differences in selected parameters between patients with ACS and stable CAD.

	ACS	Stable CAD	*p*-Value
N	67	66	-
Sex (♀/♂)	16/51	21/45	0.307
Age (years)	66 (37–95)	63 (51–85)	0.345
BMI (kg/m^2)	27.7 (16.9–45.3)	27.6 (19.3–54.1)	0.791
t2DM (no/yes/pre-diabetes) **	44/19/4	40/23/3	—
t2DM (no/yes)	44/19	40/23	0.449
Hyperlipidaemia (no/yes) *	35/26	33/29	0.643
HDL (mg/dL) *	46.6 (25.7–81.3)	46.0 (18.6–97.4)	0.279
LDL (mg/dL) *	101.5 (26.6–244.3)	78.6 (26.2–204.0)	0.008
TG (mg/dL) *	103.3 (42.6–367.8)	124.7 (66.6–357.6)	0.008
TC (mg/dL) *	168.3 (70.0–334.1)	159.6 (96.4–291.1)	0.142
Hypertension (no/yes)	8/59	11/55	0.436
History of MI (no/yes)	49/18	44/22	0.416
Smoking (no/yes/ex-smokers) **	43/22/2	33/18/15	—
Smoking (no/yes)	43/22	33/18	0.871
Zn (ppm)	166.0 (39.0–285.0)	166.5 (25.0–495.0)	0.937

*—assessed in 123 patients; **—three subgroups of patients due to the low number of patients in the selected subgroup; data for three and two groups are presented separately for statistical purposes.

A lack of significant association between Zn level and CASSS in groups of patients without (H = 2.076 p = 0.557; Figure 1A) and with a history of MI (H = 0.000 p = 1.000; Figure 1B) was observed.

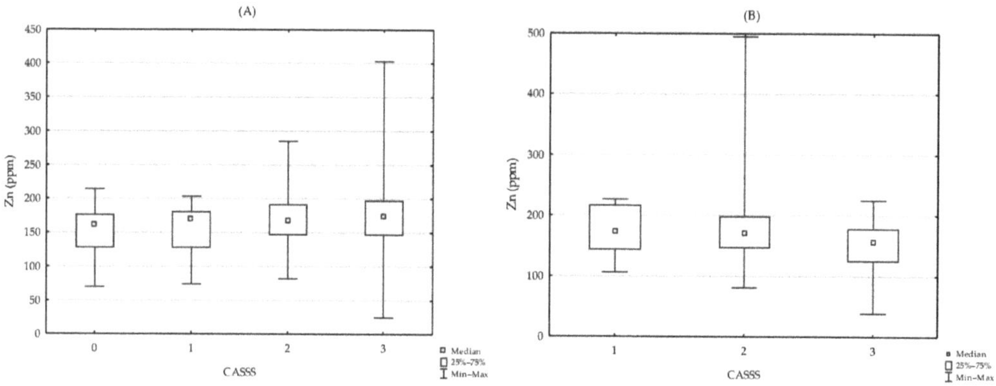

Figure 1. Association between Zn level and CASSS in group of patients without (**A**) and with history of MI (**B**).

There were no significant differences in Zn level between patients with stable CAD and ACS in groups of patients without ($p = 0.159$; Figure 2A) and with a history of MI ($p = 0.084$; Figure 2B).

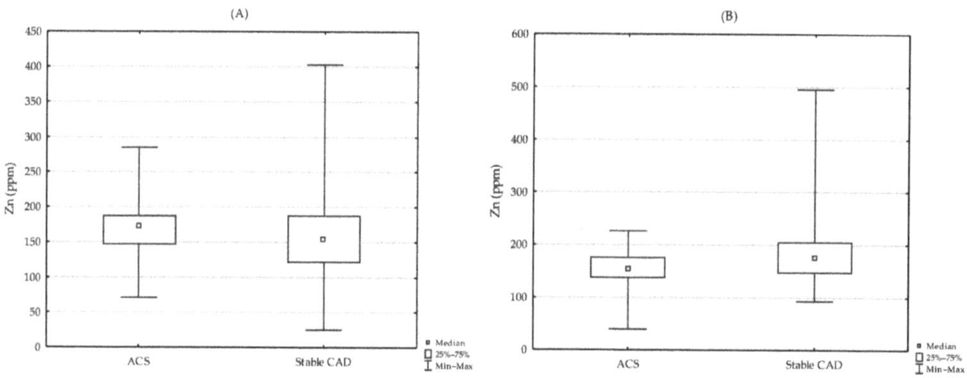

Figure 2. Differences in Zn level between patients with stable CAD and ACS in group of patients without (**A**) and with history of MI (**B**).

3.4. Association between Zn Level and Selected Parameters

There was no significant correlation between Zn and age (R = −0.005, $p = 0.952$; Figure 3F) or BMI (R = −0.15, $p = 0.084$; Figure 3G). There were no significant differences in Zn between males and females ($p = 0.218$; Figure 3A), patients with different smoking status (H 2, (N = 133) = 2073; $p = 0.355$; Figure 3D), patients with and without hypertension ($p = 0.384$; Figure 3B) and patients with or without hyperlipidaemia ($p = 0.335$; Figure 3C). Significant association was observed between Zn level and t2DM (H 2, (N = 133) = 10,952; $p = 0.004$; Figure 3E). Patients with t2DM presented significantly lower Zn values than patients without t2DM ($p = 0.006$). A lack of significant correlation was observed between Zn and TC (R = −0.070, $p = 0.440$), HDL (R = 0.091, $p = 0.320$), LDL (R = −0.020, $p = 0.829$) (Figure 4). A significant correlation was noted between Zn and TG (R = −0.193, $p = 0.032$) (Figure 4).

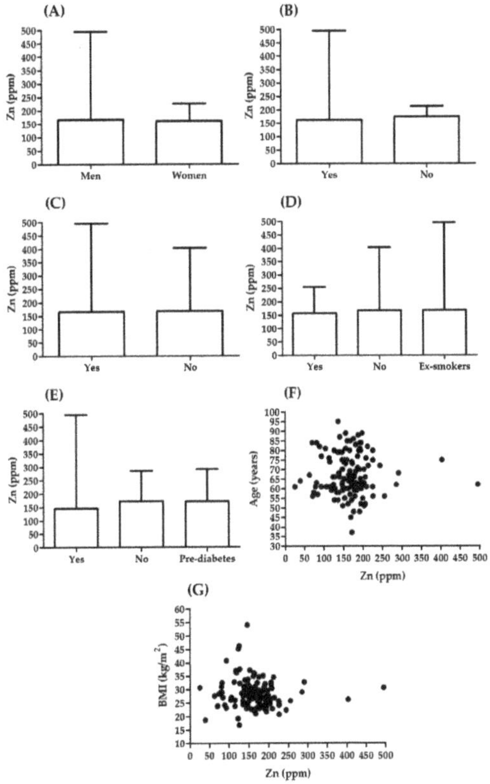

Figure 3. Association between Zn level and selected parameters: (**A**) sex, (**B**) hypertension, (**C**) hyperlipidaemia, (**D**) smoking status, (**E**) t2DM, (**F**) age, (**G**) BMI.

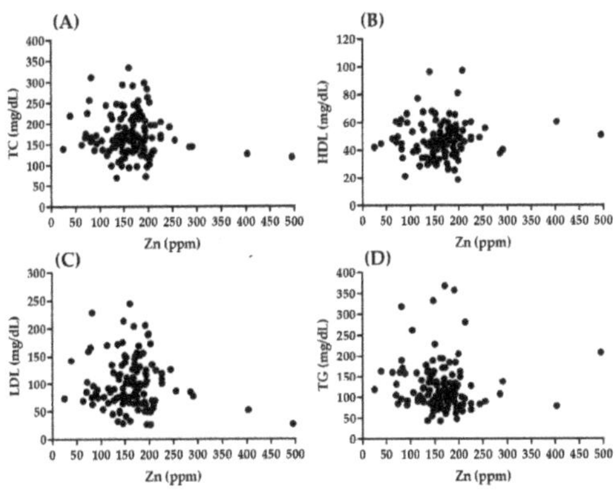

Figure 4. Correlation between Zn level and lipid profile: (**A**) TC, (**B**) HDL, (**C**) LDL, (**D**) TG.

4. Discussion

This study analysed the nutritional status of Zn among patients with angiographically confirmed CAD. Intake was assessed in hair samples using ICP-OES. There was no association between the bioelement and the advancement of CAD and episodes of myocardial infarction in the analysed cohort of patients. Moreover, patients with t2DM had significantly lower Zn content in comparison to non-diabetic individuals. In this cohort, serum TG concentration was found to negatively correlate with Zn content in hair.

Zinc (Zn) is one of the key microelements of the human body [2]. Of all the transition metals, it is second only to haemoglobin-bound iron in terms of prevalence in humans [25]. It is widespread in all tissues and bodily fluids [26]. Zn deficiency has been diagnosed in 17% of the world population. This percentage increases to 35% in developing countries [27]. As Zn influences the activity of more than 300 enzymes [28], it contributes to the stabilization of many protein structures, including more than 2000 transcription factors, most notably Zn finger proteins [29]. In the human heart, more than 24 Zn transporting proteins have been identified, indicating the important role of Zn in the homeostasis of the cardiovascular system [30].

The results of previous studies indicate that Zn has a protective role in atherosclerosis, a fundamental process in CVDs. Zn and Zn-transporting proteins are important for the function of the vessel wall and its integrity. Zn is essential for the superoxide dismutase function, as it is involved in the dimerization of endothelial NO synthase and nitric oxide production. NO enables labile Zn supplies to be released from endothelial cells, which dilate the vessels and protect endothelial cells [31]. Zn deficiency allows atherosclerotic plaque to build through the aggravation of oxidative stress; destruction of NO, NF-kB and the endothelium; as well as the production of pro-inflammatory cytokines [32]. An inverse correlation was found between Zn blood concentration and the risk of CVDs [33], atherosclerosis progression [34] and CVDs complications, such as acute coronary syndrome and heart failure [35,36]. Zn also plays a key role in the immune response, as its deficiency causes a weak cellular and humoral response [37]. By regulation of the oxidation–reduction balance of cells [32], this micronutrient facilitates the integrity of the endothelial cell membrane [38] and protects it from oxidation stress [39,40]. In animal models, Zn deficiency resulted in increased levels of reactive oxygen species [41], decreased levels of glutathione [42] and superoxide dismutase 1 [43]. Supplementation reduced serum lipid peroxidation [44] and normalized inducible nitric oxide synthase activity [45]. In both animal and human models, it also suppressed the expression of pro-inflammatory cytokines regulated by NF-kB [46,47]. These effects are components of the anti-inflammatory action, which modulates the development of atherosclerotic plaques. In summary, the experimental data indicate that low levels of Zn correlate with endothelium dysfunction [47], high levels of oxidation stress and vessel wall inflammation [48]—all of them being well-established risk factors for atherosclerosis. The importance of adequate Zn intake is also supported by data suggesting a relationship between Zn deficiency and subclinical inflammation [9,49].

In this study, we did not observe a statistically significant difference in Zn levels in hair samples between subgroups of patients with different severity of CAD (CASSS 0–3). At present, the relationship between this bioelement and CAD has not yet been determined, similar to the data on CAD risk in patients and their nutritional status of Zn being equivocal. Although Islamoglu et al. suggested a direct relationship between Zn level and CAD diagnosis [10] and El-Mahdy et al. correlated lower concentrations of this microelement with a higher SYNTAX score [11], de Paula et al. suggested that Zn levels have no association with CAD course [12]. On the other hand, data from epidemiological studies show a relationship of micronutrient levels, including Zn, on the presence and progression of atherosclerosis [50,51] and CAD [9]. The ratio of Zn excretion in urine to serum concentration is well-correlated with CAD and its severity [52]. Liu et al. found that levels of serum zinc-α2-glycoprotein were decreased in patients with premature CAD [14], while Meng et al. observed that Zn levels were higher in ACS groups than in CAD groups [53]. Furthermore, their analysis revealed that serum Zn level is an independent risk factor for the development of CAD [13]. Gao et al. demonstrated an inverse correlation of Zn levels with the progression of atherosclerotic calcification in CAD [54] and the carotid intima-media thickness test as a subclinical marker of atherosclerosis [34,55]. The aforementioned articles, presenting a significant difference in Zn levels in CAD patients compared to healthy controls, are not in line with our results. This discrepancy may be due to the characteristics of the patient group whose CAD is angiographically confirmed.

In our cohort, in addition to the lack of a relationship between Zn and CAD, we did not find significant differences in the measured concentrations in patients with ACS and stable CAD. Just as the association of Zn levels with CAD is not well-established, the relationship of this micronutrient with ACS is yet to be conclusively settled. The available data show a significant decrease in Zn levels in serum and hair of patients after ACS compared to healthy controls [11,56,57]. Lower amounts of this bioelement were also observed in patients with ST-elevation myocardial infarction compared to those with non-ST-elevation myocardial infarction [11]. An inverse correlation of serum Zn concentration with the concentration of myocardial necrosis markers and clinical predictors of ACS was found [56,58]. Furthermore, the higher the concentration of Zn, the less frequently ACS was observed in a group of patients [58]. In addition to the suggested use of Zn as the prognostic marker for ACS, Lal et al. focused on the cardioprotective role of this microelement after myocardial infarction [59]. In animal models, Zn administration caused a halving of the area of the induced infarction. In addition, a reduction in the frequency of arrythmia was observed [60,61]. Taking into account the discrepancies in the results of a few randomized trials [62], some cohort research articles [33] and our results showing the lack of association of Zn in acute CAD complications, we suggest further research on this topic. A meta-analysis of prospective cohort studies by Chu et al. suggests the possibility of a correlation of higher Zn concentrations with lower CVDs risk observed in three of the five included papers. The effect of this micronutrient was more prominent in patients with t2DM who underwent angiography due to chest pain compared to the healthy population [33]. In addition, the Iowa Women's Health Study observed an inverse correlation of dietary supplementation of Zn with CVD mortality (CAD and stroke included) in more than 30,000 postmenopausal women observed over more than ten years [62]. This result was, however, not confirmed by the recent metanalysis (49 studies with more than 300,000 participants) by Schwingshackl et al. [63]. It is worth noting that both studies used a multinutrient supplement formula with quite a small dose of Zn ($\leq$20 mg/d), and most of the participants were healthy.

The significantly lower hair Zn content observed in patients with t2DM may indicate a correlation of this element with glucose metabolism. Experimental studies have shown that Zn plays a role in the regulation of synthesis, storage and insulin excretion from β-cells in the pancreas [64], as well as improving tissue insulin sensitivity [65,66] and regulating the activity of gluconeogenetic enzymes [67]. In diabetic patients, both hypozincaemia and hyperzincuria are frequent findings [68,69]. Supplementation with Zn reduces the risk of t2DM by up to 40% [70] and improves glucose control [71].

Our data, collected from patients with CAD treated with comparable statin doses, reveal a negative correlation between Zn content and serum TG concentration. A similar trend is observed without statistical significance for total cholesterol and LDL. These findings seem to be consistent with other studies since Zn deficiency is found to have several effects on lipid metabolism. Firstly, it decreases the TG absorption [72]. Secondly, it causes a decrease in levels of Zn-α2-glycoprotein, which in turn increases lipogenesis [73]. Supplementation with Zn decreases total cholesterol levels and increases high-density lipoprotein levels [74]. Compared to the healthy control group, lower levels of this micronutrient were found in obese patients [75]. Zn supplementation also decreases insulin resistance and inflammation marker concentrations [76].

Since almost 86% of our patients were normotensive, we were unable to prove or disprove the existence of a correlation between the nutritional status of Zn and hypertension. However, previous studies on this matter showed contradictory results [55,77,78]. Similarly, we did not find any correlation between hair Zn levels and BMI or smoking; this subject needs further in-depth research, as previous research was mainly conducted on healthy individuals [75,79].

There are a few notable limitations to this study. In particular, we did not take into consideration factors interfering with Zn levels, e.g., the potential interactions of nutritional ingredients and other microelements, which can modulate the biological effects of Zn. We omitted the differentiation of the source of Zn, which could have an influence on metabolic syndrome and the risk of CVDs [80]. We neglected the effect of drugs commonly used in CVDs, such as beta-blockers, angiotensin receptor blockers, angiotensin-converting enzyme inhibitors and diuretics. Our analysis was based on a single method of quantifying the Zn level, namely hair concentration. This omits the fact that Zn can be in different forms and its concentration changes are dynamic. The use of multiple measurement methods (hair, serum, erythrocytes, urine concentration) may result in a better understanding of the relationship between Zn and CVDs. The degree of progression of CAD was determined on the basis of the results of coronary angiography using the CASSS; perhaps using the SYNTAX or Gensini score would be more optimal.

We did not confirm an association of Zn nutritional status with CAD severity and ACS in the cohort of patients with CAD. We found a statistically significant, negative, correlation between Zn levels in hair and serum TG concentration. Furthermore, patients with t2DM had a lower Zn hair content compared to non-diabetic individuals. Evaluation of the impact of Zn on cardiovascular health requires more well-designed randomized studies to specify the advantages, dangers and contraindications for different levels of Zn dietary intake and supplementation. It is necessary to define and implement global dietary recommendations and food fortification strategies, particularly in developing countries, to achieve optimal Zn intake and reduce CVD risk.

5. Conclusions

In patients with angiographically confirmed CAD, there are no significant statistical differences in Zn levels between groups with different severity of CAD (CASSS 0–3). Differences in the concentration of the analysed element in hair samples were not statistically significant in patients with ACS compared to those with stable CAD. Significantly lower levels of Zn were found in the hair of patients with t2DM. A negative correlation was identified between Zn content in the hair and serum TG concentration. The role of Zn in the pathogenesis of CAD and its complications requires further well-designed research.

Author Contributions: Conceptualization, E.A.D.; methodology, E.A.D. and M.K.; statistical analysis, J.S.G.; investigation, E.A.D.; data curation, E.A.D.; writing—original draft preparation, E.A.D., J.S.G., A.T. and J.P.; writing—review and editing, all authors; funding acquisition, E.A.D. and M.D. All authors have read and agreed to the published version of the manuscript.

Funding: This research was partly supported by a statutory grant to the Cardiology Clinic of Physiotherapy Division from the 2nd Faculty of Medicine, Medical University of Warsaw, Poland (grant number: 2F5/PM2/16). The APC was funded by the Lazarski University in Warsaw.

Institutional Review Board Statement: The study was conducted according to the guidelines of the Declaration of Helsinki and approved by the Institutional Review Board (or Ethics Committee) of the Medical University of Warsaw (KB/124/2014).

Informed Consent Statement: Informed consent was obtained from all subjects involved in the study.

Data Availability Statement: Data can be provided by the corresponding author upon reasonable request.

Conflicts of Interest: The authors declare no conflict of interest.

References

1. World Health Organisation. Global Health Estimates: Life Expectancy and Leading Causes of Death and Disability. Available online: https://www.who.int/data/gho/data/themes/mortality-and-global-health-estimates (accessed on 11 May 2022).
2. Knez, M.; Glibetic, M. Zinc as a Biomarker of Cardiovascular Health. *Front. Nutr.* **2021**, *8*, 686078. [CrossRef] [PubMed]
3. Hashemian, M.; Poustchi, H.; Mohammadi-Nasrabadi, F.; Hekmatdoost, A. Systematic Review of Zinc Biochemical Indicators and Risk of Coronary Heart Disease. *ARYA Atheroscler.* **2015**, *11*, 357–365.

4. Alexanian, I.; Parissis, J.; Farmakis, D.; Athanaselis, S.; Pappas, L.; Gavrielatos, G.; Mihas, C.; Paraskevaidis, I.; Sideris, A.; Kremastinos, D.; et al. Clinical and Echocardiographic Correlates of Serum Copper and Zinc in Acute and Chronic Heart Failure. *Clin. Res. Cardiol. Off. J. Ger. Card. Soc.* **2014**, *103*, 938–949. [CrossRef] [PubMed]
5. Yu, X.; Huang, L.; Zhao, J.; Wang, Z.; Yao, W.; Wu, X.; Huang, J.; Bian, B. The Relationship between Serum Zinc Level and Heart Failure: A Meta-Analysis. *BioMed Res. Int.* **2018**, *2018*, 2739014. [CrossRef] [PubMed]
6. Alcantara, E.H.; Shin, M.Y.; Feldmann, J.; Nixon, G.F.; Beattie, J.H.; Kwun, I.S. Long-Term Zinc Deprivation Accelerates Rat Vascular Smooth Muscle Cell Proliferation Involving the down-Regulation of JNK1/2 Expression in MAPK Signaling. *Atherosclerosis* **2013**, *228*, 46–52. [CrossRef]
7. Yoshihisa, A.; Abe, S.; Kiko, T.; Kimishima, Y.; Sato, Y.; Watanabe, S.; Kanno, Y.; Miyata-Tatsumi, M.; Misaka, T.; Sato, T.; et al. Association of Serum Zinc Level with Prognosis in Patients with Heart Failure. *J. Card. Fail.* **2018**, *24*, 375–383. [CrossRef]
8. Huang, L.; Teng, T.; Bian, B.; Yao, W.; Yu, X.; Wang, Z.; Xu, Z.; Sun, Y. Zinc Levels in Left Ventricular Hypertrophy. *Biol. Trace Elem. Res.* **2017**, *176*, 48–55. [CrossRef]
9. Beattie, J.H.; Kwun, I.-S. Is Zinc Deficiency a Risk Factor for Atherosclerosis? *Br. J. Nutr.* **2004**, *91*, 177–181. [CrossRef]
10. Islamoglu, Y.; Evliyaoglu, O.; Tekbas, E.; Cil, H.; Elbey, M.A.; Atilgan, Z.; Kaya, H.; Bilik, Z.; Akyuz, A.; Alan, S. The Relationship between Serum Levels of Zn and Cu and Severity of Coronary Atherosclerosis. *Biol. Trace Elem. Res.* **2011**, *144*, 436–444. [CrossRef]
11. El-Mahdy, R.I.; Mostafa, M.M.; El-Deen, H.S. Serum Zinc Measurement, Total Antioxidant Capacity, and Lipid Peroxide Among Acute Coronary Syndrome Patients with and Without ST Elevation. *Appl. Biochem. Biotechnol.* **2019**, *188*, 208–224. [CrossRef]
12. de Paula, R.C.S.; Aneni, E.C.; Costa, A.P.R.; Figueiredo, V.N.; Moura, F.A.; Freitas, W.M.; Quaglia, L.A.; Santos, S.N.; Soares, A.A.; Nadruz, W.; et al. Low Zinc Levels Is Associated with Increased Inflammatory Activity but Not with Atherosclerosis, Arteriosclerosis or Endothelial Dysfunction among the Very Elderly. *BBA Clin.* **2014**, *2*, 1–6. [CrossRef] [PubMed]
13. Meng, H.; Wang, Y.; Zhou, F.; Ruan, J.; Duan, M.; Wang, X.; Yu, Q.; Yang, P.; Chen, W.; Meng, F. Reduced Serum Zinc Ion Concentration Is Associated with Coronary Heart Disease. *Biol. Trace Elem. Res.* **2021**, *199*, 4109–4118. [CrossRef] [PubMed]
14. Liu, M.; Zhu, H.; Zhai, T.; Pan, H.; Wang, L.; Yang, H.; Yan, K.; Zeng, Y.; Gong, F. Serum Zinc-A2-Glycoprotein Levels Were Decreased in Patients with Premature Coronary Artery Disease. *Front. Endocrinol.* **2019**, *10*, 197. [CrossRef] [PubMed]
15. Arsenault, J.E.; Wuehler, S.E.; de Romaña, D.L.; Penny, M.E.; Sempértegui, F.; Brown, K.H. The Time of Day and the Interval since Previous Meal Are Associated with Plasma Zinc Concentrations and Affect Estimated Risk of Zinc Deficiency in Young Children in Peru and Ecuador. *Eur. J. Clin. Nutr.* **2011**, *65*, 184–190. [CrossRef] [PubMed]
16. Katz, S.A.; Chatt, A. *Hair Analysis: Applications in the Biomedical and Environmental Sciences*; VCH Publishers: New York, NY, USA, 1988; ISBN 978-0-89573-310-8.
17. Pozebon, D.; Scheffler, G.L.; Dressler, V.L. Elemental Hair Analysis: A Review of Procedures and Applications. *Anal. Chim. Acta* **2017**, *992*, 1–23. [CrossRef] [PubMed]
18. Mesko, M.F.; Henn, A.S.; Novo, D.L.R.; Wilhelm, E.A.; Flores, E.M.M. Elemental Determination for Clinical Diagnosis and Prognosis: Challenges and Trends in Sample Preparation. In *Comprehensive Analytical Chemistry*; Elsevier: Amsterdam, The Netherlands, 2022; Volume 97, pp. 1–52. ISBN 978-0-444-64347-6.
19. Suliburska, J. A Comparison of Levels of Select Minerals in Scalp Hair Samples with Estimated Dietary Intakes of These Minerals in Women of Reproductive Age. *Biol. Trace Elem. Res.* **2011**, *144*, 77–85. [CrossRef] [PubMed]
20. Klevay, L.M.; Christopherson, D.M.; Shuler, T.R. Hair as a Biopsy Material: Trace Element Data on One Man over Two Decades. *Eur. J. Clin. Nutr.* **2004**, *58*, 1359–1364. [CrossRef]
21. Gellein, K.; Lierhagen, S.; Brevik, P.S.; Teigen, M.; Kaur, P.; Singh, T.; Flaten, T.P.; Syversen, T. Trace Element Profiles in Single Strands of Human Hair Determined by HR-ICP-MS. *Biol. Trace Elem. Res.* **2008**, *123*, 250–260. [CrossRef]
22. Khan, S.R.; Sharma, B.; Chawla, P.A.; Bhatia, R. Inductively Coupled Plasma Optical Emission Spectrometry (ICP-OES): A Powerful Analytical Technique for Elemental Analysis. *Food Anal. Methods* **2022**, *15*, 666–688. [CrossRef]
23. Lawton, J.S.; Tamis-Holland, J.E.; Bangalore, S.; Bates, E.R.; Beckie, T.M.; Bischoff, J.M.; Bittl, J.A.; Cohen, M.G.; DiMaio, J.M.; Don, C.W.; et al. 2021 ACC/AHA/SCAI Guideline for Coronary Artery Revascularization: Executive Summary: A Report of the American College of Cardiology/American Heart Association Joint Committee on Clinical Practice Guidelines. *Circulation* **2022**, *145*, e4–e17. [CrossRef]
24. Collet, J.-P.; Thiele, H.; Barbato, E.; Barthélémy, O.; Bauersachs, J.; Bhatt, D.L.; Dendale, P.; Dorobantu, M.; Edvardsen, T.; Folliguet, T.; et al. 2020 ESC Guidelines for the Management of Acute Coronary Syndromes in Patients Presenting without Persistent ST-Segment Elevation. *Eur. Heart J.* **2021**, *42*, 1289–1367. [CrossRef] [PubMed]
25. Vasak, M. Metallothioneins: New Functional and Structural Insights. *Curr. Opin. Chem. Biol.* **2000**, *4*, 177–183. [CrossRef]
26. Chasapis, C.T.; Loutsidou, A.C.; Spiliopoulou, C.A.; Stefanidou, M.E. Zinc and Human Health: An Update. *Arch. Toxicol.* **2012**, *86*, 521–534. [CrossRef] [PubMed]
27. Wessells, K.R.; Brown, K.H. Estimating the Global Prevalence of Zinc Deficiency: Results Based on Zinc Availability in National Food Supplies and the Prevalence of Stunting. *PLoS ONE* **2012**, *7*, e50568. [CrossRef] [PubMed]
28. Rink, L.; Gabriel, P. Zinc and the Immune System. *Proc. Nutr. Soc.* **2000**, *59*, 541–552. [CrossRef] [PubMed]
29. Beyersmann, D. Homeostasis and Cellular Functions of Zinc. *Mater. Werkst.* **2002**, *33*, 764–769. [CrossRef]
30. Fagerberg, L.; Hallström, B.M.; Oksvold, P.; Kampf, C.; Djureinovic, D.; Odeberg, J.; Habuka, M.; Tahmasebpoor, S.; Danielsson, A.; Edlund, K.; et al. Analysis of the Human Tissue-Specific Expression by Genome-Wide Integration of Transcriptomics and Antibody-Based Proteomics. *Mol. Cell. Proteom. MCP* **2014**, *13*, 397–406. [CrossRef]

31. Zalewski, P.D.; Beltrame, J.F.; Wawer, A.A.; Abdo, A.I.; Murgia, C. Roles for Endothelial Zinc Homeostasis in Vascular Physiology and Coronary Artery Disease. *Crit. Rev. Food Sci. Nutr.* **2019**, *59*, 3511–3525. [CrossRef]
32. Choi, S.; Liu, X.; Pan, Z. Zinc Deficiency and Cellular Oxidative Stress: Prognostic Implications in Cardiovascular Diseases. *Acta Pharmacol. Sin.* **2018**, *39*, 1120–1132. [CrossRef]
33. Chu, A.; Foster, M.; Samman, S. Zinc Status and Risk of Cardiovascular Diseases and Type 2 Diabetes Mellitus-A Systematic Review of Prospective Cohort Studies. *Nutrients* **2016**, *8*, 707. [CrossRef]
34. Masley, S.C.; Roetzheim, R.; Masley, L.V.; McNamara, T.; Schocken, D.D. Emerging Risk Factors as Markers for Carotid Intima Media Thickness Scores. *J. Am. Coll. Nutr.* **2015**, *34*, 100–107. [CrossRef] [PubMed]
35. Shokrzadeh, M.; Ghaemian, A.; Salehifar, E.; Aliakbari, S.; Saravi, S.S.S.; Ebrahimi, P. Serum Zinc and Copper Levels in Ischemic Cardiomyopathy. *Biol. Trace Elem. Res.* **2009**, *127*, 116–123. [CrossRef] [PubMed]
36. Little, P.J.; Bhattacharya, R.; Moreyra, A.E.; Korichneva, I.L. Zinc and Cardiovascular Disease. *Nutr. Burbank Los Angel. Cty. Calif* **2010**, *26*, 1050–1057. [CrossRef] [PubMed]
37. Pae, M.; Meydani, S.N.; Wu, D. The Role of Nutrition in Enhancing Immunity in Aging. *Aging Dis.* **2012**, *3*, 91–129. [PubMed]
38. Korichneva, I. Zinc Dynamics in the Myocardial Redox Signaling Network. *Antioxid. Redox Signal.* **2006**, *8*, 1707–1721. [CrossRef]
39. Allen-Redpath, K.; Ou, O.; Beattie, J.H.; Kwun, I.-S.; Feldmann, J.; Nixon, G.F. Marginal Dietary Zinc Deficiency In Vivo Induces Vascular Smooth Muscle Cell Apoptosis in Large Arteries. *Cardiovasc. Res.* **2013**, *99*, 525–534. [CrossRef]
40. Prasad, A.S.; Beck, F.W.J.; Bao, B.; Fitzgerald, J.T.; Snell, D.C.; Steinberg, J.D.; Cardozo, L.J. Zinc Supplementation Decreases Incidence of Infections in the Elderly: Effect of Zinc on Generation of Cytokines and Oxidative Stress. *Am. J. Clin. Nutr.* **2007**, *85*, 837–844. [CrossRef]
41. Eide, D.J. The Oxidative Stress of Zinc Deficiency. *Met. Integr. Biomet. Sci.* **2011**, *3*, 1124–1129. [CrossRef]
42. Cortese, M.M.; Suschek, C.V.; Wetzel, W.; Kröncke, K.-D.; Kolb-Bachofen, V. Zinc Protects Endothelial Cells from Hydrogen Peroxide via Nrf2-Dependent Stimulation of Glutathione Biosynthesis. *Free Radic. Biol. Med.* **2008**, *44*, 2002–2012. [CrossRef]
43. Li, H.-T.; Jiao, M.; Chen, J.; Liang, Y. Roles of Zinc and Copper in Modulating the Oxidative Refolding of Bovine Copper, Zinc Superoxide Dismutase. *Acta Biochim. Biophys. Sin.* **2010**, *42*, 183–194. [CrossRef]
44. Jenner, A.; Ren, M.; Rajendran, R.; Ning, P.; Huat, B.T.K.; Watt, F.; Halliwell, B. Zinc Supplementation Inhibits Lipid Peroxidation and the Development of Atherosclerosis in Rabbits Fed a High Cholesterol Diet. *Free Radic. Biol. Med.* **2007**, *42*, 559–566. [CrossRef] [PubMed]
45. Cortese-Krott, M.M.; Kulakov, L.; Opländer, C.; Kolb-Bachofen, V.; Kröncke, K.-D.; Suschek, C.V. Zinc Regulates INOS-Derived Nitric Oxide Formation in Endothelial Cells. *Redox Biol.* **2014**, *2*, 945–954. [CrossRef] [PubMed]
46. Wong, C.P.; Ho, E. Zinc and Its Role in Age-Related Inflammation and Immune Dysfunction. *Mol. Nutr. Food Res.* **2012**, *56*, 77–87. [CrossRef] [PubMed]
47. Beattie, J.H.; Gordon, M.-J.; Duthie, S.J.; McNeil, C.J.; Horgan, G.W.; Nixon, G.F.; Feldmann, J.; Kwun, I.-S. Suboptimal Dietary Zinc Intake Promotes Vascular Inflammation and Atherogenesis in a Mouse Model of Atherosclerosis. *Mol. Nutr. Food Res.* **2012**, *56*, 1097–1105. [CrossRef]
48. Jurowski, K.; Szewczyk, B.; Nowak, G.; Piekoszewski, W. Biological Consequences of Zinc Deficiency in the Pathomechanisms of Selected Diseases. *J. Biol. Inorg. Chem. JBIC Publ. Soc. Biol. Inorg. Chem.* **2014**, *19*, 1069–1079. [CrossRef]
49. Bao, B.; Prasad, A.S.; Beck, F.W.J.; Fitzgerald, J.T.; Snell, D.; Bao, G.W.; Singh, T.; Cardozo, L.J. Zinc Decreases C-Reactive Protein, Lipid Peroxidation, and Inflammatory Cytokines in Elderly Subjects: A Potential Implication of Zinc as an Atheroprotective Agent. *Am. J. Clin. Nutr.* **2010**, *91*, 1634–1641. [CrossRef]
50. Kritchevsky, D. Diet and Atherosclerosis. *J. Nutr. Health Aging* **2001**, *5*, 155–159.
51. Singh, R.B.; Niaz, M.A.; Rastogi, S.S.; Bajaj, S.; Gaoli, Z.; Shoumin, Z. Current Zinc Intake and Risk of Diabetes and Coronary Artery Disease and Factors Associated with Insulin Resistance in Rural and Urban Populations of North India. *J. Am. Coll. Nutr.* **1998**, *17*, 564–570. [CrossRef]
52. Giannoglou, G.D.; Konstantinou, D.M.; Kovatsi, L.; Chatzizisis, Y.S.; Mikhailidis, D.P. Association of Reduced Zinc Status with Angiographically Severe Coronary Atherosclerosis: A Pilot Study. *Angiology* **2010**, *61*, 449–455. [CrossRef]
53. Meng, H.; Wang, Y.; Ruan, J.; Chen, Y.; Wang, X.; Zhou, F.; Meng, F. Decreased Iron Ion Concentrations in the Peripheral Blood Correlate with Coronary Atherosclerosis. *Nutrients* **2022**, *14*, 319. [CrossRef]
54. Gao, J.-W.; Zhang, S.-L.; Hao, Q.-Y.; Huang, F.-F.; Liu, Z.-Y.; Zhang, H.-F.; Yan, L.; Wang, J.-F.; Liu, P.-M. Association of Dietary Zinc Intake with Coronary Artery Calcium Progression: The Multi-Ethnic Study of Atherosclerosis (MESA). *Eur. J. Nutr.* **2021**, *60*, 2759–2767. [CrossRef] [PubMed]
55. Yang, Y.J.; Choi, B.Y.; Chun, B.-Y.; Kweon, S.-S.; Lee, Y.-H.; Park, P.S.; Kim, M.K. Dietary Zinc Intake Is Inversely Related to Subclinical Atherosclerosis Measured by Carotid Intima-Media Thickness. *Br. J. Nutr.* **2010**, *104*, 1202–1211. [CrossRef] [PubMed]
56. Low, W.I.; Ikram, H. Plasma Zinc in Acute Myocardial Infarction. Diagnostic and Prognostic Implications. *Br. Heart J.* **1976**, *38*, 1339–1342. [CrossRef] [PubMed]
57. Liu, B.; Cai, Z.-Q.; Zhou, Y.-M. Deficient Zinc Levels and Myocardial Infarction: Association between Deficient Zinc Levels and Myocardial Infarction: A Meta-Analysis. *Biol. Trace Elem. Res.* **2015**, *165*, 41–50. [CrossRef]
58. Huang, L.; Teng, T.; Zhao, J.; Bian, B.; Yao, W.; Yu, X.; Wang, Z.; Xu, Z.; Sun, Y. The Relationship Between Serum Zinc Levels, Cardiac Markers and the Risk of Acute Myocardial Infarction by Zinc Quartiles. *Heart Lung Circ.* **2018**, *27*, 66–72. [CrossRef]
59. Lal, A. Effect of Zinc Sulphate on Infarct Size in Experimental Myocardial Infarction in Dogs. *Indian J. Med. Res.* **1991**, *94*, 316–319.

60. Karagulova, G.; Yue, Y.; Moreyra, A.; Boutjdir, M.; Korichneva, I. Protective Role of Intracellular Zinc in Myocardial Ischemia/Reperfusion Is Associated with Preservation of Protein Kinase C Isoforms. *J. Pharmacol. Exp. Ther.* **2007**, *321*, 517–525. [CrossRef]
61. Xu, Z.; Kim, S.; Huh, J. Zinc Plays a Critical Role in the Cardioprotective Effect of Postconditioning by Enhancing the Activation of the RISK Pathway in Rat Hearts. *J. Mol. Cell. Cardiol.* **2014**, *66*, 12–17. [CrossRef]
62. Lee, D.-H.; Folsom, A.R.; Jacobs, D.R. Iron, Zinc, and Alcohol Consumption and Mortality from Cardiovascular Diseases: The Iowa Women's Health Study. *Am. J. Clin. Nutr.* **2005**, *81*, 787–791. [CrossRef]
63. Schwingshackl, L.; Boeing, H.; Stelmach-Mardas, M.; Gottschald, M.; Dietrich, S.; Hoffmann, G.; Chaimani, A. Dietary Supplements and Risk of Cause-Specific Death, Cardiovascular Disease, and Cancer: A Systematic Review and Meta-Analysis of Primary Prevention Trials. *Adv. Nutr. Bethesda Md* **2017**, *8*, 27–39. [CrossRef]
64. Ruz, M.; Carrasco, F.; Rojas, P.; Basfi-Fer, K.; Hernández, M.C.; Pérez, A. Nutritional Effects of Zinc on Metabolic Syndrome and Type 2 Diabetes: Mechanisms and Main Findings in Human Studies. *Biol. Trace Elem. Res.* **2019**, *188*, 177–188. [CrossRef] [PubMed]
65. Sujatha, P. Trace Elements in Diabetes Mellitus. *J. Clin. Diagn. Res.* **2013**, *7*, 1863–1865. [CrossRef] [PubMed]
66. Maret, W. Zinc in Pancreatic Islet Biology, Insulin Sensitivity, and Diabetes. *Prev. Nutr. Food Sci.* **2017**, *22*, 1–8. [CrossRef] [PubMed]
67. Chabosseau, P.; Rutter, G.A. Zinc and Diabetes. *Arch. Biochem. Biophys.* **2016**, *611*, 79–85. [CrossRef] [PubMed]
68. Garg, V.K.; Gupta, R.; Goyal, R.K. Hypozincemia in Diabetes Mellitus. *J. Assoc. Physicians India* **1994**, *42*, 720–721.
69. Pidduck, H.G.; Wren, P.J.J.; Price Evans, D.A. Hyperzincuria of Diabetes Mellitus and Possible Genetical Implications of This Observation. *Diabetes* **1970**, *19*, 240–247. [CrossRef]
70. Fernández-Cao, J.C.; Warthon-Medina, M.; H Moran, V.; Arija, V.; Doepking, C.; Serra-Majem, L.; Lowe, N.M. Zinc Intake and Status and Risk of Type 2 Diabetes Mellitus: A Systematic Review and Meta-Analysis. *Nutrients* **2019**, *11*, 1027. [CrossRef]
71. Capdor, J.; Foster, M.; Petocz, P.; Samman, S. Zinc and Glycemic Control: A Meta-Analysis of Randomised Placebo Controlled Supplementation Trials in Humans. *J. Trace Elem. Med. Biol. Organ Soc. Miner. Trace Elem. GMS* **2013**, *27*, 137–142. [CrossRef]
72. Desmarchelier, C.; Borel, P.; Lairon, D.; Maraninchi, M.; Valéro, R. Effect of Nutrient and Micronutrient Intake on Chylomicron Production and Postprandial Lipemia. *Nutrients* **2019**, *11*, 1299. [CrossRef]
73. Olechnowicz, J.; Tinkov, A.; Skalny, A.; Suliburska, J. Zinc Status Is Associated with Inflammation, Oxidative Stress, Lipid, and Glucose Metabolism. *J. Physiol. Sci. JPS* **2018**, *68*, 19–31. [CrossRef]
74. Ranasinghe, P.; Wathurapatha, W.S.; Ishara, M.H.; Jayawardana, R.; Galappatthy, P.; Katulanda, P.; Constantine, G.R. Effects of Zinc Supplementation on Serum Lipids: A Systematic Review and Meta-Analysis. *Nutr. Metab.* **2015**, *12*, 26. [CrossRef]
75. Knez, M.; Pantovic, A.; Zekovic, M.; Pavlovic, Z.; Glibetic, M.; Zec, M. Is There a Link between Zinc Intake and Status with Plasma Fatty Acid Profile and Desaturase Activities in Dyslipidemic Subjects? *Nutrients* **2019**, *12*, 93. [CrossRef] [PubMed]
76. Kelishadi, R.; Hashemipour, M.; Adeli, K.; Tavakoli, N.; Movahedian-Attar, A.; Shapouri, J.; Poursafa, P.; Rouzbahani, A. Effect of Zinc Supplementation on Markers of Insulin Resistance, Oxidative Stress, and Inflammation among Prepubescent Children with Metabolic Syndrome. *Metab. Syndr. Relat. Disord.* **2010**, *8*, 505–510. [CrossRef] [PubMed]
77. Afridi, H.; Kazi, T.; Kazi, N.; Talpur, F.; Naeemullah; Arain, S.; Brahman, K.; Wadhwa, S.; Shah, F. Distribution of Copper, Iron, and Zinc in Biological Samples of Pakistani Hypertensive Patients and Referent Subjects of Different Age Groups. *Clin. Lab.* **2013**, *59*. [CrossRef]
78. Ghayour-Mobarhan, M.; Shapouri-Moghaddam, A.; Azimi-Nezhad, M.; Esmaeili, H.; Parizadeh, S.M.R.; Safarian, M.; Kazemi-Bajestani, S.M.R.; Khodaei, G.H.; Hosseini, S.J.; Parizadeh, S.M.J.; et al. The Relationship between Established Coronary Risk Factors and Serum Copper and Zinc Concentrations in a Large Persian Cohort. *J. Trace Elem. Med. Biol.* **2009**, *23*, 167–175. [CrossRef] [PubMed]
79. Knez, M.; Nikolic, M.; Zekovic, M.; Stangoulis, J.C.; Gurinovic, M.; Glibetic, M. The Influence of Food Consumption and Socio-Economic Factors on the Relationship between Zinc and Iron Intake and Status in a Healthy Population. *Public Health Nutr.* **2017**, *20*, 2486–2498. [CrossRef] [PubMed]
80. De Oliveira Otto, M.C.; Alonso, A.; Lee, D.-H.; Delclos, G.L.; Bertoni, A.G.; Jiang, R.; Lima, J.A.; Symanski, E.; Jacobs, D.R.; Nettleton, J.A. Dietary Intakes of Zinc and Heme Iron from Red Meat, but Not from Other Sources, Are Associated with Greater Risk of Metabolic Syndrome and Cardiovascular Disease. *J. Nutr.* **2012**, *142*, 526–533. [CrossRef]

biomolecules

MDPI

Review

Zinc Homeostasis: An Emerging Therapeutic Target for Neuroinflammation Related Diseases

Shunfeng Liu [1,2,†], Nan Wang [2,3,†], Yaqian Long [4], Zhuan Wu [5] and Shouhong Zhou [2,3,*]

1 College of Pharmacy, Guilin Medical College, Guilin 541199, China
2 Guangxi Key Laboratory of Brain and Cognitive Neuroscience, Guilin Medical College, Guilin 541199, China
3 Basic Medical College, Guilin Medical College, Guilin 541199, China
4 Department of Radiology, Fifth Clinical Medical College, Guilin Medical College, Guilin 541002, China
5 Central Laboratory, Guangxi Health Commission Key Laboratory of Glucose and Lipid Metabolism Disorders, Second Affiliated Hospital of Guilin Medical College, Guilin 541199, China
* Correspondence: 112019023@glmc.edu.cn
† These authors contributed equally to this work.

Abstract: Zinc is an indispensable trace element in the human body and plays an important role in regulating normal growth and development. Zinc homeostasis in the central nervous system is closely related to the development of neuroinflammation, and synaptic zinc homeostasis disorders affect zinc homeostasis in the brain. Under the condition of synaptic zinc homeostasis, proper zinc supplementation improves the body's immunity and inhibits neuroinflammation. Synaptic zinc homeostasis disorder in the brain promotes the occurrence and development of neuroinflammation. Cerebral ischemia and hypoxia cause a massive release of synaptic Zn^{2+} into the synaptic cleft, resulting in neurotoxicity and neuroinflammation. Synaptic zinc homeostasis disorder is a high-risk factor for neurodegenerative diseases. Maintaining cerebral zinc homeostasis suppresses the progression of neuroinflammation-mediated neurodegenerative diseases. This article reviews the relationship between brain zinc homeostasis and neuroinflammation and proposes that maintaining synaptic zinc homeostasis prevents neuroinflammation.

Keywords: zinc homeostasis; Zn^{2+}; neuroinflammation; synaptic zinc; neuroinflammation-related diseases

1. Introduction

Neuroinflammation is an inflammatory response that occurs in the central nervous system (CNS). Because the blood-brain barrier (BBB) prevents a large number of peripheral immune cells from entering the brain, neuroinflammation is different from peripheral inflammation, and microglia are the main functional cells in the CNS involved in the inflammatory response. Resting-state microglia are highly branched and monitor the state of other cells in the brain [1]. Under stimulation, resting microglia become polarized to form an amoeba-like shape and become activated [1]. Proinflammatory stimulation induces the polarization of microglia to the M1 type, which induces microglial expression and release of interleukin (IL)-1β, IL-6, and tumor necrosis factor-α (TNF-α) [2]. Anti-inflammatory stimulation induces microglial polarization to the M2 type, which induces microglia to express and release IL-4, IL-10 and transforms growth factor-β [2].

Zinc is the second most abundant trace element in the human body, and it exists stably in the form of Zn^{2+}. Zinc promotes neurogenesis [3], improves immunity [4], acts as a cofactor for enzymes [5], and participates in synaptic transmission [6] and cell apoptosis [7]. Many proteins in the human body, such as the zinc finger protein, contain Zn^{2+}; many enzymes in the body need to combine with Zn^{2+} to play a catalytic role [5]. Therefore, maintaining zinc homeostasis is important for the normal functioning of the human body, and zinc homeostasis disorder may be one of the causes of many diseases, such as cerebral ischemia and hypoxia [8], epilepsy [9], depression [10], and neurodegenerative diseases [11]. However, the mechanism by which zinc homeostasis disorders cause these diseases needs to be further researched.

Synaptic zinc homeostasis disorder affects brain zinc homeostasis. When synaptic zinc is in a steady state, appropriate supplement of zinc improves the body's immunity and inhibits neuroinflammation [12,13]. When synaptic zinc is disordered, abnormal release of synaptic Zn^{2+} leads to severe neuroinflammation. Severe stresses, such as cerebral ischemia and hypoxia [8], and epilepsy [14], lead to a large amount of synaptic Zn^{2+} being released into the synaptic cleft, which mediates neuronal death and severe neuroinflammation. Neurodegenerative diseases are chronic and progressive diseases with widespread neurotransmitter imbalance. Therefore, maintaining synaptic zinc homeostasis plays an important role in delaying the progression of neurodegenerative diseases [11,15,16]. Neurotoxicity and neuroinflammation caused by disorders of synaptic zinc homeostasis promote the progression of neurodegenerative diseases. This article outlines the relationship between brain zinc homeostasis disorder and neuroinflammation and suggests that maintaining synaptic zinc homeostasis prevents neuroinflammation.

2. Zinc Homeostasis and Its Regulatory Mechanism

2.1. Body Zinc Homeostasis

Zinc homeostasis in the body is regulated by the gastrointestinal tract by absorption of dietary zinc or excretion of zinc by the kidney based on existing levels [17]. More than 90% of the zinc in the body is stored in the skeletal muscle and skeleton, and less than 3% of the total zinc is found in the brain [17]. Zinc in the body exists stably in the form of Zn^{2+}, with one part of the Zn^{2+} being in the binding state and the other part being in the free state. Zinc in cells is divided into three parts: 1. Zn^{2+} combined with proteins, including metallothionein (MT); 2. Zn^{2+} in organelles and vesicles; and 3. Free Zn^{2+} in the cytoplasm [15,18]. Zrt/Irt-like protein (ZIP) (encoded by the SLC39A gene family) has 14 types of zinc transporters. ZIP transports extracellular Zn^{2+} as well as Zn^{2+} in organelles and vesicles to the cytoplasm [17]. ZIP plays an important role in maintaining intracellular zinc homeostasis, and the zinc transporter (ZnT) (encoded by the SLC30A gene family) has 10 ZnT members, among which ZnT9 does not possess the function of zinc transport; hence, only nine types of ZnT transport zinc [17]. The outflow of Zn^{2+} from the cytoplasm to the extracellular space is mainly mediated by ZnT1, while other zinc transporters mediate the transport of Zn^{2+} from the cytoplasm to organelles or vesicles [17]. ZnT in the cell membrane is mainly distributed in the organelle membrane, which mediates the uptake of Zn^{2+} by Golgi bodies, lysosomes, and vesicles [17]. MT is a protein rich in cysteine, which scavenges free radicals [19]. MT reduces the free zinc content in cells by combining with Zn^{2+} in the cytoplasm and also releases Zn^{2+} when needed [19]. Therefore, MT is a key protein that regulates cellular zinc homeostasis.

2.2. Brain Zinc Homeostasis and Synaptic Zinc

Zinc in the brain is mainly located in glutamate neurons, most of which are highly enriched in Zn^{2+} [6]. Highly enriched Zn^{2+} in the nerve endings of glutamate neurons is released into the synaptic cleft when nerve impulses arrive and enter the postsynaptic membrane, thus regulating the physiological function of the brain [6]. Therefore, an imbalance in synaptic zinc homeostasis significantly affects zinc homeostasis in the brain. Zn^{2+} binds to N-methyl-d-aspartate (NMDA) and γ-aminobutyric acid type A (GABA-A) receptors, thereby affecting excitatory and inhibitory synaptic transmission in the brain [6]. Synaptic transport of Zn^{2+} involves the release of Zn^{2+} from the presynaptic membrane to the postsynaptic membrane and into postsynaptic neurons.

ZnT3 mediates the transfer of Zn^{2+} from cells to synaptic vesicles [20,21], and synaptic vesicles containing Zn^{2+} combine with the presynaptic membrane to release Zn^{2+} into the synaptic cleft. The pathways of Zn^{2+} entering the postsynaptic neurons in the synaptic cleft include the following [6]: 1. ZIP-mediated transport; 2. Voltage-gated calcium channel (VGCC)-mediated transport; 3. Ca^{2+}-and Zn^{2+}-permeable GluR2-lacking AMPA (α-amino-3-hydroxy-5-methyl-4-isoxazole propionic acid) receptors ($AMPAR_{Ca-Zn}$)-mediated transport; 4. NMDA receptor-mediated transport; and 5. Na^+/Zn^{2+} exchanger-mediated transport [22]. The main transport protein of Zn^{2+} efflux is ZnT1, whereas MT stores Zn^{2+} and dynamically regulates zinc homeostasis in cells. The release of zinc from presynaptic to postsynaptic neurons increases the level of oxidative stress in postsynaptic neurons, leading to neuronal death [16,23] (as shown in Figure 1).

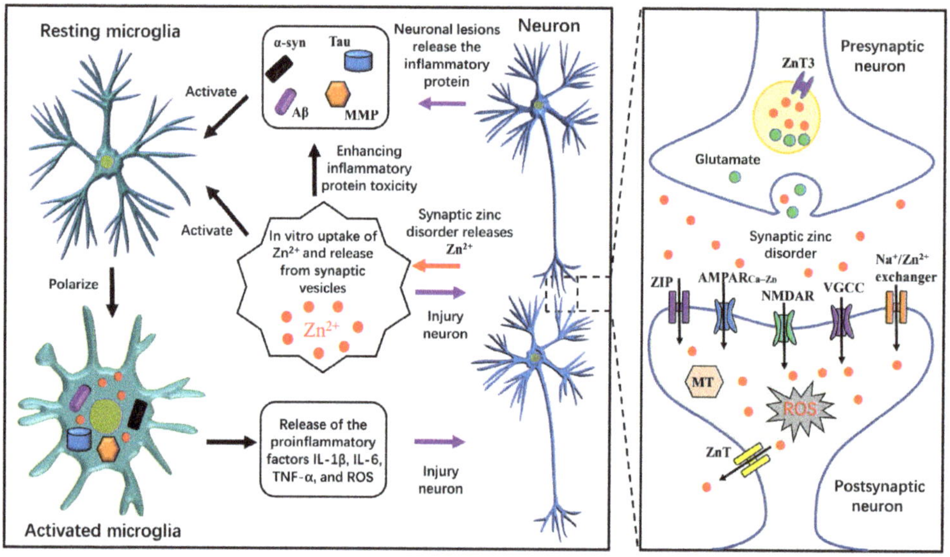

Figure 1. Zinc homeostasis disorder in the brain promotes neuroinflammation. When diseases, such as AD and PD, and conditions, such as cerebral ischemia and hypoxia, lead to synaptic function disorder, synaptic zinc levels are in a state of fragile balance or imbalance. In this case, zinc supplementation increases the extracellular Zn^{2+} and synaptic Zn^{2+} content in the brain, leading to the uncontrolled release of a large amount of Zn^{2+} from the synapse to the extracellular space. Large amounts of extracellular Zn^{2+} damage neurons in several ways: 1. Zn^{2+} flows directly into neurons, thus promoting oxidative stress and causing neuronal death; 2. Zn^{2+} damages neurons by activating microglia to produce proinflammatory factors and ROS; 3. Zn^{2+} damages neurons by enhancing the expression and neurotoxicity of inflammatory proteins (α-syn, Tau, Aβ, and MMP). Therefore, patients with disorders involving abnormal synaptic function should strictly control zinc uptake. α-synuclein (α-syn); Alzheimer's disease (AD); Parkinson's disease (PD); β-amyloid (Aβ); matrix metalloproteinase (MMP); interleukin (IL); tumor necrosis factor-α (TNF-α); reactive oxygen species (ROS); zinc transporter (ZnT); Zrt/Irt-like protein (ZIP); Ca^{2+}-and Zn^{2+}-permeable GluR2-lacking AMPA (α-amino-3-hydroxy-5-methyl-4-isoxazole propionic acid) receptors ($AMPAR_{Ca-Zn}$); N-methyl-d-aspartate receptor (NMDAR); voltage-gated calcium channel (VGCC); metallothionein (MT).

3. Matrix Metalloproteinases and Neuroinflammation

The catalytically active region of matrix metalloproteinase (MMP) contains a zinc-binding site, which is a zinc-dependent endopeptidase. Under the regulation of inflammatory cytokines and reactive oxygen species (ROS), neurons, astrocytes, and microglia express and release MMP [24,25]. MMP's function has two sides. MMP promotes tissue remodeling and repair. However, excessive activation also leads to tissue damage. Furthermore, overactivation of MMP destroys the BBB, leading to peripheral immune cells entering the brain and participating in the inflammatory response in the CNS. There is an interaction between neuroinflammation and MMP. IL-1β induces pericyte expression and secretion of MMP-9 by activating the nuclear factor-κB (NF-κB) signaling pathway, and high expression of MMP-9 damages the integrity of the BBB [26]. An in vitro study in which neurons and astrocytes were isolated from mice with Alzheimer's disease (AD) showed that knockout of the MMP-24 gene significantly reduced IL-1β and β-amyloid (Aβ) in cells [27]. Lipopolysaccharide (LPS) activates the NF-κB signaling pathway by increasing ROS content in astrocytes, thereby upregulating MMP-9 expression [28].

Zinc promotes MMP-mediated neuroinflammation. MMP activation requires Zn^{2+} binding to the catalytically active region of MMP. Treatment with the Zn^{2+} chelator clioquinol (CQ) significantly inhibited MMP-2 and MMP-9 activity in hippocampal neurons, indicating that Zn^{2+} is critical for MMP activation [29]. Treatment of mice with minocycline, an inhibitor of microglial activation, inhibits rotenone-induced MMP-2 and MMP-9 activation, indicating that microglial activation is a key factor mediating MMP activation [30]. Zinc chloride ($ZnCl_2$) treatment enhanced the effect of LPS-induced microglia to release proinflammatory factors, but ROS scavenger inhibited the effect of $ZnCl_2$, suggesting that Zn^{2+} promoted microglial expression and release of proinflammatory factors by mediating ROS generation [31]. Zn^{2+} increased the intracellular ROS content by activating reduced nicotinamide adenine dinucleotide phosphate oxidase (NOX) in microglia [8,32]. Zn^{2+} treatment significantly increased the content of ROS in neurons [33] and enhanced the effect of LPS-induced ROS production in astrocytes [34]. Persistently high concentrations of intracellular and extracellular Zn^{2+} would promote increased production of ROS and proinflammatory factors, especially in microglia, which mediate immune responses [32,35,36]. The generated ROS and proinflammatory factors increase the activity and expression of MMP, which further promotes neuroinflammation-mediated CNS injury (Figure 1).

4. Zinc Homeostasis and Neuroinflammation

4.1. Zinc Homeostasis Has a Role in Regulating Neuroinflammatory Responses

Zinc sulfate ($ZnSO_4$) downregulated the expression of inducible NO synthase (iNOS), TNF-α, and IL-6 induced by LPS (100 ng/mL) and significantly reduced the content of ROS in cells [37]. A20 is a negative regulator of NF-κB signaling and contains a zinc-finger domain. $ZnSO_4$ inhibits NF-κB signaling by upregulating A20 expression in BV2 cells, thereby inhibiting LPS-induced neuroinflammation [37]. Arginase 1 (Arg-1) is a marker of M2 polarization in microglia. IL-4 induces microglia to express Arg-1 mRNA and increases the cellular zinc in cells [38]. The intracellular Zn^{2+} chelator TPEN upregulated the expression of Arg-1 mRNA in microglia induced by IL-4, indicating that intracellular free Zn^{2+} negatively regulates the expression of Arg-1 [38]. In summary, intracellular Zn^{2+}, in cases of zinc homeostasis, inhibits microglial polarization to the M2 type. Under conditions of zinc homeostasis, Zn^{2+} not only inhibited the expression of proinflammatory factors in microglia but also inhibited the expression of anti-inflammatory factors, playing a role in regulating the neuroinflammatory response.

Zinc supplementation inhibited the activation of microglia in the cerebral cortex of rats fed a high sugar and fat diet, and downregulated the toll-like receptor 4 expression in the rat hippocampus. A high sugar and fat diet damaged the cognitive function of rats, and zinc supplementation significantly improved their cognitive function [4]. Therefore, zinc supplementation improved cognitive function in rats by reducing neuroinflammation. A high-fat diet induced high expression of IL-1β in the cerebral cortex and hippocampus of rats, while zinc supplementation significantly reduced the expression of IL-1β and enhanced novel object recognition in the rats [4]. Thus, zinc supplementation inhibited high-fat diet induced neuroinflammation. Maternal zinc supplementation downregulated the expression of the proinflammatory factors IL-6, IL-1β, and TNF-α in the prefrontal cortex (PFC) of LPS-exposed adult male offspring and decreased the expression of NF-κB in the PFC [13].

4.2. Zinc Homeostasis Imbalance

4.2.1. Low Zinc Promotes Neuroinflammation

TPEN significantly upregulated IL-1β expression in macrophages treated with LPS (1 μg/mL), whereas $ZnCl_2$ treatment reversed the effects of TPEN [39]. TPEN enhanced the release of IL-1β from macrophages induced by LPS [40]. Compared to monocytes cultured at normal Zn^{2+} concentrations, zinc-deficient monocytes treated with LPS showed significantly upregulated TNF-α and IL-1β expressions, indicating that zinc-deficient monocytes were more sensitive to LPS-induced inflammation [41]. A low-zinc diet or inhibition of zinc absorption changed the intestinal microbiota composition and intestinal function of pregnant mice, upregulating the expression of IL-6 in the brain and activating astrocytes in the hippocampus [42]. Therefore, zinc deficiency may contribute to neuroinflammation by altering intestinal function and intestinal microbiota composition. Prenatal zinc-deficient mice exhibit autism spectrum disorder. A study found that the brains of prenatal zinc-deficient mice showed a neuroinflammatory response, suggesting that prenatal zinc deficiency-mediated neuroinflammation may be the cause of autism in mice [43].

4.2.2. Zinc Accumulation Accelerates Neuroinflammation

Zn^{2+} increased the content of ROS in microglia by activating NOX in microglia; then, ROS activated the extracellular signal-regulated kinase (ERK) signaling pathway to upregulate the expression of transient receptor potential melastatin 2 (TRPM2) on the cell membrane, which mediates Ca^{2+} influx [32]. Therefore, microglia were overloaded with Ca^{2+}, leading to cell death. Zeolite-based nanomaterials (Zbn) adsorbed excessive Zn^{2+} and cleared ROS in the brain of rats after cerebral ischemia, inhibiting the activation of microglia [44]. Therefore, Zbn provided neuroprotection by reducing the content of Zn^{2+} in the brain. TPEN chelation of Zn^{2+} inhibited the activation of microglia induced by $ZnCl_2$, indicating that Zn^{2+} activated microglia and promoted neuroinflammation [35]. Hypoosmotic stress increased the level of extracellular free Zn^{2+} by upregulating the expression of ZnT1 in astrocytes. Microglia were activated by culture medium obtained by treating astrocytes with hypoosmotic stress, whereas extracellular Zn^{2+} chelator CaEDTA inhibited microglial activation [45]. Therefore, hypoosmotic stress promoted the release of Zn^{2+} from astrocytes by upregulating the expression of ZnT1 in astrocytes, thus activating microglia. Zn^{2+} promoted LPS-induced production of ROS in astrocytes. Zn^{2+} significantly increased NO production in astrocytes induced by LPS (1 μg/mL) through the p38-mitogen-activated protein kinase signaling pathway [34]. Zn^{2+} treatment reduced the levels of glutathione (GSH), superoxide dismutase (SOD), and catalase (CAT) in neurons; thus, leading to an increased level of oxidative stress [46]. In conclusion, Zn^{2+} promoted oxidative stress and neuroinflammation (Figure 1).

4.3. Severe Inflammatory Stress Leads to Zinc Homeostasis Imbalance

Intraperitoneal injection of N-acetyl-L-cysteine (NAC) in rats significantly reduced neuronal death and zinc accumulation induced by cerebral ischemia. NAC downregulated the expression of the non-selective cationic channel TRPM2 in neurons, thereby inhibiting the activation of astrocytes and microglia [47]. Therefore, NAC may reduce the entry of Zn^{2+} into neurons by downregulating the expression of TRPM2, thus reducing the neuroinflammation mediated by Zn^{2+}. Another study showed that treatment with interferon γ (INFγ) increased Zn^{2+} content in neurons and upregulated IL-1β, IL-6, and TNF-α expressions in microglia, whereas TRPM2 channel inhibitors reversed the effects of INF γ, indicating that INFγ increased intracellular Zn^{2+} content by activating TRPM2 channels and promoted neuroinflammation [48]. Therefore, severe inflammatory stress led to an excessive influx of Zn^{2+} into cells by upregulating the expression of TRPM2, leading to neuroinflammation.

5. Zinc Homeostasis and Neuroinflammatory Diseases

5.1. Zinc Supplementation Promotes the Recovery of Spinal Cord Injury (SCI)

Intraperitoneal injection of zinc gluconate (ZnG) in mice upregulated the expression of Bcl-2 and downregulated the expression of Bax in the injured spinal cord, indicating that ZnG inhibited cell apoptosis in the injured spinal cord [49]. TPEN chelating Zn^{2+} promoted apoptosis of hippocampal neurons, ERK inhibitor significantly increased TPEN-induced apoptosis of neurons, and ERK agonist reversed the effect of TPEN [50]. Therefore, Zn^{2+} reduced neuronal apoptosis by activating the ERK signaling pathway. ZnG promoted macrophages to express and release granulocyte colony stimulating factor (G-CSF) [7]. Intraperitoneal injection of ZnG promoted the production of G-CSF in the spinal cord tissue of spinal cord injury (SCI) mice and inhibited the apoptosis of nerve cells, but G-CSF neutralizing antibody reversed this effect [7]. Therefore, ZnG inhibited neuronal apoptosis in the injured spinal cord by promoting G-CSF expression. Treatment with methylprednisolone (MP) reduced neuronal death in rats with SCI, downregulated iNOS, IL-6, and IL-10 expressions in the spinal cord, and inhibited microglial activation. In vitro experiments showed that MP treatment significantly increased Zn^{2+} and ZIP8 levels in microglia and reversed LPS-induced NF-κB expression in microglia [51]. Therefore, MP may inhibit the NF-κB pathway by promoting the accumulation of Zn^{2+} in microglia, reducing the release of inflammatory factors and subsequently reducing SCI.

Intraperitoneal injection of ZnG into SCI mice downregulated the expression of NOD-like receptor protein 3 (NLRP3), apoptosis-associated speck-like protein containing a CARD (ASC), caspase-1, and IL-1β in the spinal cord, indicating that zinc treatment inhibited the NLRP3 inflammatory pathway. An in vitro study showed that ZnG inhibited the activation of the LPS-induced NLRP3 inflammatory pathway in microglia [49]. Therefore, Zn^{2+} reduced the expression and release of inflammatory factors by inhibiting the NLRP3 inflammatory pathway in microglia and accelerating the repair of spinal cord tissues. Intraperitoneal injection of ZnG into SCI mice increased the content of Beclin-1 and the ratio of microtubule-associated protein 1 light chain 3-II/I and reduced the expression of P62. Therefore, ZnG promoted autophagy in nerve cells in the injured spinal cord [49]. The expressions of IL-1β, IL-6, and the NLRP3 inflammasome upregulated in the injured spinal cord, and intraperitoneal injection of ZnG promoted autophagy and inhibited the expression of the NLRP3 inflammasome in SCI mice [52]. An in vitro study showed that ZnG promoted autophagy of microglia and inhibited the expression of the NLRP3 inflammasome, and an autophagy inhibitor reversed the effect of zinc on inhibiting the expression of the NLRP3 inflammasome in microglia. A ubiquitin inhibitor reversed the effect of zinc in promoting NLRP3 inflammasome degradation in microglia, suggesting that zinc promotes NLRP3 degradation through ubiquitination modification [52]. In conclusion, zinc reduced NLRP3 inflammasome content through autophagy and ubiquitination, subsequently promoting repair in SCI mice (Figure 2).

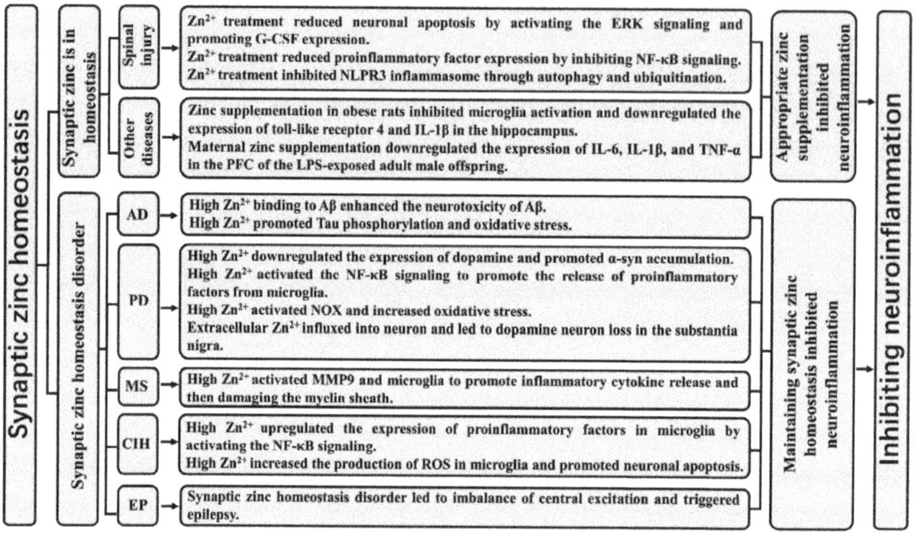

Figure 2. Relationship between synaptic zinc and neuroinflammation. Zinc is an indispensable trace element for maintaining life-preserving activities of organisms. When synaptic zinc is in homeostasis, the appropriate zinc supplementation enhances the body's utilization of zinc, thereby enhancing the body's ability to resist neuroinflammation. However, synaptic zinc disorder further damages the synaptic function when suffering from synapse damage-related diseases. Zn^{2+} uptake in vitro and the large amount of Zn^{2+} released by impaired synapses promote neuroinflammation, so maintaining synaptic zinc homeostasis inhibits neuroinflammation. ERK (extracellular signal-regulated kinase); granulocyte colony stimulating factor (G-CSF); nuclear factor-κB (NF-κB); NOD-like receptor protein 3 (NLRP3); interleukin (IL); tumor necrosis factor-α (TNF-α); prefrontal cortex (PFC); lipopolysaccharide (LPS); Alzheimer's disease (AD); Parkinson's disease (PD); multiple sclerosis (MS); cerebral ischemia and hypoxia (CIH); epilepsy (EP); β-amyloid (Aβ); α-synuclein (α-syn); reduced nicotinamide adenine dinucleotide phosphate oxidase (NOX); matrix metalloproteinase (MMP); reactive oxygen species (ROS).

5.2. Zinc Supplement Has Antidepressant Effect

Clinical research has shown that the decrease in zinc concentration in peripheral blood is related to the occurrence and development of depression [53], and zinc supplementation enhances the therapeutic effect of antidepressants [10,54,55]. A low-zinc diet significantly increased the immobility time of mice in the forced swim test [56] and the tail suspension test [57], and decreased the preference for sugar and water [58]. Therefore, the reduction of zinc intake in the body could lead to depression. The combination of $ZnCl_2$ and the natural peptide Cyclo-(His Pro) generated zinc plus Cyclo-(His-Pro) (ZC). Feeding rats with ZC enhanced the body's absorption of zinc and significantly increased the level of vesicular zinc and neurogenesis in the hippocampus [59]. $ZnCl_2$ treatment upregulated the expression of nestin in induced pluripotent stem cells (iPSCs), whereas TPEN treatment inhibited the expression of nestin in iPSCs, indicating that Zn^{2+} promoted the differentiation of iPSCs. $ZnCl_2$ treatment promoted ERK and signal transducer and activator of transcription (STAT) phosphorylation, but ERK inhibitors inhibited STAT3 phosphorylation and downregulated nestin expression, indicating that Zn^{2+} promoted iPSCs differentiation through the ERK-STAT3 signal pathway [60]. Patients with depression show decreased hippocampal neurogenesis, and zinc supplementation promotes hippocampal neurogenesis and has anti-depressive effects. Therefore, zinc supplementation may play an antidepressant role by enhancing neurogenesis.

5.3. Zinc Accumulation Promotes the Development of Acute Cerebral Ischemia

After global cerebral ischemia, the release of Zn^{2+} from the presynaptic membrane to the postsynaptic membrane led to increased Zn^{2+} content in postsynaptic neurons, followed by neuronal degeneration, while Zn^{2+} chelation reduced neuronal degeneration, indicating that the massive release of Zn^{2+} from the presynaptic membrane is neurotoxic after global cerebral ischemia [61]. Hypobaric hypoxia led to a protein imbalance related to zinc homeostasis, upregulated the expression of TNF-α and iNOS, and also led to neuronal apoptosis in the hippocampus. Zn^{2+} chelators significantly reversed these effects, indicating that zinc promoted neuroinflammation and apoptosis during hypobaric hypoxia [62]. Chelating Zn^{2+} by intravenous etidronate-zinc complex (Eti-Zn) inhibited the activation of microglia and decreased the expressions of IL-1β, IL-6, and TNF-α after cerebral ischemia and hypoxia. Eti–Zn also activated the NF-κB signaling pathway, suggesting that Eti–Zn reduced microglial activation by inhibiting the NF-κB signaling pathway, thereby inhibiting neuroinflammation [63]. $ZnCl_2$ significantly induced ROS production in microglia and also enhanced the effects of LPS on the expression of the proinflammatory factors IL-1β, IL-6, and TNF-α [8]. Cerebral ischemia induced microglia to transform into the M1 type and upregulated proinflammatory factor expression in the mouse hippocampus, whereas CaEDTA inhibited its expression in mice after cerebral ischemia, suggesting that chelating Zn^{2+} attenuated post-ischemic brain injury by inhibiting neuroinflammation [8]. Pretreatment with a high concentration of zinc enhanced the effect of LPS-induced proinflammatory factor secretion in microglia, but this effect was reversed by TPEN and ROS scavengers [31]. Preinjection of CaEDTA into the lateral ventricle inhibited M1 type polarization of microglia in the hippocampus, downregulated the expression of IL-1β, IL-6, and TNF-α, and improved cognitive function in mice after ischemia and reperfusion [31]. In conclusion, zinc accumulation promoted transformation of microglia into a proinflammatory M1 type via ROS, thereby impairing cognitive function in mice after acute cerebral ischemia (Figure 2).

5.4. Zinc Homeostasis Disorder Promotes the Development of Epilepsy

Synaptic zinc homeostasis disorder increased susceptibility to epilepsy [9,64]. Synaptic zinc homeostasis disorder led to an imbalance in central excitation and inhibition, which causes abnormal nerve conduction and triggers epilepsy [9,65]. Knockout of MT3 in mice significantly reduced kainate-induced zinc accumulation, neuronal death, and epileptic seizures in the hippocampal CA1 region [14]. Intracerebroventricular injection of CaEDTA reduced kainate-induced hippocampal neuronal death and epileptic seizures in wild-type mice, indicating that Zn^{2+} overload is a vital cause of neuronal death and epilepsy [14]. Knockout of Zip1 and Zip3 genes inhibited zinc absorption by pyramidal neurons in the hippocampus and kainic acid-induced epilepsy in mice, thus reducing neuronal damage [66]. Therefore, chelating Zn^{2+} may be a measure to prevent epileptic seizures. Transient receptor potential cation channel 5 (TRPC5) was involved in abnormal epileptic discharges and mediated neuronal death. Treatment of cortical neurons with hydrogen peroxide led to zinc-dependent TRPC5 opening, which triggered Ca^{2+} influx and caused neuronal death and seizures [67]. Therefore, zinc plays an important role in the process of seizure-mediated neuronal death. Ganaxolone (GX) inhibited seizures by acting on extrasynaptic GABA-A receptors. Zn^{2+} selectively blocked extrasynaptic GABA-A receptors in the hippocampus, thereby attenuating the antiepileptic effect of GX [68] (Figure 2).

5.5. Zinc Homeostasis Imbalance Promotes Alzheimer's Disease Progression

Daily intraperitoneal injections of the Zn^{2+} chelator CQ for 2 weeks significantly decreased cognitive function in young mice [29]. Knockdown of ZnT3 significantly impaired cognitive function in mice [69]. Knockout of ZnT3 significantly reduced zinc levels in the mouse hippocampus, led to abnormal pre- and post-synaptic protein expressions, and affected learning and memory. CQ treatment reversed these phenomena in ZnT3 knockout mice, indicating that zinc homeostasis disorder led to cognitive impairment [70]. Intracellular and extracellular ROS increased the content of free Zn^{2+} [71]. ROS increased the content of Zn^{2+} in the neurons of AD mice. Intraperitoneal injection of pyruvate reduced the expression of ROS and the content of Zn^{2+} in neurons of AD mice and improved their cognitive function [72]. Therefore, pyruvate reduced the content of Zn^{2+} in the brain by downregulating the level of ROS in neurons, thereby improving the cognitive function of AD mice.

ZnAβ generated by the combination of Zn^{2+} and Aβ significantly inhibited long-term potentiation in the mouse hippocampus and activated microglia, and Zn^{2+} also enhanced the toxicity of Aβ to neuroblastoma cells [73]. Thus, ZnAβ is more neurotoxic than Aβ. Tau mice are transgenic mice containing human tau protein gene. High concentrations of zinc significantly increased the phosphorylation level of tau in tau mice, the number of tau tangles in the hippocampus, and memory impairment in tau mice [74]. Intracerebroventricular injection of $ZnSO_4$ caused hyperphosphorylation of tau in the hippocampus, promoted oxidative stress injury, and impaired learning and memory function in rats. The nuclear factor E2 related factor 2 (Nrf2)/ heme oxygenase-1 (HO-1) pathway was inhibited by intracerebroventricular injection of $ZnSO_4$ in rats and $ZnSO_4$ treatment of SH-SY5Y cells [33]. In conclusion, $ZnSO_4$ promoted tau phosphorylation by inhibiting the Nrf2/HO-1 pathway, thereby damaging the cognitive function of rats. Zn^{2+} enhanced tau neurotoxicity by inducing tau hyperphosphorylation and conformational changes. CQ reversed the effect of Zn^{2+} on tau, indicating that Zn^{2+} is the key to tau neurotoxicity [75]. Zn^{2+} enhanced the toxic effects of Aβ and promoted tau phosphorylation, indicating that restoring zinc homeostasis in the brain can prevent and treat AD [76] (Figure 2).

5.6. Zinc Led to Parkinson's Disease Mediated by Inflammatory Response and Oxidative Stress

Intraperitoneal injection of high concentrations of $ZnSO_4$ increased the content of Zn^{2+} in the substantia nigra and striatum, promoted oxidative stress, reduced the level of dopamine in the striatum, and reduced the spontaneous locomotor activity of rats [77]. Intraperitoneal injection of high concentrations of $ZnSO_4$ in rats caused α-synuclein (α-syn) aggregation in the substantia nigra dopaminergic neurons [78]. Kufor-Rakeb syndrome (KRS) is a genetic disease caused by ATP13A2 (PARK9) mutations that manifests as Parkinson's disease (PD). Knockdown of the PARK9 gene in primary cortical neurons caused the accumulation of α-syn, and Zn^{2+} treatment further increased α-syn accumulation, indicating that Zn^{2+} promoted the progression of PD [79]. Intraperitoneal injection of high concentrations of $ZnSO_4$ in rats reduced the number of dopamine neurons and dopamine content, and increased the activity of NOX in the substantia nigra and striatum. The inhibition of microglial activation by minocycline reversed the effects of $ZnSO_4$, suggesting that a high concentration of Zn^{2+} caused dopamine neuron damage by activating microglia to release inflammatory factors; thus, participating in the occurrence of PD [80]. The intraperitoneal injection of high concentrations of $ZnSO_4$ in adult rats reduced the number of substantia nigra dopamine neurons and the content of striatal dopamine, increased the expression of oxidative stress and inflammatory factors, and caused PD symptoms. $ZnSO_4$ treatment in newborn and adult rats further aggravated the above effects, indicating that a high concentration of Zn^{2+} promoted the occurrence and development of PD [81]. It was also found that high concentrations of Zn^{2+} promoted the release of proinflammatory factors from microglia in the substantia nigra by activating the NF-κB pathway and also promoted apoptosis [81] (Figure 2).

AMPA injection into the substantia nigra significantly increased the amount of free Zn^{2+} and the loss of dopamine neurons in the substantia nigra and caused movement disorder in rats. Zn^{2+} chelator reversed these effects, suggesting that $AMPAR_{Ca\text{-}Zn}$ mediated Zn^{2+} influx and damaged nigral dopamine neurons, leading to PD [82]. 6-OHDA impaired dopamine neurons in the substantia nigra and striatum by promoting extracellular Zn^{2+} to flow into dopamine neurons [83], while blocking Zn^{2+} influx into dopamine neurons inhibited paraquat-induced dopamine neuron loss [84]. In conclusion, a large influx of extracellular Zn^{2+} into dopaminergic neurons promoted the development of PD. Synaptic

Zn^{2+} release led to a rapid increase in extracellular Zn^{2+} concentration, so maintaining synaptic zinc homeostasis is a potential strategy to combat PD [15,82].

5.7. Zinc Accumulation Promotes the Occurrence and Development of Multiple Sclerosis

ZnT3 on the synaptic vesicle membrane transports zinc into the vesicles for storage [20,21]. Knockdown of ZnT3 reduced Zn^{2+} release into the synaptic cleft, significantly inhibited spinal white matter demyelination, reduced the inflammatory response in mice with experimental autoimmune encephalomyelitis (EAE), and significantly inhibited the activation of MMP-9 and the formation of zinc plaques [3]. 1H10 is an inhibitor of AMP-activated protein kinase (AMPK) phosphorylation, and it was experimentally demonstrated that 1H10 also has the ability to chelate Zn^{2+}. Intraperitoneal injection of 1H10 in mice inhibited EAE-induced demyelination and microglia/macrophage activation, inhibited EAE-induced MMP-9 activation and zinc plaque formation and alleviated multiple sclerosis (MS) [85]. Therefore, inhibition of AMPK phosphorylation or chelation of Zn^{2+} may be an effective target for MS treatment. Zn^{2+} damaged myelin sheath by activating MMP-9 to destroy the BBB and activating microglia to release proinflammatory cytokines [16] (Figure 2).

6. Summary and Outlook

Synaptic zinc homeostasis plays an important role in maintaining brain zinc homeostasis and CNS stability. When synaptic zinc is in a homeostatic state, preventive zinc supplementation can improve the body's immunity and reduce the production of inflammatory factors in the brain and is conducive to the utilization of zinc in the body [12,13]. Under intense stress conditions, such as cerebral ischemia [8], epilepsy [14], and hypoglycemia [86], the presynaptic membrane Zn^{2+} is released at a great rate into the synaptic cleft, resulting in Zn^{2+} overload in the postsynaptic neurons and neurotoxicity. Excessive Zn^{2+} levels in the synaptic cleft also activate extracellular MMP and microglia, promoting neuroinflammation and causing tissue damage [16]. Thus, the use of extracellular Zn^{2+} chelators, reduces neuronal death caused by ischemic hypoxia and epilepsy, and resists neuroinflammation.

Neurodegenerative diseases are closely related to synaptic transmission, such as the imbalance of acetylcholine and dopamine levels in AD and PD patients. Zn^{2+} uptake in vitro and a large amount of Zn^{2+} released by impaired synaptic function can activate microglia and also enhance inflammatory protein (α-syn, tau, Aβ, and MMP) expression and neurotoxicity, thereby driving neuroinflammation-mediated neuronal damage [16,33,73,78]. Therefore, maintaining synaptic zinc homeostasis plays an important role in treating neurodegenerative diseases [11,15,16]. Synaptic zinc homeostasis disorders cause neurotoxicity and further affect nerve transmission, thereby promoting the progression of neurodegenerative diseases. Regulating zinc homeostasis in the early onset of neurodegenerative diseases may delay the disease progression. When neurodegenerative disease progresses to a moderate or severe degree, nerve transmission may become impaired, and zinc intake should be strictly controlled.

Preparations for zinc supplementation have evolved from inorganic zinc agents to organic zinc agents [87]. Inorganic zinc agents have the disadvantages of strong intestinal irritation and low bioavailability [88]. Now, an organic zinc agent is mainly used to improve zinc content in the body, including zinc gluconate, zinc citrate, zinc-enriched yeast and protein-chelated zinc [88–90]. The clinical application of zinc chelating agent is chloroiodohydroxyquine, which is mainly used in the treatment of dermatological diseases [91,92]. Some studies showed the neuroprotective effect of chelating Zn^{2+} after cerebral ischemia and hypoxia [8,31,63], and the development of a clinical drug to Zn^{2+} in the brain may have a high application prospect.

Neuroinflammation is a common pathophysiological basis of neurodegenerative diseases and is a major research topic in the field of neurodegenerative diseases [93]. Regulation of neuroinflammation is an important way to prevent and treat neurodegenerative diseases. Zinc homeostasis disorders affect the occurrence and development of neuroinflammation. Therefore, restoring zinc homeostasis by targeting specific proteins that regulate zinc homeostasis may be vital for the prevention and treatment of neuroinflammation-related neurodegenerative diseases. Hence, the role of zinc homeostasis disorders in the development and progression of neuroinflammation has received increasing attention. Further clarification of the molecular mechanism of neuroinflammation caused by zinc homeostasis disorders has profound scientific and clinical significance for the prevention and treatment of neuroinflammation-related neurodegenerative diseases.

Funding: This work was supported by the National Natural Science Foundation of China (Grant No. 82060250), the Guangxi Science and Technology Plan Project (Grant No. AD20238035), the Natural Science Foundation of Hunan Province (Grant Nos. 2021JJ30614 and 2021JJ70036), and the Hunan Clinical Medical Technology Innovation and Guidance Project (Grant No. 2020SK517).

Institutional Review Board Statement: Not applicable.

Informed Consent Statement: Not applicable.

Data Availability Statement: Not applicable.

Acknowledgments: The authors are grateful to all members of the research group for their discussions.

Conflicts of Interest: The authors declare no conflict of interest.

References

1. Colonna, M.; Butovsky, O. Microglia function in the central nervous system during health and neurodegeneration. *Annu. Rev. Immunol.* **2017**, *35*, 441–468. [CrossRef] [PubMed]
2. Orihuela, R.; McPherson, C.A.; Harry, G.J. Microglial M1/M2 polarization and metabolic states. *Br. J. Pharmacol.* **2016**, *173*, 649–665. [CrossRef] [PubMed]
3. Choi, B.Y.; Kim, I.Y.; Kim, J.H.; Kho, A.R.; Lee, S.H.; Lee, B.E.; Sohn, M.; Koh, J.Y.; Suh, S.W. Zinc transporter 3 (ZnT3) gene deletion reduces spinal cord white matter damage and motor deficits in a murine MOG-induced multiple sclerosis model. *Neurobiol. Dis.* **2016**, *94*, 205–212. [CrossRef] [PubMed]
4. Feijó, G.D.S.; Jantsch, J.; Correia, L.L.; Eller, S.; Furtado-Filho, O.V.; Giovenardi, M.; Porawski, M.; Braganhol, E.; Guedes, R.P. Neuroinflammatory responses following zinc or branched-chain amino acids supplementation in obese rats. *Metab. Brain Dis.* **2022**, *37*, 1875–1886. [CrossRef]
5. Lin, P.H.; Sermersheim, M.; Li, H.; Lee, P.H.U.; Steinberg, S.M.; Ma, J. Zinc in wound healing modulation. *Nutrients* **2017**, *10*, 16. [CrossRef]
6. Sensi, S.L.; Paoletti, P.; Bush, A.I.; Sekler, I. Zinc in the physiology and pathology of the CNS. *Nat. Rev. Neurosci.* **2009**, *10*, 780–791. [CrossRef]
7. Li, X.; Chen, S.; Mao, L.; Li, D.; Xu, C.; Tian, H.; Mei, X. Zinc Improves Functional recovery by regulating the secretion of granulocyte colony stimulating factor from microglia/macrophages after spinal cord injury. *Front. Mol. Neurosci.* **2019**, *12*, 18. [CrossRef] [PubMed]
8. Ueba, Y.; Aratake, T.; Onodera, K.I.; Higashi, Y.; Hamada, T.; Shimizu, T.; Shimizu, S.; Yawata, T.; Nakamura, R.; Akizawa, T.; et al. Attenuation of zinc-enhanced inflammatory M1 phenotype of microglia by peridinin protects against short-term spatial-memory impairment following cerebral ischemia in mice. *Biochem. Biophys. Res. Commun.* **2018**, *507*, 476–483. [CrossRef]
9. Doboszewska, U.; Młyniec, K.; Wlaź, A.; Poleszak, E.; Nowak, G.; Wlaź, P. Zinc signaling and epilepsy. *Pharmacol. Ther.* **2019**, *193*, 156–177. [CrossRef]
10. da Silva, L.E.M.; de Santana, M.L.P.; Costa, P.R.F.; Pereira, E.M.; Nepomuceno, C.M.M.; Queiroz, V.A.O.; de Oliveira, L.P.M.; Machado, M.; de Sena, E.P. Zinc supplementation combined with antidepressant drugs for treatment of patients with depression: A systematic review and meta-analysis. *Nutr. Rev.* **2021**, *79*, 1–12. [CrossRef]
11. Szewczyk, B. Zinc homeostasis and neurodegenerative disorders. *Front. Aging Neurosci.* **2013**, *5*, 33. [CrossRef] [PubMed]
12. De Oliveira, S.; Feijó, G.D.S.; Neto, J.; Jantsch, J.; Braga, M.F.; Castro, L.; Giovenardi, M.; Porawski, M.; Guedes, R.P. Zinc supplementation decreases obesity-related neuroinflammation and improves metabolic function and memory in rats. *Obesity* **2021**, *29*, 116–124. [CrossRef] [PubMed]
13. Mousaviyan, R.; Davoodian, N.; Alizadeh, F.; Ghasemi-Kasman, M.; Mousavi, S.A.; Shaerzadeh, F.; Kazemi, H. Zinc supplementation during pregnancy alleviates lipopolysaccharide-induced glial activation and inflammatory markers expression in a rat model of maternal immune activation. *Biol. Trace Elem. Res.* **2021**, *199*, 4193–4204. [CrossRef] [PubMed]
14. Lee, J.Y.; Kim, J.H.; Palmiter, R.D.; Koh, J.Y. Zinc released from metallothionein-iii may contribute to hippocampal CA1 and thalamic neuronal death following acute brain injury. *Exp. Neurol.* **2003**, *184*, 337–347. [CrossRef] [PubMed]
15. Sikora, J.; Ouagazzal, A.M. Synaptic zinc: An emerging player in parkinson's disease. *Int. J. Mol. Sci.* **2021**, *22*, 4724. [CrossRef]
16. Choi, B.Y.; Jung, J.W.; Suh, S.W. The Emerging role of zinc in the pathogenesis of multiple sclerosis. *Int J Mol Sci.* **2017**, *18*, 2070. [CrossRef] [PubMed]

17. Kambe, T.; Tsuji, T.; Hashimoto, A.; Itsumura, N. The physiological, biochemical, and molecular roles of zinc transporters in zinc homeostasis and metabolism. *Physiol. Rev.* **2015**, *95*, 749–784. [CrossRef]
18. Kimura, T.; Kambe, T. The functions of metallothionein and ZIP and ZnT transporters: An overview and perspective. *Int. J. Mol. Sci.* **2016**, *17*, 336. [CrossRef]
19. Baltaci, A.K.; Yuce, K.; Mogulkoc, R. Zinc metabolism and metallothioneins. *Biol. Trace Elem. Res.* **2018**, *183*, 22–31. [CrossRef]
20. Wenzel, H.J.; Cole, T.B.; Born, D.E.; Schwartzkroin, P.A.; Palmiter, R.D. Ultrastructural localization of zinc transporter-3 (ZnT-3) to synaptic vesicle membranes within mossy fiber boutons in the hippocampus of mouse and monkey. *Proc. Natl. Acad. Sci. USA* **1997**, *94*, 12676–12681. [CrossRef]
21. Palmiter, R.D.; Cole, T.B.; Quaife, C.J.; Findley, S.D. ZnT-3, a putative transporter of zinc into synaptic vesicles. *Proc. Natl. Acad. Sci. USA* **1996**, *93*, 14934–14939. [CrossRef] [PubMed]
22. Ohana, E.; Segal, D.; Palty, R.; Ton-That, D.; Moran, A.; Sensi, S.L.; Weiss, J.H.; Hershfinkel, M.; Sekler, I. A sodium zinc exchange mechanism is mediating extrusion of zinc in mammalian cells. *J. Biol. Chem.* **2004**, *279*, 4278–4284. [CrossRef] [PubMed]
23. Granzotto, A.; Sensi, S.L. Intracellular zinc is a critical intermediate in the excitotoxic cascade. *Neurobiol. Dis.* **2015**, *81*, 25–37. [CrossRef] [PubMed]
24. Bitanihirwe, B.K.Y.; Woo, T.W. A conceptualized model linking matrix metalloproteinase-9 to schizophrenia pathogenesis. *Schizophr. Res.* **2020**, *218*, 28–35. [CrossRef]
25. Ciccone, L.; Vandooren, J.; Nencetti, S.; Orlandini, E. Natural marine and terrestrial compounds as modulators of matrix metalloproteinases-2 (MMP-2) and MMP-9 in Alzheimer's disease. *Pharmaceuticals* **2021**, *14*, 86. [CrossRef] [PubMed]
26. Qin, W.; Li, J.; Zhu, R.; Gao, S.; Fan, J.; Xia, M.; Zhao, R.C.; Zhang, J. Melatonin protects blood-brain barrier integrity and permeability by inhibiting matrix metalloproteinase-9 via the NOTCH3/NF-κB pathway. *Aging* **2019**, *11*, 11391–11415. [CrossRef]
27. Pilat, D.; Paumier, J.M.; García-González, L.; Louis, L.; Stephan, D.; Manrique, C.; Khrestchatisky, M.; Di Pasquale, E.; Baranger, K.; Rivera, S. MT5-MMP promotes neuroinflammation, neuronal excitability and Aβ production in primary neuron/astrocyte cultures from the 5xFAD mouse model of Alzheimer's disease. *J. Neuroinflamm.* **2022**, *19*, 65. [CrossRef] [PubMed]
28. Yang, C.C.; Hsiao, L.D.; Tseng, H.C.; Kuo, C.M.; Yang, C.M. Pristimerin inhibits MMP-9 expression and cell migration through attenuating NOX/ROS-dependent NF-κB activation in rat brain astrocytes challenged with LPS. *J. Inflamm. Res.* **2020**, *13*, 325–341. [CrossRef] [PubMed]
29. Frazzini, V.; Granzotto, A.; Bomba, M.; Massetti, N.; Castelli, V.; d'Aurora, M.; Punzi, M.; Iorio, M.; Mosca, A.; Delli Pizzi, S.; et al. The pharmacological perturbation of brain zinc impairs BDNF-related signaling and the cognitive performances of young mice. *Sci. Rep.* **2018**, *8*, 9768. [CrossRef]
30. Ruan, Z.; Zhang, D.; Huang, R.; Sun, W.; Hou, L.; Zhao, J.; Wang, Q. Microglial activation damages dopaminergic neurons through MMP-2/-9-mediated increase of blood-brain barrier permeability in a Parkinson's disease mouse model. *Int. J. Mol. Sci.* **2022**, *23*, 2793. [CrossRef]
31. Higashi, Y.; Aratake, T.; Shimizu, S.; Shimizu, T.; Nakamura, K.; Tsuda, M.; Yawata, T.; Ueba, T.; Saito, M. Influence of extracellular zinc on M1 microglial activation. *Sci. Rep.* **2017**, *7*, 43778. [CrossRef] [PubMed]
32. Mortadza, S.S.; Sim, J.A.; Stacey, M.; Jiang, L.H. Signalling mechanisms mediating $Zn^{(2+)}$-induced TRPM2 channel activation and cell death in microglial cells. *Sci. Rep.* **2017**, *7*, 45032. [CrossRef] [PubMed]
33. Lai, C.; Chen, Z.; Ding, Y.; Chen, Q.; Su, S.; Liu, H.; Ni, R.; Tang, Z. Rapamycin attenuated zinc-induced tau phosphorylation and oxidative stress in rats: Involvement of dual mTOR/p70S6K and Nrf2/HO-1 pathways. *Front. Immunol.* **2022**, *13*, 782434. [CrossRef] [PubMed]
34. Moriyama, M.; Fujitsuka, S.; Kawabe, K.; Takano, K.; Nakamura, Y. Zinc potentiates lipopolysaccharide-induced nitric oxide production in cultured primary rat astrocytes. *Neurochem. Res.* **2018**, *43*, 363–374. [CrossRef] [PubMed]
35. Higashi, Y.; Segawa, S.; Matsuo, T.; Nakamura, S.; Kikkawa, Y.; Nishida, K.; Nagasawa, K. Microglial zinc uptake via zinc transporters induces ATP release and the activation of microglia. *Glia* **2011**, *59*, 1933–1945. [CrossRef] [PubMed]
36. Doi, T.; Hara, H.; Kajita, M.; Kamiya, T.; Adachi, T. Zinc regulates expression of IL-23 p19 mRNA via activation of eIF2α/ATF4 axis in HAPI cells. *Biometals* **2015**, *28*, 891–902. [CrossRef]
37. Hongxia, L.; Yuxiao, T.; Zhilei, S.; Yan, S.; Yicui, Q.; Jiamin, S.; Xin, X.; Jianxin, Y.; Fengfeng, M.; Hui, S. Zinc inhibited LPS-induced inflammatory responses by upregulating A20 expression in microglia BV2 cells. *J. Affect Disord.* **2019**, *249*, 136–142. [CrossRef]
38. Aratake, T.; Higashi, Y.; Ueba, Y.; Hamada, T.; Shimizu, T.; Shimizu, S.; Yawata, T.; Ueba, T.; Saito, M. The inhibitory role of intracellular free zinc in the regulation of Arg-1 expression in interleukin-4-induced activation of M2 microglia. *Metallomics* **2018**, *10*, 1501–1509. [CrossRef]
39. Summersgill, H.; England, H.; Lopez-Castejon, G.; Lawrence, C.B.; Luheshi, N.M.; Pahle, J.; Mendes, P.; Brough, D. Zinc depletion regulates the processing and secretion of IL-1β. *Cell Death Dis.* **2014**, *5*, e1040. [CrossRef]
40. Rivers-Auty, J.; Tapia, V.S.; White, C.S.; Daniels, M.J.D.; Drinkall, S.; Kennedy, P.T.; Spence, H.G.; Yu, S.; Green, J.P.; Hoyle, C.; et al. Zinc status alters Alzheimer's disease progression through NLRP3-dependent inflammation. *J. Neurosci.* **2021**, *41*, 3025–3038. [CrossRef]
41. Wong, C.P.; Magnusson, K.R.; Ho, E. Increased inflammatory response in aged mice is associated with age-related zinc deficiency and zinc transporter dysregulation. *J. Nutr. Biochem.* **2013**, *24*, 353–359. [CrossRef] [PubMed]
42. Sauer, A.K.; Grabrucker, A.M. Zinc deficiency during pregnancy leads to altered microbiome and elevated inflammatory markers in mice. *Front. Neurosci.* **2019**, *13*, 1295. [CrossRef] [PubMed]

43. Sauer, A.K.; Malijauskaite, S.; Meleady, P.; Boeckers, T.M.; McGourty, K.; Grabrucker, A.M. Zinc is a key regulator of gastrointestinal development, microbiota composition and inflammation with relevance for autism spectrum disorders. *Cell Mol. Life Sci.* **2021**, *79*, 46. [CrossRef] [PubMed]
44. Huang, Z.; Qian, K.; Chen, J.; Qi, Y.; E, Y.; Liang, J.; Zhao, L. A biomimetic zeolite-based nanoenzyme contributes to neuroprotection in the neurovascular unit after ischaemic stroke via efficient removal of zinc and ROS. *Acta Biomater.* **2022**, *144*, 142–156. [CrossRef]
45. Segawa, S.; Nishiura, T.; Furuta, T.; Ohsato, Y.; Tani, M.; Nishida, K.; Nagasawa, K. Zinc is released by cultured astrocytes as a gliotransmitter under hypoosmotic stress-loaded conditions and regulates microglial activity. *Life Sci.* **2014**, *94*, 137–144. [CrossRef]
46. Olasehinde, T.A.; Olaniran, A.O.; Okoh, A.I. Neuroprotective effects of some seaweeds against Zn-induced neuronal damage in HT-22 cells via modulation of redox imbalance, inhibition of apoptosis and acetylcholinesterase activity. *Metab. Brain Dis.* **2019**, *34*, 1615–1627. [CrossRef]
47. Hong, D.K.; Kho, A.R.; Lee, S.H.; Jeong, J.H.; Kang, B.S.; Kang, D.H.; Park, M.K.; Park, K.H.; Lim, M.S.; Choi, B.Y.; et al. Transient receptor potential melastatin 2 (TRPM2) inhibition by antioxidant, N-acetyl-l-cysteine, reduces global cerebral ischemia-induced neuronal death. *Int. J. Mol. Sci.* **2020**, *21*, 6026. [CrossRef]
48. Güzel, M.; Nazıroğlu, M.; Akpınar, O.; Çınar, R. Interferon gamma-mediated oxidative stress induces apoptosis, neuroinflammation, zinc ion influx, and TRPM2 channel activation in neuronal cell line: Modulator role of curcumin. *Inflammation* **2021**, *44*, 1878–1894. [CrossRef]
49. Zhao, X.; Sun, J.; Yuan, Y.; Lin, S.; Lin, J.; Mei, X. Zinc promotes microglial autophagy through NLRP3 inflammasome inactivation via XIST/miR-374a-5p axis in spinal cord injury. *Neurochem. Res.* **2022**, *47*, 372–381. [CrossRef]
50. Pang, W.; Leng, X.; Lu, H.; Yang, H.; Song, N.; Tan, L.; Jiang, Y.; Guo, C. Depletion of intracellular zinc induces apoptosis of cultured hippocampal neurons through suppression of ERK signaling pathway and activation of caspase-3. *Neurosci. Lett.* **2013**, *552*, 140–145. [CrossRef] [PubMed]
51. Li, D.; Wang, G.; Han, D.; Bi, J.; Li, C.; Wang, H.; Liu, Z.; Gao, W.; Gao, K.; Yao, T.; et al. MP resulting in autophagic cell death of microglia through zinc changes against spinal cord injury. *Biomed Res. Int.* **2016**, *2016*, 6090316. [CrossRef] [PubMed]
52. Lin, J.Q.; Tian, H.; Zhao, X.G.; Lin, S.; Li, D.Y.; Liu, Y.Y.; Xu, C.; Mei, X.F. Zinc provides neuroprotection by regulating NLRP3 inflammasome through autophagy and ubiquitination in a spinal contusion injury model. *CNS Neurosci. Ther.* **2021**, *27*, 413–425. [CrossRef] [PubMed]
53. Swardfager, W.; Herrmann, N.; Mazereeuw, G.; Goldberger, K.; Harimoto, T.; Lanctôt, K.L. Zinc in depression: A meta-analysis. *Biol. Psychiatry* **2013**, *74*, 872–878. [CrossRef] [PubMed]
54. Siwek, M.; Dudek, D.; Paul, I.A.; Sowa-Kućma, M.; Zieba, A.; Popik, P.; Pilc, A.; Nowak, G. Zinc supplementation augments efficacy of imipramine in treatment resistant patients: A double blind, placebo-controlled study. *J. Affect Disord.* **2009**, *118*, 187–195. [CrossRef] [PubMed]
55. Nowak, G.; Siwek, M.; Dudek, D.; Zieba, A.; Pilc, A. Effect of zinc supplementation on antidepressant therapy in unipolar depression: A preliminary placebo-controlled study. *Pol. J. Pharmacol.* **2003**, *55*, 1143–1147. [PubMed]
56. Młyniec, K.; Davies, C.L.; Budziszewska, B.; Opoka, W.; Reczyński, W.; Sowa-Kućma, M.; Doboszewska, U.; Pilc, A.; Nowak, G. Time course of zinc deprivation-induced alterations of mice behavior in the forced swim test. *Pharmacol. Rep.* **2012**, *64*, 567–575. [CrossRef]
57. Młyniec, K.; Nowak, G. Zinc deficiency induces behavioral alterations in the tail suspension test in mice. Effect of antidepressants. *Pharmacol. Rep.* **2012**, *64*, 249–255. [CrossRef]
58. Tassabehji, N.M.; Corniola, R.S.; Alshingiti, A.; Levenson, C.W. Zinc deficiency induces depression-like symptoms in adult rats. *Physiol. Behav.* **2008**, *95*, 365–369. [CrossRef]
59. Choi, B.Y.; Kim, I.Y.; Kim, J.H.; Lee, B.E.; Lee, S.H.; Kho, A.R.; Sohn, M.; Suh, S.W. Zinc plus cyclo-(His-Pro) promotes hippocampal neurogenesis in rats. *Neuroscience* **2016**, *339*, 634–643. [CrossRef]
60. Yang, M.; Bao, D.; Shi, A.; Yuan, H.; Wang, J.; He, W.; Tong, X.; Qin, H. Zinc promotes patient-derived induced pluripotent stem cell neural differentiation via ERK-STAT signaling. *Stem. Cells Dev.* **2020**, *29*, 863–875. [CrossRef]
61. Koh, J.Y.; Suh, S.W.; Gwag, B.J.; He, Y.Y.; Hsu, C.Y.; Choi, D.W. The role of zinc in selective neuronal death after transient global cerebral ischemia. *Science* **1996**, *272*, 1013–1016. [CrossRef] [PubMed]
62. Malairaman, U.; Dandapani, K.; Katyal, A. Effect of Ca2EDTA on zinc mediated inflammation and neuronal apoptosis in hippocampus of an in vivo mouse model of hypobaric hypoxia. *PLoS ONE* **2014**, *9*, e110253. [CrossRef]
63. Feng, L.; Gao, J.; Wang, Y.; Cheong, Y.K.; Ren, G.; Yang, Z. Etidronate-zinc complex ameliorated cognitive and synaptic plasticity impairments in 2-vessel occlusion model rats by reducing neuroinflammation. *Neuroscience* **2018**, *390*, 206–217. [CrossRef] [PubMed]
64. Saghazadeh, A.; Mahmoudi, M.; Meysamie, A.; Gharedaghi, M.; Zamponi, G.W.; Rezaei, N. Possible role of trace elements in epilepsy and febrile seizures: A meta-analysis. *Nutr. Rev.* **2015**, *73*, 760–779. [CrossRef]
65. Staley, K. Molecular mechanisms of epilepsy. *Nat. Neurosci.* **2015**, *18*, 367–372. [CrossRef] [PubMed]
66. Qian, J.; Xu, K.; Yoo, J.; Chen, T.T.; Andrews, G.; Noebels, J.L. Knockout of Zn transporters Zip-1 and Zip-3 attenuates seizure-induced CA1 neurodegeneration. *J. Neurosci.* **2011**, *31*, 97–104. [CrossRef]

67. Park, S.E.; Song, J.H.; Hong, C.; Kim, D.E.; Sul, J.W.; Kim, T.Y.; Seo, B.R.; So, I.; Kim, S.Y.; Bae, D.J.; et al. Contribution of zinc-dependent delayed calcium influx via TRPC5 in oxidative neuronal death and its prevention by novel TRPC antagonist. *Mol. Neurobiol.* **2019**, *56*, 2822–2835. [CrossRef]
68. Chuang, S.H.; Reddy, D.S. Zinc reduces antiseizure activity of neurosteroids by selective blockade of extrasynaptic GABA-A receptor-mediated tonic inhibition in the hippocampus. *Neuropharmacology* **2019**, *148*, 244–256. [CrossRef] [PubMed]
69. Adlard, P.A.; Parncutt, J.M.; Finkelstein, D.I.; Bush, A.I. Cognitive loss in zinc transporter-3 knock-out mice: A phenocopy for the synaptic and memory deficits of Alzheimer's disease? *J. Neurosci.* **2010**, *30*, 1631–1636. [CrossRef]
70. Adlard, P.A.; Parncutt, J.; Lal, V.; James, S.; Hare, D.; Doble, P.; Finkelstein, D.I.; Bush, A.I. Metal chaperones prevent zinc-mediated cognitive decline. *Neurobiol. Dis.* **2015**, *81*, 196–202. [CrossRef]
71. Aras, M.A.; Aizenman, E. Redox regulation of intracellular zinc: Molecular signaling in the life and death of neurons. *Antioxid. Redox. Signal* **2011**, *15*, 2249–2263. [CrossRef]
72. Isopi, E.; Granzotto, A.; Corona, C.; Bomba, M.; Ciavardelli, D.; Curcio, M.; Canzoniero, L.M.; Navarra, R.; Lattanzio, R.; Piantelli, M.; et al. Pyruvate prevents the development of age-dependent cognitive deficits in a mouse model of Alzheimer's disease without reducing amyloid and tau pathology. *Neurobiol. Dis.* **2015**, *81*, 214–224. [CrossRef] [PubMed]
73. Lee, M.C.; Yu, W.C.; Shih, Y.H.; Chen, C.Y.; Guo, Z.H.; Huang, S.J.; Chan, J.C.C.; Chen, Y.R. Zinc ion rapidly induces toxic, off-pathway amyloid-β oligomers distinct from amyloid-β derived diffusible ligands in Alzheimer's disease. *Sci. Rep.* **2018**, *8*, 4772. [CrossRef]
74. Craven, K.M.; Kochen, W.R.; Hernandez, C.M.; Flinn, J.M. Zinc exacerbates tau pathology in a tau mouse model. *J. Alzheimers Dis.* **2018**, *64*, 617–630. [CrossRef]
75. Huang, Y.; Wu, Z.; Cao, Y.; Lang, M.; Lu, B.; Zhou, B. Zinc binding directly regulates tau toxicity independent of tau hyperphosphorylation. *Cell Rep.* **2014**, *8*, 831–842. [CrossRef]
76. Huat, T.J.; Camats-Perna, J.; Newcombe, E.A.; Valmas, N.; Kitazawa, M.; Medeiros, R. Metal toxicity links to Alzheimer's disease and neuroinflammation. *J. Mol. Biol.* **2019**, *431*, 1843–1868. [CrossRef] [PubMed]
77. Kumar, A.; Ahmad, I.; Shukla, S.; Singh, B.K.; Patel, D.K.; Pandey, H.P.; Singh, C. Effect of zinc and paraquat co-exposure on neurodegeneration: Modulation of oxidative stress and expression of metallothioneins, toxicant responsive and transporter genes in rats. *Free Radic Res.* **2010**, *44*, 950–965. [CrossRef]
78. Kumar, V.; Singh, D.; Singh, B.K.; Singh, S.; Mittra, N.; Jha, R.R.; Patel, D.K.; Singh, C. Alpha-synuclein aggregation, Ubiquitin proteasome system impairment, and L-Dopa response in zinc-induced Parkinsonism: Resemblance to sporadic Parkinson's disease. *Mol. Cell Biochem.* **2018**, *444*, 149–160. [CrossRef]
79. Tsunemi, T.; Krainc, D. Zn^{2+} dyshomeostasis caused by loss of ATP13A2/PARK9 leads to lysosomal dysfunction and alpha-synuclein accumulation. *Hum. Mol. Genet.* **2014**, *23*, 2791–2801. [CrossRef]
80. Kumar, V.; Singh, B.K.; Chauhan, A.K.; Singh, D.; Patel, D.K.; Singh, C. Minocycline rescues from zinc-induced nigrostriatal dopaminergic neurodegeneration: Biochemical and molecular interventions. *Mol. Neurobiol.* **2016**, *53*, 2761–2777. [CrossRef] [PubMed]
81. Mittra, N.; Chauhan, A.K.; Singh, G.; Patel, D.K.; Singh, C. Postnatal zinc or paraquat administration increases paraquat or zinc-induced loss of dopaminergic neurons: Insight into augmented neurodegeneration. *Mol. Cell Biochem.* **2020**, *467*, 27–43. [CrossRef] [PubMed]
82. Tamano, H.; Morioka, H.; Nishio, R.; Takeuchi, A.; Takeda, A. AMPA-induced extracellular $Zn^{(2+)}$ influx into nigral dopaminergic neurons causes movement disorder in rats. *Neurotoxicology* **2018**, *69*, 23–28. [CrossRef]
83. Tamano, H.; Nishio, R.; Morioka, H.; Takeda, A. Extracellular $Zn^{(2+)}$ influx into nigral dopaminergic neurons plays a key role for pathogenesis of 6-Hydroxydopamine-induced Parkinson's disease in rats. *Mol. Neurobiol.* **2019**, *56*, 435–443. [CrossRef] [PubMed]
84. Tamano, H.; Morioka, H.; Nishio, R.; Takeuchi, A.; Takeda, A. Blockade of rapid influx of extracellular $Zn^{(2+)}$ into nigral dopaminergic neurons overcomes paraquat-induced Parkinson's disease in rats. *Mol. Neurobiol.* **2019**, *56*, 4539–4548. [CrossRef] [PubMed]
85. Choi, B.Y.; Jeong, J.H.; Eom, J.W.; Koh, J.Y.; Kim, Y.H.; Suh, S.W. A novel zinc chelator, 1H10, ameliorates experimental autoimmune encephalomyelitis by modulating zinc toxicity and AMPK activation. *Int. J. Mol. Sci.* **2020**, *21*, 3375. [CrossRef] [PubMed]
86. Kho, A.R.; Choi, B.Y.; Kim, J.H.; Lee, S.H.; Hong, D.K.; Lee, S.H.; Jeong, J.H.; Sohn, M.; Suh, S.W. Prevention of hypoglycemia-induced hippocampal neuronal death by N-acetyl-L-cysteine (NAC). *Amino Acids* **2017**, *49*, 367–378. [CrossRef] [PubMed]
87. Zhang, L.; Guo, Q.; Duan, Y.; Lin, X.; Ni, H.; Zhou, C.; Li, F. Comparison of the Effects of Inorganic or Amino Acid-Chelated Zinc on Mouse Myoblast Growth in vitro and Growth Performance and Carcass Traits in Growing-Finishing Pigs. *Front. Nutr.* **2022**, *9*, 857393. [CrossRef]
88. Fazel Torshizi, F.; Chamani, M.; Khodaei, H.R.; Sadeghi, A.A.; Hejazi, S.H.; Majidzadeh Heravi, R. Therapeutic effects of organic zinc on reproductive hormones, insulin resistance and mTOR expression, as a novel component, in a rat model of Polycystic ovary syndrome. *Iran J. Basic Med. Sci.* **2020**, *23*, 36–45. [CrossRef]
89. Maares, M.; Keil, C.; Pallasdies, L.; Schmacht, M.; Senz, M.; Nissen, J.; Kieserling, H.; Drusch, S.; Haase, H. Zinc availability from zinc-enriched yeast studied with an in vitro digestion/Caco-2 cell culture model. *J. Trace Elem. Med. Biol.* **2022**, *71*, 126934. [CrossRef] [PubMed]

90. Zhang, Z.; Cheng, Q.; Liu, Y.; Peng, C.; Wang, Z.; Ma, H.; Liu, D.; Wang, L.; Wang, C. Zinc-Enriched Yeast May Improve Spermatogenesis by Regulating Steroid Production and Antioxidant Levels in Mice. *Biol. Trace Elem. Res.* **2022**, *200*, 3712–3722. [CrossRef]
91. Da Costa, B.; Pippi, B.; Berlitz, S.J.; Carvalho, A.R.; Teixeira, M.L.; Külkamp-Guerreiro, I.C.; Andrade, S.F.; Fuentefria, A.M. Evaluation of activity and toxicity of combining clioquinol with ciclopirox and terbinafine in alternative models of dermatophytosis. *Mycoses* **2021**, *64*, 727–733. [CrossRef] [PubMed]
92. Da Costa, B.; Pippi, B.; Andrzejewski Kaminski, T.F.; Andrade, S.F.; Fuentefria, A.M. In vitro antidermatophytic synergism of double and triple combination of clioquinol with ciclopirox and terbinafine. *Mycoses* **2020**, *63*, 993–1001. [CrossRef] [PubMed]
93. Kwon, H.S.; Koh, S.H. Neuroinflammation in neurodegenerative disorders: The roles of microglia and astrocytes. *Transl. Neurodegener* **2020**, *9*, 42. [CrossRef] [PubMed]

biomolecules

MDPI

Review

The Role of Zinc in Modulating Acid-Sensing Ion Channel Function

Amber W. Sun †, Michelle H. Wu †, Madhumathi Vijayalingam †, Michael J. Wacker and Xiang-Ping Chu *

Department of Biomedical Sciences, School of Medicine, University of Missouri-Kansas City, Kansas City, MO 64108, USA

* Correspondence: chux@umkc.edu; Tel.: +1-816-235-2248; Fax: +1-816-235-6517

† These authors contributed equally to this work.

Abstract: Acid-sensing ion channels (ASICs) are proton-gated, voltage-independent sodium channels widely expressed throughout the central and peripheral nervous systems. They are involved in synaptic plasticity, learning/memory, fear conditioning and pain. Zinc, an important trace metal in the body, contributes to numerous physiological functions, with neurotransmission being of note. Zinc has been implicated in the modulation of ASICs by binding to specific sites on these channels and exerting either stimulatory or inhibitory effects depending on the ASIC subtype. ASICs have been linked to several neurological and psychological disorders, such as Alzheimer's disease, Parkinson's disease, ischemic stroke, epilepsy and cocaine addiction. Different ASIC isoforms contribute to the persistence of each of these neurological and psychological disorders. It is critical to understand how various zinc concentrations can modulate specific ASIC subtypes and how zinc regulation of ASICs can contribute to neurological and psychological diseases. This review elucidates zinc's structural interactions with ASICs and discusses the potential therapeutic implications zinc may have on neurological and psychological diseases through targeting ASICs.

Keywords: acid-sensing ion channels; zinc; function; neuron; modulation; neurological diseases

1. Introduction

Acid-sensing ion channels (ASICs) are proton-gated, voltage-independent Na^+ channels found predominantly in the central and peripheral nervous systems [1,2]. ASICs are part of the degenerin/epithelial sodium channel (DEG/ENaC) superfamily of amiloride-sensitive ion channels [3]. To date, at least six ASIC isoforms (ASIC1a, 1b, 2a, 2b, 3a, and 4) encoded on four genes (*ACCN1–ACCN4)* have been cloned [2,4]. These ASIC isoforms form homo- and heterotrimers with different physiological and pharmacological properties [3,4]. ASIC subunits, mainly those present in the brain, are activated by rapid drops in pH [3]. Transient pH drops lead to rapid desensitization of ASICs, producing a detectable current [3]. On the contrary, gradual acidification causes gradual desensitization. Thus, no detectable current is produced from ASICs with slow drops in pH [3]. Most ASIC subunits are located in the brain; however, ASIC3 is a calcium-insensitive channel that is predominantly found peripherally in the dorsal root ganglion (DRG) neurons and other locations in the peripheral nervous system (PNS) [2,3]. Moreover, ASIC4 is unlike other ASIC channels in that, despite its name, ASIC4 does not induce currents when protonated [3,5]. Rather than a protein channel, it is hypothesized to be a modulator, specifically downregulating ASIC1a and ASIC3 surface expression. In addition, ASIC4 is unique from other ASIC channels in that it is found intracellularly, mainly in endosome-related vacuoles [6,7].

In concurrence with the major physiological roles of ASICs, studies have indicated that the potentiation, desensitization, and inactivation of ASICs play a role in pathological processes such as brain ischemia [8], Parkinson's disease [9], multiple sclerosis [10], and cocaine addiction [11]. It has been discovered that at both physiological and pathological levels, zinc serves as an important modulator of ASICs and can cause their activation or inhibition, depending on the specific ASIC isoform [4,12]. For example, the finger domain binds zinc, which has a potentiating effect on ASIC2a and an inhibiting effect on ASIC1b [13]. Beyond these findings, there are few studies reviewing the exact relationship of zinc with the different isoforms of ASICs. This review serves to elucidate zinc's structural interactions with ASICs and discuss the potential therapeutic implications zinc may have on neurological and psychological diseases through targeting ASICs.

1.1. ASIC Structure

As a part of the DEG/ENaC superfamily, ASIC subunits are composed of intracellular NH_2 and COOH terminals and two hydrophobic transmembrane domains (TMD1 and TMD2) separated by a large extracellular domain of approximately 370 residues [14–17]. The grossly viewed structure of an individual ASIC subunit's extracellular domain resembles a "clenched fist" with wrist, palm, finger, knuckle, thumb, and β-ball domains [15,17]. The palm domain serves as the central structure within each extracellular ASIC subunit and has direct connections to the transmembrane domains (TMD1 and TMD2) and the thumb domain [17]. The knuckle domain and its attached finger domain lie superior to the palm domain [17]. The outer edges of the finger domain come in contact with the thumb domain, and altogether the palm, thumb, finger, and knuckle domains surround the small β-ball domain [17]. Between the β-ball, thumb, and finger domains lies a highly negatively charged cavity called the acidic pocket [17]. Within this acidic pocket are three pairs of carboxyl–carboxylate interactions between the side chains of aspartate or glutamate residues [17]. These carboxyl–carboxylate interactions are responsible for the ion and pH-sensing capabilities of the acidic pocket [15,17]. This extracellular "clenched fist" domain is rich in cysteine residues, and studies have indicated that mutating various cysteine residues in the extracellular domain of ASICs play a critical role in zinc's potentiation and inhibition effects [15,17]. The zinc-binding site is located within the extracellular domain of the ASIC channel [4,12,18]. The specific extracellular domain zinc binds to varies on the ASIC subtype (see Figure 1). Zinc's effects on the activation or inactivation of ASICs can be further regulated by extracellular residues, such as histidine and cysteine [4,12,18].

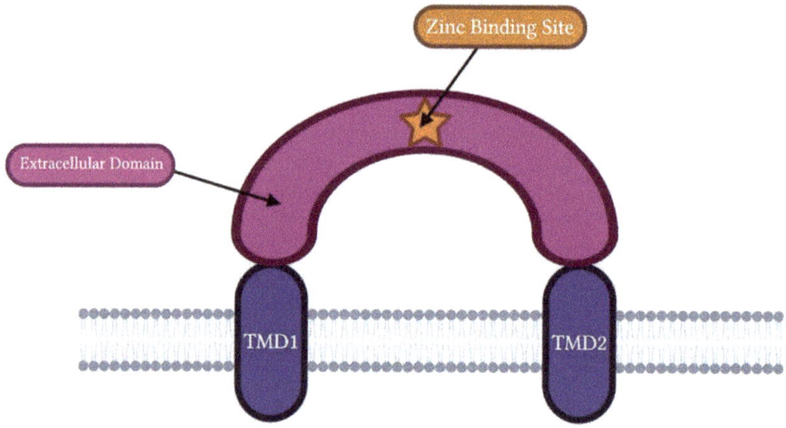

Figure 1. General ASIC structure with the zinc-binding site. Acid-sensing ion channels are composed of two transmembrane domains (TMD1 and TMD2) separated by a large extracellular domain. The *C* and *N* terminal face inside of the membrane. The binding site of zinc is contained within the ASIC extracellular domain. The exact location of zinc's binding site within this extracellular domain varies depending on the ASIC subtype. Adapted from "Transporters" by BioRender.com (2022 and ac) (https://app.biorender.com/biorender-templates, accessed on 10 December 2022).

1.2. Zinc Physiology

Zinc is essential to the growth and development of living organisms, and after iron, zinc is the most abundant trace metal within the human body. The human body contains approximately 2–4 grams of zinc, and the majority of zinc is distributed within the testes, muscles, liver, and prostate [19–21]. Over 300 enzymes rely on zinc as their cofactor, and zinc plays an important structural role for various proteins [21]. Furthermore, zinc is an essential ion for neurotransmission and is widely distributed within the presynaptic vesicles [22]. Zinc plays a critical role in neurogenesis, as it controls the cell cycle, apoptosis, and the binding of DNA and several proteins [22]. In addition, zinc plays a significant role in inhibiting growth within the prostate gland [23]. For example, zinc inhibits the enzyme in the first step of the Krebs cycle, accumulating citrate within the prostate gland and preventing further downstream energy production [24,25]. Moreover, high levels of zinc induce mitochondrial apoptosis, dampening prostatic tissue's growth and proliferation [25].

1.3. Zinc and Disease

The importance of zinc is underscored during the adverse effects that arise from a state of zinc deficiency. A lack of zinc in the body can lead to impaired immune function, growth retardation, delayed sexual maturation, poor wound healing, prostate cancer, and neurodegenerative disease [19,22,25–27]. Studies have indicated that men with prostate cancer have markedly decreased zinc levels by up to 80% compared to healthy controls. No study has reported a case of prostate cancer without decreased zinc levels [24,25].

Furthermore, zinc is involved in the pathogenesis of Alzheimer's disease, ischemic stroke, traumatic brain injury, epilepsy, and Parkinson's disease [26,27]. Most of these diseases are caused by zinc deficiency or intracellular zinc overload that disrupts multiple signaling pathways. A myriad of studies have been conducted to evaluate the concentration levels at which zinc develops neuroprotective or neurotoxic effects [27,28].

1.4. Zinc and Ion Channel Regulation

Ions cannot pass freely across cell membranes due to their charge; therefore, ion channels or transporters are required to transport ions between intra- and extracellular compartments [29]. Ion channels are mainly either voltage-gated or ligand-gated. Voltage-gated ion channels open and close based on changes in the membrane potential [30]. Ligand-gated ion channels rely on the release of molecules such as glutamate, glycine, acetylcholine, GABA, ATP, and serotonin [31]. These messengers selectively open cationic or anionic channels to depolarize or hyperpolarize the cell, respectively [31]. If a channel allows for the influx of sodium or calcium, it is classified as excitatory. Likewise, if a channel allows for the influx of chloride ions or efflux of potassium, it is classified as inhibitory. Zinc has been studied for the regulation or modulation of several ion channels, including the K^+ channel, Ca^{2+} channel, *N*-methyl-*D*-aspartate (NMDA) receptors, α-amino-3-hydroxy-5-methyl-4-isoxazole propionic acid (AMPA) receptors, kainate receptors, transient receptor potential (TRP) channels, and γ-aminobutyric acid (GABA) receptors [31,32].

K^+ channels selectively transport potassium from inside the cell to the extracellular space [33]. There are four main classes of K^+ channels: calcium-activated K^+ channels, inward-rectifying K^+ channels, tandem pore domain K^+ channels, and voltage-gated K^+ channels [33]. Each K^+ channel is divided into a pore-forming and regulatory domains [33]. The pore-forming domain allows for the passage of K^+ ions, and its structure is conserved among the different types of K^+ channels [33]. The regulatory domain varies in structure depending on the K^+ channel type [33]. Potassium channels are widely expressed in the peripheral and central nervous system cell membranes and play a major role in multiple cellular physiological processes as they control the resting membrane potential, repolarization rate of action potentials, and spike frequency adaptation [33]. Therefore, potassium channel dysfunction is associated with multiple neurological disorders such as epilepsy [34], Huntington's disease [35], and Parkinson's disease [36,37]. At micromolar concentrations, zinc is a negative modulator for most K^+ channels [32]. The exceptions are that zinc positively modulates TWIK-related potassium channels subtype 2 (TREK-2) at an EC_{50} of 659 μM and voltage-gated potassium channels subfamily Q member 5 (KCNQ5) at an EC_{50} of 22 μM [32]. Furthermore, zinc activates Slo1 K^+ channels at an EC_{50} of 34 μM [32].

There are two types of Ca^{2+} channels: high voltage-activated (Ca_V1.1–1.4, Ca_V2.1–2.3) and low voltage-activated (Ca_V3.1–3.3) channels [38]. These voltage-gated calcium channels (VGCC) open for the entry of Ca^{2+} into the cell in response to membrane depolarization [38]. Zinc serves as a negative modulator for all calcium channels [32]. Ca_V1.1–1.4 channels conduct L-type calcium currents, distinguished by slow voltage-dependent inactivation [38]. The L-type Ca^{2+} channels are mainly expressed in smooth muscle cells [32]. These channels are involved in initiating contraction, hormone secretion, and local calcium signaling to gene transcription [38]. Ca_V1.1 has a zinc IC_{50} of 11 μM or 18 μM, and Ca_V1.2 has a zinc IC_{50} of 34 μM [32]. Ca_V2.1–2.3 channels conduct P/Q-, N-, and R-type calcium currents, respectively [38]. These currents contain faster voltage-dependent inactivation and are located in neurons [38]. Ca_V2.1 has a zinc IC_{50} of 110 μM, Ca_V2.2 has an IC_{50} of 98 μM, and Ca_V2.3 has an IC_{50} of 32 μM [31]. Ca_V3.1–3.3 channels conduct T-type calcium currents, which are activated at negative membrane potentials [38]. T-type calcium currents also have fast deactivation upon repolarization and fast voltage-dependent inactivation during sustained depolarizations [38]. Ca_V3.1–3.3 channels are prominent within cardiac myocytes in the sino-atrial node and neurons within the thalamus [38]. They have zinc IC_{50}s of 82, 0.8, and 159 μM or 196, 24, and 152 μM for Ca_v3.1, Ca_v3.2, and Ca_v3.3, respectively [32]. Mutations in calcium channels are responsible for neuropsychiatric diseases, migraines, hypertension, heart failure, and chronic pain syndromes [39].

Glutamatergic neurons are involved in important roles in the CNS, such as learning/memory and synaptic plasticity [40]. Glutamatergic neurotransmission is primarily modulated through ionotropic and metabotropic glutamate receptors [40]. The ionotropic glutamate receptors are ligand-gated ion channels that are permeable to Na^+ and K^+. Certain ionotropic glutamate receptors have Ca^{2+} permeability, such as GluN2-containing subunits [40]. There are three groups of ionotropic glutamate receptors: *N*-methyl-*D*-aspartate (NMDA) receptors, α-amino-3-hydroxy-5-methyl-4-isoxazolepropionic acid (AMPA) receptors, and kainate receptors [40]. At normal physiological conditions, NMDA receptors are blocked by Mg^{2+}. To function properly, NMDA receptors first require membrane depolarization via an influx of Na^+ from AMPA receptors to remove this Mg^{2+} block and allow subsequent ion permeability [40]. Once the Mg^{2+} block is removed, NMDA receptors allow the influx of Na^+ and Ca^{2+} [40]. This influx of Ca^{2+} activates intracellular mechanisms that lead to the phosphorylation and subsequent upregulation of AMPA receptors [40]. AMPA and NMDA receptors play a major role in synaptic plasticity and are involved in pathological processes such as Alzheimer's disease [41]. Kainate receptors and AMPA receptors are inhibited by zinc [31,32]. The regulation of zinc on these ionotropic glutamate receptors heavily depends on the subunit composition [31,32]. For example, the co-expression of GluN1 and GluN2A subunits in HEK293 cells increased the negative zinc modulation by 1000-fold compared to the co-expression of GluN1 and GluN2B subunits in HEK293 cells [31]. Ionotropic glutamate receptors, particularly NMDA and AMPA receptors, have been linked to neurological disorders such as Alzheimer's disease, ischemic stroke and schizophrenia [41–43].

TRP channels are a large family of channels that are activated by a variety of sensory stimuli [44]. Out of the 28 members, there are six TRP subfamilies: canonical (TRPC), vanilloid (TRPV), melastatin (TRPM), polycystin (TRPP), mucolipin (TRPML), and ankyrin (TRPA) [44]. These channels are integral in evaluating environmental stimuli and act as signal transducers via altering the membrane potential and intracellular Ca^{2+} levels [44]. TRP channels are involved in a large number of physiological processes and are potential therapeutic targets. For example, TRPA contains a single chemo-nociceptor that may be a potential analgesic target, and TRPP is involved in autosomal dominant polycystic kidney disease [44]. Zinc serves as a negative modulator for TRPM2 and TRPM5 [31,32]. When

TRPM2 is activated by ADP ribose, zinc negatively modulates the channel because of inhibition of the channel above levels of 30 μM of extracellular zinc [31,32]. When TRPM5 is activated by 500 nM of intracellular Ca^{2+}, it has a zinc IC_{50} of 4.3 μM [31,32].

GABA is a fast inhibitory neurotransmitter predominant within the CNS and acts on $GABA_A$ and $GABA_B$ receptors [45]. A dysfunction with these receptors can lead to a wide array of neurological problems [45]. $GABA_A$ receptors are the target for multiple therapeutic drugs, including those for epilepsy, anxiety, insomnia, and panic disorder [45]. $GABA_A$ receptors are composed of 19 subunits, with the predominant presynaptic isoform composed of two α1 subunits, two β2 subunits and one γ2 subunit [45]. GABA receptors are negatively modulated by zinc [32]. The GABAρ1 subunit has a zinc IC_{50} of 22 μM or 20 μM, the GABAα1β2γ2 subunit has a zinc IC_{50} of 441 μM, and $GABA_A$ has a zinc IC_{50} of 7 μM [32].

Despite the research into zinc and other membrane channels, few studies or reviews have linked zinc's pathological role in these neurodegenerative disorders with its effects on ASICs [13]. Both zinc and ASICs are involved in the pathophysiology of multiple neurological and psychological disorders. Depending on the ASIC subtype and composition, zinc can serve as a potentiator or inhibitor of that channel by binding to the ASIC extracellular domain. Further study of the connection between zinc and ASICs may reveal a critical point for the therapeutic treatment of several of these "incurable" neurodegenerative disorders. Thus, the next section of the review details the relationship between zinc and ASICs and how these two substances correlate with several neurological and psychological diseases.

2. Zinc's Effects on Different Types of ASICs

2.1. Zinc and ASIC1a

Although ASIC1a and ASIC1b are transcribed from the *ACCN2* gene [2,16], their function and location vary significantly. ASIC1a is widely expressed in the CNS and PNS. Centrally located homotrimeric ASIC1a plays a vital role in synaptic plasticity, learning, and memory [3]. ASIC1a also plays a role in fear and anxiety [3]. Unlike the other ASIC channels, ASIC1a is uniquely permeable to Ca^{2+} and is heavily involved in acidosis-induced injuries such as ischemic brain injury [46–50]. Though their calcium permeability is poor, activation of ASIC1a channels may induce damage through secondary mechanisms that increase intracellular Ca^{2+} levels, such as the activation of voltage-gated Ca^{2+} channels and intracellular storage release [50]. Zinc chelation potentiates ASIC1a-mediated currents, consequently increasing intracellular calcium levels and inducing membrane depolarization [46].

Our laboratory's previous studies have shown that zinc has a high-affinity binding site on the lysine-133 residue in the extracellular domain of ASIC1a channels, where it exhibits inhibitory behavior. Consequently, mutations of K133 rendered zinc's function to be obsolete [46]. The presence of 0.3 μM of zinc displayed reversible inhibition of ASIC1a [50]. The effects of *N,N,N′,N′*-tetrakis(2-pyridinylmethyl)-1,2-ethanediamine (TPEN), a high-affinity zinc chelator, on closed-state ASIC1a revealed current potentiation with an EC_{50} value of 2 μM in mouse cortical neurons [46]. The maximum potentiation induced by TPEN can be achieved in cells within 2–4 min of perfusion, and the effect is reversible following a washout of TPEN. Adding TPEN alone in the solution does not trigger any current in pH 7.4 solution, and co-applicating TPEN to pH 6.5 solution for 10 seconds does not induce any significant potentiation. Another study revealed that the co-application of 300 μM of zinc to *Xenopus* oocytes displayed slight inhibition of the ASIC1a current that was not significant [18], suggesting that zinc only has significant inhibitory effects on the closed state of ASIC1a channels.

Zinc-mediated inhibition is pH- and dose-dependent based on zinc chelation using 10 μM of TPEN in various concentrations of buffered free zinc ions with an IC_{50} value of 14 nM in mouse cortical neurons [46]. Consequently, a zinc dose-inhibition curve yielded an IC_{50} value of 7.0 ± 0.35 nM for buffered zinc solutions. An increase of zinc concentration up to 30 μM does not display additional zinc-mediated inhibition [46].

2.2. Zinc and ASIC1b

ASIC1b and ASIC3 channels normally possess a transient current followed by a sustained current, which contributes to prolonged acidosis and pain sensation [12,51,52]. Zinc binds to the extracellular domain of ASIC1b channels and displays inhibitory mechanisms [12,51]. Zinc inhibits the peak amplitude in both ASIC1b and ASIC3 channels, but unlike ASIC1b, zinc inhibits both the peak and sustained component in ASIC3, thereby indicating that zinc has differing mechanisms on ASIC1b and ASIC3 [12,51,52].

Zinc concentrations of 1 or 3 μM harbor no significant inhibition on ASIC1b channels [12,51] while concentrations of 10, 30, 100, or 300 μM exhibit a profound and concentration-dependent inhibitory effect of the peak amplitude [12,51], revealing that zinc has a low-affinity binding site of ASIC1b channels. Furthermore, pretreatment of zinc possesses inhibitory behavior with an IC_{50} of 36.5 ± 1.5 μM while co-application harbors no effects on ASIC1b currents [12,51]. Additionally, pH activation, steady-state desensitization, and extracellular calcium concentrations harbor no effect on zinc-mediated inhibition of ASIC1b channels, indicating non-competitive processes [12,51]. Therefore, zinc reveals a strong pH- and calcium-independent inhibitory effect on the peak component of ASIC1b channels in closed states only.

Studies from our laboratory have shown that mutation of the extracellular cysteine-149 residue in mouse ASIC1b did not reveal any zinc-mediated inhibition, indicating a lack of zinc-binding site in rat ASIC1b when cysteine 149 residue was replaced [51]. Patch-clamp recordings of zinc and human ASIC1b with a mutation of the extracellular cysteine-196 residue demonstrated no inhibition of the current due to the lack of zinc binding [12]. Consequently, mutating other cysteine or non-cysteine residues on ASIC1b channels exhibited a reduction of ASIC current in peak amplitude by a drop in pH [12], revealing that the extracellular cysteine-196 residue of human ASIC1b is the low-affinity binding site of zinc responsible for mediating inhibitory mechanisms [12].

2.3. Zinc and ASIC1a/3

ASIC channels can form homomeric and heteromeric channels [4,46]. ASIC1a is predominantly located centrally, whereas ASIC3 is primarily expressed in the peripheral sensory neurons. The coexistence of the ASIC1a/3 heterodimer is currently discovered to be located mostly in skeletal muscle and plays a role in pain management [53,54]. Recently, we found that systematically mutated histidine residues 72 and 73 in both ASIC1a and ASIC3 and histidine residue 83 in ASIC3 were responsible for the dual effects of zinc on heteromeric hamster ovary ASIC1a/3 channels [4].

Co-application of zinc dose-dependently potentiated the peak and sustained component of ASIC1a/3 channels. Concentrations between 1 and 100 μM displayed an EC_{50} of 26 μM whereas concentrations between 100 and 1000 μM displayed an EC_{50} of 343 μM [4]. Overall, zinc has a low-affinity binding site on ASIC1a/3 that mediates potentiation of both the peak and sustained components of open-state ASIC1a/3 heteromeric channels in a dose- and pH-dependent manner.

Pretreatment with zinc between 3 to 100 μM exerted the same potentiation as co-application [4]. Concentrations between 1 and 100 μM exhibited an EC_{50} of 24 μM, and concentrations between 100 and 250 μM exhibited an EC_{50} of 128 μM. In contrast, concentrations above 250 μM exerted profound inhibition on the peak amplitude with an IC_{50} of 306 μM [4]. Taken together, zinc displays dual effects on the closed state of ASIC1a/3 channels in a dose- and pH-dependent manner [4].

2.4. Zinc and ASIC1a/2b

Homomeric ASIC2b channels do not produce currents independently, but ASIC2b, associated with other ASIC subunits, can form functional heteromeric ASIC channels [55]. The ASIC2b subunit enables the heteromeric channel ASIC1a/2b to harbor unique channel properties different from homomeric ASIC1a [55]. ASIC1a/2b undergoes steady-state desensitization at more basic pH values (pH 7.4) than other ASIC channels [55]. In *Xenopus* oocytes, co-application of 300 μM of zinc on ASIC1a/2b heteromers and ASIC1a homomers displays reversible inhibition. This is in contrast to ASIC1a/2a, which demonstrates profound potentiation. However, zinc modulation has reduced effects in the presence of ASIC2b with ASIC1a compared to ASIC1a homomers [55]. The same study also utilized 10 μM TPEN and assessed that nanomolar chelating concentrations of zinc enhance the ASIC1a/2b current amplitude. In fact, ASIC1a, ASIC1a/2a, and ASIC1a/2b are all inhibited by low nanomolar concentrations of zinc [55]. These studies monitored all currents with zinc modulation after the voltage ramp, indicating that these channels were assessed in an open state. The exact zinc-binding site of ASIC1a/2b is unknown, although it is strongly hypothesized that it is located on the ASIC2 subunit based on the differing current modulation of zinc when comparing homomeric ASIC1a and ASIC1a/2b [55]. Further research is required to identify the exact binding site for zinc modulation.

2.5. *Zinc and ASIC1a/2a*

Zinc has a dual dose-dependent effect on ASIC1a/2a heteromers. At high micromolar concentrations (100–300 μM), zinc binds to the ASIC1a/2a channels with low affinity and potentiates the effect of the ASIC1a/2a heteromers as it does with ASIC2a [18,46]. Zinc potentiates the ASIC1a/2a current at an EC_{50} of 111 μM upon co-application with an acidic pH [18]. Zinc cannot potentiate the channel unless the ASIC1a/2a heteromer is in an open state. Normally, ASIC1a/2a requires a more acidic pH to be activated in the CNS. However, the potentiation of ASIC1a/2a by zinc at high micromolar concentrations leftward shifts the pH dependence to a pH closer to physiological pH. Rather than becoming activated at a $pH_{0.5}$ of 5.5 as the channel usually does, the $pH_{0.5}$ for ASIC1a/2a channels activated by zinc is higher, at 6.0 [18]. The greatest potentiation by zinc was between a pH of 6.9 and 6.3 [46]. These heteromers increase the Hill coefficient by shifting the activation curve to the left, demonstrating that one ASIC2a subunit is enough for Zn^{2+} to potentiate the heteromer [18]. In homomeric ASIC2a channels, mutation of either the histidine residue H162 or H339 to alanine inhibits zinc coactivation of the channel [18]. Conversely, zinc coactivation of the heteromeric ASIC1a/2a channel is inhibited by H339A mutations but not by H162A mutations. This indicates a difference between the zinc binding sites of the ASIC2a containing homomeric and heteromeric channels that can alter the capability of zinc as a potentiator [14]. An increasing number of 2a subunits in comparison to 1a subunits also does not increase its potentiation by Zn^{2+}. Unexpectedly, the ASIC1a-2a-1a concatemer with two 1a subunits had greater potentiation by zinc than the ASIC1a-2a-2a concatemer [56]. This suggests that the cooperation of the three subunits contributes to its ability to be potentiated by Zn^{2+} rather than simply the presence of ASIC2a.

At nanomolar concentrations, Zn^{2+} inhibits the ASIC1a/2a heteromer as it does with ASIC1a [46]. Chelation of zinc by TPEN potentiates the current through ASIC1a/2a in a dose-dependent manner [46]. The IC_{50} value for inhibition of ASIC1a/2a by zinc is 10 nM [46]. While the absence of ASIC2a did not impact the inhibition of zinc, the absence of the ASIC1a subunit did eliminate the inhibition of zinc, confirming that the presence of ASIC1a is necessary for the inhibitory effects of zinc [46]. The high-affinity site K133 located extracellularly on the ASIC1a subunit is highly indicated in the binding of zinc to the channel. This positively charged lysine residue could be involved in the inhibitory effect of zinc at nanomolar concentrations on ASIC1a/2a heteromers since ASIC1a mediates zinc inhibition [46]. However, the exact binding site for the inhibitory action of zinc on the ASIC1a/2a channel is currently unknown.

2.6. *Zinc and ASIC2a*

Unlike ASIC1a channels, high micromolar concentrations of zinc (e.g., 100 or 300 μM) potentiate ASIC2a and ASIC2a-containing channels [18,46]. Upon mutation of the extracellular His-162 (H162) and His-339 (H339) residues to alanine, zinc was no longer able to potentiate ASIC2a-containing channels [18]. Specifically, zinc binds with low affinity to the H339 and H162 residues at the interface between the upper palm, finger, and ball domains [18]. This is different from heteromeric ASIC2a-containing channels in which H339 is necessary, but H162 has either moderate or no effect on zinc sensitivity, indicating a difference in binding sites between the homomers and heteromers [18]. ASIC2a is potentiated by zinc at an EC_{50} of 120 μM upon co-application with an acidic pH [18]. Thus, zinc potentiates the ASIC2a homomer when the channel is in an open state. Homomeric ASIC2a normally has low acid sensitivity and cannot be activated at a pH between 7.4 and 5.5. It requires a pH of 4.5 to activate a large current [18,46]. However, the potentiation of ASIC2a by zinc occurs at a much lower concentration of H^+ between a pH of 6.9 and 5.0 since zinc induces an alkaline shift of the pH dependence [18]. The pH sensitivity of ASIC2a is dependent on the extracellular His-72 residue found immediately after the first transmembrane domain and is abolished if this is mutated [18].

2.7. Zinc and ASIC2a/3

Another function of ASIC2a is to increase the conductance sensitivity of the cell membrane to protons by increasing the expression of ASIC3 at the cell surface and assembling heteromers with ASIC3 [56,57]. ASIC2a/3, like the previous ASIC2a-containing channels, is also potentiated by zinc [18]. The potentiation of this channel by micromolar concentrations of zinc mimics that of ASIC1a/2a, with the greatest potentiation being at a pH of 6 [18]. Zinc acts on the ASIC2a/3 channel in an open state, requiring co-application with an acidic pH for potentiation to occur [18]. Similar to ASIC1a/2a, potentiation by zinc decreased with increasing extracellular acidity [18]. While the effect of zinc on the potentiation of ASIC1a/2a was unchanged with a mutation of H162, for ASIC2a/3 channels, both H162 and H339 mutations decreased the effects of zinc [18]. This indicates that zinc likely binds with low affinity to both the H162 and H339 residues to potentiate the heteromeric ASIC2a/3. However, the H339 residue is likely more significant because an H339 mutation had a greater zinc potentiation reduction than the H162 mutation [18].

2.8. Zinc and ASIC3

Zinc binds to the extracellular domain of ASIC3 channels and exhibits inhibitory behavior [52]. Similar to ASIC1b, ASIC3 channels possess a transient current followed by a sustained current, thereby prolonging acidosis and further contributing to pain perception [54]. Nanomolar concentrations of zinc harbor no involvement in ASIC3 channels, while micromolar concentrations exhibit inhibitory behavior on both the transient and sustained components of ASIC3 currents, indicating that zinc has a low-affinity binding site on ASIC3 channels [52]. Additionally, our studies have shown that the co-application of zinc has no effects on ASIC3 currents, whereas pretreatment of zinc displays dose-dependent inhibition with an IC_{50} of 61 ± 3.2 μM [52]. Zinc-mediated inhibition occurs rapidly and at a narrow concentration range between 30 to 300 μM [52]. Thus, zinc reveals a strong inhibitory effect on ASIC3 channels in closed states within a narrow micromolar concentration range.

Administration of intracellular zinc does not diminish zinc-mediated inhibition of ASIC3 currents, confirming that the zinc-binding site is located outside the cell [52]. When administering zinc on ASIC3 channels with calcium concentrations at 2, 5, or 10 mM, the degree of zinc inhibition remains unchanged, concluding that zinc and calcium do not share the same binding site on ASIC3 channels [52]. Therefore, zinc-mediated inhibition is calcium-independent for this particular isoform. When comparing zinc inhibition at various pH values, the percent inhibition does not deviate significantly, suggesting that this inhibition is pH-independent [52]. With consistent inhibitory behavior and pH activation acuity, zinc may be a crucial ASIC3 channel regulator in pathophysiological conditions associated with pH changes, such as epilepsy [58,59], myocardial ischemia [60], rheumatoid arthritis [61,62], Alzheimer's disease [63], and trauma [64].

While zinc commonly binds to histidine or cysteine residues in other ASIC channels, modification of these two amino acid residues on ASIC3 did not affect zinc inhibition, suggesting that zinc binds to a site unrelated to histidine or cysteine [52]. Further examination of other extracellular residues like glutamate may help to identify the specific zinc-binding site of the ASIC3 channels. Table 1 shows zinc's effects on different type of ASICs.

Table 1. Summary of zinc's effects on different ASIC subtypes.

ASIC Subtype	Zinc Binding Site	Zinc Effect	EC_{50}/IC_{50}	Binding State of Channel	References
ASIC1a	K133$^{\uparrow}$	Inhibitory	IC_{50}: 7.0 ± 0.35 nM	Closed	[46]
ASIC1b	C149$^{\downarrow}$	Inhibitory	IC_{50}: 36.5 ± 1.5 μM	Closed	[12,51]
ASIC1a/3	H72$^{\downarrow}$, H73$^{\downarrow}$, H83$^{\downarrow}$	Excitatory* [μM] Inhibitory$^{\#}$ [1–250 μM] Excitatory$^{\#}$ [> 250 μM]	$EC_{50}{}^{*}$: 26 μM $EC_{50}{}^{\#}$ [1–100 μM]: 24 μM $EC_{50}{}^{\#}$ [100–250 μM]: 128 μM $IC_{50}{}^{\#}$ [> 250 μM]: 206 μM	Closed^{1,2} Open^{1}	[4]
ASIC1a/2b	Unknown	Inhibitory	Unknown	Open	[50]
ASIC1a/2a	H339$^{\downarrow,1}$, K133$^{\uparrow,2}$	Excitatory [μM] Inhibitory [1–250 μM]	EC_{50}: 111 μM IC_{50}: 10.04 ± 1.23 nM	Open	[18,46]
ASIC2a	H339$^{\downarrow}$, H162$^{\downarrow}$	Excitatory	EC_{50}: 120 μM	Open	[18]
ASIC2a/3	H339$^{\downarrow}$, H162$^{\downarrow}$	Excitatory	Unknown	Open	[18]
ASIC3	Unknown$^{\downarrow}$	Inhibitory	IC_{50}: 61 ± 3.2 μM	Closed	[52]

[] The values in brackets represent the zinc concentrations used to determine the subsequent findings. $^{\uparrow}$—High affinity; $^{\downarrow}$—Low affinity; *—Open state; #—Closed state; ^{1}—Excitatory; ^{2}—Inhibitory.

3. Zinc Regulation of ASICs in Neurological and Psychological Diseases

3.1. *ASIC1a*

3.1.1. Epilepsy

Epilepsy is a neurological disorder characterized by abnormal, excessive, or synchronized neuronal activity [65]. Various studies reveal the significance of zinc homeostasis in seizures and epilepsy. Zinc is necessary for proper neural signaling, whereas zinc dyshomeostasis leads to an improper balance of neural excitation and inhibition, resulting in seizures [20]. Increased zinc serves as a protective tool in preventing ASIC1a- and NMDA-mediated excitotoxicity in neuropathological conditions like epilepsy [59,66]. Activating ASIC1a channels from decreased pH levels in the brain leads to acidosis-mediated neurological damage [67]. The zinc-mediated inhibition of ASIC1a channels may reduce acidosis and thus prevent brain injury from seizures [67].

Further research on administering zinc as a therapy to epileptic patients is required to document the specific relationship between zinc and epilepsy. Notably, increased zinc concentrations beyond physiological concentrations were found to be toxic due to their entry into neurons [34,68]; yet another study found that a moderate increase of zinc in the extracellular space is neuroprotective against pathological conditions with severe acidosis. The corresponding study used HEK 293 cells, so future research is needed on primary neurons to study the neuroprotection of different concentrations of extracellular zinc [67].

3.1.2. Migraines

The pathophysiology of migraines is poorly understood, although one hypothesis suggests extracellular acidification [69]. Calcium ions play a vital role in the human body and can contribute to non-mitochondrial reactive oxygen species (ROS) production [70] and induce acidosis-mediated injury [46]. Various studies reveal a correlation between migraine headaches and zinc deficiency [71]. A 2020 randomized 8-week clinical trial revealed that 220 mg of zinc sulfate per day reduced the frequency of migraine attacks in comparison to the placebo group. However, other factors such as headache severity, migraine duration, and presence of auras were not affected by zinc supplementation [72]. Similarly, a 2021 randomized 12-week clinical trial revealed that zinc glucose supplementation not only significantly reduced the frequency but also the periods and severity of migraine attacks in comparison to the control group [73]. Taken together, zinc supplementation has an overall positive impact on migraine attacks by reducing the frequency and severity. ASIC1a is vital for normal brain function, though due to its calcium permeability, excessive ASIC1a signaling can contribute to acidosis-mediated injury [50] and cortical spreading depression in migraines [71]. Administration of the inhibitors mambalgin-1 and amiloride to ASIC1a channels revealed significant efficacy as acute and prophylactic treatment options for migraines [71]. Most therapeutic options for chronic migraines are notoriously difficult and commonly fail [74,75]. Further research is necessary to analyze the potential therapeutic effects of zinc administration and chronic migraines. Thus, zinc supplementation might reveal a potential therapeutic option for the treatment of acute migraine attacks due to its inhibitory modulator on ASIC1a.

3.1.3. Alzheimer's Disease

Alzheimer's disease (AD) is a progressive neurodegenerative disorder and is one of the leading causes of dementia [76]. It is characterized by advanced cognitive impairment associated with behavioral changes, memory loss, and learning and orientation complications [76,77]. Diagnostic histopathologic features of AD include amyloid plaques aggregated by β-amyloid (Aβ) peptides and neuronal fibrillary tangles (NFTs) by hyperphosphorylated tau protein in the brain [78,79].

The imbalance of zinc in the brain is one of the pathological features of AD. Zinc concentrations are elevated in specific regions of the brain affected by AD, which may be a result of the Aβ peptides capturing zinc ions. Additionally, zinc directly promotes the aggregation of Aβ peptides and tau hyperphosphorylation, thereby exacerbating the advancement of AD [80–84]. Furthermore, the quantity of zinc transporters decreases as the disease progresses, correlating to increased disease severity and cognitive impairment. It is unclear whether zinc concentrations or zinc transporters are the ultimate cause of AD [84,85].

ASIC1a channels may play a role in the pathogenesis of AD through their involvement in the Aβ-mediated effect on metabotropic glutamate (mGlu) receptor-dependent transmission. Consequently, utilizing the ASIC1a-selective inhibitor, psalmotoxin-1, restored the intrinsic excitability of mGlu in the hippocampus, revealing that ASIC1a channels play a role in the Aβ-related depolarizing response and long-term depression [77]. Taken together, the relationship between ASIC1a channels and mGlu potentially suggests their significant role in the pathogenesis of AD.

Because both zinc and ASIC1a channels play a role in the pathogenesis of AD, targeting zinc and ASIC1a may have therapeutic potential for AD. Various studies revealed potential therapeutic effects of zinc and copper balance on the early stages of AD [86], unrelated to ASIC channels. Further research is required to analyze the relationship between zinc and ASIC1a channels in patients with AD.

3.1.4. Parkinson's Disease

Parkinson's disease (PD) is a neurodegenerative disorder characterized by Lewy body inclusions and the degeneration of the dopaminergic neurons in the substantia nigra pars compacta [87], leading to dopamine deficiency in the striatal pathway and, ultimately, basal ganglia deterioration [88,89]. Clinical presentation of PD includes olfactory dysfunction, tremor, cogwheel rigidity, cognitive impairment, and more [87].

It is uncertain how zinc dyshomeostasis and PD are related. Some studies have found lower serum zinc levels in PD patients [90,91], while others have found higher serum zinc levels. Studies reveal that methamphetamine causes dopaminergic cell death by generating reactive oxygen species and increasing the total amount of α-synuclein, a key element of Lewy bodies [85]. Zinc pretreatment reverses the aforementioned phenomena by increasing metallothionein expression in vitro, attenuating the accumulation of ROS in neurons [92,93]. By pretreating the cells with 50 μM of zinc chloride, methamphetamine-induced expression of α-synuclein was significantly reduced [92]. This observation supports the potential relationship between low zinc levels and α-synuclein production in PD. Contrastingly, numerous studies have demonstrated the detrimental effects of an aberrant accumulation of zinc in the substantia nigra and striatum, leading to dopaminergic neuronal cell death [88,94]. These findings are reinforced by the observation of 1-methyl- 4-phenyl-1, 2, 3, 6-tetrahydropyridine (MPTP)-induced neuronal cell death in mice when zinc is pretreated [85]. In principle, these research findings imply that zinc dyshomeostasis may harbor dual involvement in the pathogenesis of PD contingent on zinc-modulated signaling pathways at specific stages of the disease.

Aberrant excess of neural inflammation and lactic acidosis contribute to neurodegeneration in PD [87,95]. Observation of lactic acidosis in the animal model of PD reveals that ASIC1a may also play a role in dopaminergic neuronal cell death. Amiloride (non-selective ASIC inhibitor) and psalmotoxin-1 (selective inhibitor of ASIC1a) were revealed to attenuate neurodegeneration in the substantia nigra pars compacta [95]. This observation is supported by the conclusion that ASIC1a may play a significant role in the pathophysiology of PD [77], either by mutations in the Parkin gene associated with the autosomal recessive juvenile-onset of PD [96] or by the absence of the Parkin gene, which promotes hippocampal ASIC1a currents [97]. The role of ASIC1a as a therapeutic target for PD has not been the subject of many research investigations.

The significance of zinc and ASIC1a channels as a potential treatment of PD demands further investigation. Zinc-mediated inhibition of ASIC1a channels [46] is a probable therapeutic target to prevent neurodegeneration, but future research is required to detail zinc's precise effects on ASIC1a channels and how it impacts PD.

3.1.5. Depression

Major depressive disorder (MDD) is a prevalent mental illness with unclear etiology and poor effective therapies [98,99]. A key contributing factor to depression may be alterations to serotonin levels in the brain [99]. The glutamatergic theory, which postulates that depression results from an imbalance between the excitatory effect of glutamate and the inhibitory action of γ-aminobutyric acid (GABA), is the widely accepted approach to understanding the pathophysiology of depression. However, the glutamatergic theory of depression does not account for the large spectrum of symptoms observed in MDD [99]. Consequently, there is potential significance between zinc and its receptor, GPR39, that bridges the gaps in understanding depression. Therefore, it is crucial to investigate potential interactions between the brain's monoaminergic, glutamatergic, and zincergic systems.

Zinc can act as an inhibitory neuromodulator of NMDA channels, a major pharmacotherapeutic target in depression, [100] or as a neurotransmitter [101]. Neural transmission is disrupted in zinc deficiency, clinically manifesting as cognitive, emotional, and behavioral impairment [102]. In rats, low levels of zinc cause elevated cortisol, enhancing the hypothalamic–pituitary–adrenal (HPA) axis and ultimately facilitating the pathogenesis of depression [103]. GPR39 also plays a significant role relevant to cognition, emotions, and memory processing [101,102]. When bound to zinc, GPR39 is hypothesized to participate in serotonin synthesis [104], serotonin receptor signaling [105], and higher brain-derived neurotrophic factor (BDNF) [106]. Consequently, GPR39 knock-out mice are resistant to traditional antidepressants [104]. Zinc is also hypothesized to directly affect serotonin signaling through agonistic binding to the 5-HT1A receptor and antagonistic binding to the 5-HT7. Likewise, zinc transporters play a role in depression. Mice with absent zinc transporter-3 exhibit reduced proliferating progenitor neurons [107] and decreased hippocampal volume [108], suggesting that alterations to zinc transporter-3 contribute to the pathogenesis of depression. The impacts of zinc are comparable to that of typical antidepressants, and prior research has demonstrated that antidepressant treatment of depression restores low BDNF levels [109].

Fear, addiction, and depression are attributed to ASIC channels [110,111]. ASIC1a channels located in the amygdala are hypothesized to play a significant role in depression [110–112]. ASIC1a channel disturbance is theorized to disrupt the fear circuit, leading to deficits in fear-related behavior [112]. A 2009 study investigated the antidepressant-like effects when disrupting ASIC1a channels in mice. Using the forced swim test and tail suspension test, mice with absent ASIC1a channels displayed antidepressant-like findings compared to mice with normal ASIC1a. Administration of PcTx1 and amiloride also produced antidepressant-like effects. Overall, findings from this study reveal that ASIC1a contributes to depression in mice [110]. Surprisingly, a 2017 study revealed normal nucleus accumbens ASIC1a expression in knock-out mice with absent ASIC1a alleles. The same mice also demonstrated a normal forced swim test. The discrepancy may be due to the newer study using SynAsic1a KO mice generated by floxed ASIC1a alleles disrupted by Cre recombinase driven by the neuron-specific synapsin I promoter, whereas the 2009 study used ASIC1a$^{-/-}$ mice. Compared to ASIC1a$^{-/-}$ mice, SynAsic1a KO mice do not display identical behavioral changes but have similar deficits in fear conditioning. Furthermore, not all neurons had disrupted ASIC1a expression in SynAsic1a KO mice [109]. Therefore, ASIC1a channels impact fear-related behaviors in mice, but further research is required to assess the specific relationship between ASIC1a and depression. With zinc intrinsically harboring antidepressant effects, targeting ASIC1a channels using zinc-mediated inhibition may be another potential antidepressant. Future investigation is warranted to identify a potential correlation between ASIC1a channels, zinc, and depression in humans.

3.1.6. Stroke

Ischemic stroke is characterized as a thrombo-inflammatory condition that induces a pro-inflammatory state at the site of vascular injury, consequently compromising the blood–brain barrier (BBB) and inducing neuronal cell death [113,114]. Under physiological conditions, zinc and other divalent cation transport are maintained by the BBB [115]. Some research utilizing animal models of global ischemia in cortical neurons reveals that zinc accumulation drives the progression of brain infarction. Additionally, pathological zinc concentrations in the synaptic cleft of ischemic neurons also elicit cell death, revealing that zinc toxicity may be an independent risk factor for ischemic stroke [115,116]. Consequently, chelating zinc using EDTA in rats with ischemia revealed neuroprotection through enhanced cognitive function and deterred apoptosis of ischemic cells [116]. Surprisingly, other studies have revealed that zinc administration in rats with cerebral ischemia guards the hippocampus against neuronal injury during the reperfusion phase [115,117].

Furthermore, patients with zinc deficiency have a greater likelihood of ischemic strokes and an enhanced rehabilitation of neurological deficits with zinc supplementation [115,118]. These contradictory findings suggest that zinc plays a variety of functions in both the early and late stages of ischemic strokes. The laboratory settings may also substantially influence the beneficial or detrimental effects of zinc addressed above.

ASIC1a channels are suggested to play a role in the progression of ischemic strokes due to their activation during hypoxia [119,120]. Lactic acid production from increased anaerobic glycolysis promotes an acidic pH in the brain, thereby activating ASIC1a and ASIC1a-containing channels [119]. ASIC1a-mediated neuronal ischemic injury is further enhanced by concurrent induction of other elements such as Ca^{2+}/calmodulin kinase II and NMDA receptors [121]. Consequently, the administration of amiloride, a high-affinity inhibitor of ASIC1a and most other ASIC subtypes, demonstrated a reduced cerebral ischemic cell injury [120]. Taken together, the inhibition of ASIC1a channels may have significant therapeutic potential in ischemic strokes. Within the current literature, the effect of zinc on ischemic stroke is inconclusive, as different studies have seen the neuroprotective effects of both zinc chelation and administration. However, further research on the effect of zinc chelation and administration in the early versus late stages of ischemic stroke may ascertain whether the neuroprotective effects of zinc chelation are time-dependent and/or superior to the effects of zinc administration.

3.1.7. Cocaine Addiction

Cocaine triggers drug-seeking behavior by binding to the dopamine transporter at the synapse [122,123]. A 2021 study revealed that increased zinc concentrations in mice enhance cocaine binding to the dopamine transporter (DAT) protein. Repeated cocaine administration increased zinc concentrations in the caudate putamen (CPu) and nucleus accumbens (NAc). Conversely, low levels of zinc revealed decreased zinc content and cocaine sensitivity in the brain, confirming that zinc plays a role in cocaine-seeking behavior [123]. Additionally, ASIC channels are abundantly expressed in the NAc [122,124]. Previous studies discovered that mice deficient in the ASIC1a gene had increased cocaine-conditioned place preference (CPP); consequently, the effect is abolished when ASIC1a is restored in the NAc [11,125]. More recently, cocaine priming-induced reinstatement and drug-seeking were amplified when ASIC1a was overexpressed in mice NAc [122]. Additionally, our studies have indicated that, at various dosages of cocaine (5, 10, 20, and 30 mg/kg), cocaine drastically reduced acute cocaine-induced motor responses in ASIC1a$^{-/-}$ mice [124]. Behavioral sensitization in chronic cocaine addiction was also prevalent in the same ASIC1a$^{-/-}$ mice, suggesting that ASIC1a plays a role in chronic cocaine-induced behavioral changes [124]. These findings demonstrate that ASIC1a channels partake in both acute and chronic cocaine addiction. Taken together, it is unclear whether zinc-mediated inhibition of ASIC1a would attenuate or stimulate cocaine addiction. Further research is required to investigate the specific therapeutic potential of zinc modulation of ASIC1a channels in cocaine addiction in humans.

Zinc regulation of ASIC1a in certain neurological and psychological disorders has been shown in Figure 2.

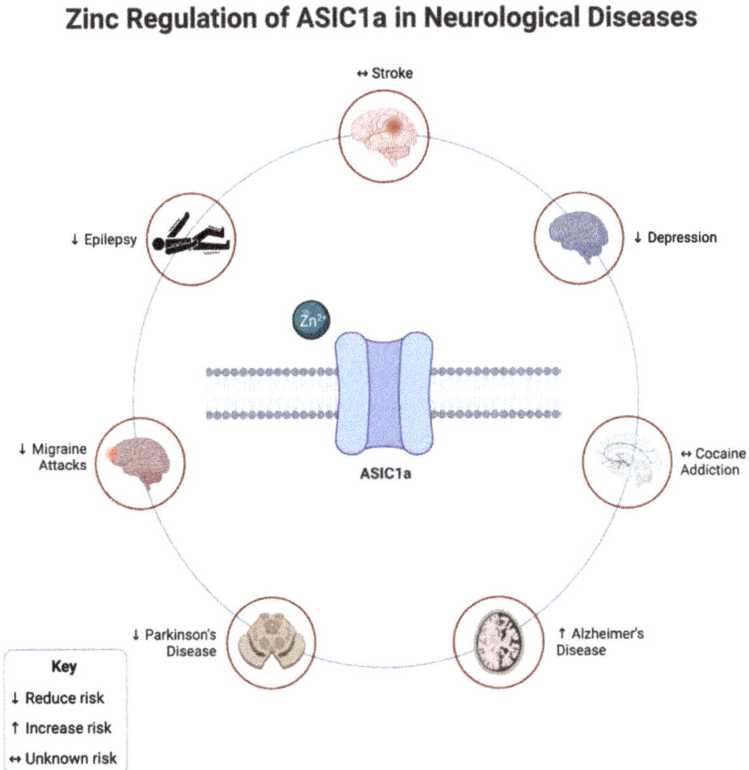

Figure 2. Zinc Regulation of ASIC1a in Neurological Diseases. Zinc-mediated inhibition of ASIC1a channels is theorized to reduce the risk of epilepsy, migraine attacks, Parkinson's disease, and depression and increase the risk of Alzheimer's disease. Its effects on cocaine addiction and strokes are unknown based on current literature. Adapted from "Transporters" by BioRender.com (2022) (https://app.biorender.com/biorender-templates, accessed on 28 December 2022).

3.2. ASIC1b

Zinc's anti-inflammatory qualities from zinc-mediated enhancement/induction of metallothionein are thought to be the reason for its therapeutic effectiveness in the treatment of pain [126], which is significant given inflammation is a primary contributor to the onset of chronic pain, including neuropathic pain [127]. However, the exact role that zinc plays in nociception is unknown. Studies show a diet lacking in zinc enables mice to experience less of the antinociceptive effects of morphine [128]. In addition, zinc was found to inhibit paclitaxel-induced mechanical hypersensitivity, the capsaicin response in DRG neurons, and TRPV1 channels [129]. These are several mechanisms of action that potentially reveal how zinc plays a role in pain sensation. Zinc therapy significantly reduced inflammatory hyperalgesia in a rat model of generated neuropathic pain and levels of the inflammatory biomarker IL-1B and nerve growth factor (NGF) [130].

Consequently, zinc chelation is found to cause hyperalgesia, and a zinc injection leads to pain relief [131,132]. Through TRPV inhibition, zinc may also be used therapeutically to treat chemotherapy-induced peripheral neuropathy [133]. Additionally, the depletion of vesicular zinc in the dorsal root ganglion of mice led to an increase in neuropathic pain due to a lower pain threshold [134]. In mice with fibromyalgia, it was found that when inducing hyperalgesia in both ASIC1b knockout mice and wild-type mice, the knockout mice had a shorter response [132]. Studies also revealed that human and mouse ASIC1b channels exhibit a small sustained current when activated [132,135]. The physiological significance of the sustained current is unclear but has been implicated in persistent pain [136]. Taken together, zinc-mediated inhibition of ASIC1b may play a key role in pain management therapy.

3.3. ASIC1a/3

Studies from ASIC knock-out mice have indicated that ASIC1a/3 channels could be one of the main ASIC components within skeletal muscle afferents [137]. ASIC channels play a significant role in detecting protons within sensory muscle neurons, particularly with sensing drops in pH [54,58]. Skeletal muscle afferents are predominantly composed of ASIC heteromers, ASIC1a/2a/3 or ASIC1a/3 [137]. Additional studies have hypothesized that ASIC1a/3 heterotrimers are involved with muscle pain [138]. This study found that PcTx1 significantly inhibited pH 6-evoked currents in ASIC1a/3 heteromeric channels in CHO cells [138]. PcTx1 binds a location on the ASIC1a/3 extracellular domain that controls pH-dependent channel desensitization [138]. With the presence of PcTx1, ASIC1a/3 channels are desensitized at a neutral pH and are, therefore, unable to be opened by acidic pH changes [138]. Thus, targeted inhibition of ASIC1a/3 can be a therapeutic option for patients with activity-induced hyperalgesia [138]. Zinc exhibits dual effects on ASIC1a/3 heterotrimers, causing inhibition at higher concentrations [4]. For example, the pretreatment of zinc of 1–100 μM has an EC_{50} of 24 μM, 100–250 μM has an EC_{50} of 128 μM, and 300 μM has an IC_{50} of 306 μM [4]. Furthermore, it has been shown that systemic zinc administration reduces hyperalgesia during early inflammation by decreasing cytokine IL-beta and growth factor NGF [130]. Therefore, zinc-mediated inhibition of ASIC1a/3 in the DRG of skeletal muscle cells may be a method of treatment for activity-induced pain, and further investigation should be done to explore this relationship.

Furthermore, ASIC1a and ASIC3 channels have a significant expression in the retina and subsequently can form functional ASIC1a/3 channels. In addition, zinc is released into the retina during neurotransmission. Multiple studies have shown that exogenous dietary supplementation of zinc is vital in preventing retinal aging [139,140], age-related macular degeneration [141,142], and maintaining the taurine system [143]. Hence, we hypothesize that zinc may be a modulator of ASIC1a/3 activity under physiological and pathological conditions. Prospective studies are required to investigate the relationship between zinc and ASIC1a/3 channels on the retina and explore zinc's effects on retinal pathological conditions.

3.4. ASIC1a/2b

ASIC1a/2b expression in dorsal horn neurons is upregulated by peripheral inflammation, thereby suggesting a vital role in central pain perception in physiological and pathological circumstances [144]. Similar to homomeric ASIC1a channels, ASIC1a/2b channels are calcium-permeable [55,145]. ASIC1a/2b calcium permeability plays a significant role in mediating neuronal death under various pathological states [55]. When the *ACCN1* (ASIC2) gene is mutated, ASIC1a/2a and ASIC1a/2b heteromers display significantly reduced acidosis-induced calcium rise after an ischemic event [146,147]. Current studies indicate that barium can reduce acid-induced neuronal death by inhibiting ASIC1a/2b [55]. Surprisingly, although ASIC1a homomers resist barium, barium could still profoundly inhibit neuronal death [55]. These results could be due to differences in the expression of ASIC1a/2b channels compared to ASIC1a homomeric channels in the neurons studied. However, this could also be evidence that ASIC1a/2b can be specifically targeted for treatment in neuron populations that are not affected by inhibitors of ASIC1a homomers. No current studies suggest the association between zinc regulation of ASIC1a/2b and tissue acidosis. Since millimolar and nanomolar concentrations of zinc can inhibit ASIC1a/2b as well as ASIC1a, additional research regarding the effects of zinc-mediated inhibition on neuronal cell death under ischemic conditions is necessary to determine whether zinc could be an advantageous therapy.

3.5. ASIC1a/2a

ASIC1a-containing channels, such as ASIC1a/2a heterotrimers, are indicated in neuronal injury after an ischemic stroke [49,147]. Although ASIC1a/2a channels are calcium impermeable, in acidic conditions, they can increase calcium levels and contribute to neuronal or axonal degeneration through secondary mechanisms [147,148]. This is true for calcium permeable ASIC1a channels as well. In fact, the activation of VGCC and release of calcium stores in acidotic and ischemic conditions have a higher contribution to the total increase in calcium than the calcium influx through ASIC1a [147]. While the activation of ASIC1a/2a channels may not directly contribute to injury, the greater the number of ASIC1a-containing channels, the greater the magnitude of the acid-induced calcium release during acidosis. Thus, the inhibition of the ASIC1a/2a heterotrimers with nanomolar concentrations of zinc could potentially attenuate the effect of the channel on a neuronal injury during acidotic and ischemic conditions [148,149]. However, as previously mentioned with ASIC1a, current research demonstrates that zinc plays a controversial role in the pathophysiology of ischemic stroke. In one study, the accumulation of zinc in the brains of adult rats was shown to be an independent risk factor for ischemic stroke [116]. In another study on rats, zinc supplementation after ischemia protected the hippocampus from neuronal injury [117]. With these conflicting findings, it is difficult to theorize the effects of zinc modulation of ASIC1a/2a on reducing the likelihood of neuronal degeneration after an ischemic stroke without additional research.

ASIC1a/2a heteromers and ASIC1a homomers are the most commonly expressed ASICs in the neurons of the olfactory bulb and are necessary for odorant stimulation and synaptic transmission [150]. A study proposed that a reduced ASIC function could contribute to the loss of olfactory perception in PD [150]. Additional research on the pathophysiology of ASICs and PD could be beneficial in determining whether micromolar concentrations of zinc could help potentiate ASIC1a/2a and attempt to restore olfactory perception. Currently, studies show that zinc treatment has controversial roles in PD depending on the affinity at which it binds to proteins in different signaling pathways. In one study on Drosophila flies, zinc treatment increased the lifespan and motility of the flies with PD [151]. In another study on mice, zinc supplementation resulted in methamphetamine-induced dopaminergic neuronal loss [94]. Thus, further research is necessary to elucidate if ASIC1a/2a is a channel that zinc will preferentially bind to for the restoration of olfactory perception in patients with PD or if it is more likely to bind to another protein that could result in neurotoxicity.

3.6. ASIC2a

ASIC2a homomers are indicated in numerous pathological conditions that affect different parts of the body. While ASIC1a, ASIC2a, and ASIC2b are highly expressed in the CNS, ASIC2a expression is thought to increase susceptibility to temporal lobe epilepsy [59]. There is greater ASIC2a expression post-seizure, and overexpression of ASIC2a hastened the onset of the first epileptic episode as well as increased the occurrence of status epilepticus episodes reaching Racine stage IV [59]. In those with focal brain ischemia due to a stroke, the deletion of ASIC2 channels has a protective effect on hippocampal, cortical, and striatal neurons by decreasing the effects of acidosis-induced injury [147,152]. ASIC2 deletion contributes to this protective effect by deleting heteromeric 1a/2a channels and decreasing the expression of ASIC1a [147]. It also eliminates ASIC2a, and this will result in a reduction of neuronal cell death under pathological conditions [147]. Since zinc helps potentiate ASIC2a currents and could increase seizure susceptibility and acid-induced injury after a stroke, systemic treatment with zinc may result in increased adverse effects for these patients.

In the eyes, ASIC2a is important for retinal function due to its expression in retinal ganglion cells. Zinc, abundant in the retina, can potentiate ASIC2a and consequently protect the eye from light-induced retinal degeneration [153]. Zinc supplementation has been studied to slow the progression of age-related macular degeneration [141,142]. However, in conditions such as glaucoma, in which there is optic nerve injury, retinal ganglion cells cannot regenerate and begin to die. In this circumstance, zinc accumulates in the retina and further prevents the axon regeneration of the retinal ganglion cells [154,155]. Studies propose that zinc chelation in patients with glaucoma could promote the regenerative capacity of the retinal ganglion cells [154,155]. The role of ASIC2a in the progressive damage of the retina after injury to the optic nerve is currently unclear. Further investigation is necessary to elucidate whether the modulation of ASIC2a by zinc plays a role in patients with glaucoma.

3.7. ASIC2a/3

ASIC2a/3 is a heteromeric channel found in cardiac dorsal root ganglion neurons [156]. Potentiation of these channels increases the firing of action potentials and could cause persistent angina during myocardial infarction [156]. Previous studies have shown that zinc's antioxidative and anti-inflammatory nature can protect against myocardial infarction [157], attenuate cardiac remodeling after infarction [158,159], improve the rate of contraction, and increase myocardial flow rate [160]. Although zinc supplementation has proven to be an effective cardioprotective treatment, since it potentiates ASIC2a/3, it should be studied further to determine if it could cause angina as an adverse effect.

3.8. ASIC3

ASIC3 is predominantly expressed within nociceptors and is a target of therapy for acid-induced pain [161]. In animal and human studies, selective ASIC3 drug antagonists are effective in relieving acid-evoked pain [161]. Furthermore, data has indicated that ASIC3 is involved in the maintenance of inflammatory pain [162]. One study utilized CFA-treated rats to demonstrate the role of ASIC3 in inflammatory pain [162]. Injection of APETx2, a potent peptide inhibitor of ASIC3, into a CFA-induced inflamed paw of the rat resulted in a complete reversal of mechanical hypersensitivity levels before the CFA treatment, reaching maximal efficacy 30 min post-APETx2 injection [162]. Therefore, inhibition of ASIC3 serves as a powerful therapeutic agent for both acid-induced and inflammatory nociception. Zinc inhibits the peak and sustained components of ASIC3 channels; hence, zinc may play a role in treating chronic pain [9]. Furthermore, ASIC3 is expressed within neuronal cells in the brain and adipocytes, and ASIC3 knockout mice show reduced anxiety levels and enhanced insulin sensitivity [161]. Thus, inhibition of ASIC3, potentially via zinc, may have additional analeptic implications outside of pain management.

ASIC3 has profound expression in the retina, and zinc is released during retinal neurotransmission; therefore, zinc may influence ASIC3 and ASIC1a/3 activity during normal or pathological conditions [4]. As previously stated, multiple studies have shown that exogenous dietary supplementation of zinc is vital in preventing retinal aging [140], age-related macular degeneration [141,142], and maintaining the taurine system [143]. Hence, we hypothesize that zinc may be a modulator of ASIC1a/3 activity under physiological and pathological conditions. Additional research is warranted regarding the zinc regulation on ASIC3 and ASIC1a/3 channels in the retina.

ASIC3 was also discovered to have an influential expression in the bladder. Subsequently, mutations of this channel can cause micturition-affiliated pathophysiological conditions such as urinary incontinence [163]. ASIC3 channels are significantly acid-sensitive and therefore play a vital role in tissue acidosis, nociception, and mechanosensation [52,164]. Knowing that zinc exhibits inhibitory behavior on ASIC3 channels, zinc may consequently play a vital role in tissue acidosis, nociception, and other underlying pathological conditions. A recent study from 2019 discovered that zinc is a protective biomolecule in ischemia/reperfusion injury in various organs [165]. However, no current studies support this relationship between ASIC3 and zinc and warrant future research.

3.9. ASIC4

ASIC4, unlike most ASIC channels, does not induce currents when protonated [5]. It is also mainly found intracellularly in endosome-related vacuoles [7]. ASIC4 is hypothesized to be a modulator that downregulates ASIC1a and ASIC3 surface expression [6]. This ultimately reduces the production of currents that trigger acidosis. Furthermore, ASIC4 channels may play a role in reducing fear and anxiety by modulating ASIC1a channels. One study found that knockout mice with ASIC4 mutations demonstrated heightened fear and anxious behavior [47]. Chemo-sensing and mechano-sensing roles of ASIC4 channels are still unclear since they do not form functional heteromeric or homomeric channels [164].

There is a potential linkage between zinc and ASIC4 channels found in zebrafish. Zebrafish ASIC4 (zASIC4) shares characteristics of a cytoplasmic N terminal domain with mammalian ASIC4 that is completely conserved in human ASIC4 channels [166]. zASIC4.1 has a transient component and a sustained component induced by calcium influx. In the study, the administration of 0.5 mM of zinc blocked the sustained component of zASIC4.1 channels but not the transient component, indicating a potential inhibitory modulation by zinc [166,167]. Further research is needed to analyze a possible relationship between zinc and human ASIC4 channels.

3.10. ASIC5

Similar to ASIC4, ASIC5 channels are also not activated by protons [168,169]. Recent studies have indicated that ASIC5 is important for type II Unipolar brush cells activity and that disruption of ASIC5 contributes to impaired movement, likely, at least in part, due to altered temporal processing of vestibular input [170]. In fact, ASIC5 has been discovered to be sensitive to bile acids. The closed state of ASIC5 is destabilized in the presence of bile acids, thus activating the channel [169]. It was subsequently named bile acid-sensitive ion channel (BASIC) [168,169]. In addition, physiological concentrations of magnesium and calcium strongly inhibited ASIC5 channels, indicating that extracellular divalent cations stabilize the inactive state of ASIC5 channels [168,169]. Therefore, bile acids and the displacement of extracellular divalent cations activate ASIC5 channels. There is currently no relevant literature discussing the relationship between zinc and ASIC5 channels, but since zinc is also an extracellular divalent cation with a history of effects on other ASIC channels, it is essential to rule out the possible correlation between zinc and ASIC5.

4. Conclusions

In conclusion, there has been significant interest in the function of zinc as a modulator of ASICs. ASICs have been implicated in numerous neurological disorders such as ischemic stroke, epilepsy, PD, and AD. Depending on the concentration of zinc and the ASIC isoform, zinc can exert either a stimulatory or inhibitory effect on the ion channels. With further research, zinc has promising potential to provide therapeutic benefits for various neuropathologies through the modulation of ASICs.

Author Contributions: Conceptualization, supervision, methodology, project administration, formal analysis, investigation, funding acquisition, validation, writing—review and editing, X.-P.C.; methodology, formal analysis, investigation, writing—original draft preparation, formal analysis, A.W.S., M.H.W. and M.V.; editing, M.J.W. All authors have read and agreed to the published version of the manuscript.

Funding: This work was partly supported by the American Heart Association (19AIREA34470007), USA (X.-P.C.).

Acknowledgments: We would like to thank the University of Missouri–Kansas City School of Medicine's student research program for supporting A.W.S., M.H.W., and M.V.'s professional studies. X.-P.C. acknowledges the support from the American Heart Association (grant number: 19AIREA34470007). We would also like to thank BioRender for its application in developing the figures and table.

Conflicts of Interest: The authors declare no conflict of interest.

References

1. Waldmann, R.; Champigny, G.; Bassilana, F.; Heurteaux, C.; Lazdunski, M. A proton-gated cation channel involved in acid-sensing. *Nature* **1997**, *386*, 173–177. [CrossRef]
2. Ruan, N.; Tribble, J.; Peterson, A.M.; Jiang, Q.; Wang, J.Q.; Chu, X.P. Acid-Sensing Ion Channels and Mechanosensation. *Int. J. Mol. Sci.* **2021**, *22*, 4810. [CrossRef]
3. Storozhuk, M.; Cherninskyi, A.; Maximyuk, O.; Isaev, D.; Krishtal, O. Acid-Sensing Ion Channels: Focus on Physiological and Some Pathological Roles in the Brain. *Curr. Neuropharmacol.* **2021**, *19*, 1570–1589.
4. Jiang, Q.; Peterson, A.M.; Chu, Y.; Yao, X.; Zha, X.M.; Chu, X.P. Histidine Residues Are Responsible for Bidirectional Effects of Zinc on Acid-Sensing Ion Channel 1a/3 Heteromeric Channels. *Biomolecules* **2020**, *10*, 1264. [CrossRef] [PubMed]
5. Chen, Y.J.; Deng, S.M.; Chen, H.W.; Tsao, C.H.; Chen, W.T.; Cheng, S.J.; Huang, H.S.; Tan, B.C.; Matzuk, M.M.; Flint, J.; et al. Follistatin mediates learning and synaptic plasticity via regulation of Asic4 expression in the hippocampus. *Proc. Natl. Acad. Sci. USA* **2021**, *118*, e2109040118. [CrossRef] [PubMed]
6. Schwartz, V.; Friedrich, K.; Polleichtner, G.; Gründer, S. Acid-sensing ion channel (ASIC) 4 predominantly localizes to an early endosome-related organelle upon heterologous expression. *Sci. Rep.* **2015**, *5*, 18242. [CrossRef] [PubMed]
7. Hoshikawa, M.; Kato, A.; Hojo, H.; Shibata, Y.; Kumamoto, N.; Watanabe, M.; Ugawa, S. Distribution of ASIC4 transcripts in the adult wild-type mouse brain. *Neurosci. Lett.* **2017**, *651*, 57–64. [CrossRef]
8. Zhou, R.; Leng, T.; Yang, T.; Chen, F.; Hu, W.; Xiong, Z.G. β-Estradiol protects against acidosis-mediated and ischemic neuronal injury by promoting ASIC1a (acid-sensing ion channel 1a) protein degradation. *Stroke* **2019**, *50*, 2902–2911. [CrossRef]
9. Arias, R.L.; Sung, M.L.; Vasylyev, D.; Zhang, M.Y.; Albinson, K.; Kubek, K.; Kagan, N.; Beyer, C.; Lin, Q.; Dwyer, J.M.; et al. Amiloride is neuroprotective in an MPTP model of Parkinson's disease. *Neurobiol. Dis.* **2008**, *31*, 334–341. [CrossRef] [PubMed]
10. Kweon, H.J.; Suh, B.C. Acid-sensing ion channels (ASICs): Therapeutic targets for neurological diseases and their regulation. *BMB Rep.* **2013**, *46*, 295–304. [CrossRef]
11. Kreple, C.J.; Lu, Y.; Taugher, R.J.; Schwager-Gutman, A.L.; Du, J.; Stump, M.; Wang, Y.; Ghobbeh, A.; Fan, R.; Cosme, C.V.; et al. Acid-sensing ion channels contribute to synaptic transmission and inhibit cocaine-evoked plasticity. *Nat. Neurosci.* **2014**, *17*, 1083–1091. [CrossRef]
12. Jiang, Q.; Zha, X.M.; Chu, X.P. Inhibition of human acid-sensing ion channel 1b by zinc. *Int. J. Physiol. Pathophysiol. Pharmacol.* **2012**, *4*, 84–93.
13. Chu, X.P.; Papasian, C.J.; Wang, J.Q.; Xiong, Z.G. Modulation of acid-sensing ion channels: Molecular mechanisms and therapeutic potential. *Int. J. Physiol. Pathophysiol. Pharmacol.* **2011**, *3*, 288–309. [PubMed]
14. Sivils, A.; Yang, F.; Wang, J.Q.; Chu, X.P. Acid-Sensing Ion Channel 2: Function and Modulation. *Membranes* **2022**, *12*, 113. [CrossRef]
15. Gonzales, E.B.; Kawate, T.; Gouaux, E. Pore architecture and ion sites in acid-sensing ion channels and P2X receptors. *Nature* **2009**, *460*, 599–604. [CrossRef]

16. Cristofori-Armstrong, B.; Rash, L.D. Acid-sensing ion channel (ASIC) structure and function: Insights from spider, snake and sea anemone venoms. *Neuropharmacology* **2017**, *127*, 173–184. [CrossRef] [PubMed]
17. Jasti, J.; Furukawa, H.; Gonzales, E.B.; Gouaux, E. Structure of acid-sensing ion channel 1 at 1.9 A resolution and low pH. *Nature* **2007**, *449*, 316–323. [CrossRef] [PubMed]
18. Baron, A.; Schaefer, L.; Lingueglia, E.; Champigny, G.; Lazdunski, M. Zn2+ and H+ are coactivators of acid-sensing ion channels. *J. Biol. Chem.* **2001**, *276*, 35361–35367. [CrossRef]
19. Hassan, A.; Sada, K.K.; Ketheeswaran, S.; Dubey, A.K.; Bhat, M.S. Role of Zinc in Mucosal Health and Disease: A Review of Physiological, Biochemical, and Molecular Processes. *Cureus* **2020**, *12*, 8197. [CrossRef]
20. Doboszewska, U.; Młyniec, K.; Wlaź, A.; Poleszak, E.; Nowak, G.; Wlaź, P. Zinc signaling and epilepsy. *Pharmacol. Ther.* **2019**, *193*, 156–177. [CrossRef]
21. Mezzaroba, L.; Alfieri, D.F.; Colado Simão, A.N.; Vissoci Reiche, E.M. The role of zinc, copper, manganese and iron in neurodegenerative diseases. *Neurotoxicology* **2019**, *74*, 230–241. [CrossRef] [PubMed]
22. Kumar, V.; Kumar, A.; Singh, K.; Avasthi, K.; Kim, J.J. Neurobiology of zinc and its role in neurogenesis. *Eur. J. Nutr.* **2021**, *60*, 55–64. [CrossRef] [PubMed]
23. Sauer, A.K.; Vela, H.; Vela, G.; Stark, P.; Barrera-Juarez, E.; Grabrucker, A.M. Zinc Deficiency in Men Over 50 and Its Implications in Prostate Disorders. *Front. Oncol.* **2020**, *10*, 1293. [CrossRef]
24. Christudoss, P.; Selvakumar, R.; Fleming, J.J.; Gopalakrishnan, G. Zinc status of patients with benign prostatic hyperplasia and prostate carcinoma. *Ind. J. Urol.* **2011**, *27*, 14–18. [CrossRef]
25. Costello, L.C.; Franklin, R.B. A comprehensive review of the role of zinc in normal prostate function and metabolism; and its implications in prostate cancer. *Arch. Biochem. Biophys.* **2016**, *611*, 100–112. [CrossRef]
26. Portbury, S.D.; Adlard, P.A. Zinc Signal in Brain Diseases. *Int. J. Mol. Sci.* **2017**, *18*, 2506. [CrossRef]
27. Bu, S.; Lv, Y.; Liu, Y.; Qiao, S.; Wang, H. Zinc Finger Proteins in Neuro-Related Diseases Progression. *Front. Neurosci.* **2021**, *15*, 760567. [CrossRef]
28. Mammadova-Bach, E.; Braun, A. Zinc Homeostasis in Platelet-Related Diseases. *Int. J. Mol. Sci.* **2019**, *20*, 5258. [CrossRef] [PubMed]
29. Kefauver, J.M.; Ward, A.B.; Patapoutian, A. Discoveries in structure and physiology of mechanically activated ion channels. *Nature* **2020**, *587*, 567–576. [CrossRef] [PubMed]
30. Bennett, D.L.; Clark, A.J.; Huang, J.; Waxman, S.G.; Dib-Hajj, S.D. The Role of Voltage-Gated Sodium Channels in Pain Signaling. *Physiol. Rev.* **2019**, *99*, 079–1151. [CrossRef]
31. Peralta, F.A.; Huidobro-Toro, J.P. Zinc as Allosteric Ion Channel Modulator: Ionotropic Receptors as Metalloproteins. *Int. J. Mol. Sci.* **2016**, *17*, 1059. [CrossRef]
32. Noh, S.; Lee, S.R.; Jeong, Y.J.; Ko, K.S.; Rhee, B.D.; Kim, N.; Han, J. The direct modulatory activity of zinc toward ion channels. *Integr. Med. Res.* **2015**, *4*, 142–146. [CrossRef] [PubMed]
33. Kuang, Q.; Purhonen, P.; Hebert, H. Structure of potassium channels. *Cell. Mol. Life Sci.* **2015**, *72*, 3677–3693. [CrossRef]
34. Niday, Z.; Tzingounis, A.V. Potassium Channel Gain of Function in Epilepsy: An Unresolved Paradox. *Neuroscientist* **2018**, *24*, 368–380. [CrossRef] [PubMed]
35. Zhang, X.; Wan, J.Q.; Tong, X.P. Potassium channel dysfunction in neurons and astrocytes in Huntington's disease. *CNS Neurosci. Ther.* **2018**, *24*, 311–318. [CrossRef] [PubMed]
36. Wie, J.; Liu, Z.; Song, H.; Tropea, T.F.; Yang, L.; Wang, H.; Liang, Y.; Cang, C.; Aranda, K.; Lohmann, J.; et al. A growth-factor-activated lysosomal K+ channel regulates Parkinson's pathology. *Nature* **2021**, *591*, 431–437. [CrossRef] [PubMed]
37. Sarkar, S.; Nguyen, H.M.; Malovic, E.; Luo, J.; Langley, M.; Palanisamy, B.N.; Singh, N.; Manne, S.; Neal, M.; Gabrielle, M.; et al. Kv1.3 modulates neuroinflammation and neurodegeneration in Parkinson's disease. *J. Clin. Investig.* **2020**, *130*, 4195–4212. [CrossRef] [PubMed]
38. Hering, S.; Zangerl-Plessl, E.M.; Beyl, S.; Hohaus, A.; Andranovits, S.; Timin, E.N. Calcium channel gating. *Pflug. Arch.* **2018**, *470*, 1291–1309. [CrossRef]
39. Nanou, E.; Catterall, W.A. Calcium Channels, Synaptic Plasticity, and Neuropsychiatric Disease. *Neuron* **2018**, *98*, 466–481. [CrossRef] [PubMed]
40. Meldrum, B.S. Glutamate as a neurotransmitter in the brain: Review of physiology and pathology. *J. Nutr.* **2000**, *130*, 1007S–1015S. [CrossRef]
41. Cuestas Torres, D.M.; Cardenas, F.P. Synaptic plasticity in Alzheimer's disease and healthy aging. *Rev. Neurosci.* **2020**, *31*, 245–268. [CrossRef]
42. Nakazawa, K.; Sapkota, K. The origin of NMDA receptor hypofunction in schizophrenia. *Pharmacol. Ther.* **2020**, *205*, 107426. [CrossRef] [PubMed]
43. Babaei, P. NMDA and AMPA receptors dysregulation in Alzheimer's disease. *Eur. J. Pharmacol.* **2021**, *908*, 174310. [CrossRef] [PubMed]
44. Samanta, A.; Hughes, T.E.T.; Moiseenkova-Bell, V.Y. Transient Receptor Potential (TRP) Channels. *Subcell Biochem.* **2018**, *87*, 141–165. [PubMed]
45. Zhu, S.; Noviello, C.M.; Teng, J.; Walsh, R.M., Jr.; Kim, J.J.; Hibbs, R.E. Structure of a human synaptic GABAA receptor. *Nature* **2018**, *559*, 67–72. [CrossRef] [PubMed]

46. Chu, X.P.; Wemmie, J.A.; Wang, W.Z.; Zhu, X.M.; Saugstad, J.A.; Price, M.P.; Simon, R.P.; Xiong, Z.G. Subunit-Dependent High-Affinity Zinc Inhibition of Acid-Sensing Ion Channels. *J. Neurosci.* **2004**, *24*, 8678–8689. [CrossRef] [PubMed]
47. Lin, S.H.; Chien, Y.C.; Chiang, W.W.; Liu, Y.Z.; Lien, C.C.; Chen, C.C. Genetic mapping of ASIC4 and contrasting phenotype to ASIC1a in modulating innate fear and anxiety. *Eur. J. Neurosci.* **2015**, *41*, 1553–1568. [CrossRef]
48. Sun, C.; Wang, S.; Hu, W. Acid-sensing ion channel 1a mediates acid-induced inhibition of matrix metabolism of rat articular chondrocytes via the MAPK signaling pathway. *Mol. Cell. Biochem.* **2018**, *443*, 81–91. [CrossRef] [PubMed]
49. Pignataro, G.; Simon, R.P.; Xiong, Z.G. Prolonged activation of ASIC1a and the time window for neuroprotection in cerebral ischaemia. *Brain A J. Neurol.* **2007**, *130*, 151158. [CrossRef]
50. Sherwood, T.W.; Lee, K.G.; Gormley, M.G.; Askwith, C.C. Heteromeric acid-sensing ion channels (ASICs) composed of ASIC2b and ASIC1a display novel channel properties and contribute to acidosis-induced neuronal death. *J. Neurosci.* **2011**, *31*, 9723–9734. [CrossRef]
51. Jiang, Q.; Inoue, K.; Wu, X.; Papasian, C.J.; Wang, J.Q.; Xiong, Z.G.; Chu, X.P. Cysteine 149 in the extracellular finger domain of acid-sensing ion channel 1b subunit is critical for zinc-mediated inhibition. *Neuroscience* **2011**, *193*, 89–99. [CrossRef] [PubMed]
52. Jiang, Q.; Papasian, C.J.; Wang, J.Q.; Xiong, Z.G.; Chu, X.P. Inhibitory regulation of acid-sensing ion channel 3 by zinc. *Neuroscience* **2010**, *169*, 574–583. [CrossRef]
53. Zaremba, M.; Ruiz-Velasco, V. Opioid-mediated modulation of acid-sensing ion channel currents in adult rat sensory neurons. *Mol. Pharmacol.* **2019**, *95*, 519–527. [CrossRef] [PubMed]
54. Lee, C.H.; Chen, C.C. Roles of ASICs in nociception and proprioception. *Adv. Exp. Med. Biol.* **2018**, *1099*, 37–47.
55. Chen, X.; Kalbacher, H.; Grunder, S. The tarantula toxin psalmotoxin 1 inhibits acid-sensing ion channel (ASIC) 1a by increasing its apparent H+ affinity. *J. Gen. Physiol.* **2005**, *126*, 71–79. [CrossRef]
56. Babinski, K.; Catarsi, S.; Biagini, G.; Séguéla, P. Mammalian ASIC2a and ASIC3 subunits co-assemble into heteromeric proton-gated channels sensitive to Gd3+. *J. Biol. Chem.* **2000**, *275*, 28519–28525. [CrossRef]
57. Kweon, H.J.; Cho, J.H.; Jang, I.S.; Suh, B.C. ASIC2a-dependent increase of ASIC3 surface expression enhances the sustained component of the currents. *BMB Rep.* **2016**, *49*, 542–547. [CrossRef] [PubMed]
58. Heusser, S.A.; Pless, S.A. Acid-sensing ion channels as potential therapeutic targets. *Trends Pharmacol. Sci.* **2021**, *42*, 1035–1050. [CrossRef] [PubMed]
59. Zhang, H.; Gao, G.; Zhang, Y.; Sun, Y.; Li, H.; Dong, S.; Ma, W.; Liu, B.; Wang, W.; Wu, H.; et al. Glucose Deficiency Elevates Acid-Sensing Ion Channel 2a Expression and Increases Seizure Susceptibility in Temporal Lobe Epilepsy. *Sci. Rep.* **2017**, *7*, 5870. [CrossRef]
60. Yagi, J.; Wenk, H.N.; Naves, L.A.; McCleskey, E.W. Sustained currents through ASIC3 ion channels at the modest pH changes that occur during myocardial ischemia. *Circ. Res.* **2006**, *99*, 501–509. [CrossRef]
61. González-Garrido, A.; Vega, R.; Mercado, F.; López, I.A.; Soto, E. Acid-Sensing Ion Channels Expression, Identity and Role in the Excitability of the Cochlear Afferent Neurons. *Front. Cell. Neurosci.* **2015**, *9*, 483. [CrossRef]
62. Goldie, I.; Nachemson, A. Synovial pH in rheumatoid knee-joints I. The effect of synovectomy. *Acta Orthop.* **1969**, *40*, 634–641. [CrossRef] [PubMed]
63. Decker, Y.; Németh, E.; Schomburg, R.; Chemla, A.; Fülöp, L.; Menger, M.D.; Liu, Y.; Fassbender, K. Decreased pH in the aging brain and Alzheimer's disease. *Neurobiol. Aging* **2021**, *101*, 40–49. [CrossRef] [PubMed]
64. Dulai, J.S.; Smith, E.S.J.; Rahman, T. Acid-sensing ion channel 3: An analgesic target. *Channels* **2021**, *15*, 94–127. [CrossRef]
65. Fisher, R.S.; Acevedo, C.; Arzimanoglou, A.; Bogacz, A.; Cross, J.H.; Elger, C.E.; Engel, J., Jr.; Forsgren, L.; French, J.A.; Glynn, M.; et al. ILAE official report: A practical clinical definition of epilepsy. *Epilepsia* **2014**, *55*, 475–482. [CrossRef]
66. Amar, M.; Perin-Dureau, F.; Neyton, J. High-affinity Zn block in recombinant N-methyl-D-aspartate receptors with cysteine substitutions at the Q/R/N site. *Biophys. J.* **2001**, *81*, 107–116. [CrossRef]
67. Hey, J.G.; Chu, X.P.; Seeds, J.; Simon, R.P.; Xiong, Z.G. Extracellular zinc protects against acidosis-induced injury of cells expressing Ca2+-permeable acid-sensing ion channels. *Stroke* **2007**, *38* (Suppl. 2), 670–673. [CrossRef]
68. Koh, J.Y.; Choi, D.W. Zinc toxicity on cultured cortical neurons: Involvement of N-methyl-d-aspartate receptors. *Neuroscience* **1994**, *60*, 1049–1057. [CrossRef] [PubMed]
69. Goadsby, P.J.; Holland, P.R.; Martins-Oliveira, M.; Hoffmann, J.; Schankin, C.; Akerman, S. Pathophysiology of Migraine: A Disorder of Sensory Processing. *Physiol. Rev.* **2017**, *97*, 553–622. [CrossRef]
70. Azoulay, I.S.; Qi, X.; Rozenfeld, M.; Liu, F.; Hu, Q.; Ben-Kasus Nissim, T.; Stavsky, A.; Zhu, M.X.; Xu, T.L.; Sekler, I. ASIC1a senses lactate uptake to regulate metabolism in neurons. *Redox Biol.* **2022**, *51*, 102253. [CrossRef]
71. Verkest, C.; Piquet, E.; Diochot, S.; Dauvois, M.; Lanteri-Minet, M.; Lingueglia, E.; Baron, A. Effects of systemic inhibitors of acid-sensing ion channels 1 (ASIC1) against acute and chronic mechanical allodynia in a rodent model of migraine. *Br. J. Pharmacol.* **2018**, *175*, 4154–4166. [CrossRef]
72. Ahmadi, H.; Mazloumi-Kiapey, S.S.; Sadeghi, O.; Nasiri, M.; Khorvash, F.; Mottaghi, T.; Askari, G. Zinc supplementation affects favorably the frequency of migraine attacks: A double-blind randomized placebo-controlled clinical trial. *Nutr. J.* **2020**, *19*, 101. [CrossRef] [PubMed]
73. Mazaheri, M.; Aghdam, A.M.; Heidari, M.; Zarrin, R. Assessing the Effect of Zinc Supplementation on the Frequency of Migraine Attack, Duration, Severity, Lipid Profile and hs-CRP in Adult Women. *Clin. Nutr. Res.* **2021**, *10*, 127–139. [CrossRef] [PubMed]

74. Schytz, H.W.; Hargreaves, R. Ashina, M. Challenges in developing drugs for primary headaches. *Prog. Neurobiol.* **2017**, *152*, 70–88. [CrossRef] [PubMed]
75. Lipton, R.B.; Munjal, S.; Buse, D.C.; Bennett, A.; Fanning, K.M.; Burstein, R.; Reed, M.L. Allodynia Is Associated With Initial and Sustained Response to Acute Migraine Treatment: Results from the American Migraine Prevalence and Prevention Study. *Headache* **2017**, *57*, 1026–1040. [CrossRef]
76. Chapman, P.F.; White, G.L.; Jones, M.W.; Cooper-Blacketer, D.; Marshall, V.J.; Irizarry, M.; Younkin, L.; Good, M.A.; Bliss, T.V.; Hyman, B.T.; et al. Impaired synaptic plasticity and learning in aged amyloid precursor protein transgenic mice. *Nat. Neurosci.* **1999**, *2*, 271–276. [CrossRef]
77. Mango, D.; Nisticò, R. Acid-Sensing ion channel 1a is involved in N-Methyl D-aspartate receptor-dependent long-term depression in the hippocampus. *Front. Pharmacol.* **2019**, *10*, 555. [CrossRef]
78. Bush, A.I.; Pettingell, W.H.; Multhaup, G.; Paradis, M.; Vonsattel, J.P.; Gusella, J.F.; Beyreuther, K.; Masters, C.L.; Tanzi, R.E. Rapid induction of Alzheimer A beta amyloid formation by zinc. *Science* **1994**, *265*, 1464–1467. [CrossRef]
79. Lashley, T.; Schott, J.M.; Weston, P.; Murray, C.E.; Wellington, H.; Keshavan, A.; Foti, S.C.; Foiani, M.; Toombs, J.; Rohrer, J.D.; et al. Molecular biomarkers of Alzheimer's disease: Progress and prospects. *Dis. Model. Mech.* **2018**, *11*, dmm031781. [CrossRef]
80. Lee, M.C.; Yu, W.C.; Shih, Y.H.; Chen, C.Y.; Guo, Z.H.; Huang, S.J.; Chan, J.C.C.; Chen, Y.R. Zinc ion rapidly induces toxic, off-pathway amyloid-beta oligomers distinct from amyloid-beta derived diffusible ligands in Alzheimer's disease. *Sci. Rep.* **2018**, *8*, 4772. [CrossRef]
81. James, S.A.; Churches, Q.I.; de Jonge, M.D.; Birchall, I.E.; Streltsov, V.; McColl, G.; Adlard, P.A.; Hare, D.J. Iron, copper, and zinc concentration in abeta plaques in the APP/PS1 mouse model of Alzheimer's disease correlates with metal levels in the surrounding neuropil. *ACS Chem. Neurosci.* **2017**, *8*, 629–637. [CrossRef]
82. Ayton, S.; Wang, Y.; Diouf, I.; Schneider, J.A.; Brockman, J.; Morris, M.C.; Bush, A.I. Brain iron is associated with accelerated cognitive decline in people with Alzheimer pathology. *Mol. Psychiatry* **2019**, *45*, 358–367. [CrossRef]
83. Datki, Z.; Galik-Olah, Z.; Janosi-Mozes, E.; Szegedi, V.; Kalman, J.; Hunya, A.G.; Fulop, L.; Tamano, H.; Takeda, A.; Adlard, P.A.; et al. Alzheimer risk factors age and female sex induce cortical Aβ aggregation by raising extracellular zinc. *Mol. Psychiatry* **2020**, *25*, 2728–2741. [CrossRef]
84. Kepp, K.P. Alzheimer's disease: How metal ions define β-amyloid function. *Coord. Chem. Rev.* **2017**, *351*, 127–157. [CrossRef]
85. Zhang, H.L.; Wang, X.C.; Liu, R. Zinc in Regulating Protein Kinases and Phosphatases in Neurodegenerative Diseases. *Biomolecules* **2022**, *12*, 785. [CrossRef] [PubMed]
86. Squitti, R.; Pal, A.; Picozza, M.; Avan, A.; Ventriglia, M.; Rongioletti, M.C.; Hoogenraad, T. Zinc Therapy in Early Alzheimer's Disease: Safety and Potential Therapeutic Efficacy. *Biomolecules* **2020**, *10*, 1164. [CrossRef]
87. Gelders, G.; Baekelandt, V.; Van der Perren, A. Linking Neuroinflammation and Neurodegeneration in Parkinson's Disease. *J. Immunol. Res.* **2018**, *2018*, 4784268. [CrossRef]
88. Granzotto, A.; Sensi, S.L. Intracellular zinc is a critical intermediate in the excitotoxic cascade. *Neurobiol. Dis.* **2015**, *81*, 25–37. [CrossRef] [PubMed]
89. Calabresi, P.; Di Filippo, M.; Gallina, A.; Wang, Y.; Stankowski, J.N.; Picconi, B.; Dawson, V.L.; Dawson, T.M. New synaptic and molecular targets for neuroprotection in Parkinson's disease. *Mov. Disord.* **2013**, *28*, 51–60. [CrossRef]
90. Ahmed, S.S.; Santosh, W. Metallomic profiling and linkage map analysis of early Parkinson's disease: A new insight to aluminum marker for the possible diagnosis. *PLoS ONE* **2010**, *5*, e11252. [CrossRef]
91. Hegde, M.L.; Shanmugavelu, P.; Vengamma, B.; Rao, T.S.; Menon, R.B.; Rao, R.V.; Rao, K.S. Serum trace element levels and the complexity of inter-element relations in patients with Parkinson's disease. *J. Trace Elem. Med. Biol.* **2004**, *18*, 163–171. [CrossRef]
92. Ajjimaporn, A.; Phansuwan-Pujito, P.; Ebadi, M.; Govitrapong, P. Zinc protects SK-N-SH cells from methamphetamine-induced alpha-synuclein expression. *Neurosci. Lett.* **2007**, *419*, 59–63. [CrossRef]
93. Ajjimaporn, A.; Shavali, S.; Ebadi, M.; Govitrapong, P. Zinc rescues dopaminergic SK-N-SH cell lines from methamphetamine-induced toxicity. *Brain Res. Bull.* **2008**, *77*, 361–366. [CrossRef]
94. Hussain, S.; Ali, S.F. Zinc potentiates 1-methyl-4-phenyl-1,2,3,6-tetrahydropyridine induced dopamine depletion in caudate nucleus of mice brain. *Neurosci. Lett.* **2002**, *335*, 25–28. [CrossRef]
95. Raj, K.; Kaur, P.; Gupta, G.D.; Singh, S. Metals associated neurodegeneration in Parkinson's disease: Insight to physiological, pathological mechanisms and management. *Neurosci. Lett.* **2021**, *753*, 135383. [CrossRef]
96. Kitada, T.; Asakawa, S.; Hattori, N.; Matsumine, H.; Yamamura, Y.; Minoshima, S.; Yokochi, M.; Mizuno, Y.; Shimizu, N. Mutations in the parkin gene cause autosomal recessive juvenile parkinsonism. *Nature* **1998**, *392*, 605–608. [CrossRef]
97. Joch, M.; Ase, A.R.; Chen, C.X.; MacDonald, P.A.; Kontogiannea, M.; Corera, A.T.; Brice, A.; Séguéla, P.; Fon, E.A. Parkin-mediated monoubiquitination of the PDZ protein PICK1 regulates the activity of acid-sensing ion channels. *Mol. Biol. Cell* **2007**, *18*, 3105–3118. [CrossRef]
98. GBD 2017 Disease and Injury Incidence and Prevalence Collaborators. Global, regional, and national incidence, prevalence, and years lived with disability for 354 diseases and injuries for 195 countries and territories, 1990–2017: A systematic analysis for the Global Burden of Disease Study 2017. *Lancet* **2018**, *392*, 1789–1858. [CrossRef] [PubMed]
99. Siodłak, D.; Nowak, G.; Mlyniec, K. Interaction between zinc, the GPR39 zinc receptor and the serotonergic system in depression. *Brain Res. Bull.* **2021**, *170*, 146–154. [CrossRef] [PubMed]

100. Westbrook, G.L.; Mayer, M.L. Micromolar concentrations of Zn2+ antagonize NMDA and GABA responses of hippocampal neurons. *Nature* **1987**, *328*, 640–643. [CrossRef] [PubMed]
101. Assaf, S.Y.; Chung, S.H. Release of endogenous Zn2+ from brain tissue during activity. *Nature* **1984**, *308*, 734–736. [CrossRef] [PubMed]
102. Prasad, A.S. Zinc deficiency. *Bmj* **2003**, *326*, 409–410. [CrossRef] [PubMed]
103. Jokinen, J.; Nordström, P. HPA axis hyperactivity and attempted suicide in young adult mood disorder inpatients. *J. Affect. Disord.* **2009**, *116*, 117–120. [CrossRef] [PubMed]
104. Młyniec, K.; Gaweł, M.; Librowski, T.; Reczyński, W.; Bystrowska, B.; Holst, B. Investigation of the GPR39 zinc receptor following inhibition of monoaminergic neurotransmission and potentialization of glutamatergic neurotransmission. *Brain Res. Bull.* **2015**, *115*, 23–29. [CrossRef]
105. Satała, G.; Duszyńska, B.; Lenda, T.; Nowak, G.; Bojarski, A.J. Allosteric Inhibition of Serotonin 5-HT7 Receptors by Zinc Ions. *Mol. Neurobiol.* **2018**, *55*, 2897–2910. [CrossRef]
106. Mo, F.; Tang, Y.; Du, P.; Shen, Z.; Yang, J.; Cai, M.; Zhang, Y.; Li, H.; Shen, H. GPR39 protects against corticosterone-induced neuronal injury in hippocampal cells through the CREB-BDNF signaling pathway. *J. Affect. Disord.* **2020**, *272*, 474–484. [CrossRef]
107. Suh, S.W.; Won, S.J.; Hamby, A.M.; Yoo, B.H.; Fan, Y.; Sheline, C.T.; Tamano, H.; Takeda, A.; Liu, J. Decreased brain zinc availability reduces hippocampal neurogenesis in mice and rats. *J. Cereb. Blood Flow Metab.* **2009**, *29*, 1579–1588. [CrossRef]
108. Wang, J.; Um, P.; Dickerman, B.A.; Liu, J. Zinc, Magnesium, Selenium and Depression: A Review of the Evidence, Potential Mechanisms and Implications. *Nutrients* **2018**, *10*, 584. [CrossRef]
109. Malykhin, N.; Coupland, N. Hippocampal neuroplasticity in major depressive disorder. *Neuroscience* **2015**, *309*, 200–213. [CrossRef]
110. Taugher, R.J.; Lu, Y.; Fan, R.; Ghobbeh, A.; Kreple, C.J.; Faraci, F.M.; Wemmie, J.A. ASIC1A in neurons is critical for fear-related behaviors. *Genes Brain Behav.* **2017**, *16*, 745–755. [CrossRef]
111. Coryell, M.W.; Wunsch, A.M.; Haenfler, J.M.; Allen, J.E.; Schnizler, M.; Ziemann, A.E.; Cook, M.N.; Dunning, J.P.; Price, M.P.; Rainier, J.D.; et al. Acid-sensing ion channel-1a in the amygdala, a novel therapeutic target in depression-related behavior. *J. Neurosci.* **2009**, *29*, 5381–5388. [CrossRef] [PubMed]
112. Wemmie, J.A.; Askwith, C.C.; Lamani, E.; Cassell, M.D.; Freeman, J.H., Jr.; Welsh, M.J. Acid-sensing ion channel 1 is localized in brain regions with high synaptic density and contributes to fear conditioning. *J. Neurosci.* **2003**, *23*, 5496–5502. [CrossRef]
113. Zhu, H.; Hu, S.; Li, Y.; Sun, Y.; Xiong, X.; Hu, X.; Chen, J.; Qiu, S. Interleukins and Ischemic Stroke. *Front. Immunol.* **2022**, *13*, 828447. [CrossRef] [PubMed]
114. Maida, C.D.; Norrito, R.L.; Daidone, M.; Tuttolomondo, A.; Pinto, A. Neuroinflammatory Mechanisms in Ischemic Stroke: Focus on Cardioembolic Stroke, Background, and Therapeutic Approaches. *Int. J. Mol. Sci.* **2020**, *21*, 6454. [CrossRef]
115. Mammadova-Bach, E.; Nagy, M.; Heemskerk, J.W.M.; Nieswandt, B.; Braun, A. Store-operated calcium entry in thrombosis and thrombo-inflammation. *Cell Calcium* **2019**, *77*, 39–48. [CrossRef]
116. Lee, J.M.; Zipfel, G.J.; Park, K.H.; He, Y.Y.; Hsu, C.Y.; Choi, D.W. Zinc translocation accelerates infarction after mild transient focal ischemia. *Neuroscience* **2002**, *115*, 871–878. [CrossRef] [PubMed]
117. Helal, G.K. Systemic administration of Zn2+ during the reperfusion phase of transient cerebral ischaemia protects rat hippocampus against iron-catalysed postischaemic injury. *Clin. Exp. Pharmacol. Physiol.* **2008**, *35*, 775–781. [CrossRef]
118. Aquilani, R.; Baiardi, P.; Scocchi, M.; Iadarola, P.; Verri, M.; Sessarego, P.; Boschi, F.; Pasini, E.; Pastoris, O.; Viglio, S. Normalization of zinc intake enhances neurological retrieval of patients suffering from ischemic strokes. *Nutr. Neurosci.* **2009**, *12*, 219–225. [CrossRef]
119. Xiong, Z.G. Neuroprotection in ischemia: Blocking calcium-permeable acid-sensing ion channels. *Cell* **2004**, *118*, 687–698. [CrossRef]
120. O'Bryant, Z.; Vann, K.T.; Xiong, Z.G. Translational strategies for neuroprotection in ischemic stroke–focusing on acid-sensing ion channel 1a. *Transl. Stroke Res.* **2014**, *5*, 59–68. [CrossRef]
121. Gao, J.; Duan, B.; Wang, D.G.; Deng, X.H.; Zhang, G.Y.; Xu, L.; Xu, T.L. Coupling between NMDA receptor and acid-sensing ion channel contributes to ischemic neuronal death. *Neuron* **2005**, *48*, 635–646. [CrossRef]
122. Gutman, A.L.; Cosme, C.V.; Noterman, M.F.; Worth, W.R.; Wemmie, J.A.; LaLumiere, R.T. Overexpression of ASIC1A in the nucleus accumbens of rats potentiates cocaine-seeking behavior. *Addict. Biol.* **2020**, *25*, e12690. [CrossRef] [PubMed]
123. Gomez, J.L.; Bonaventura, J.; Keighron, J.; Wright, K.M.; Marable, D.L.; Rodriguez, L.A.; Lam, S.; Carlton, M.L.; Ellis, R.J.; Jordan, C.J.; et al. Synaptic Zn2+ potentiates the effects of cocaine on striatal dopamine neurotransmission and behavior. *Transl. Psychiatry* **2021**, *11*, 570. [CrossRef] [PubMed]
124. Jiang, Q.; Li, M.H.; Papasian, C.J.; Branigan, D.; Xiong, Z.G.; Wang, J.Q.; Chu, X.P. Characterization of acid-sensing ion channels in medium spiny neurons of mouse striatum. *Neuroscience* **2009**, *162*, 55–66. [CrossRef] [PubMed]
125. Kreple, C.J.; Lu, Y.; LaLumiere, R.T.; Wemmie, J.A. Drug abuse and the simplest neurotransmitter. *ACS Chem. Neurosci.* **2014**, *5*, 746–748. [CrossRef]
126. Hennigar, S.R.; Kelley, A.M.; Anderson, B.J.; Armstrong, N.J.; McClung, H.L.; Berryman, C.E.; Karl, J.P.; McClung, J.P. Sensitivity and reliability of zinc transporter and metallothionein gene expression in peripheral blood mononuclear cells as indicators of zinc status: Responses to ex vivo zinc exposure and habitual zinc intake in humans. *Br. J. Nutr.* **2021**, *125*, 361–368. [CrossRef]

127. Dursun, N.; Erenmemisoglu, A.; Cem, S.; Gogusten, B. The effect of zinc deficiency on morphine antinociception. *Res. Commun. Alcohol. Subst. Abus.* **1995**, *16*, 47–52.
128. Malcangio, M. Role of the immune system in neuropathic pain. *Scand. J. Pain* **2019**, *20*, 33–37. [CrossRef]
129. Li, L.; Chen, C.; Chiang, C.; Xiao, T.; Chen, Y.; Zhao, Y.; Zheng, D. The Impact of TRPV1 on Cancer Pathogenesis and Therapy: A Systematic Review. *Int. J. Biol. Sci.* **2021**, *17*, 2034–2049. [CrossRef]
130. Safieh-Garabedian, B.; Poole, S.; Allchorne, A.; Kanaan, S.; Saade, N.; Woolf, C.J. Zinc reduces the hyperalgesia and upregulation of NGF and IL-1 beta produced by peripheral inflammation in the rat. *Neuropharmacology* **1996**, *35*, 599–603. [CrossRef]
131. Larson, A.A.; Kitto, K.F. Manipulations of zinc in the spinal cord, by intrathecal injection of zinc chloride, disodium-calcium-EDTA, or dipicolinic acid, alter nociceptive activity in mice. *J. Pharmacol. Exp. Ther.* **1997**, *282*, 1319–1325. [PubMed]
132. Chang, C.T.; Fong, S.W.; Lee, C.H.; Chuang, Y.C.; Lin, S.H.; Chen, C.C. Involvement of Acid-Sensing Ion Channel 1b in the Development of Acid-Induced Chronic Muscle Pain. *Front. Neurosci.* **2019**, *13*, 1247. [CrossRef] [PubMed]
133. Luo, J.; Bavencoffe, A.; Yang, P.; Feng, J.; Yin, S.; Qian, A.; Yu, W.; Liu, S.; Gong, X.; Cai, T.; et al. Zinc Inhibits TRPV1 to Alleviate Chemotherapy-Induced Neuropathic Pain. *J. Neurosci.* **2018**, *38*, 474–483. [CrossRef] [PubMed]
134. Jo, S.M.; Danscher, G.; Schrøder, H.D.; Suh, S.W. Depletion of vesicular zinc in dorsal horn of spinal cord causes increased neuropathic pain in mice. *Biometals* **2008**, *21*, 151–158. [CrossRef]
135. Hoagland, E.N.; Sherwood, T.W.; Lee, K.G.; Walker, C.J.; Askwith, C.C. Identification of a calcium permeable human acid-sensing ion channel 1 transcript variant. *J. Biol. Chem.* **2010**, *285*, 41852–41862. [CrossRef]
136. Dubé, G.R.; Lehto, S.G.; Breese, N.M.; Baker, S.J.; Wang, X.; Matulenko, M.A.; Honoré, P.; Stewart, A.O.; Moreland, R.B.; Brioni, J.D. Electrophysiological and in vivo characterization of A-317567, a novel blocker of acid sensing ion channels. *Pain* **2005**, *117*, 88–96. [CrossRef]
137. Gautam, M.; Benson, C.J. Acid-sensing ion channels (ASICs) in mouse skeletal muscle afferents are heteromers composed of ASIC1a, ASIC2, and ASIC3 subunits. *FASEB J.* **2013**, *27*, 793–802. [CrossRef]
138. Gregory, N.S.; Gautam, M.; Benson, C.J.; Sluka, K.A. Acid Sensing Ion Channel 1a (ASIC1a) Mediates Activity-induced Pain by Modulation of Heteromeric ASIC Channel Kinetics. *Neuroscience* **2018**, *386*, 166–174. [CrossRef]
139. Ugarte, M.; Osborne, N.N.; Brown, L.A.; Bishop, P.N. Iron, zinc, and copper in retinal physiology and disease. *Surv. Ophthalmol.* **2013**, *58*, 585–609. [CrossRef]
140. Gilbert, R.; Peto, T.; Lengyel, I.; Emri, E. Zinc Nutrition and Inflammation in the Aging Retina. *Mol. Nutr. Food Res.* **2019**, *63*, 15. [CrossRef]
141. Blasiak, J.; Pawlowska, E.; Chojnacki, J.; Szczepanska, J.; Chojnacki, C.; Kaarniranta, K. Zinc and Autophagy in Age-Related Macular Degeneration. *Int. J. Mol. Sci.* **2020**, *21*, 4994. [CrossRef]
142. Evans, J.R.; Lawrenson, J.G. Antioxidant vitamin and mineral supplements for slowing the progression of age-related macular degeneration. *Cochrane Database Syst. Rev.* **2017**, *7*, CD000254. [CrossRef]
143. Márquez, A.G.; Salazar, V.; Lima, L.P. Consequences of zinc deficiency on zinc localization, taurine transport, and zinc transporters in rat retina. *Microsc. Res. Tech.* **2022**, *85*, 3382–3390. [CrossRef] [PubMed]
144. Wu, L.J.; Duan, B.; Mei, Y.D.; Gao, J.; Chen, J.G.; Zhuo, M.; Xu, L.; Wu, M.; Xu, T.L. Characterization of acid-sensing ion channels in dorsal horn neurons of rat spinal cord. *J. Biol. Chem.* **2004**, *279*, 43716–43724. [CrossRef]
145. Papalampropoulou-Tsiridou, M.; Labrecque, S.; Godin, A.G.; De Koninck, Y.; Wang, F. Differential Expression of Acid—Sensing Ion Channels in Mouse Primary Afferents in Naïve and Injured Conditions. *Front. Cell. Neurosci.* **2020**, *14*, 103. [CrossRef]
146. Zha, X.M.; Costa, V.; Harding, A.M.S.; Reznikov, L.; Benson, C.J.; Welsh, M.J. ASIC2 Subunits target acid-sensing ion channels to the synapse via an association with PSD-95. *J. Neurosci.* **2009**, *29*, 8438–8446. [CrossRef]
147. Jiang, N.; Wu, J.; Leng, T.; Yang, T.; Zhou, Y.; Jiang, Q.; Wang, B.; Hu, Y.; Ji, Y.; Simon, R.P.; et al. Region specific contribution of ASIC2 to acidosis-and ischemia-induced neuronal injury. *J. Cereb. Blood Flow Metab.* **2017**, *37*, 528–540. [CrossRef]
148. Huang, Y.; Jiang, N.; Li, J.; Ji, Y.H.; Xiong, Z.G.; Zha, X.M. Two aspects of ASIC function: Synaptic plasticity and neuronal injury. *Neuropharmacology.* **2015**, *94*, 42–48. [CrossRef]
149. Bartoi, T.; Augustinowski, K.; Polleichtner, G.; Gründer, S.; Ulbrich, M.H. Acid-sensing ion channel (ASIC) 1a/2a heteromers have a flexible 2:1/1:2 stoichiometry. *Proc. Natl. Acad. Sci. USA* **2014**, *111*, 8281–8286. [CrossRef] [PubMed]
150. Li, M.H.; Liu, S.Q.; Inoue, K.; Lan, J.; Simon, R.P.; Xiong, Z.G. Acid-sensing ion channels in mouse olfactory bulb M/T neurons. *J. Gen. Physiol.* **2014**, *143*, 719–731. [CrossRef] [PubMed]
151. Saini, N.; Schaffner, W. Zinc supplement greatly improves the condition of parkin mutant Drosophila. *Biol. Chem.* **2010**, *391*, 513–518. [CrossRef] [PubMed]
152. Askwith, C.C.; Wemmie, J.A.; Price, M.P.; Rokhlina, T.; Welsh, M.J. Acid-sensing ion channel 2 (ASIC2) modulates ASIC1 H+-activated currents in hippocampal neurons. *J. Biol. Chem.* **2004**, *279*, 18296–18305. [CrossRef] [PubMed]
153. Ettaiche, M.; Guy, N.; Hofman, P.; Lazdunski, M.; Waldmann, R. Acid-sensing ion channel 2 is important for retinal function and protects against light-induced retinal degeneration. *J. Neurosci.* **2004**, *24*, 1005–1012. [CrossRef] [PubMed]
154. Tang, J.; Zhuo, Y.; Li, Y. Effects of Iron and Zinc on Mitochondria: Potential Mechanisms of Glaucomatous Injury. *Front. Cell. Dev. Biol.* **2021**, *9*, 720288. [CrossRef] [PubMed]
155. Li, Y.; Andereggen, L.; Yuki, K.; Omura, K.; Yin, Y.; Gilbert, H.Y.; Erdogan, B.; Asdourian, M.S.; Shrock, C.; de Lima, S.; et al. Mobile zinc increases rapidly in the retina after optic nerve injury and regulates ganglion cell survival and optic nerve regeneration. *Proc. Natl. Acad. Sci. USA* **2017**, *114*, E209–E218. [CrossRef] [PubMed]

156. Hattori, T.; Chen, J.; Harding, A.M.; Price, M.P.; Lu, Y.; Abboud, F.M.; Benson, C.J. ASIC2a and ASIC3 heteromultimerize to form pH-sensitive channels in mouse cardiac dorsal root ganglia neurons. *Circ. Res.* **2009**, *105*, 279–286. [CrossRef]
157. Choi, S.; Liu, X.; Pan, Z. Zinc deficiency and cellular oxidative stress: Prognostic implications in cardiovascular diseases. *Acta Pharmacol. Sin.* **2018**, *39*, 1120–1132. [CrossRef]
158. Li, L.; Niu, P.; Wang, X.; Bing, F.; Tan, W.; Huo, Y. Short-Term Inhalation of Ultrafine Zinc Particles Could Alleviate Cardiac Dysfunctions in Rats of Myocardial Infarction. *Front. Bioeng. Biotechnol.* **2021**, *9*, 646533. [CrossRef]
159. Gonçalves, A.F.; Polegato, B.F.; Fernandes, A.A.; Ishikawa, L.L.; Okoshi, K.; Bazan, S.G.Z.; Minicucci, M.F.; Azevedo, P.S.; Ikoma, M.R.; Penitenti, M.; et al. Zinc Supplementation Attenuates Cardiac Remodeling After Experimental Myocardial Infarction. *Cell. Physiol. Biochem.* **2018**, *50*, 353–362. [CrossRef]
160. Abdel-Hady, E.; Mohamed, F.; Ahmed, M.; Abdel-Salam, M.; Ayobe, M. Supplementation of Lipoic Acid, Zinc and Clopidogrel Reduces Mortality Rate and Incidence of Ventricular Arrhythmia in Experimental Myocardial Infarction. *Front Physiol.* **2021**, *12*, 582223. [CrossRef]
161. Wu, W.L.; Cheng, C.F.; Sun, W.H.; Wong, C.W.; Chen, C.C. Targeting ASIC3 for pain, anxiety, and insulin resistance. *Pharmacol. Ther.* **2012**, *134*, 127–138. [CrossRef]
162. Karczewski, J.; Spencer, R.H.; Garsky, V.M.; Liang, A.; Leitl, M.D.; Cato, M.J.; Cook, S.P.; Kane, S.; Urban, M.O. Reversal of acid-induced and inflammatory pain by the selective ASIC3 inhibitor, APETx2. *Br. J. Pharmacol.* **2010**, *161*, 950–960. [CrossRef]
163. Montalbetti, N.; Rooney, J.G.; Marciszyn, A.L.; Carattino, M.D. ASIC3 fine-tunes bladder sensory signaling. *Am. J. Physiol. Renal. Physiol.* **2018**, *315*, F870–F879. [CrossRef] [PubMed]
164. Cheng, Y.R.; Jiang, B.Y.; Chen, C.C. Acid-sensing ion channels: Dual function proteins for chemo-sensing and mechano-sensing. *J. Biomed Sci.* **2018**, *25*, 46. [CrossRef]
165. Akbari, G. Role of Zinc Supplementation on Ischemia/Reperfusion Injury in Various Organs. *Biol. Trace Elem. Res.* **2020**, *196*, 1–9. [CrossRef]
166. Chen, X.; Polleichtner, G.; Kadurin, I.; Gründer, S. Zebrafish acid-sensing ion channel (ASIC) 4, characterization of homo- and heteromeric channels, and identification of regions important for activation by H+. *J. Biol. Chem.* **2007**, *282*, 30406–30413. [CrossRef]
167. Paukert, M.; Sidi, S.; Russell, C.; Siba, M.; Wilson, S.W.; Nicolson, T.; Gründer, S. A family of acid-sensing ion channels from the zebrafish: Widespread expression in the central nervous system suggests a conserved role in neuronal communication. *J. Biol. Chem.* **2004**, *279*, 18783–18791. [CrossRef] [PubMed]
168. Lefèvre, C.M.T.; Diakov, A.; Haerteis, S.; Korbmacher, C.; Gründer, S.; Wiemuth, D. Pharmacological and electrophysiological characterization of the human bile acid-sensitive ion channel (hBASIC). *Pflug. Arch.–Eur. J. Physiol.* **2014**, *466*, 253–263. [CrossRef]
169. Wiemuth, D.; Assmann, M.; Gründer, S. The bile acid-sensitive ion channel (BASIC), the ignored cousin of ASICs and ENaC. *Channels* **2014**, *8*, 29–34. [CrossRef]
170. Kreko-Pierce, T.; Boiko, N.; Harbidge, D.G.; Marcus, D.C.; Stockand, J.D.; Pugh, J.R. Cerebellar Ataxia Caused by Type II Unipolar Brush Cell Dysfunction in the Asic5 Knockout Mouse. *Sci. Rep.* **2020**, *10*, 2168. [CrossRef] [PubMed]

biomolecules

MDPI

Review

The Important Role of Zinc in Neurological Diseases

Zhe Li [1,2,3], Yang Liu [1,2,3], Ruixue Wei [1,2,3], V. Wee Yong [4,*] and Mengzhou Xue [1,2,3,*]

1 Department of Cerebrovascular Diseases, The Second Affiliated Hospital of Zhengzhou University, Zhengzhou 450001, China
2 Academy of Medical Science, Zhengzhou University, Zhengzhou 450001, China
3 Henan Medical Key Laboratory of Translational Cerebrovascular Diseases, Zhengzhou 450001, China
4 Department of Clinical Neurosciences, Hotchkiss Brain Institute, University of Calgary, Calgary, AB T2N 1N4, Canada
* Correspondence: vyong@ucalgary.ca (V.W.Y.); xuemengzhou@zzu.edu.cn (M.X.); Tel.: +86-0371-63974732 (M.X.)

Abstract: Zinc is one of the most abundant metal ions in the central nervous system (CNS), where it plays a crucial role in both physiological and pathological brain functions. Zinc promotes antioxidant effects, neurogenesis, and immune system responses. From neonatal brain development to the preservation and control of adult brain function, zinc is a vital homeostatic component of the CNS. Molecularly, zinc regulates gene expression with transcription factors and activates dozens of enzymes involved in neuronal metabolism. During development and in adulthood, zinc acts as a regulator of synaptic activity and neuronal plasticity at the cellular level. There are several neurological diseases that may be affected by changes in zinc status, and these include stroke, neurodegenerative diseases, traumatic brain injuries, and depression. Accordingly, zinc deficiency may result in declines in cognition and learning and an increase in oxidative stress, while zinc accumulation may lead to neurotoxicity and neuronal cell death. In this review, we explore the mechanisms of brain zinc balance, the role of zinc in neurological diseases, and strategies affecting zinc for the prevention and treatment of these diseases.

Keywords: zinc; brain; stroke; neurotoxicity; brain injury

1. Introduction

Zinc is a trace element essential for human survival, and its deficiency has been linked to various adverse effects, such as growth retardation, impaired functioning of the immune system, and cognitive dysfunction [1,2]. Zinc is associated with the synthesis and activity of numerous proteins and enzymes, including matrix metalloproteinases (MMPs), deoxyribonucleic acid (DNA) and ribonucleic acid (RNA) polymerases, and insulin [3]. Zinc insufficiency is one of the world's major public health issues since it is linked to several diseases [4–7]. Maintaining zinc homeostasis is crucial for normal brain function, and zinc deficiency or overload may contribute to brain injury and exacerbate neurological conditions [8].

Zinc plays a role in both physiological and pathological processes in CNS [9,10]. A major site of storage of zinc is the telencephalon, especially in the cortical areas, hippocampus, and amygdala [11]. Accordingly, zinc has far-reaching effects on cognition, emotional stability, and memory [12]. Therefore, maintaining zinc homeostasis is essential for brain health, and it appears relevant to investigate its potential contributions to many neurological diseases. In this review, we provide a quick overview of zinc regulation and then focus on the roles that zinc plays in diverse brain disorders, as well as recent advances in zinc-related therapies.

2. The Homeostasis of Zinc in Physiological Conditions

The adult human brain contains approximately 0.04 g of zinc, accounting for 1.5% of the total body zinc [13,14]. The majority of zinc in the human body is located in zinc-containing proteins. A bioinformatics analysis in humans found that over 2800 proteins are potentially zinc-binding, constituting about 10% of all proteins in the human proteome; these zinc-binding proteins fulfill signaling, catalytic and structural functions [15]. Most zinc-binding proteins feature zinc-finger motifs, which occur when a zinc ion binds four invariant cysteines and/or histidine residues to form a stable structure, regulating protein-DNA [16], protein-RNA, and protein-protein interactions [17]. The physiological roles of zinc in cells and organs are summarized in Figure 1.

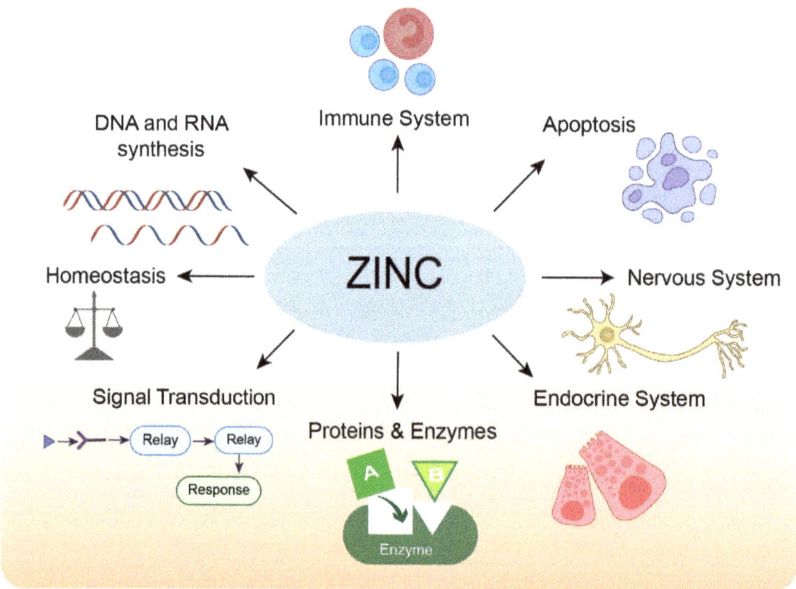

Figure 1. The physiological roles of zinc.

Zinc ions (Zn^{2+}), which can be detected histochemically and eliminated by chelating agents, are abundant in all organs, tissues, body fluids, and secretions in addition to being tightly linked to proteins [1,18]. Zinc is highly concentrated (>95%) in fat-free body masses, mainly in intracellular compartments [19]. Under physiological conditions, zinc ions and zinc-binding proteins comprise the zinc pool in human tissues and maintain zinc homeostasis [20].

Zinc is a structural or functional component of many proteins [21]. It is an essential cofactor in protein structure stabilization and enzyme catalyses, such as DNA synthesis, brain development, and neurotransmission [22,23]. Because of its multifaceted role in biological processes, changes in zinc concentrations from normal levels can contribute to a variety of devastating diseases. For instance, zinc deficiency is associated with decreased immunological responses and growth retardation, in addition to neurological disorders [24–26].

3. Distribution of Zinc in the Brain

At 2–3 g in total, zinc is an abundant transition metal found in high concentrations in the mammalian brain and is distributed unequally throughout different organs and tissues [26]. In adults, 60% of zinc is found in skeletal muscles; 30% in bones; 5% in the liver and skin; 1.5% in the brain, kidney, and heart; and less than 2% in other tissues [27]. Only a small fraction of zinc circulates in the blood, approximately 80% of which is loosely bound to albumin, and 20% is tightly bound to α2-macroglobulin [28]. Zinc is crucial for the development and physiology of the mammalian brain [29]. The amount of zinc in brain tissue is second only to iron in terms of trace metal concentration [20]. In this regard, while the iron content in the normal brain is around 0.04 mg/g with a concentration of ~ 720 μM, that of zinc is about 10 μg/g with a concentration of 150 μM [30,31]. Timm-Danscher and Nissl stainings were used to analyze the distribution of zinc in the rat brain and revealed that zinc was highly deposited in neuron-rich areas, primarily the hippocampus and cerebral cortex [32,33].

The brain stores zinc in three different forms: vesicular zinc, protein-bound zinc, and zinc ions (Zn^{2+}) [32]. In the CNS, around 85% of zinc is tightly linked to proteins that perform both functional and structural purposes, such as zinc-related metalloenzymes, MMPs, and zinc transcription factors [34–36]. Vesicular zinc is mostly present in the synaptic vesicles at the axon terminals of glutamatergic neurons, where it is released in a calcium- and the impulse-dependent way [37]. The concentration of Zn^{2+} in the brain is roughly ten times higher than that in serum, and it is required for neural regulation, synaptic plasticity, learning, and memory [34].

Given that zinc is unable to diffuse across the cell membrane, specialized transporters or membrane channels are required for its entry into cells [38]. Zinc transporters mainly include metallothioneins (MTs), Zrt-Irt-like proteins (ZIPs), and the zinc transporter family (ZnTs). Zinc enters neurons through three main pathways: α-amino-3-hydroxyl-5-methyl-4-isoxazolepropionate-acid /kainite (Ca^{2+}-A/K) AMPA/KA channels [39], voltage-dependent channels, and N-methyl-D-aspartate (NMDA) receptor-gated channels [40]. In particular, AMPA/KA gated channels can pass more zinc in neurons [41–43].

3.1. Role of Zinc in Neurogenesis

Neurogenesis is particularly robust in embryonic and neonatal periods. Zinc plays critical roles in neuronal proliferation and differentiation, neuronal migration, and axonal growth during neurodevelopment [30]. It is necessary for the synthesis of a wide range of proteins, hormones, and growth factors [44]. During embryogenesis, stem cells and neuronal progenitor cells proliferate and differentiate in the neural tube, one of the first brain structures. Zinc has been recognized to be essential for DNA polymerase and other crucial enzymes in the developing brain [33,45]. As a result, zinc shortage during pregnancy leads to abnormal hippocampal proliferation and neuronal differentiation in the fetus, which appears to impact future memory and learning processes and even neurological abnormalities [46,47].

However, neurogenesis is not limited to the developmental phase, and zinc is required for the proliferation, differentiation, and death of neurons throughout the human brain's life cycle [44]. Wistar females, during pregnancy and lactation, received a drinking water solution of ZnSO4 at an estimated dose of 16 mg/kg. Behavioral tests showed enhanced spatial memory and increased blood zinc levels in the pups [48]. The US recommended dietary allowance for zinc for women during pregnancy is 11.5 ± 1.75 mg/d [49]. Zinc deficiency (<56 μg/dL) during pregnancy and breastfeeding can impede the growth of newborns, increasing the risk of dwarfism, impaired learning memory function, and delayed mental development, whereas zinc supplementation can improve the development of premature newborns [50,51].

3.2. Role of Zinc in Promoting Redox Homeostasis

Several age-related chronic illnesses, including atherosclerosis, cancer, and dementia, display oxidative stress as a contributing component. Reactive oxygen species (ROS) are chemically reactive chemical entities containing oxygen, such as $O_2{\cdot}^-$, H_2O_2, and OH, which are constantly produced in vivo under aerobic conditions [52,53]. Zinc is a redox-inert metal that participates in redox-regulated signaling. It acts as an antioxidant by accelerating the action of copper/zinc superoxide dismutase [54], maintaining membrane structure, and promoting the synthesis of metallothionein (MT), a metal-binding protein [55]. Zinc helps to maintain cellular redox equilibrium via a variety of processes, including zinc's dynamic interaction with sulfur in protein cysteine clusters [56], control of oxidant generation and metal-induced oxidative damage [57], and controlling redox signaling by altering enzyme activity, binding interactions, and molecular chaperone activity [58].

Given the complexity of the pathways and events that disruption of zinc homeostasis might influence, the negative effects of these changes are not to be underestimated. The events involved in the regulation of cellular oxidative/antioxidant homeostasis by zinc are diverse and interrelated, including (i) regulation of oxidant production and metal-induced oxidative damage; (ii) regulation of glutathione (GSH) metabolism and overall thiol redox status by zinc; and (iii) direct or indirect regulation of redox signaling [52]. Thus, the dysregulation of zinc homeostasis may have significant adverse effects. Deficits in zinc are associated with oxidative stress, impaired GSH metabolism, tubulin oxidation, and disruptions in redox-sensitive signaling in the developing nervous system [59]. These alterations may lead to altered organ cellularity, organization, and connectivity, thereby increasing the risk of diseases later in life [60,61].

3.3. Role of Zinc on Immunity in the CNS

Patients with zinc deficiency show increased susceptibility to various pathogens [62]. Regarding the human immune system, the innate and adaptive immune systems are both regulated by zinc-finger-bearing transcription factors. Hence, a direct, as well as an indirect role of zinc in altering intracellular signaling, can be anticipated [63]. The role of zinc in immune function is also connected to zinc transporter proteins. Overall, even a minor zinc shortage may elevate pro-inflammatory cytokines, including tumor necrosis factor-alpha (TNF-α), interleukin 1 beta (IL-1β), and IL-6 [64–66].

Microglia are an innate immune cell type found in the CNS. In response to insults, such as infection or other detrimental molecules, microglia are activated and secrete MMPs, ROS, and other pro-inflammatory molecules [67,68]. In models of ischemic stroke and neurodegenerative diseases, zinc chelator TPEN has been shown to prevent neuronal loss by inhibiting microglial activation, which is triggered when neurons release zinc [69]. PARP-1, which can chelate zinc, has been shown to minimize neuronal death in rats with ischemic stroke [70] and other neurodegenerative diseases [71].

4. The Role of Zinc in Stroke

4.1. Ischemic Stroke

As the primary cause of disability, stroke ranks just behind heart disease as the second biggest cause of mortality globally [67]. Ischemic stroke accounts for approximately 80% of all strokes and may be caused by a combination of events such as cardiogenic embolism, obstruction of tiny blood arteries in the brain, and atherosclerosis influencing cerebral circulation [72]. Ischemic stroke is a thrombo-inflammatory disease in which platelets and immune cells accumulate at sites of ischemic vascular damage, disrupting the permeability of the BBB, triggering other processes such as neuroinflammation, microglia activation, and excitotoxicity; these collectively contribute to neuronal death [67,73].

Zinc may provide anti-atherogenic properties by preventing metabolic and physiological dysregulation of the vascular endothelium due to its antioxidant and membrane-stabilizing characteristics [74,75]. The supplement, along with zinc in zinc-deficient endothelial cells, induces a partial repair of the endothelial cell barrier, but supplementation with calcium and magnesium does not achieve the same effect [76]. As a consequence, zinc appears to be required for endothelial integrity, and a deficit might compromise endothelial barrier function [77].

During ischemic stroke, zinc is released from synaptic vesicles of glutamatergic neurons, and the abnormal accumulation of zinc stimulates the activation of microglia [78] and ultimately influences cell survival and function in part through mitochondrial dysfunction [79]. In ischemic stroke, zinc release can directly produce ROS or activate NADPH oxidase to produce ROS, leading to brain injury [17] (Figure 2). Zinc is considered an independent risk factor for ischemic stroke [75,80], and it has been observed to accumulate in the synaptic gap of neurons in animal models of ischemic stroke, speeding up the development of cerebral infarction [81]. As a result, inhibiting zinc over-release may prevent brain injury after ischemic stroke.

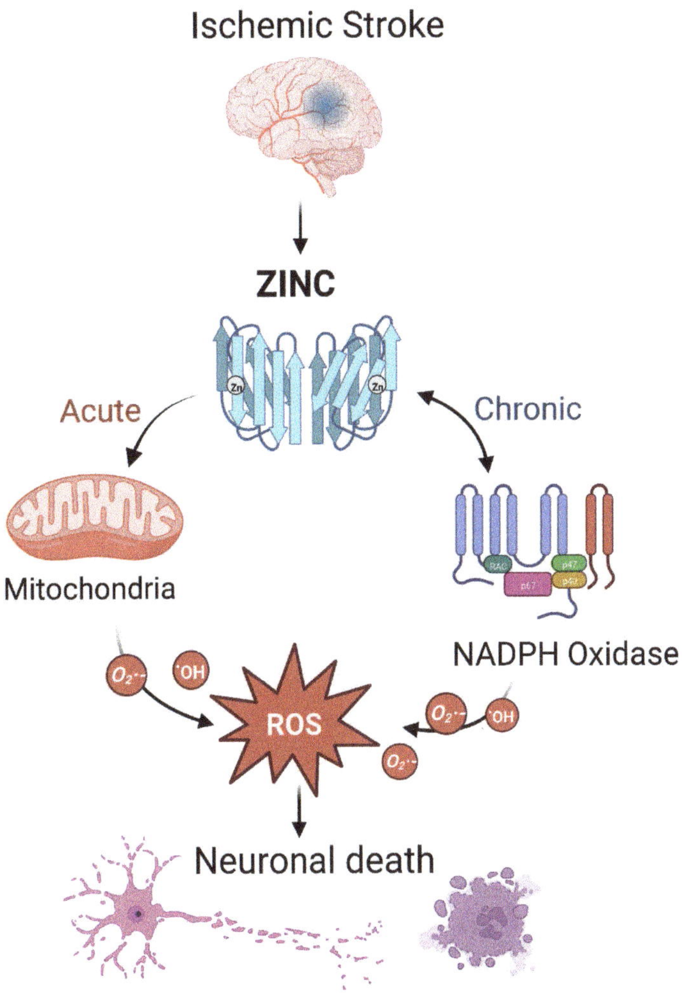

Figure 2. The interaction between zinc and ROS in ischemic stroke. In ischemic stroke, zinc is released from sources such as the synaptic vesicles of glutamatergic neurons. Acutely, zinc may elevate ROS through action on mitochondria. In the chronic phase, zinc accumulation can activate neuronal NADPH oxidases to generate additional oxidative species. The resultant ROS may produce neurotoxicity, further increasing zinc accumulation and ischemic brain injury. Note that while we show a central role for ROS, other mechanisms of zinc toxicity exist and are described in the text.

In vitro and animal experiments support a causal relationship between zinc dysregulation and neuronal damage after ischemic stroke. The injection of the zinc chelator ZnEDTA 14 days after middle cerebral artery occlusion (MCAO) in adult male rats led to a significant decrease in infarct volume and neuronal damage and improvement in neurological function [82]. In MCAO rats, zinc-induced CDK5-Tyr15 phosphorylation activates CDK5 in the hippocampus, which exacerbates neuronal death in ischemic stroke [83]. In an experimental model of ischemic stroke, normobaric hyperoxia treatment reduces zinc accumulation in penumbral tissues, thus reducing ischemic injury [84].

Some noteworthy findings about zinc levels have been reported from clinical investigations of patients diagnosed with an ischemic stroke. Serum zinc concentrations were considerably lower in ischemic stroke patients compared to age- and sex-matched healthy controls [85]. Calcium, copper, and iron levels did not change significantly between patients with ischemic stroke and healthy controls; the researchers postulated that low blood zinc concentration is associated with an increased risk of ischemic stroke based on their findings [86]. In a case-cohort study, blood zinc concentrations were inversely linked with the incidence of ischemic stroke, especially in women. This might be related to the function of zinc in the metabolism of sex hormones and the reproductive cycle, but the mechanism underlying the possible interaction is still unclear and warrants further investigation [87]. It appears that ischemic stroke could be a condition that would benefit from preventative zinc supplementation [86].

4.2. Intracerebral Hemorrhage (Hemorrhagic Stroke)

Intracerebral hemorrhage (ICH) is a non-traumatic hemorrhage in the brain parenchyma. ICH is particularly catastrophic, with a high mortality rate of up to 50% of the survivors. Over 70% are dependent on functioning aids a year after the event [88–90]. Despite clinical advances, including the continued evolution of minimally invasive surgical procedures to remove the blood clot, the prognosis of ICH remains poor [91–93]. A number of factors play a role in the pathogenesis of ICH, including hypertension, diabetes, lipid metabolism problems, and genetics, but also changes in the levels of essential trace elements and heavy metals [80,94–96].

Zinc is involved in coagulation's intrinsic pathway, where it binds directly to the XII factor and increases its sensitivity to enzymatic activation [97]. Therefore, both animal models and patients with zinc deficiency have been shown to have prolonged bleeding times as a result of impaired coagulation cascades and fibrin formation [98]. Similarly, after platelets are activated at the site of injury, zinc binds to fibrinogen and fibrin, promoting the formation of a fibrin network [98,99]. Patients with ICH had lower zinc levels than normal controls (0.13 ± 0.02 vs. 3.17 ± 0.74 μg/dl); $p < 0.001$) [100]. It has been determined that plasma zinc levels and the risk of first-ever stroke in hypertensive patients are significantly correlated with the risk of ICH in a recent study [101]. Previously, a retrospective cohort study examined the relationship between hypozincemia and the severity of aneurysmal subarachnoid hemorrhage. The findings indicated that zinc deficiency (plasma zinc concentration < 10 μmol/L) is associated with a more severe course of the disease. As a result, the authors concluded that zinc deficiency independently contributes to an unfavorable outcome [102].

5. Zinc and Other Disorders of the CNS

5.1. Alzheimer's Disease

Alzheimer's disease (AD) is a neurodegenerative condition that affects over 37 million individuals throughout the globe and is triggered by a wide range of risk factors [103]. Neuropathological diagnosis of AD requires the presence of two primary indicators: the presence of amyloid-beta (Aβ) protein plaques outside neurons and the presence of deposits of neurofibrillary tangles (NFTs) within neurons, both of which lead to neuronal degeneration in the hippocampus, amygdala, and neocortex [104,105].

AD is a neurodegenerative condition characterized by a gradual but steady deterioration in cognitive abilities, such as cognitive decline, language impairment, mental and physical disorientation, and altered behavior [106]. Depression, anxiety, and agitation are just some of the non-memory-related behavioral symptoms it produces [107]. There was a twofold increase (25 μg/g) in zinc levels in AD brain tissue as compared with normal age-matched control subjects [108]. Brain zinc accumulation is a prominent feature of advanced AD. During the development of AD, there are unexpected changes in the expression levels of the zinc transporter, which includes six families of the zinc transporter (ZnT) and one family of the Zrt-/Irt-like protein (ZIP) transporter. There were changes in ZnT proteins (ZnT-1, ZnT-4, and ZnT-6) observed in AD patients [109]. ZnT1 and ZnT4 were expressed across the whole senile plaque, whereas ZnT3, ZnT5, and ZnT6 were localized to the plaque's perimeter [110]. Genetic ablation of ZnT3 may reflect a phenotype of synaptic and memory deficiencies in AD since ZnT3 knockout animals have been shown to have problems in learning and memory [111].

Amyloid deposits of aggregated Aβ protein and neurofibrillary tangles (NFT) are pathological characteristics of AD. Zinc-binding sites may be found in both the amyloid precursor protein (APP) and Aβ [112]. Zinc transporters have been implicated in the development of AD senile plaques by several studies [106,109,113]. In senile plaques of the human Alzheimer's disease brain and APP or APP/presenilin (PS) animal models, zinc transporters were discovered to be overexpressed and distributed differently; an extracellular elevation of zinc concentration may initiate the deposition of Aβ and lead to the formation of senile plaques [106,114]. Excess chronic zinc supplementation is also linked to hyperphosphorylated tau aggregation in AD pathogenesis at the molecular level [115].

The effect of excessive chronic zinc supplementation on tau pathology was assessed using a transgenic mouse model containing the human tau (P301L) gene [116]. In tauopathic mice, zinc supplementation aggravated circadian rhythm deficits, nesting behavior, and the Morris water maze deficits, suggesting zinc exacerbates tauopathic and biochemical deficits [116]. Tau hyperphosphorylation and other metabolic abnormalities have been attributed to zinc's inhibition of proteins such as protein phosphatase 2A (PP2A), followed by activation of extracellular signal-regulated kinase (ERK), glycogen synthase kinase-3 (GSK-3) and other pathways [117]; these culminate in a characteristic reduction in microtubule stability in neurons and ultimately neurodegenerative morphological abnormalities [118].

Overall, zinc dyshomeostasis may play a crucial role in AD, making the discovery of an appropriate agent for zinc chelation a promising therapeutic target.

5.2. Traumatic Brain Injury

Traumatic brain injury (TBI) is a widespread public health and socioeconomic issue that affects about 70 million people worldwide annually and is a leading cause of disability and fatality [119]. TBI patients have a variety of behavioral abnormalities, including impairments in motor, sensory, attention, memory, and executive skills [120,121]. The major cause of TBI is mechanical damage to brain tissue induced by external forces (impact, nerve axon rupture, vascular injury, etc.), and the initial phases are associated with reduced cerebral blood flow, oxygen consumption, and metabolism [122]. Secondary injuries may cause brain edema, tissue atrophy and death, neuroinflammation, oxidative stress, mitochondrial dysfunction, and other harmful chain reactions [123]. The degree of damage is determined by the ultrastructure and mechanical injury to the brain and can occur from minutes to months [121].

The function of zinc in TBI has been the subject of numerous ideas throughout the last three decades of study. Following TBI, glutamate is released from presynaptic terminals, and intracellular zinc accumulates inappropriately in postsynaptic neurons, resulting in excitotoxicity and cell death [124]. As a result, researchers have started investigating whether zinc chelators could be useful in reducing the severity of neuronal damage caused by TBI. Previous research has demonstrated that as zinc chelators, clioquinol (CQ) [125] and N,N,N′,N′-tetrakis-(2-pyridylmethyl) ethylenediamine (TPEN) [126] may regulate zinc overload after TBI and have the potential to be therapeutic or preventative for TBI.

During pathological conditions, excess zinc is rapidly released from presynaptic neuronal vesicles, crosses the postsynaptic membrane via channels or transporters (translocation), and causes neuronal damage and death. Recent studies have shifted the focus from this hypothesis to zinc chelation, where zinc is released and induces neuronal death when hippocampal neurons are overexcited [127]. According to preliminary in vitro investigations, zinc deficiency inhibits the proliferation of neural precursor cells and triggers death through p53 [128] and cysteine-mediated pathways [129]. Similar to this, dietary zinc deficiency (zinc contents in the zinc-deficient and zinc-adequate control diets were 0.85 ppm and 30 ppm, respectively) in rodents for 5 weeks inhibited the proliferation of neural precursor cells and hippocampus neurogenesis in animal studies [33,130]. Furthermore, as mentioned above, neurogenesis was considerably decreased in ZnT-3 deficient animals lacking presynaptic vesicle zinc [111]. As a consequence, research in recent years has focused on the possible neuroprotective effects of zinc supplementation. In a clinical trial, 68 patients with severe brain injury were randomly allocated to the zinc supplementation (12 mg elemental zinc) versus the routine zinc therapy group (2.5 mg elemental zinc), both as the sulfate salt in their total parenteral nutrition solution. One month after injury, the Glasgow Coma Scale (GCS) scores considerably improved, and the death rate was cut in half (26% vs. 12%) in the zinc supplementation group [131]. Similarly, animal studies have demonstrated that dietary zinc therapy may boost TBI resistance and enhance behavioral outcomes such as anxiety, sadness, spatial learning, and memory [132].

5.3. Epilepsy

Epilepsy is a common neurological disorder characterized by chronic brain seizures caused primarily by abnormal, excessive, or synchronized neuronal activity [133]. More than 60 million people worldwide suffer from epilepsy, which can be brought on by a variety of factors such as genetics, abnormal brain development, medications, and head trauma [134,135]. According to epilepsy research data, zinc's neuroprotective and neurotoxic effects appear to be dose-dependent, and it can have both pro- and anticonvulsant effects [136,137]. Postnatal day 27 (P27) male Sprague-Dawley rats pretreatment with a zinc-deficient diet (2.7 mg/kg) for 4 weeks had long-term adverse effects of seizures. In contrast, zinc supplementation (246 mg/kg) for 4 weeks significantly mitigated the severity of pilocarpine-induced limbic seizures and prolonged the time it took for forelimb contracture to occur. High-zinc dietary treatment improved cognitive impairment and reduced regenerative sprouting of hippocampal mossy fibers [138,139].

In addition, zinc chelation reduced the frequency of EEG spikes and the duration of both behavioral seizures and electrical after-discharges. While this did not alter the severity of behavioral seizures, it does suggest that zinc may play a facilitative role in the onset of epilepsy [140]. Treatment with zinc with cyclo-(His-Pro) enhances vesicular zinc levels as well as hippocampus neurogenesis in rats [141]. According to a case-control study, children with intractable epilepsy had considerably lower blood zinc levels than healthy children. One possible explanation for this is that low zinc levels increase oxidative stress and the incidence of seizures since zinc is a component of glutathione peroxidase and superoxide dismutase [142].

Consequently, before considering zinc supplementation as a therapeutic method for the treatment of epilepsy, it is essential to investigate the anticonvulsant or antiepileptic effects of proteins related to zinc signaling.

5.4. Depression

Depression is a common chronic mental disorder that has risen to become the world's second most common disorder, with symptoms primarily manifesting as changes in mood, sleep, cognition, and appetite [143,144]. Overall social functioning is impaired in patients with depression, and the vast majority of them do not receive an accurate and timely diagnosis, making depression one of the leading causes of suicide and disability [145,146]. Despite the lack of clarity surrounding the pathogenesis of depression, numerous studies have revealed that the majority of patients with major depressive disorder (MDD) have decreased serum zinc levels [147]. One possible mechanism involved in the changes in hippocampal functions, behavior, and pathophysiology observed in zinc deficiency maybe the activation of the hypothalamo-pituitary-adrenocortical (HPA) system [148].

Three decades ago, the association between zinc and depression was hypothesized for the first time [149]. In subsequent animal experiments, zinc deficiency was found to increase depression-like behavior in rodents during forced swimming and tail suspension tests [150–152]. Recent research using the olfactory bulbectomy model (OB) of depression in rats has revealed interesting results, whereby a zinc-deficient diet (3 mg Zn/kg) exacerbated depression. Importantly, treatment with escitalopram, venlafaxine 10 mg/kg, i.p., or combined escitalopram/venlafaxine (1 mg/kg, i.p.) with zinc (5 mg/kg) for 3 weeks reversed the behavioral changes of depression, indicating that zinc deficiency may lower the efficacy of anti-depressant drugs [153].

Human studies corroborated the above finding, showing an inverse relationship between serum zinc levels and the severity of depression. A meta-analysis revealed that zinc concentrations were approximately 1.85 μmol/L lower in depressed patients, and the severity of depression correlates with the relative zinc deficiency [154]. Furthermore, similar findings were observed in depressed patients such as postmenopausal women (72 ± 14 μg/dL) [155], college-aged women (79.6 ± 30.7 μg/dL) [156], and hemodialysis patients with end-stage renal disease (67.46 ± 29.7 μg/dL) [157]. Additionally, women with poor dietary or supplementary zinc consumption were more likely to suffer depressive symptoms, whereas men had no significant correlation [158].

6. Summary and Conclusions

Zinc plays a critical role in all life forms, in which zinc homeostasis is required for catalysis, structure, and regulatory processes. Zinc deficiency is commonly caused by poor dietary intake, while abnormal zinc accumulation is associated with the inhibition of zinc transporters or the activation of oxidative stress. Hence, zinc homeostasis and its regulation may affect a wide range of molecular functions and pathological outcomes, and different environmental and dietary habits and ecological conditions may influence zinc homeostasis.

Zinc has a profound effect on the brain throughout human life, from the development of the neonatal brain to neurological conditions. Zinc homeostasis in the CNS is regulated by a complex interplay of components, including the permeability of the BBB, the activity of zinc-binding proteins and ion channels, and the activity of various transporters (particularly ZIPs). Zinc is a signaling molecule in numerous metabolic pathways, the coordination of which occurs through the activity of zinc transporters. The surge in the development of a full suite of new fluorescent Zn^{2+} sensors, chelators, and genetic mouse models allows us to gain a greater understanding of neuronal signaling of Zn^{2+} in live primary neuron culture, brain slices, and live animals. In this review, we look at the evidence linking zinc metabolism to common neurological diseases such as stroke, Alzheimer's disease, epilepsy, and TBI in animal and human studies.

Regulating zinc levels in the brain has emerged as a potential treatment target for neurological disorders. It is now clear that the development of new strategies for the treatment of CNS insults needs to take into account the role of zinc in neuronal function, damage, repair, apoptosis, and necrosis. Excessive zinc is released following brain injury in conditions including ischemic stroke and epilepsy, causing neurotoxic damage to neurons and glial cells. Given this, zinc chelators may be used as a preventative therapy for diseases exacerbated by excess zinc, although removing zinc from its dependent enzymes may be undesirable. Regardless of the approach, systemic zinc supplementation or central zinc chelation, future work that seeks to explore the possible efficacy of zinc modulation in stroke, TBI, and seizure disorders should employ clinically relevant models to permit not only the development of drug therapies but also the clarification of dietary recommendations after injury.

Nevertheless, there is still room for improvement in clinical research on zinc chelators and zinc supplements. While nanoparticle delivery holds the best promise for restoring neural zinc imbalances, methods to therapeutically manipulate zinc levels in the brain remain elusive. These limitations are contributed by several factors, such as serum zinc measurements being non-indicative of brain zinc concentrations, the efficacy of supplemental zinc absorption and transport, the lack of therapeutic ways to manipulate zinc levels in the brain, and most importantly, nanoparticle compounds must be directed past the BBB in order to reach their intended targets.

It is important to take note that zinc's functions in the central nervous system are quite diverse, but all of its actions are linked to homeostasis. Thus, in order to utilize zinc as a therapeutic agent, it is necessary to have a detailed understanding of the factors that influence zinc homeostasis as well as the ability to optimize the relevant concentrations. There is an ongoing research effort investigating the tight regulatory balance between bound zinc and labile zinc, which can influence many molecular functions and pathological outcomes. Even though significant progress has been achieved in deciphering the processes that underlie zinc's metabolism and function, there is still a significant amount of information that has to be uncovered at the frontier of zinc neuroscience.

Author Contributions: Conceptualization, formal analysis, writing—original draft and writing—review and editing, Z.L.; writing—review and editing, Y.L. and R.W.; supervision and writing—review and editing, M.X. and V.W.Y. All authors have read and agreed to the published version of the manuscript.

Funding: The authors acknowledge operating grant support from the National Key Research and Development Program of China (grant no. 2018YFC1312200), National Natural Science Foundation of China (grants nos. 82071331 and 81870942), China Postdoctoral Science Foundation (grants nos. 2020TQ0289, 2020M672291), and from the Canadian Institutes of Health Sciences (V.W.Y.).

Institutional Review Board Statement: Not applicable.

Informed Consent Statement: Not applicable.

Data Availability Statement: The data in this study could be available from the corresponding author.

Acknowledgments: We appreciate the coworkers who participated in this study.

Conflicts of Interest: The authors declare no conflict of interest.

References

1. Prasad, A.S. Discovery of human zinc deficiency: Its impact on human health and disease. *Adv. Nutr.* **2013**, *4*, 176–190. [CrossRef] [PubMed]
2. Sanna, A.; Firinu, D.; Zavattari, P.; Valera, P. Zinc Status and Autoimmunity: A Systematic Review and Meta-Analysis. *Nutrients* **2018**, *10*, 68. [CrossRef]
3. McAllister, B.B.; Dyck, R.H. A new role for zinc in the brain. *eLife* **2017**, *6*, e31816. [CrossRef] [PubMed]
4. Nakatani, S.; Mori, K.; Shoji, T.; Emoto, M. Association of Zinc Deficiency with Development of CVD Events in Patients with CKD. *Nutrients* **2021**, *13*, 1680. [CrossRef] [PubMed]
5. Begum, F.; Me, H.M.; Christov, M. The Role of Zinc in Cardiovascular Disease. *Cardiol. Rev.* **2022**, *30*, 100–108. [CrossRef]
6. Vujasinovic, M.; Hedstrom, A.; Maisonneuve, P.; Valente, R.; von Horn, H.; Lohr, J.M.; Haas, S.L. Zinc deficiency in patients with chronic pancreatitis. *World J. Gastroenterol.* **2019**, *25*, 600–607. [CrossRef]
7. Swardfager, W.; Herrmann, N.; McIntyre, R.S.; Mazereeuw, G.; Goldberger, K.; Cha, D.S.; Schwartz, Y.; Lanctot, K.L. Potential roles of zinc in the pathophysiology and treatment of major depressive disorder. *Neurosci. Biobehav. Rev.* **2013**, *37*, 911–929. [CrossRef]
8. Wessells, K.R.; Brown, K.H. Estimating the global prevalence of zinc deficiency: Results based on zinc availability in national food supplies and the prevalence of stunting. *PLoS ONE* **2012**, *7*, e50568. [CrossRef]
9. Adamo, A.M.; Zago, M.P.; Mackenzie, G.G.; Aimo, L.; Keen, C.L.; Keenan, A.; Oteiza, P.I. The role of zinc in the modulation of neuronal proliferation and apoptosis. *Neurotox. Res.* **2010**, *17*, 1. [CrossRef]
10. Portbury, S.D.; Adlard, P.A. Zinc Signal in Brain Diseases. *Int. J. Mol. Sci.* **2017**, *18*, 2506. [CrossRef]
11. Choi, S.; Hong, D.K.; Choi, B.Y.; Suh, S.W. Zinc in the Brain: Friend or Foe? *Int. J. Mol. Sci.* **2020**, *21*, 8941. [CrossRef] [PubMed]
12. Bertoni-Freddari, C.; Fattoretti, P.; Casoli, T.; Di Stefano, G.; Giorgetti, B.; Balietti, M. Brain aging: The zinc connection. *Exp. Gerontol.* **2008**, *43*, 389–393. [CrossRef] [PubMed]
13. Takeda, A. Zinc homeostasis and functions of zinc in the brain. *Biometals* **2001**, *14*, 343–351. [CrossRef] [PubMed]
14. Sekler, I.; Silverman, W.F. Zinc homeostasis and signaling in glia. *Glia* **2012**, *60*, 843–850. [CrossRef]

15. Andreini, C.; Banci, L.; Bertini, I.; Rosato, A. Counting the zinc-proteins encoded in the human genome. *J. Proteome Res.* **2006**, *5*, 196–201. [CrossRef]
16. Khmeleva, S.A.; Mezentsev, Y.V.; Kozin, S.A.; Tsvetkov, P.O.; Ivanov, A.S.; Bodoev, N.V.; Makarov, A.A.; Radko, S.P. Zinc-induced interaction of the metal-binding domain of amyloid-beta peptide with DNA. *J. Alzheimers Dis.* **2013**, *36*, 633–636. [CrossRef]
17. Zhou, X.; Cooper, K.L.; Sun, X.; Liu, K.J.; Hudson, L.G. Selective Sensitization of Zinc Finger Protein Oxidation by Reactive Oxygen Species through Arsenic Binding. *J. Biol. Chem.* **2015**, *290*, 18361–18369. [CrossRef]
18. Stefanidou, M.; Maravelias, C.; Dona, A.; Spiliopoulou, C. Zinc: A multipurpose trace element. *Arch. Toxicol.* **2006**, *80*, 1–9. [CrossRef]
19. Micheletti, A.; Rossi, R.; Rufini, S. Zinc status in athletes: Relation to diet and exercise. *Sports Med.* **2001**, *31*, 577–582. [CrossRef]
20. Marger, L.; Schubert, C.R.; Bertrand, D. Zinc: An underappreciated modulatory factor of brain function. *Biochem. Pharmacol.* **2014**, *91*, 426–435. [CrossRef]
21. Oteiza, P.I.; Mackenzie, G.G. Zinc, oxidant-triggered cell signaling, and human health. *Mol. Asp. Med.* **2005**, *26*, 245–255. [CrossRef] [PubMed]
22. Barceloux, D.G. Zinc. *J. Toxicol. Clin. Toxicol.* **1999**, *37*, 279–292. [CrossRef] [PubMed]
23. Nakashima, A.S.; Dyck, R.H. Zinc and cortical plasticity. *Brain Res. Rev.* **2009**, *59*, 347–373. [CrossRef]
24. Plum, L.M.; Rink, L.; Haase, H. The essential toxin: Impact of zinc on human health. *Int. J. Environ. Res. Public Health* **2010**, *7*, 1342–1365. [CrossRef] [PubMed]
25. Read, S.A.; Obeid, S.; Ahlenstiel, C.; Ahlenstiel, G. The Role of Zinc in Antiviral Immunity. *Adv. Nutr.* **2019**, *10*, 696–710. [CrossRef] [PubMed]
26. Kambe, T.; Tsuji, T.; Hashimoto, A.; Itsumura, N. The Physiological, Biochemical, and Molecular Roles of Zinc Transporters in Zinc Homeostasis and Metabolism. *Physiol. Rev.* **2015**, *95*, 749–784. [CrossRef]
27. Solomons, N.W. Update on zinc biology. *Ann Nutr Metab* **2013**, *62* (Suppl. 1), 8–17. [CrossRef]
28. King, J.C.; Shames, D.M.; Woodhouse, L.R. Zinc homeostasis in humans. *J. Nutr.* **2000**, *130*, 1360S–1366S. [CrossRef]
29. Bitanihirwe, B.K.; Cunningham, M.G. Zinc: The brain's dark horse. *Synapse* **2009**, *63*, 1029–1049. [CrossRef]
30. Sensi, S.L.; Paoletti, P.; Bush, A.I.; Sekler, I. Zinc in the physiology and pathology of the CNS. *Nat. Rev. Neurosci.* **2009**, *10*, 780–791. [CrossRef]
31. Li, Z.; Liu, Y.; Wei, R.; Khan, S.; Zhang, R.; Zhang, Y.; Yong, V.W.; Xue, M. Iron Neurotoxicity and Protection by Deferoxamine in Intracerebral Hemorrhage. *Front. Mol. Neurosci.* **2022**, *15*, 927334. [CrossRef] [PubMed]
32. Frederickson, C.J. Neurobiology of zinc and zinc-containing neurons. *Int. Rev. Neurobiol.* **1989**, *31*, 145–238. [CrossRef] [PubMed]
33. Suh, S.W.; Won, S.J.; Hamby, A.M.; Yoo, B.H.; Fan, Y.; Sheline, C.T.; Tamano, H.; Takeda, A.; Liu, J. Decreased brain zinc availability reduces hippocampal neurogenesis in mice and rats. *J. Cereb. Blood Flow Metab.* **2009**, *29*, 1579–1588. [CrossRef] [PubMed]
34. Koh, J.Y. Zinc and disease of the brain. *Mol. Neurobiol.* **2001**, *24*, 99–106. [CrossRef] [PubMed]
35. Parisi, A.F.; Vallee, B.L. Zinc metalloenzymes: Characteristics and significance in biology and medicine. *Am. J. Clin. Nutr.* **1969**, *22*, 1222–1239. [CrossRef] [PubMed]
36. Liu, Y.; Mu, Y.; Li, Z.; Yong, V.W.; Xue, M. Extracellular matrix metalloproteinase inducer in brain ischemia and intracerebral hemorrhage. *Front. Immunol.* **2022**, *13*, 986469. [CrossRef]
37. Toth, K. Zinc in neurotransmission. *Annu. Rev. Nutr.* **2011**, *31*, 139–153. [CrossRef] [PubMed]
38. Gibson, R.S.; Ferguson, E.L. Assessment of dietary zinc in a population. *Am. J. Clin. Nutr.* **1998**, *68*, 430S–434S. [CrossRef]
39. Kerchner, G.A.; Canzoniero, L.M.; Yu, S.P.; Ling, C.; Choi, D.W. Zn^{2+} current is mediated by voltage-gated Ca^{2+} channels and enhanced by extracellular acidity in mouse cortical neurones. *J. Physiol.* **2000**, *528 Pt 1*, 39–52. [CrossRef]
40. Koh, J.Y.; Choi, D.W. Zinc toxicity on cultured cortical neurons: Involvement of N-methyl-D-aspartate receptors. *Neuroscience* **1994**, *60*, 1049–1057. [CrossRef]
41. Marin, P.; Israel, M.; Glowinski, J.; Premont, J. Routes of zinc entry in mouse cortical neurons: Role in zinc-induced neurotoxicity. *Eur. J. Neurosci.* **2000**, *12*, 8–18. [CrossRef]
42. Karakas, E.; Simorowski, N.; Furukawa, H. Structure of the zinc-bound amino-terminal domain of the NMDA receptor NR2B subunit. *EMBO J.* **2009**, *28*, 3910–3920. [CrossRef] [PubMed]
43. Weiss, J.H.; Sensi, S.L. Ca^{2+}-Zn^{2+} permeable AMPA or kainate receptors: Possible key factors in selective neurodegeneration. *Trends Neurosci.* **2000**, *23*, 365–371. [CrossRef] [PubMed]
44. Kumar, V.; Kumar, A.; Singh, K.; Avasthi, K.; Kim, J.J. Neurobiology of zinc and its role in neurogenesis. *Eur. J. Nutr.* **2021**, *60*, 55–64. [CrossRef] [PubMed]
45. Levenson, C.W.; Morris, D. Zinc and neurogenesis: Making new neurons from development to adulthood. *Adv. Nutr.* **2011**, *2*, 96–100. [CrossRef] [PubMed]
46. Chudasama, Y.; Wright, K.S.; Murray, E.A. Hippocampal lesions in rhesus monkeys disrupt emotional responses but not reinforcer devaluation effects. *Biol. Psychiatr.* **2008**, *63*, 1084–1091. [CrossRef]
47. Goeldner, C.; Reiss, D.; Wichmann, J.; Meziane, H.; Kieffer, B.L.; Ouagazzal, A.M. Nociceptin receptor impairs recognition memory via interaction with NMDA receptor-dependent mitogen-activated protein kinase/extracellular signal-regulated kinase signaling in the hippocampus. *J. Neurosci.* **2008**, *28*, 2190–2198. [CrossRef]
48. Piechal, A.; Blecharz-Klin, K.; Pyrzanowska, J.; Widy-Tyszkiewicz, E. Maternal zinc supplementation improves spatial memory in rat pups. *Biol. Trace Elem. Res.* **2012**, *147*, 299–308. [CrossRef]

49. National Research Council. *Recommended Dietary Allowances*, 10th ed.; Reports funded by National Institutes of Health; The National Academies Collection: Washington, DC, USA, 1989. [PubMed]
50. Yu, X.; Chen, W.; Wei, Z.; Ren, T.; Yang, X.; Yu, X. Effects of maternal mild zinc deficiency and different ways of zinc supplementation for offspring on learning and memory. *Food Nutr. Res.* **2016**, *60*, 29467. [CrossRef]
51. Wang, H.; Hu, Y.F.; Hao, J.H.; Chen, Y.H.; Su, P.Y.; Wang, Y.; Yu, Z.; Fu, L.; Xu, Y.Y.; Zhang, C.; et al. Maternal zinc deficiency during pregnancy elevates the risks of fetal growth restriction: A population-based birth cohort study. *Sci. Rep.* **2015**, *5*, 11262. [CrossRef]
52. Oteiza, P.I. Zinc and the modulation of redox homeostasis. *Free Radic. Biol. Med.* **2012**, *53*, 1748–1759. [CrossRef] [PubMed]
53. Zhao, X.; Drlica, K. Reactive oxygen species and the bacterial response to lethal stress. *Curr. Opin. Microbiol.* **2014**, *21*, 1–6. [CrossRef] [PubMed]
54. Maret, W. The redox biology of redox-inert zinc ions. *Free Radic. Biol. Med.* **2019**, *134*, 311–326. [CrossRef] [PubMed]
55. Hernandez-Camacho, J.D.; Vicente-Garcia, C.; Parsons, D.S.; Navas-Enamorado, I. Zinc at the crossroads of exercise and proteostasis. *Redox Biol.* **2020**, *35*, 101529. [CrossRef] [PubMed]
56. Hempe, J.M.; Cousins, R.J. Cysteine-rich intestinal protein binds zinc during transmucosal zinc transport. *Proc. Natl. Acad. Sci. USA* **1991**, *88*, 9671–9674. [CrossRef]
57. Baltaci, A.K.; Yuce, K.; Mogulkoc, R. Zinc Metabolism and Metallothioneins. *Biol. Trace Elem. Res.* **2018**, *183*, 22–31. [CrossRef]
58. Maret, W. Zinc coordination environments in proteins as redox sensors and signal transducers. *Antioxid. Redox Sign.* **2006**, *8*, 1419–1441. [CrossRef]
59. Mehta, A.J.; Yeligar, S.M.; Elon, L.; Brown, L.A.; Guidot, D.M. Alcoholism causes alveolar macrophage zinc deficiency and immune dysfunction. *Am. J. Respir. Crit. Care Med.* **2013**, *188*, 716–723. [CrossRef]
60. Joshi, P.C.; Mehta, A.; Jabber, W.S.; Fan, X.; Guidot, D.M. Zinc deficiency mediates alcohol-induced alveolar epithelial and macrophage dysfunction in rats. *Am. J. Respir. Cell Mol. Biol.* **2009**, *41*, 207–216. [CrossRef]
61. Ayton, S.; Lei, P.; Appukuttan, A.T.; Renoir, T.; Foliaki, S.; Chen, F.; Adlard, P.A.; Hannan, A.J.; Bush, A.I. Brain Zinc Deficiency Exacerbates Cognitive Decline in the R6/1 Model of Huntington's Disease. *Neurotherapeutics* **2020**, *17*, 243–251. [CrossRef]
62. Ibs, K.H.; Rink, L. Zinc-altered immune function. *J. Nutr.* **2003**, *133*, 1452S–1456S. [CrossRef] [PubMed]
63. Wessels, I.; Maywald, M.; Rink, L. Zinc as a Gatekeeper of Immune Function. *Nutrients* **2017**, *9*, 1286. [CrossRef] [PubMed]
64. Gammoh, N.Z.; Rink, L. Zinc in Infection and Inflammation. *Nutrients* **2017**, *9*, 624. [CrossRef] [PubMed]
65. Wessels, I.; Haase, H.; Engelhardt, G.; Rink, L.; Uciechowski, P. Zinc deficiency induces production of the proinflammatory cytokines IL-1beta and TNFalpha in promyeloid cells via epigenetic and redox-dependent mechanisms. *J. Nutr. Biochem.* **2013**, *24*, 289–297. [CrossRef] [PubMed]
66. Kahmann, L.; Uciechowski, P.; Warmuth, S.; Plumakers, B.; Gressner, A.M.; Malavolta, M.; Mocchegiani, E.; Rink, L. Zinc supplementation in the elderly reduces spontaneous inflammatory cytokine release and restores T cell functions. *Rejuvenation Res.* **2008**, *11*, 227–237. [CrossRef]
67. Xue, M.; Yong, V.W. Neuroinflammation in intracerebral haemorrhage: Immunotherapies with potential for translation. *Lancet Neurol.* **2020**, *19*, 1023–1032. [CrossRef]
68. Liu, Y.; Li, Z.; Khan, S.; Zhang, R.; Wei, R.; Zhang, Y.; Xue, M.; Yong, V.W. Neuroprotection of minocycline by inhibition of extracellular matrix metalloproteinase inducer expression following intracerebral hemorrhage in mice. *Neurosci. Lett.* **2021**, *764*, 136297. [CrossRef]
69. Haase, H.; Rink, L. Functional significance of zinc-related signaling pathways in immune cells. *Annu. Rev. Nutr.* **2009**, *29*, 133–152. [CrossRef]
70. Chen, J.; Li, X.; Xu, S.; Zhang, M.; Wu, Z.; Zhang, X.; Xu, Y.; Chen, Y. Delayed PARP-1 Inhibition Alleviates Post-stroke Inflammation in Male Versus Female Mice: Differences and Similarities. *Front. Cell. Neurosci* **2020**, *14*, 77. [CrossRef]
71. Kauppinen, T.M.; Suh, S.W.; Higashi, Y.; Berman, A.E.; Escartin, C.; Won, S.J.; Wang, C.; Cho, S.H.; Gan, L.; Swanson, R.A. Poly(ADP-ribose)polymerase-1 modulates microglial responses to amyloid beta. *J. Neuroinflamm.* **2011**, *8*, 152. [CrossRef]
72. Maida, C.D.; Norrito, R.L.; Daidone, M.; Tuttolomondo, A.; Pinto, A. Neuroinflammatory Mechanisms in Ischemic Stroke: Focus on Cardioembolic Stroke, Background, and Therapeutic Approaches. *Int. J. Mol. Sci.* **2020**, *21*, 6454. [CrossRef] [PubMed]
73. Zhang, X.; Zhang, Y.; Wang, F.; Liu, Y.; Yong, V.W.; Xue, M. Necrosulfonamide Alleviates Acute Brain Injury of Intracerebral Hemorrhage via Inhibiting Inflammation and Necroptosis. *Front. Mol. Neurosci.* **2022**, *15*, 916249. [CrossRef] [PubMed]
74. Hennig, B.; Toborek, M.; McClain, C.J. Antiatherogenic properties of zinc: Implications in endothelial cell metabolism. *Nutrition* **1996**, *12*, 711–717. [CrossRef] [PubMed]
75. Grungreiff, K.; Gottstein, T.; Reinhold, D. Zinc Deficiency-An Independent Risk Factor in the Pathogenesis of Haemorrhagic Stroke? *Nutrients* **2020**, *12*, 3548. [CrossRef]
76. Hennig, B.; Wang, Y.; Ramasamy, S.; McClain, C.J. Zinc deficiency alters barrier function of cultured porcine endothelial cells. *J. Nutr.* **1992**, *122*, 1242–1247. [CrossRef]
77. Shuttleworth, C.W.; Weiss, J.H. Zinc: New clues to diverse roles in brain ischemia. *Trends Pharmacol. Sci.* **2011**, *32*, 480–486. [CrossRef]
78. Bu, S.; Lv, Y.; Liu, Y.; Qiao, S.; Wang, H. Zinc Finger Proteins in Neuro-Related Diseases Progression. *Front. Neurosci.* **2021**, *15*, 760567. [CrossRef]
79. Mammadova-Bach, E.; Braun, A. Zinc Homeostasis in Platelet-Related Diseases. *Int. J. Mol. Sci.* **2019**, *20*, 5258. [CrossRef]

80. Zhang, Y.; Khan, S.; Liu, Y.; Siddique, R.; Zhang, R.; Yong, V.W.; Xue, M. Gap Junctions and Hemichannels Composed of Connexins and Pannexins Mediate the Secondary Brain Injury Following Intracerebral Hemorrhage. *Biology* **2021**, *11*, 27. [CrossRef]
81. Weiss, J.H.; Sensi, S.L.; Koh, J.Y. Zn($^{2+}$): A novel ionic mediator of neural injury in brain disease. *Trends Pharmacol. Sci.* **2000**, *21*, 395–401. [CrossRef]
82. Lee, J.M.; Zipfel, G.J.; Park, K.H.; He, Y.Y.; Hsu, C.Y.; Choi, D.W. Zinc translocation accelerates infarction after mild transient focal ischemia. *Neuroscience* **2002**, *115*, 871–878. [CrossRef] [PubMed]
83. Tuo, Q.Z.; Liuyang, Z.Y.; Lei, P.; Yan, X.; Shentu, Y.P.; Liang, J.W.; Zhou, H.; Pei, L.; Xiong, Y.; Hou, T.Y.; et al. Zinc induces CDK5 activation and neuronal death through CDK5-Tyr15 phosphorylation in ischemic stroke. *Cell Death Dis.* **2018**, *9*, 870. [CrossRef] [PubMed]
84. Dong, W.; Qi, Z.; Liang, J.; Shi, W.; Zhao, Y.; Luo, Y.; Ji, X.; Liu, K.J. Reduction of zinc accumulation in mitochondria contributes to decreased cerebral ischemic injury by normobaric hyperoxia treatment in an experimental stroke model. *Exp. Neurol.* **2015**, *272*, 181–189. [CrossRef] [PubMed]
85. Munshi, A.; Babu, S.; Kaul, S.; Shafi, G.; Rajeshwar, K.; Alladi, S.; Jyothy, A. Depletion of serum zinc in ischemic stroke patients. *Methods Find. Exp. Clin. Pharmacol.* **2010**, *32*, 433–436. [CrossRef] [PubMed]
86. Mattern, L.; Chen, C.; McClure, L.A.; Brockman, J.; Cushman, M.; Judd, S.; Kahe, K. Serum Zinc Levels and Incidence of Ischemic Stroke: The Reasons for Geographic and Racial Differences in Stroke Study. *Stroke* **2021**, *52*, 3953–3960. [CrossRef]
87. Howard, V.J.; Madsen, T.E.; Kleindorfer, D.O.; Judd, S.E.; Rhodes, J.D.; Soliman, E.Z.; Kissela, B.M.; Safford, M.M.; Moy, C.S.; McClure, L.A.; et al. Sex and Race Differences in the Association of Incident Ischemic Stroke With Risk Factors. *JAMA Neurol.* **2019**, *76*, 179–186. [CrossRef]
88. Shao, Z.; Tu, S.; Shao, A. Pathophysiological Mechanisms and Potential Therapeutic Targets in Intracerebral Hemorrhage. *Front. Pharmacol.* **2019**, *10*, 1079. [CrossRef]
89. Keep, R.F.; Hua, Y.; Xi, G. Intracerebral haemorrhage: Mechanisms of injury and therapeutic targets. *Lancet Neurol.* **2012**, *11*, 720–731. [CrossRef]
90. Li, Z.; Liu, Y.; Wei, R.; Khan, S.; Xue, M.; Yong, V.W. The combination of deferoxamine and minocycline strengthens neuroprotective effect on acute intracerebral hemorrhage in rats. *Neurol. Res.* **2021**, *43*, 854–864. [CrossRef]
91. Hanley, D.F.; Thompson, R.E.; Rosenblum, M.; Yenokyan, G.; Lane, K.; McBee, N.; Mayo, S.W.; Bistran-Hall, A.J.; Gandhi, D.; Mould, W.A.; et al. Efficacy and safety of minimally invasive surgery with thrombolysis in intracerebral haemorrhage evacuation (MISTIE III): A randomised, controlled, open-label, blinded endpoint phase 3 trial. *Lancet* **2019**, *393*, 1021–1032. [CrossRef]
92. Bai, Q.; Xue, M.; Yong, V.W. Microglia and macrophage phenotypes in intracerebral haemorrhage injury: Therapeutic opportunities. *Brain* **2020**, *143*, 1297–1314. [CrossRef] [PubMed]
93. Liu, Y.; Bai, Q.; Yong, V.W.; Xue, M. EMMPRIN Promotes the Expression of MMP-9 and Exacerbates Neurological Dysfunction in a Mouse Model of Intracerebral Hemorrhage. *Neurochem. Res.* **2022**, *47*, 2383–2395. [CrossRef] [PubMed]
94. Levenson, C.W. Trace metal regulation of neuronal apoptosis: From genes to behavior. *Physiol. Behav.* **2005**, *86*, 399–406. [CrossRef] [PubMed]
95. Zhang, Y.; Khan, S.; Liu, Y.; Wu, G.; Yong, V.W.; Xue, M. Oxidative Stress Following Intracerebral Hemorrhage: From Molecular Mechanisms to Therapeutic Targets. *Front. Immunol.* **2022**, *13*, 847246. [CrossRef]
96. Bai, Q.; Sheng, Z.; Liu, Y.; Zhang, R.; Yong, V.W.; Xue, M. Intracerebral haemorrhage: From clinical settings to animal models. *Stroke Vasc. Neurol.* **2020**, *5*, 388–395. [CrossRef]
97. Bernardo, M.M.; Day, D.E.; Halvorson, H.R.; Olson, S.T.; Shore, J.D. Surface-independent acceleration of factor XII activation by zinc ions. II. Direct binding and fluorescence studies. *J. Biol. Chem.* **1993**, *268*, 12477–12483. [CrossRef]
98. Taylor, K.A.; Pugh, N. The contribution of zinc to platelet behaviour during haemostasis and thrombosis. *Metallomics* **2016**, *8*, 144–155. [CrossRef]
99. Vu, T.T.; Fredenburgh, J.C.; Weitz, J.I. Zinc: An important cofactor in haemostasis and thrombosis. *Thromb. Haemost.* **2013**, *109*, 421–430. [CrossRef]
100. Karadas, S.; Sayin, R.; Aslan, M.; Gonullu, H.; Kati, C.; Dursun, R.; Duran, L.; Gonullu, E.; Demir, H. Serum levels of trace elements and heavy metals in patients with acute hemorrhagic stroke. *J. Membr. Biol.* **2014**, *247*, 175–180. [CrossRef]
101. Zhang, J.; Cao, J.; Zhang, Y.; Li, H.; Zhang, H.; Huo, Y.; Li, J.; Liu, X.; Wang, X.; Qin, X.; et al. Baseline Plasma Zinc and Risk of First Stroke in Hypertensive Patients: A Nested Case-Control Study. *Stroke* **2019**, *50*, 3255–3258. [CrossRef]
102. Arleth, T.; Olsen, M.H.; Orre, M.; Rasmussen, R.; Bache, S.; Eskesen, V.; Frikke-Schmidt, R.; Moller, K. Hypozincaemia is associated with severity of aneurysmal subarachnoid haemorrhage: A retrospective cohort study. *Acta Neurochir. (Wien)* **2020**, *162*, 1417–1424. [CrossRef] [PubMed]
103. Xie, Z.; Wu, H.; Zhao, J. Multifunctional roles of zinc in Alzheimer's disease. *Neurotoxicology* **2020**, *80*, 112–123. [CrossRef] [PubMed]
104. Mangialasche, F.; Solomon, A.; Winblad, B.; Mecocci, P.; Kivipelto, M. Alzheimer's disease: Clinical trials and drug development. *Lancet Neurol.* **2010**, *9*, 702–716. [CrossRef] [PubMed]
105. Hodson, R. Alzheimer's disease. *Nature* **2018**, *559*, S1. [CrossRef]
106. Xu, Y.; Xiao, G.; Liu, L.; Lang, M. Zinc transporters in Alzheimer's disease. *Mol. Brain* **2019**, *12*, 106. [CrossRef]
107. Sensi, S.L.; Granzotto, A.; Siotto, M.; Squitti, R. Copper and Zinc Dysregulation in Alzheimer's Disease. *Trends Pharmacol. Sci.* **2018**, *39*, 1049–1063. [CrossRef]

108. Religa, D.; Strozyk, D.; Cherny, R.A.; Volitakis, I.; Haroutunian, V.; Winblad, B.; Naslund, J.; Bush, A.I. Elevated cortical zinc in Alzheimer disease. *Neurology* **2006**, *67*, 69–75. [CrossRef]
109. Lyubartseva, G.; Smith, J.L.; Markesbery, W.R.; Lovell, M.A. Alterations of zinc transporter proteins ZnT^{-1}, ZnT^{-4} and ZnT^{-6} in preclinical Alzheimer's disease brain. *Brain Pathol.* **2010**, *20*, 343–350. [CrossRef]
110. Zhang, L.H.; Wang, X.; Zheng, Z.H.; Ren, H.; Stoltenberg, M.; Danscher, G.; Huang, L.; Rong, M.; Wang, Z.Y. Altered expression and distribution of zinc transporters in APP/PS1 transgenic mouse brain. *Neurobiol. Aging* **2010**, *31*, 74–87. [CrossRef]
111. Adlard, P.A.; Parncutt, J.M.; Finkelstein, D.I.; Bush, A.I. Cognitive loss in zinc transporter-3 knock-out mice: A phenocopy for the synaptic and memory deficits of Alzheimer's disease? *J. Neurosci.* **2010**, *30*, 1631–1636. [CrossRef]
112. Lovell, M.A. A potential role for alterations of zinc and zinc transport proteins in the progression of Alzheimer's disease. *J. Alzheimers Dis.* **2009**, *16*, 471–483. [CrossRef] [PubMed]
113. Bush, A.I.; Pettingell, W.H.; Multhaup, G.; d Paradis, M.; Vonsattel, J.P.; Gusella, J.F.; Beyreuther, K.; Masters, C.L.; Tanzi, R.E. Rapid induction of Alzheimer A beta amyloid formation by zinc. *Science* **1994**, *265*, 1464–1467. [CrossRef] [PubMed]
114. Zhang, L.H.; Wang, X.; Stoltenberg, M.; Danscher, G.; Huang, L.; Wang, Z.Y. Abundant expression of zinc transporters in the amyloid plaques of Alzheimer's disease brain. *Brain Res. Bull.* **2008**, *77*, 55–60. [CrossRef]
115. 2020 Alzheimer's disease facts and figures. *Alzheimers Deme.* **2020**, *16*, 391–460. [CrossRef] [PubMed]
116. Craven, K.M.; Kochen, W.R.; Hernandez, C.M.; Flinn, J.M. Zinc Exacerbates Tau Pathology in a Tau Mouse Model. *J. Alzheimers Dis.* **2018**, *64*, 617–630. [CrossRef] [PubMed]
117. Kim, I.; Park, E.J.; Seo, J.; Ko, S.J.; Lee, J.; Kim, C.H. Zinc stimulates tau S214 phosphorylation by the activation of Raf/mitogen-activated protein kinase-kinase/extracellular signal-regulated kinase pathway. *Neuroreport* **2011**, *22*, 839–844. [CrossRef]
118. Wang, L.; Yin, Y.L.; Liu, X.Z.; Shen, P.; Zheng, Y.G.; Lan, X.R.; Lu, C.B.; Wang, J.Z. Current understanding of metal ions in the pathogenesis of Alzheimer's disease. *Transl. Neurodegener.* **2020**, *9*, 10. [CrossRef]
119. Roozenbeek, B.; Maas, A.I.; Menon, D.K. Changing patterns in the epidemiology of traumatic brain injury. *Nat. Rev. Neurol.* **2013**, *9*, 231–236. [CrossRef]
120. Selassie, A.W.; Zaloshnja, E.; Langlois, J.A.; Miller, T.; Jones, P.; Steiner, C. Incidence of long-term disability following traumatic brain injury hospitalization, United States, 2003. *J. Head Trauma Rehabil.* **2008**, *23*, 123–131. [CrossRef]
121. Salmond, C.H.; Sahakian, B.J. Cognitive outcome in traumatic brain injury survivors. *Curr. Opin. Crit. Care* **2005**, *11*, 111–116. [CrossRef]
122. Capizzi, A.; Woo, J.; Verduzco-Gutierrez, M. Traumatic Brain Injury: An Overview of Epidemiology, Pathophysiology, and Medical Management. *Med. Clin. N. Am.* **2020**, *104*, 213–238. [CrossRef] [PubMed]
123. Sandsmark, D.K.; Diaz-Arrastia, R. Advances in traumatic brain injury research in 2020. *Lancet Neurol.* **2021**, *20*, 5–7. [CrossRef] [PubMed]
124. Suh, S.W.; Listiack, K.; Bell, B.; Chen, J.; Motamedi, M.; Silva, D.; Danscher, G.; Whetsell, W.; Thompson, R.; Frederickson, C. Detection of pathological zinc accumulation in neurons: Methods for autopsy, biopsy, and cultured tissue. *J. Histochem. Cytochem.* **1999**, *47*, 969–972. [CrossRef] [PubMed]
125. Choi, B.Y.; Kim, J.H.; Kim, H.J.; Lee, B.E.; Kim, I.Y.; Sohn, M.; Suh, S.W. Zinc chelation reduces traumatic brain injury-induced neurogenesis in the subgranular zone of the hippocampal dentate gyrus. *J. Trace Elem. Med. Biol.* **2014**, *28*, 474–481. [CrossRef] [PubMed]
126. Zhao, L.; Liu, Q.; Ma, S.; Zhang, Y.; Liang, P. TPEN Attenuates Neural Autophagy Induced by Synaptically-released Zinc Translocation and Improves Histological Outcomes after Traumatic Brain Injury in Rats. *Ann. Clin. Lab. Sci.* **2018**, *48*, 446–452.
127. Dominguez, M.I.; Blasco-Ibanez, J.M.; Crespo, C.; Marques-Mari, A.I.; Martinez-Guijarro, F.J. Zinc chelation during non-lesioning overexcitation results in neuronal death in the mouse hippocampus. *Neuroscience* **2003**, *116*, 791–806. [CrossRef]
128. Corniola, R.S.; Tassabehji, N.M.; Hare, J.; Sharma, G.; Levenson, C.W. Zinc deficiency impairs neuronal precursor cell proliferation and induces apoptosis via p53-mediated mechanisms. *Brain Res.* **2008**, *1237*, 52–61. [CrossRef]
129. Seth, R.; Corniola, R.S.; Gower-Winter, S.D.; Morgan, T.J., Jr.; Bishop, B.; Levenson, C.W. Zinc deficiency induces apoptosis via mitochondrial p53- and caspase-dependent pathways in human neuronal precursor cells. *J. Trace Elem. Med. Biol.* **2015**, *30*, 59–65. [CrossRef]
130. Gao, H.L.; Zheng, W.; Xin, N.; Chi, Z.H.; Wang, Z.Y.; Chen, J.; Wang, Z.Y. Zinc deficiency reduces neurogenesis accompanied by neuronal apoptosis through caspase-dependent and -independent signaling pathways. *Neurotox. Res.* **2009**, *16*, 416–425. [CrossRef]
131. Young, B.; Ott, L.; Kasarskis, E.; Rapp, R.; Moles, K.; Dempsey, R.J.; Tibbs, P.A.; Kryscio, R.; McClain, C. Zinc supplementation is associated with improved neurologic recovery rate and visceral protein levels of patients with severe closed head injury. *J. Neurotrauma* **1996**, *13*, 25–34. [CrossRef]
132. Cope, E.C.; Morris, D.R.; Scrimgeour, A.G.; Levenson, C.W. Use of zinc as a treatment for traumatic brain injury in the rat: Effects on cognitive and behavioral outcomes. *Neurorehabil. Neural Repair* **2012**, *26*, 907–913. [CrossRef] [PubMed]
133. Manford, M. Recent advances in epilepsy. *J. Neurol.* **2017**, *264*, 1811–1824. [CrossRef] [PubMed]
134. Fisher, R.S.; Cross, J.H.; French, J.A.; Higurashi, N.; Hirsch, E.; Jansen, F.E.; Lagae, L.; Moshe, S.L.; Peltola, J.; Roulet Perez, E.; et al. Operational classification of seizure types by the International League Against Epilepsy: Position Paper of the ILAE Commission for Classification and Terminology. *Epilepsia* **2017**, *58*, 522–530. [CrossRef] [PubMed]

135. Pathak, H.R.; Weissinger, F.; Terunuma, M.; Carlson, G.C.; Hsu, F.C.; Moss, S.J.; Coulter, D.A. Disrupted dentate granule cell chloride regulation enhances synaptic excitability during development of temporal lobe epilepsy. *J. Neurosci.* **2007**, *27*, 14012–14022. [CrossRef]
136. Gower-Winter, S.D.; Levenson, C.W. Zinc in the central nervous system: From molecules to behavior. *Biofactors* **2012**, *38*, 186–193. [CrossRef]
137. Harrison, B.K.; Asplund, C. Sudden unexplained death in epilepsy during physical activity. *Curr. Sports Med. Rep.* **2007**, *6*, 13–15. [CrossRef]
138. Baraka, A.M.; Hassab El Nabi, W.; El Ghotni, S. Investigating the role of zinc in a rat model of epilepsy. *CNS Neurosci. Ther.* **2012**, *18*, 327–333. [CrossRef]
139. Chen, N.N.; Zhao, D.J.; Sun, Y.X.; Wang, D.D.; Ni, H. Long-Term Effects of Zinc Deficiency and Zinc Supplementation on Developmental Seizure-Induced Brain Damage and the Underlying GPR39/ZnT-3 and MBP Expression in the Hippocampus. *Front. Neurosci.* **2019**, *13*, 920. [CrossRef]
140. Foresti, M.L.; Arisi, G.M.; Fernandes, A.; Tilelli, C.Q.; Garcia-Cairasco, N. Chelatable zinc modulates excitability and seizure duration in the amygdala rapid kindling model. *Epilepsy Res.* **2008**, *79*, 166–172. [CrossRef]
141. Choi, B.Y.; Kim, I.Y.; Kim, J.H.; Lee, B.E.; Lee, S.H.; Kho, A.R.; Sohn, M.; Suh, S.W. Zinc plus cyclo-(His-Pro) promotes hippocampal neurogenesis in rats. *Neuroscience* **2016**, *339*, 634–643. [CrossRef]
142. Seven, M.; Basaran, S.Y.; Cengiz, M.; Unal, S.; Yuksel, A. Deficiency of selenium and zinc as a causative factor for idiopathic intractable epilepsy. *Epilepsy Res.* **2013**, *104*, 35–39. [CrossRef] [PubMed]
143. Dubovsky, S.L.; Ghosh, B.M.; Serotte, J.C.; Cranwell, V. Psychotic Depression: Diagnosis, Differential Diagnosis, and Treatment. *Psychother. Psychosom.* **2021**, *90*, 160–177. [CrossRef] [PubMed]
144. McCarron, R.M.; Shapiro, B.; Rawles, J.; Luo, J. Depression. *Ann. Intern. Med.* **2021**, *174*, ITC65–ITC80. [CrossRef] [PubMed]
145. Hammen, C. Risk Factors for Depression: An Autobiographical Review. *Annu. Rev. Clin. Psychol.* **2018**, *14*, 1–28. [CrossRef] [PubMed]
146. Fox, M.E.; Lobo, M.K. The molecular and cellular mechanisms of depression: A focus on reward circuitry. *Mol. Psychiatr.* **2019**, *24*, 1798–1815. [CrossRef]
147. Szewczyk, B.; Kubera, M.; Nowak, G. The role of zinc in neurodegenerative inflammatory pathways in depression. *Prog. Neuropsychopharmacol. Biol. Psychiatr.* **2011**, *35*, 693–701. [CrossRef]
148. Takeda, A. Zinc signaling in the hippocampus and its relation to pathogenesis of depression. *Mol. Neurobiol.* **2011**, *44*, 166–174. [CrossRef]
149. Little, K.Y.; Castellanos, X.; Humphries, L.L.; Austin, J. Altered zinc metabolism in mood disorder patients. *Biol. Psychiatr.* **1989**, *26*, 646–648. [CrossRef]
150. Tamano, H.; Kan, F.; Kawamura, M.; Oku, N.; Takeda, A. Behavior in the forced swim test and neurochemical changes in the hippocampus in young rats after 2-week zinc deprivation. *Neurochem. Int.* **2009**, *55*, 536–541. [CrossRef]
151. Tassabehji, N.M.; Corniola, R.S.; Alshingiti, A.; Levenson, C.W. Zinc deficiency induces depression-like symptoms in adult rats. *Physiol. Behav.* **2008**, *95*, 365–369. [CrossRef]
152. Whittle, N.; Lubec, G.; Singewald, N. Zinc deficiency induces enhanced depression-like behaviour and altered limbic activation reversed by antidepressant treatment in mice. *Amino Acids* **2009**, *36*, 147–158. [CrossRef] [PubMed]
153. Rafalo-Ulinska, A.; Pochwat, B.; Misztak, P.; Bugno, R.; Kryczyk-Poprawa, A.; Opoka, W.; Muszynska, B.; Poleszak, E.; Nowak, G.; Szewczyk, B. Zinc Deficiency Blunts the Effectiveness of Antidepressants in the Olfactory Bulbectomy Model of Depression in Rats. *Nutrients* **2022**, *14*, 2746. [CrossRef] [PubMed]
154. Swardfager, W.; Herrmann, N.; Mazereeuw, G.; Goldberger, K.; Harimoto, T.; Lanctot, K.L. Zinc in depression: A meta-analysis. *Biol. Psychiatr.* **2013**, *74*, 872–878. [CrossRef] [PubMed]
155. Stanislawska, M.; Szkup-Jablonska, M.; Jurczak, A.; Wieder-Huszla, S.; Samochowiec, A.; Jasiewicz, A.; Nocen, I.; Augustyniuk, K.; Brodowska, A.; Karakiewicz, B.; et al. The severity of depressive symptoms vs. serum Mg and Zn levels in postmenopausal women. *Biol. Trace Elem. Res.* **2014**, *157*, 30–35. [CrossRef] [PubMed]
156. Amani, R.; Saeidi, S.; Nazari, Z.; Nematpour, S. Correlation between dietary zinc intakes and its serum levels with depression scales in young female students. *Biol. Trace Elem. Res.* **2010**, *137*, 150–158. [CrossRef]
157. Roozbeh, J.; Sharifian, M.; Ghanizadeh, A.; Sahraian, A.; Sagheb, M.M.; Shabani, S.; Hamidian Jahromi, A.; Kashfi, M.; Afshariani, R. Association of zinc deficiency and depression in the patients with end-stage renal disease on hemodialysis. *J. Ren. Nutr.* **2011**, *21*, 184–187. [CrossRef]
158. Maserejian, N.N.; Hall, S.A.; McKinlay, J.B. Low dietary or supplemental zinc is associated with depression symptoms among women, but not men, in a population-based epidemiological survey. *J. Affect. Disord.* **2012**, *136*, 781–788. [CrossRef]

biomolecules

MDPI

Review

Zinc in Human Health and Infectious Diseases

Martina Maywald and Lothar Rink *

Institute of Immunology, Faculty of Medicine, RWTH Aachen University Hospital, 52074 Aachen, Germany
* Correspondence: LRink@ukaachen.de; Tel.: +49-241-8080-208

Abstract: During the last few decades, the micronutrient zinc has proven to be an important metal ion for a well-functioning immune system, and thus also for a suitable immune defense. Nowadays, it is known that the main cause of zinc deficiency is malnutrition. In particular, vulnerable populations, such as the elderly in Western countries and children in developing countries, are often affected. However, sufficient zinc intake and homeostasis is essential for a healthy life, as it is known that zinc deficiency is associated with a multitude of immune disorders such as metabolic and chronic diseases, as well as infectious diseases such as respiratory infections, malaria, HIV, or tuberculosis. Moreover, the modulation of the proinflammatory immune response and oxidative stress is well described. The anti-inflammatory and antioxidant properties of zinc have been known for a long time, but are not comprehensively researched and understood yet. Therefore, this review highlights the current molecular mechanisms underlying the development of a pro-/ and anti-inflammatory immune response as a result of zinc deficiency and zinc supplementation. Additionally, we emphasize the potential of zinc as a preventive and therapeutic agent, alone or in combination with other strategies, that could ameliorate infectious diseases.

Keywords: zinc; zinc homeostasis; zinc transporter; infectious diseases; pro- and anti-inflammatory signaling pathways

1. Introduction

The essential trace element zinc is extraordinarily important for a well-functioning immune system and for human health in general. Zinc has a key role in several biological processes of the human body, such as influencing cell differentiation, proliferation, and apoptosis, which affect the growth of the organism [1]. The importance of zinc quickly becomes apparent when one considers bioinformatics studies that have located nearly 3000 human proteins that are presumed to bind zinc [2]. Of these, zinc fingers and zinc finger-containing domains in particular require the divalent cation as a stabilizer. Others need zinc for catalytic and regulatory functions, as it is needed for the activity of several metalloenzymes [2,3]. It is well established that zinc is the second most abundant micronutrient in the human body after iron [4,5]. Since the first discovery of zinc being important for human health [6], much research has been performed. Nowadays it is known that zinc, and especially zinc deficiency, plays an important role in many metabolic and chronic diseases, including: diabetes, cancer (e.g., esophageal, hepatocellular, breast cancer, colon cancer), neurodegenerative diseases such as Alzheimer's disease, and intestinal diseases such as inflammatory bowel disease and irritable bowel syndrome [7–13]. There is also strong evidence between zinc deficiency and several infectious diseases such as malaria, HIV, tuberculosis, measles, and respiratory infections such as COVID-19 pneumonia [14–16]. Consequently, zinc's myriad of functions in the human immune system demonstrates how a defect in nutritional absorption may lead to the manifestation of various diseases. Nevertheless, both zinc deficiency and oversupply affect homeostasis, and consequently, cellular function. Zinc intoxication is a rare event; however, it was seen by using, e.g., denture adhesive creams that contain high amounts of zinc. Here, high amounts of zinc were reported to cause nausea, dizziness, headaches, gastric distress, vomiting, and loss of appetite [17,18]. If used for weeks, doses of 50 mg zinc or more can interfere with copper absorption, leading to a low copper status, which consequently impairs immune function [16–19]. Hence, zinc homeostasis has to be carefully monitored and regulated.

This review highlights the particularly important role of zinc for a well-functioning immune system and its versatile influence on various signaling pathways of the pro- and anti-inflammatory immune response. On the one hand, it is shown how inadequate nutrient intake can lead to the occurrence of various diseases caused by harmful pathogens such as bacteria, fungi, or viruses. On the other hand, the impact of zinc supplementation strategies that can lead to faster remission or alleviation of disease symptoms is high-lighted, illustrating the therapeutic benefits of zinc in various infectious diseases.

Search strategy: A comprehensive search of the databases Web of Science, PubMed, and Cochrane Library was conducted. The search was conducted using a combination of keywords including (1) "zinc" OR "zinc supplementation", AND (2) "infectious diseases", "viral disease" "bacterial diseases", "pneumonia", "common cold", "hepatitis", "HIV", "dengue fever", "COVID-19", "tuberculosis", "shigellosis", "H. pylori", "diarrhea", "malaria", "leishmaniosis", "fungi", or "fungal diseases", and related terms. We also manually searched references of relevant literature to identify other eligible sources.

2. Assessment of Zinc Status

Nowadays, there are a variety of methods to determine zinc status, which differ in terms of whether the question is scientific or diagnostic. Diagnostic zinc concentrations are determined as standard in whole blood, serum, plasma, or urine. For this purpose, the methods of atomic absorption spectroscopy (AAS, flame or graphite furnace), inductively coupled plasma optical (atomic) emission spectrometers (ICP-OES/ICP-AES), and ICP mass spectrometers (ICP-MS) are used often, providing excellent sensitivity, precision, and good accuracy with virtually no restriction of the element investigated [20,21].

The scientific approach is far more complex, as both the distribution and dynamics of zinc are important for biological processes that need to be specifically analyzed. Nowa-days, a growing number of fluorescent small-molecule probes and protein-based sensors exist that enable the measurement of subcellular zinc distribution in different organelles. Most of the available tools can be divided into two general classes of probes: small-molecule probes and protein sensors based on Förster resonance energy transfer (FRET). Small-molecule probes usually increase in fluorescence upon chelation of zinc, as known for the indicators Zinpyr-1, FluoZin-3, or Zinquin. They can be cell-permeable, and are therefore easy to apply. They provide a rapid means of investigating metal/protein equilibria by optical-sensing methods with both high sensitivity and very little sample preparation. They have been adapted to provide ratiometric signals that allow for the normalization of fluorescence changes that are not due to the chelation of zinc [22,23]. FRET consists of two fluorescent proteins and a zinc-coordinating site that is designed to change the relative orientation and distance of the fluorescent proteins upon binding to a target, which leads to a change in FRET signal. The sensors are genetically encoded and can be targeted to different organelles, as known for the Zap, eZinCh, or eCALWY series.

3. Zinc Intake and Distribution

Zinc needs to be properly distributed into all compartments of the human body to fulfill a multiplicity of functions. It is undisputed that there must exist a complex process for homeostasis and to ensure proper zinc allocation, because the zinc content between various organs differs greatly. Prostate, pancreas, and bone are considerably high in zinc, containing about 100 to 250 μg/g. Zinc concentrations in heart, brain, and plasma are comparatively low, at 1 to 23 μg/g. Although plasma only has a zinc content of about 1 μg/g, it is probably the most important reservoir for zinc homeostasis [24].

In total, the body of an adult human contains about 2.6 g of zinc. The largest proportion is found in bones and skeletal muscles (~86%), followed by skin (4.2%) and liver (3.4%). It should be noted that some zinc-containing entities, including the thymus and mucous membranes, were not included in the calculation, so the actual total zinc content of the human body is higher than the 2.6 g estimated [25,26]. Within body fluids, free zinc is rare, since it is predominantly bound to proteins, including albumin, α2 macroglobulin (A2M), and transferrin. Albumin binds zinc with a relatively low affinity, A2M with a medium affinity, and transferrin with a high affinity. However, although albumin's binding affinity is low, hypoalbuminemia can even result in zinc deficiency [27].

Intracellularly, zinc is distributed mainly between specific zinc-storage structures called zincosomes (~50%) and the nucleus (~30–40%). The remaining zinc is distributed between the cytoplasm and other cell organelles [28,29]. In the cytoplasm, zinc is largely bound by zinc-chelating proteins called metallothioneins (MT). Those were first discovered in 1957 [30] and play an important role in zinc homeostasis by complexing approximately 20% of intracellular zinc acting as zinc buffer [31]. The availability of cytosolic zinc is mediated by zinc storage or release from organelles (e.g., zincosomes, endoplasmic reticulum (ER), or Golgi apparatus). Zincosomes are vesicle-like storage granules for zinc that are typically visualized with zinc-binding probes Zinquin or FluoZin-3 [32,33]. These granules have different functions depending on the cell type, such as zinc loading of milk, neuron function, and insulin secretion. Influx into the cell or efflux into the extracellular space is another important aspect in zinc availability [34], which is mediated by specific zinc-transporting proteins, such as zinc importers or exporters, and membrane channels, as described later [31,35].

During evolution, plant-, animal-, and human-based cells have developed metal-binding proteins to maintain low cytosolic metal ion concentrations and thus protect themselves from high cytotoxic metal concentrations. Interestingly, the human body evolved mechanisms to compartmentalize and sequester zinc, for example, to deny vital zinc to invading microorganisms. In contrast, intracellular release of high zinc amounts, for instance from proteins and organelles, may be beneficial and desirable because transiently and locally high zinc levels also target gene expression, enzymatic activity, and cell signaling of pathogens, leading to zinc intoxication [35,36].

To maintain healthy zinc homeostasis, the body's own zinc loss needs to be compensated. Since there is no dedicated compartment for zinc storage in the human body, the trace element must be ingested daily in sufficient quantities with food [37]. Zinc loss can be of intestinal and nonintestinal origin, including fecal zinc excretions and excretions with urine, sweat, menstrual flow, and semen, as well as loss of hair, nails, and desquamated skin [38]. In addition, special life circumstances must be taken into account in which the presumed zinc loss must be individually adjusted, such as pregnancy, lactation, or even early infancy or young children in growth [38].

Zinc homeostasis in the human body is primarily regulated by absorption in the intestine, particularly in the duodenum and jejunum [39–41]. In this process, zinc transporters on the apical and basolateral membrane of enterocytes and the zinc-binding proteins (metallothioneins) regulate zinc homeostasis in the intestine and eventually in the human body [3,42]. Zinc absorption not only depends on adequate dietary intake, but is also strongly influenced by intestinal availability from food, which means that zinc bioavailability depends on dietary composition [43,44]. So far, it is known that foods prepared from unrefined cereals or legumes, or foods containing indigestible plant ligands such as phytate and lignin, form an insoluble complex with zinc, making zinc unavailable for resorption in the intestine. Lignin is important for plant structure and resilience, whereas phytate is the storage form of phosphorus in plants that binds and stores minerals. Hence, both phytate and lignin reduce the bioavailability of zinc because they bind zinc very effectively and inhibit its absorption, respectively [17,18,45,46]. Besides plant products or components, zinc absorption is also influenced by additional factors such as other trace element concentrations such as copper or iron, or mineral concentrations such as calcium, as mentioned later.

The highest zinc content is found in the following foods: red meat, offal, oysters and shellfish, fortified cereals, and whole-grain products [17,46]. Oysters contain more zinc per serving than any other food. However, in Western countries, beef contributes 20% of the zinc intakes from food because it is commonly consumed [47]. Eggs and dairy products also contain zinc [17]. Beans, nuts, and whole grains contain zinc, but the bioavailability is much lower because plant-based foods contain high amounts of phytates [17,18,46]. In summary, zinc is present in a wide variety of food classes, but its availability to the human body varies greatly. In general, zinc from animal sources has a higher bioavailability compared to zinc from plant products. People who consciously avoid eating red meat and eat mainly plant-based foods, such as vegetarians and vegans, but also people who often have to abstain from meat, such as the population living in developing countries, have a higher risk of developing zinc deficiency due to insufficient zinc intake [48].

The recommended dietary allowance (RDA) for daily zinc intake depends on various factors, such as age, gender, weight, and phytate content of the diet. The recommended levels vary slightly from country to country: the National Institute of Health recommends an intake of 11 mg/day for adult men and 8 mg/day for adult women [18]. During pregnancy and lactation 11 mg/day to 12 mg/day are recommended. For infants and children the RDA is 3 mg/day for both sexes [18]. The German Nutrition Society (DGE) recommends 11–16 mg/day for adult men and 7–10 mg/day for adult women, depending on phythate intake. During pregnancy and lactation up to 13–14 mg/day are recommended. For infants and children the RDA is 2.5 mg/day for both sexes [49]. Due to the inhibitory effect of phytate on zinc absorption from food, phytate levels are taken into account by both the World Health Organization (WHO) and the European Food Safety Authority (EFSA) when establishing recommendations for daily zinc intake. However, there are different classifications of the WHO and the EFSA: WHO divides diets into three groups according to their potential absorption capacity of zinc per phytate-zinc molar ratio: high (<5), moderate (5–15), and low (>15) zinc bioavailability [50]. EFSA and (DGE) provides several zinc reference recommendations for diets with phytate intakes of about 300–(330), 600–(660), 900–(990), and 1200 mg/day [49,51]. Oral zinc supplements are readily available, but not all offer the same zinc bioavailability. Zinc bound to amino acids such as aspartate, cysteine, and histidine shows the highest absorption concentration, followed by zinc chloride, sulfate, and acetate, whereas zinc oxide shows the lowest bioavailability [52].

The importance of zinc homeostasis is underlined by the extraordinarily high number of diseases associated with changes in zinc content in human organism. Most of these diseases are associated with zinc deficiency [24,53,54]. About two billion people worldwide suffer from zinc deficiency, which has serious consequences for their health, affecting every solid organ in the human organism as well as the immune system [6,24,55]. Nowadays, it is known that zinc deficiency is very common worldwide, especially in developing countries. Here, it is the fifth leading cause of loss of healthy life years [56]. In Western countries, the elderly population is particularly affected by zinc deficiency, as nearly 30% of them are considered to be zinc-deficient [24,57]. Moreover, vegetarian or vegan people are also concerned [58,59]. In general, the WHO estimates that about one-third of the world's population is at risk for zinc deficiency [56]. Globally, it is responsible for approximately 16% of lower respiratory tract infections, 18% of malaria, and 10% of diarrheal diseases. While severe zinc deficiency is rare, mild-to-moderate deficiency is more common worldwide [54,56]. However, the lack of a suitable biomarker for physiological zinc status, and thus the poor ability to detect inadequate zinc absorption, especially in the early stages of mild zinc deficiency, is a major obstacle in this situation [48].

4. Regulation of Zinc Homeostasis

The distribution of zinc between the cytosol and cell organelles is mediated by specific zinc-transporting proteins, such as zinc importers or exporters, or by membrane channels. The mammalian zinc transporters belong to two major protein families, which are distinguished according to their specific membrane topology: Two major zinc transporter families, the SLC39 (ZIP) and SLC30 (ZnT) family, control cellular zinc homeostasis.

The first group includes 14 zinc transporters of the SLC39As family Zrt (zinc-regulated transporter)-like and Irt (iron-regulated transporter)-like proteins, called ZIP transporters. ZIPs 1-14 (SLC39A1-14) are abundantly located at the outer plasma membrane and at intracellular organelles and are responsible for transporting zinc both from the extracellular space and from intracellular compartments into the cytosol to increase cytoplasmic zinc.

The second group includes 10 zinc transporters of the SLC30As family (ZnT 1–10 (SLC30A1-10)) that decrease cytoplasmic zinc by transporting zinc either out of the cell into the extracellular space or into intracellular organelles such as Golgi apparatus, or zincosomes [60–62]. Both ZIP and ZnT transporters are expressed in a tissue-specific manner and are expressed in a variety of different cell types in the human body. They are localized to the outer plasma membrane and to organelle membranes such as mitochondria, Golgi apparatus, lysosomes, and ER. In addition, they show developmental and stimulus-dependent expression patterns, highlighting the need for adequate zinc levels for specific cellular functions [1,62].

More recently, zinc distribution has also been described in the context of receptors and ion channels other than ZIP or ZnT. These include, for example, nicotinic acetylcholine receptors, glutamatergic receptors, voltage-gated calcium channels, and transient receptor potential channels. Amino acid-bound zinc diffusion is also known, but these mechanisms remain to be further defined [1].

Interestingly, there are many zinc transporter mutations reported, to be involved in inherited diseases:

(1) Zinc importer: Mutation in the intestinal ZIP4 transporter causes the lethal autosomal-recessive inherited zinc deficiency disease, acrodermatitis enteropathica (AE). AE is characterized amongst other symptoms by gastrointestinal disturbances, retardation, and high susceptibility to infections. A complete cure can be achieved through high-dose zinc supplementation [63]. Mutation in ZIP 7 is associated with Agammaglobulinemia due to impaired B cell function and proliferation. Mutation in ZIP13 cause a novel subtype of Ehlers–Danlos Syndrome (EDS), where collagen synthesis and modification is impaired [64]. ZIP 14 mutation is associated with hypermanganism and Parkinson syndrome [65]. Additionally, various ZIP transporter modifications are addressed to different cancer types, such as ZIP 6 (esophageal squamous cell carcinoma and gastric adenocaricinoma), ZIP 9 (prostate cancer and breast cancer), and ZIP 12 (colorectal cancer and bladder cancer) [66].

(2) Zinc exporter: Mutation in ZnT2 causes a defect in zinc secretion, which leads to a reduced zinc content in the breast milk during lactation. Consequently, infants that are exclusively breast-fed develop a zinc deficiency, which can be alleviated by oral zinc supplementation [67].

Zinc homeostasis is furthermore controlled by metallothioneins (MT), which are cysteine-rich 6–7 kDa proteins that bind metal ions such as zinc [68]. So far, MTs 1–4 are known in humans to play a coordinated role in the transport, distribution, and maintenance of intracellular zinc levels [68,69]. Up to 20% of intracellular zinc is bound to MTs and can be rapidly released. MTs are known to modulate three fundamental processes: (1) release of gaseous mediators (e.g., nitric oxide); (2) apoptosis, and (3) binding/exchange of heavy metals such as zinc, cadmium, or copper. Their main functions include, amongst others, detoxification of heavy metals such as mercury and cadmium; homeostasis of essential metals, including copper and zinc; antioxidation against reactive oxygen species (ROS); protection against DNA damage; and modulating proliferation and apoptosis. Due to their ROS-scavenging properties, they help to protect against several types of environmental stress. MT-1 and MT-2 are ubiquitously expressed throughout the human body, especially in the kidney, liver, intestine, and pancreas, and are involved in heavy metal detoxification, immune function, and in a variety of G.I. tract functions [69–71]. MT-3 and MT-4 are expressed more specifically: MT-3 is mainly present in the nervous system, but also in heart, retina, kidney, breast, prostate, bladder, and reproductive organs, and is involved in the development, organization, and apoptosis of brain cells. MT-4 is found predominantly in the stratified squamous epithelium, e.g., the upper G.I. tract, where it helps to regulate stomach acid pH, taste, and texture discrimination of the tongue [71]. MT-1 and MT-2 are inducible by various stimuli such as metal ions, proinflammatory cytokines, such as interleukin (IL)-1, IL-6, interferon (IFN)-α, tumor necrosis factor (TNF)-α, glucocorticoids, and oxidative stress, whereas the expression of both MT-3 and MT-4 is very strictly controlled [69,70,72,73].

Besides intracellularly, MTs have been also found to be released in a very small amount to the extracellular environment, like it has been reported for serum, urine, bronchoalveolar spaces, liver sinusoids, and inflammatory lesions. Although MTs are present at very low levels in the extracellular environment, they support the beneficial movement of leukocytes to the site of inflammation by providing a "danger signal" to the cells and alter the character of the immune response when cells perceive cellular stress [74]. Regarding inflammation, MTs play a critical role in activating immune cells by increasing the intracellular concentration of free zinc ions and modulating proinflammatory cytokines on the one hand [55]. On the other hand, MTs themselves are synthesized in response to proinflammatory acute phase cytokines released at sites of inflammation. Glucocorticoids, a signal often associated with a stressful environment, also trigger MT expression. In addition, MTs can be induced by ROS, endotoxins, and divalent metal cations, such as zinc [74]. With all these different triggers, it is not surprising that increased MT levels are found in association with neoplastic diseases [75], autoimmune diseases [76], chronic inflammation [77], and infections [78,79]. Furthermore, it is described that MTs have immunomodulatory activities, such as reducing T-cell-dependent humoral responses, altering the proliferative capacity of lymphocytes, and decreasing cytotoxic T-cell function and macrophage effector function.

The various expression patterns and different functions illustrate the importance of MT isoforms for distinct organ and tissue function and clearly suggest a strong interplay between MTs and the immune system. Hence, insufficient expression of MTs in the context of inflammatory diseases can dramatically shorten lifespan [74].

In addition to MTs, members of the S100 protein family can also bind intracellular zinc, hence regulating zinc homeostasis [80]. All 20 S100 family proteins act in a calcium-dependent manner and have calcium-buffering capacities, whereas some proteins are also regulated by zinc and several members show higher affinity towards zinc compared to calcium, such as S100A3 [81]. Zinc-binding S100 proteins include S100B, S100A1, S100A2, S100A3, S100A5, S100A7, S100A8/9, S100A12, and S100A16. Of these, calprotectin and calgranulin (S100A8/9 and S100A12) play a crucial role in nutritional immunity, as mentioned later [82]. S100 family proteins are a broad subfamily of low-molecular-weight calcium-binding proteins (9–14 kDa) with structural similarity distributed in a cell-specific, tissue-specific, and cell-cycle-specific manner, and having functional discrepancy. Their multitude of functions range from calcium buffering, modulating enzyme activity and enzyme secretions, modulating apoptosis and transcription, proliferation and differentiation, cellular chemotaxis, and much more [82,83]. The family of S100 proteins is a critical connecting link in innate immunity. It facilitates the immune response cascade through direct participation, and provides host defense mechanisms by triggering immunological responses against invading pathogens.

Zinc sequestration is furthermore facilitated by α2-macroglobulin (A2M), which is an inhibitor of matrix metalloproteases (MMPs). A2M has a very high affinity to zinc, and zinc is needed for A2M activation and binding to cytokines [58]. A2M expression is triggered due to the expression of proinflammatory acute-phase proteins, including IL-6. Hence, zinc-binding peptides such as MTs and A2M are upregulated during inflammation, facilitating a reduction in zinc availability for invading pathogens [84]. This mechanism is beneficial during acute immune response; however, a long-term decrease in zinc availability may contribute to pathological processes in conditions of chronic inflammation, like it is seen in diabetes or dementia [11–13]. Moreover, high IL-6 expression and consequently upregulated MT and A2M expression is linked to immunosenescence and increased risk of Alzheimer's disease [58].

5. Zinc and Nutritional Immunity

Antimicrobial peptides and/or proteins (AMPs) play an essential role in the first line of defense against a wide range of pathogens [85]. In humans, many AMPs, bactericidal factors, and host defense peptides exist [82,85]. For instance, calcium-dependent proteins S100A7 (psoriasin), S100A15, the S100A8/A9 heterodimer (calprotectin), and calgranulins (calprotectin and S100A12) belong to the AMPs [82].

Calgranulins, in particular, take advantage of this mesmerizing property, and inhibit microorganism growth by essential-nutrient deprivation, thereby limiting pathogen invasion and growth [86,87].

Innate immune cells, such as neutrophil granulocytes, are the first line of cell-based immunological defense and are well known for calprotectin expression. Calprotectin is the most abundant protein in neutrophil granulocytes and function as antimicrobial protein via zinc-chelating capacity, which causes zinc depletion through sequestration. Hence, pathogens such as *Candida albicans*, *Acinetobacter baumannii*, *Klebsiella pneumoniae*, *Helicobacter pylori*, *Escherichia coli*, and *Staphylococcus aureus* are in poor condition, leading to so-called nutritional immunity [82,88]. Additionally, neutrophils are able to release neutrophil extracellular traps (NET) that consist of cellular DNA, chromatin, and granular proteins. NETs are a mechanism to capture and finally kill microorganisms, which is accompanied by simultaneous calprotectin release in a high concentration to support the elimination of the invading pathogens [89]. This mechanism is well known for host defense against bacteria, and was also recently shown for viruses, such as COVID-19 [90], and fungi, as shown for the world's three most deadly fungal pathogens (*C. albicans*, *Aspergillus fumigatus*, and *Cryptococcus neoformans*) [91]. According to the National Nosocomial Infections Surveillance System study, *Candida* species are responsible for 85% of fungal infections among patients in critical conditions. Candidemia commonly results in a high rate of morbidity and mortality, as well as high medical expense, among patients who are hospitalized. Hence, fungal control plays a decisive role in human health [92].

Interestingly, immune cells of either innate and adaptive immunity have developed two opposing strategies to eliminate invading pathogens:

On the one hand, they are able to reduce the zinc content of the phagosome or the cytoplasm to limit zinc availability, like it has been described for macrophages in *Histoplasma capsulatum* infection, resulting in nutritional immunity [93]. Additionally, zinc transporter ZIP8 is described to enable the same effect in activated T cells [94]. On the other hand, the zinc content of the phagosome is excessively increased, leading to zinc intoxication of the pathogen, as known, for instance, for *Mycobacterium tuberculosis* infection [95]. Hence, nutritional immunity and metal intoxication are feasible immune strategies to limit pathogen growth and to eventually control infection. Nutritional immunity mainly affects enzymatic and metabolic function, while metal overload contributes to reactive oxygen species and reactive nitrogen species, protein mismetallization, and subsequent respiratory arrest. Both are well described for several pathogens, including bacteria, viruses, and fungi [88,96–98].

In general, during infection, the overall plasma zinc level is reduced to limit zinc availability for pathogens. This is primarily facilitated by upregulation of the zinc transporter ZIP 14, which is induced in an IL-6-dependent manner. This occurs rapidly within only a few hours and leads to zinc accumulation in the liver and hypozincemia in the serum, respectively [99,100].

Although different strategies are known to limit zinc environment for the invading pathogen, some pathogens have developed defense strategies to overcome some mechanisms [96,101]:

Neisseria—*N. gonorrhoeae* and *N. meningitidis*—have evolved mechanisms to acquire transition metals such as manganese, iron, copper, and zinc in specific environments. Gonococcal TdfH and TdfJ are zinc-specific transporters that pirate zinc from calprotectin and psoriasin, respectively, to transport zinc across the outer cell membrane into the periplasmatic space. *N. meningitides* expresses a similar zinc transporter called CbpA. Both *N. gonorrhoeae* and *N. meningitides* additionally express a specific zinc-import system called

ZnuC, ZnuB, and ZnuA to improve intracellular zinc status. Moreover, *N. meningitidis* uses ZnuD to escape NET-mediated nutritional immunity [102]. Znu-like systems are found in a wide variety of Gram-negative and Gram-positive bacteria [103], such as *E. coli*, *M. tuberculosis*, *Bacillus subtilis*, *Salmonella enterica* and S. *typhi*, *K. pneumoniae*, *Yersinia pestis*, and much more [103–106]. Another category of zinc transporters is similar to the eukaryotic ZIP family transporters; however, ZIP homologs have only been identified in *E. coli* [103]. In summary, a lot of pathogens have evolved mechanisms to circumvent nutritional immunity as zinc deficiency.

Zinc deficiency is harmful not only for pathogens, but also for humans themselves. Low zinc is associated with an impaired immune system and poor prognosis in various diseases as common cold, sepsis, malaria, and much more [1,24,107]. Zinc supplementation can be very supportive, as shown in a multitude of studies, investigating, for example, lower respiratory tract infection, diarrhea, pneumonia, or common cold. For the latter, zinc sulfate or zinc gluconate via lozenges at concentrations of about 75 mg/day reduced the mean daily clinical score, nasal secretion weight, and viral shedding, and made more symptom-free episodes possible [108–114]. Similar results are known for pneumonia in children. Zinc supplementation of 20 mg/day enhanced recovery and reduced resistance to antimicrobials through decreasing exposure to broad-spectrum antibiotics [115]. For diarrhea, zinc supplementation was shown to reduce the incidence and prevalence [16]. This has been extensively studied in numerous investigations, and the results show that zinc reduces the overall duration, severity duration, severity, and frequency of stools in children and adults. Moreover, consequential damage of the intestine is also reduced [116]. This is of main importance, since diarrheal infections count for the second leading cause of death in children under five years of age in developing countries. Hence, the WHO and United Nation's Children Fund (UNICEF) recommend oral rehydration solutions and 10–14 days' supplemental zinc treatment in acute diarrhea, particularly in children [117].

However, supplementation during some diseases seem to be not beneficial, as shown for sepsis [118], and some contradictory results also exist showing no beneficial effect of zinc supplementation for different diseases, although this has been presented before (see Tables 1 and 2) [119–123].

It is important to consider the risk of creating a zinc microenvironment that favors pathogen growth when zinc is supplemented during infection while interfering with the efforts of the innate system to chelate available free zinc. The importance of understanding the complex interplay between different nutrients and their collective influence on health in more detail has been prominent for decades.

On the one hand, a deficiency of one micronutrient is often accompanied by a deficiency of another micronutrient, or even by a general micronutrient deficiency, as occurs particularly in the elderly or in people on a plant-based diet, often due to malnutrition [14,48,124]. On the other hand, high-dose supplementation or overdosing of one micronutrient is critical, since several micronutrients influence each other, and thus an imbalance of another micronutrient may occur. For example, iron supplementation is known to interfere with the proper absorption of zinc, resulting in zinc deficiency.

Moreover, zinc supplementation in high doses compromises copper uptake and vice versa. Adequate zinc and copper homeostasis is known to be important for decades and an altered copper/zinc ratio is found in multiple diseases, such as neurodegenerative disorders [125,126], hepatocellular and gastric carcinoma [127,128], immunological disorders such as sickle cell disease [129], and bacterial and viral infectious diseases such as leishmaniosis or COVID-19 [130,131]. In future, it may possible that the copper/zinc ratio may be considered as a biomarker for disease or for mortality in the elderly population. In advance, more research needs to be performed to understand the therapeutic potentials of micronutrient supplementation, particularly zinc supplementation, on numerous diseases.

In recent years, several clinical trials have been conducted that investigated zinc supplementation in various diseases. Studies are sorted by type and/or origin of infectious disease into viruses (Table 1), bacteria (Table 2), and parasites/fungi (Table 3). The influence of zinc depends on many factors, including the following: zinc status at baseline measurement, zinc formula administered, zinc supplementation concentration, zinc supplementation frequency and type of supplementation (preventive or therapeutic), and subject's age.

Table 1. Zinc supplementation and viral diseases.

Disease	Zinc Salt/Formulation	Period	Population	Effect of Supplementation	References
Common cold	Zinc gluconate, 23.0 mg 6 times/d	7 days	C:28 Z: 37	Shortened duration of cold, more zinc-treated subjects are asymptomatic compared to control subjects	[108]
	Zinc gluconate, 23.0 mg 6 times/d	6 days	P: 28 Z: 29	Reduced mean daily clinical score and nasal secretion weight and viral shedding	[114]
	Zinc gluconate, 23.0 mg 8 times/d	5 days	P: 16 Z: 16	No significant difference between nasal symptom scores, same median duration of viral shedding	[132]
	Zinc gluconate, 23.0 mg 6 times/d	10 days	75	Shortens duration and reduces symptom severity	[109]
	Zinc gluconate, 13.3 mg 6 times/d	8 days	P: 50 Z: 50	Shortens duration and reduces symptom severity, especially cough, headache, nasal congestion/drainage	[133]
	Zinc gluconate, 4.5 mg 4–6 times/d	10 days	P: 69 Z: 61	No benefit was observed among the groups	[119]
	Zinc acetate, 10.0 mg 4 times/d	6 days	P: 28 Z: 30	No benefit was observed among the groups	[120]
	Zinc acetate, 9.0 mg 6 times/d	14 days	P: 49 Z:52	Overall symptom duration was significantly less	[110]
	Zinc acetate, 12.8 mg 6 times/d	12 days	P: 23 Z: 25	Reduced duration and severity of cold symptoms, especially cough	[111]
	Zinc gluconate, 13.5 mg Zinc acetate, 11.5 mg or 5.0 mg 6 times/d	14 days	P: 67 ZG: 69 $ZA_{5.0}$: 66 $ZA_{11.5}$: 70	Zinc gluconate treatment: reduced median duration of symptoms Zinc acetate lozenges: no effect on the duration or severity of symptoms	[112]
	Zinc sulfate, 15.0 mg/d	7 month	P: 100 Z: 100	Mean number of colds in the zinc group was significantly fewer	[134]
	Zinc gluconate, 15.0 mg/d	7 month	P: 17 Z: 17	More symptom-free episodes	[113]
HIV/AIDS	Zinc sulfate, 45.5 mg/d	1 month	P: 29 Z: 29	Increase or stabilization in body weight, increase in plasma zinc levels, $CD4^+$ T cells and plasma active zinc-bound thymulin; reduced or delayed frequency of opportunistic infections due to *Pneumocystis jirovecii* and *C. albicans*	[135]
	Zinc gluconate, 45.0 mg 3 times/d	15 days	P: 5 Z: 5	Increased zinc concentrations in red blood cells, HLA-DR^+ cells, stimulation of lymphocyte transformation, and phagocytosis of opsonized zymosan by neutrophils	[136]
	Zinc sulfate, 10.0 mg elemental zinc/d	6 month	P: 41 Z: 44	Decreased morbidity from diarrhea	[137]
	Zinc sulfate, 50.0 mg/d	1 month	P: 34 Z: 31	No improvements in immune responses to tuberculosis, CD4/CD8 ratio, lymphocyte subsets, and viral load	[138]
	Zinc sulfate, 25.0 mg/d	6 month	P: 200 Z: 200	No effect on birth outcomes by supplementation to pregnant HIV-positive women, no effect on T-lymphocyte counts	[139]
	Zinc sulfate, 25.0 mg/d	6 month	P: 200 Z: 200	Increased risk of wasting, no effect on viral load	[140]
	Zinc gluconate, 50.0 mg/d	6 days	P: 45 Z: 44	No improvements in antibody responses to a pneumococcal conjugate vaccine	[141]
	12.0–15.0 mg zinc/d (women–men)	18 month	P: 116 Z: 115	Four-fold reduction in the likelihood of immunological failure, reduced rate of diarrhea	[142]
	Chelated zinc, 15.0 mg/d	12 month	P: 17 Z: 13	$CD4^+$ cell count significantly increased	[143]
	Daily zinc intake (not specified)	18 month	P: 128 Z: 126	Nonsignificant decrease in Veterans Aging Cohort Study (VACS) index	[144]
	Zinc sulfate, 15.0 mg/d	12 month	P: 40 Z:40	No benefit was observed among the groups	[145]
	Zinc gluconate, 12.0–15.0 mg/d (women–men)	18 month	P: 128 Z: 126	No benefit was observed among the groups	[146]
	Zinc gluconate (high (Z_{hi}) zinc): 90.0 mg elemental/d, Zinc gluconate (low (Z_{low}) zinc): 45.0 mg elemental/d	16 weeks	Z_{hi}: 27 Z_{low}: 25	Increased serum zinc, decreased biomarkers associated with clinical comorbidities (decreased systemic inflammation (c reactive protein and TNF-α), monocyte activation, and enterocyte damage)	[147]
	Zinc sulfate, 20.0 mg/d	24 weeks	P: 26 Z: 26	Decrease in viral load, anthropometric indices, and morbidity profile in HIV-infected children started on antiretroviral therapy	[148]

Table 1. *Cont.*

Disease	Zinc Salt/Formulation	Period	Population	Effect of Supplementation	References
Chronic HepC	Polaprezinc, 75.0 mg 2 times/d	24 weeks	P: 35 Z: 40	Improved response to IFN-α therapy	[149]
	Zinc sulfate, 300.0 mg/d Polaprezinc, 150.0 mg/d	24 weeks	P: 10 Z_S: 9 Z_P: 15	Normalized serum ALT levels, better eradication of HepC virus RNA	[150]
	Zinc gluconate, 78.0 mg/d 5 times/d	6 month	P: 40 Z: 18	Decreased incidences of gastrointestinal disturbances, body weight loss, and mild anemia	[151]
	Zinc gluconate, 30.0 mg/d	1 year	C: 16 Z: 16	No benefit was observed among the groups	[122]
	Polaprezinc, 75.0 mg 2 times/d	48 weeks	P: 16 Z: 16	No benefit was observed among the groups	[152]
	Polaprezinc, 75.0 mg 2 times/d	24 weeks	P: 39 Z: 39	No benefit was observed among the groups	[153]
	Polaprezinc, 75.0 mg 3 times/d	6 month	P: 12 Z: 12	Reduced serum AST, ALT, and ferritin	[154]
	Polaprezinc, 150.0 mg 2 times/d	6 years	P: 30 Z: 32	Reduced incidence of HCC (albumin-dependent)	[155]
	Polaprezinc, 150.0 mg/d	48 weeks	P: 12 Z: 11	Decrease in serum ALT levels and Th2 cells (%), decreased plasma thiobarbituric acid reactive substances, and prevented decrease in polyunsaturated fatty acids of erythrocyte membrane phospholipids	[156]
Dengue fever	Zinc bisglycinate, 15.0 mg 3 times/d	5 days	P: 25 Z: 25	Lower mean time of defervescence and shorter time of hospitalization	[157]
COVID-19	Zinc sulfate, 100.0 mg elemental/d	Until recovery or death	P: 46 Z: 196	No benefit was observed among the groups	[123]
	Zinc sulfate, 100.0 mg elemental zinc/d	5 days	P: 521 Z: 411	Increased frequency of being discharged home (OR 1.53, 95% CI 1.12–2.09) and reduction in mortality or transfer to hospice among patients who did not require ICU level of care	[158]
	Zinc sulfate, 100.0 mg elemental/d	4 days	P: 2467 Z: 1006	Increased rates of discharge home and 24% reduced risk of in-hospital mortality	[159]
	Zinc sulfate, 50.0 mg elemental/d	5 days	P: 377 Z: 141	Fewer hospitalizations	[160]
	Zinc acetate or gluconate, 2.0–2.5 mg/kg/day	10 days	Case report Z: 28	Short recovery time (~1,6 days)	[161]
	Zinc chloride, 0.24 mg/kg/d i.v.	7 days	P: 18 Z: 15	Increased zinc level but no effect on clinical outcome	[162]
	Zinc sulfate, 50.0 mg elemental/d 2 times/d	15 days	P: 95 Z: 96	Decreased duration of ventilation, decreased length of hospitalization, and reduced risk of in-hospital mortality	[163]
	Zinc bisglycinate, 15.0 mg/d	6 weeks	P: 57 Z: 59 C: 56	Significant decrease in SARS-CoV-2 infection	[164]

P: placebo, Z: zinc, C: control, ZA: sinc acetate, ZG: zinc gluconate.

Table 2. Zinc supplementation and viral diseases.

Disease	Zinc Salt/Formulation	Period	Population	Effect of Supplementation	References
Acute lower respiratory tract infection	Zinc bisglycinate, 30.0 mg elemental/d	7 days	P: 32 Z: 32	Shortened recovery time and duration of the hospital stay, and improved chest in-drawing, tachypnea, and fever	[157]
	Zinc sulfate, 20.0 mg/d	5 months	P: 124 Z: 134	Reduced acute lower respiratory tract infection morbidity	[165]
	Zinc gluconate, 10.0 mg/d	60 days	P: 48 Z: 48	Reduced episodes of acute lower respiratory infections and severe acute lower respiratory infections, increased infection-free days	[166]
	Zinc oxide, 5.0 mg/d	12 months	P: 167 Z: 162	Decreased incidence of upper respiratory tract infections and diarrheal disease episodes	[167]
	Zinc gluconate, 10.0 mg/d	6 months	P: 311 Z: 298	Increased plasma zinc level and decreased episodes of infection	[168]
	Zinc acetate, 10.0 mg 2 times/d	5 days	P: 74 Z: 76	Increased recovery rates from illness and fever in boys	[169]
	Zinc sulfate, 15.0 mg/d	6 months	P: 40 Z: 40	Increased plasma retinol concentrations, earlier sputum conversion and resolution of X-ray lesion area	[170]

Table 2. *Cont.*

Disease	Zinc Salt/Formulation	Period	Population	Effect of Supplementation	References
Pneumonia	10.0–20.0 mg zinc/d	2 weeks	P: 280 Z: 280	Acceleration in clinical resolution and shorter hospital stay	[171]
	10.0–20.0 mg zinc /d	2 weeks	610	Marginal faster recovery time	[172]
	Zinc sulfate, 10.0 mg 2 times/d	until discharge	299	No benefit was observed among the groups	[173]
	Zinc sulfate, 12.5 mg elemental/d 2 times/d	until discharge	P: 47 Z: 47	No benefit was observed among the groups	[174]
	Elemental zinc, 20.0 mg/d	until discharge	P: 84 Z: 80	Reduced duration of severe pneumonia, duration of chest in-drawing respiratory rate hypoxia, and overall hospital duration	[115]
	Zinc sulfate, 10.0–20.0 mg/d	7 days	P: 301 Z: 303	Faster recovery from lower chest wall indrawing and sternal retraction	[175]
	Zinc syrup, 20.0 mg elemental/d 2 times/d	until discharge	P: 225 Z: 225	Faster resolution of respiratory signs	[176]
	Elemental zinc, 10.0 mg 2 times/d	7 days	P: 53 Z: 64	No benefit was observed among the groups	[177]
	Zinc gluconate, 10.0 mg/d–20.0 mg/d	7 days	P: 176 Z: 176	No benefit was observed among the groups	[178]
	Zinc syrup, 40.0 mg/d	until discharge	P: 150 Z: 150	Shorter duration of relief of severe pneumonia signs and hospitalization time	[179]
	Zinc syrup, 10.0 mL/d	until discharge	P: 60 Z: 60	Faster resolution of clinical symptoms	[180]
Tuberculosis	Zinc sulfate, 220.0 mg/d	18 months	Z: 8	Reduced dose of clofazimine, withdrawal of steroids, toleration of dapsone, reduced incidence and severity of erythema nodosum leprosum, gradual decrease in the size of granuloma, and gradual increase in the number of lymphocytes	[181]
	Zinc sulfate, 15.0 mg/d, +/−VitA 5000 IU/d	6 months	P: 40 Z: 40 VitA: 40 Z+VitA: 40	Marginal earlier sputum conversion No difference in clinical, nutritional, chest X-ray, or laboratory findings	[182]
	Zinc sulfate, 30.0 mg elemental every second day	6 month	P: 37 Z: 37	Elevated plasma zinc concentrations, elevated body weight, earlier sputum smear conversion, lower SGOT and SGPT concentrations after 2 months, decreased serum levels of total protein and albumin	[183]
	Zinc sulfate, 15.0 mg/d	6 month	P: 40 Z: 40	Increased plasma retinol concentrations, earlier sputum conversion and resolution of X-ray lesion area	[170]
Shigellosis	Zinc acetate, 1.30 mg/kg 3 times/d	1 month	P: 16 Z: 16	Increased intestinal mucosal permeability and better nitrogen absorption, increased serum zinc and alkaline phosphatase activity	[184]
	Zinc acetate, 20.0 mg/d	2 weeks	P: 28 Z: 28	Increased serum zinc level, lymphocyte proliferation in response to phytohemagglutinin and plasma invasion plasmid-encoded antigen-specific IgG titers	[185]
	Zinc acetate, 20.0 mg/d	2 weeks	P: 28 Z: 28	Increased serum zinc levels, serum shigellacidal antibody titers, $CD20^+$ cells, and $CD20^+CD38^+$ cells	[186]
	Not specified, 20.0 mg/d	2 weeks	P: 16 Z: 14	Faster recovery from acute illness, increased mean body weight, and fewer episodes of diarrhea	[187]
H. pylori	Polaprezinc, 150.0 mg 2 times/d	7 days	P: 28 Z: 33	Increased cure rate of *H. pylori* infection compared to single antibiotic treatment	[188]
Diarrhea	Zinc sulfate, 3.0–7.0 mg/kg/d elemental zinc/d	4 month	P: 70 Z: 70	Decreased incidence of diarrhea, number of diarrhea episodes per child, and frequency of stools per day	[189]
	Diarrhea multiple different studies			Decreased duration, severity, and occurrence of diarrhea	[190]
	Zinc acetate, zinc gluconate, zinc sulfate ranging from 5.0–40.0 mg/d	5–15 days	P: 9353 Z: 9469	Reductions in morbidity as a result of oral therapeutic zinc supplementation for acute diarrhea among children	[191]
	Zinc sulfate, 10.0 mg/d	2 weeks	P: 536 Z: 538	No benefit was observed among the groups	[121]
	10.0–20.0 mg zinc/d	until discharge	P: 50 Z: 50	Reduced frequency of diarrheal episodes	[192]
	Zinc gluconate syrup, 20.0 mg elemental/d, +/− daily Probiotics (Pr)	7 days	P: 50 Pr: 50 Z: 46	Reduced relative risk of diarrhea persistence, decreased duration and severity, reduced post-treatment complications	[193]
	Zinc tablet, 7.0 mg/d, therapeutic zinc (TZ), 20.0 mg/d, +/− micronutrient powder (MNP)	9 month	P: 847 Z: 844 TZ: 848 MNP: 841	No benefit was observed among the groups	[194]
	Zinc sulfate, 10.0–20.0 mg/d	10 days	P: 50 Z: 53	Reduced duration of diarrhea, fewer diarrheic episodes in the next 3 months	[195]

P: placebo, Z: zinc, C: control, VitA: vitamin A, Pr: probiotics, MNP: micronutrient powder, TZ: therapeutic zinc.

Table 3. Zinc supplementation and parasitic or fungal diseases.

Disease	Zinc Salt/Formulation	Period	Population	Effect of Supplementation	References
Malaria	10.0 mg zinc 6 times/week +/− VitA single dose, 200,000 IU	6 months	P: 74 Z: 74	Decreased malaria prevalence and fewer malaria episodes, longer time to first malaria episode, and 22% fewer fever episodes	[196]
	Zinc gluconate, 10.0 mg 6 times/week	46 weeks	P: 138 Z: 136	Reduction in Plasmodium falciparum-mediated febrile episodes	[197]
	Zinc acetate or zinc gluconate, 70.0 mg 2 times/week	15 month	P: 54 Z: 55	Not statistically significant trend towards fewer malaria episodes; no effect on plasma and hair zinc, diarrhea, and respiratory illness	[198]
	Zinc sulfate, 12.5 mg 6 times/week	6 months	P: 344 Z: 336	Increased serum zinc levels and reduced prevalence of diarrhea	[199]
	Zinc sulfate, 20.0 mg or 40.0 mg/d	4 days	P: 483 Z: 473	Increased plasma zinc, no effect on fever, parasitemia, or hemoglobin concentration	[200]
	Zinc sulfate, 20.0 mg/d	7 months	P: 189 Z: 191	No significant effect on *P. vivax* incidence but significantly reduced diarrhea morbidity	[201]
	Zinc sulfate, 25.0 mg/d +/− VitA 2500 IU/d	until delivery	P: 362 VitA: 348 Z: 345 VitA+Z: 349	36% (95% CI = 9–56%) reduced risk of histopathology-positive placental infection	[202]
	5.0 mg, 10.0 mg, or 15.0 mg zinc/d	9 month	P: 785 C: 433 Z5: 429 Z10: 438 Z15: 436	No benefit was observed among the groups	[203]
	Zinc gluconate, 10.0 mg/d +/− VitA, 200,000 IU/d at the beginning and end of the study	6 month	C: 90 Z: 92	Significantly fewer (27%) malaria diagnoses	[204]
	Zinc, 10.0 mg/d +/− Multi-nutrients (M)		P: 148 M: 148 Z: 145 M + Z: 146	No benefit was observed among the groups	[205]
Leishmania infection	Zinc sulfate, 2.5 mg/kg, 5.0 mg/kg or 10.0 mg/kg 3 times/d	45 days	P: 12 Z: 92	Increased serum zinc levels and cure rate, decreased erythema and size of induration	[206]
	Zinc, 45.0 mg/d	20 days	P: 15 Z: 14	Higher expression level of transferrin receptor	[207]
	Zinc sulfate, topical 2%	3 month	P: 32 Z: 32	No benefit was observed among the groups	[208]
	Zins syrup, total dose in 2 weeks of 2 mg/kg/d	2 weeks	C: 26 Z: 26	Accelerated reduction in splenomegaly	[209]
	Zinc sulfate, 10.0 mg/kg/d	45 days	C: 50 Z: 50	Zinc supplementation is as effective as systemic meglumine antimoniate treatment	[210]
	Zinc sulfate, itralesional injections of 2% zinc solution	6 weeks	C: 35 Z: 31	Higher efficacy after the second and fourth weeks	[211]
C. albicans	Zinc syrup, 20.0 mg elemental/d	2 weeks	P: 366 Z: 358	Increased blood zinc concentration, reduced prevalence of candidemia and candiduria by 50%, lesser nosocomial urinary tract infection and bloodstream infection, shorter treatment with broad-spectrum antibiotics, shorter length of hospital stay	[212]

P: placebo, Z: zinc, C: control, VitA: vitamin A, M: multinutrients.

6. Zinc and the Inflamed Immune System

Invading pathogens put the healthy human body into a state of inflammation by alerting and activating the immune system to fight those pathogens and to eventually restore the healthy state. Inflammation is a natural and necessary process that is required to protect the host from tissue damage; however, sometimes overshooting immune reactions occur that cause severe tissue damage or inflammation processes, which are not resolved and later become chronic, leading to loss of tissue function [213]. Nonresolving inflammation contributes to many diseases, including COVID-19 in its fatal and long forms, sepsis, or multiple organ failure, which are a major cause of admission to the intensive care unit and high mortality rates [214].

Since 2010, the toll that inflammation takes on human health has not abated, despite major advances in understanding the underlying biology. Nevertheless, a lot of research has been performed regarding drug development, such as biologics that block signaling through IL-1β and TNF-α, or regarding nutritional supplements, including zinc, that improve the overall immune status of patients suffering of severe diseases [13,214,215].

In immunological pathogen defense, neutrophil leukocytes (polymorphonuclear leukocytes (PMNs)) and macrophages are among the first responders to infection or tissue damage [216]. The first signals responsible for the inflammatory processes are initially activated by damage-associated molecular patterns (DAMPs), such as DNA and extracellular components or ATP, at the tissue injury site. After cellular activation through pattern recognition receptors (PRR), immune cells migrate to the site of injury, which is called chemotaxis. PRR can be subdivided into three groups: toll-like receptors, C-type lectin receptors, and retinoic acid-inducible gene 1-like receptors [217]. The early recruitment of PMN can be mediated by a complex interplay of chemokines, proinflammatory cytokines, complement factors, and bacterial cell wall compounds, which is highly affected by the patient's zinc status. It is reported that during zinc deficiency, chemotaxis is reduced, and can be restored by physiological zinc supplementation [218–220]. Early activation of PMN functions, amongst others, through the SCR family kinase LYN, which is known to be regulated in a zinc-dependent manner. In all leukocytes, the SCR family kinase is prevalent. Activation and inhibition are mediated on the molecular level by phosphorylation, which is facilitated by kinases such as protein tyrosine phosphatases (PTP). For example, CD45 is a transmembrane PTP that is an important regulator of SCR family kinase activity. Since the inhibition constants of zinc for PTPs are very low (varying from micromolar to picomolar concentrations [221]), cellular zinc levels are sufficient to alter enzymatic activity of PTP that eventually regulate cellular activation and function [222].

Besides chemotaxis, phagocytosis and pathogen killing are controlled by zinc status. Following phagocytosis, pathogens are killed by the activity of nicotinamide adenine dinucleotide phosphate (NADPH) oxidases that have been shown to be inhibited by zinc deficiency and zinc excess, respectively [223,224]. Alternatively, PMNs kill pathogens by releasing neutrophil extracellular traps (NETs), which are abolished during zinc deficiency [218]. NETs are web-like structures composed of DNA, chromatin, fibers, and various neutrophil granule proteins [225]. NETosis is a double-edged sword for the immune system. On the one hand, it is known that NETs can encapsulate, capture, and kill pathogens efficiently, especially large microorganisms such as *C. albicans* and *Mycobacteria* aggregates that are difficult to phagocytose. On the other hand, the excessive release of NETs can trigger and amplify inflammatory diseases such as sepsis, rheumatoid arthritis, or inflammatory bowel disease [225]. However, NETosis is an indispensable method to protect humans of invading harmful pathogens.

Due to pathogens, monocytes circulating in the blood may be activated and migrate to the damaged tissue. Therefore, they need to adhere to endothelial cells, which can be intensified by zinc supplementation. During zinc deficiency, this mechanism is impaired, as is, for example, well reported for the elderly population suffering of zinc deficiency very often [226]. Moreover, proinflammatory cytokine production, such as IL-1, IL-6, and TNF-α, is increased. In line with that, studies uncovered a rapid decline of plasma zinc concentrations during the acute phase of infection, leading to hypozincemia [227–229]. Hence, redistribution of zinc during inflammation seems to be mediated by proinflammatory cytokine production. Zinc is shuttled into cellular compartments to assist protein synthesis, neutralization of free radicals, and to prevent microbial survival.

One of the most important mediators in inflammation is the nuclear factor kappa-light-chain-enhancer of activated B cells (NFκB) (Figure 1). Besides inflammatory process regulation, it controls apoptosis, cell adhesion, proliferation, tissue remodeling, the innate and adaptive immune responses, and cellular stress responses. NFκB activation requires the phosphorylation of IKKα/β by the IκB kinase (IKK) complex, which degrades the IκB and releases NFκB and allows it to translocate freely to the nucleus to induce targeted gene expression [230,231].

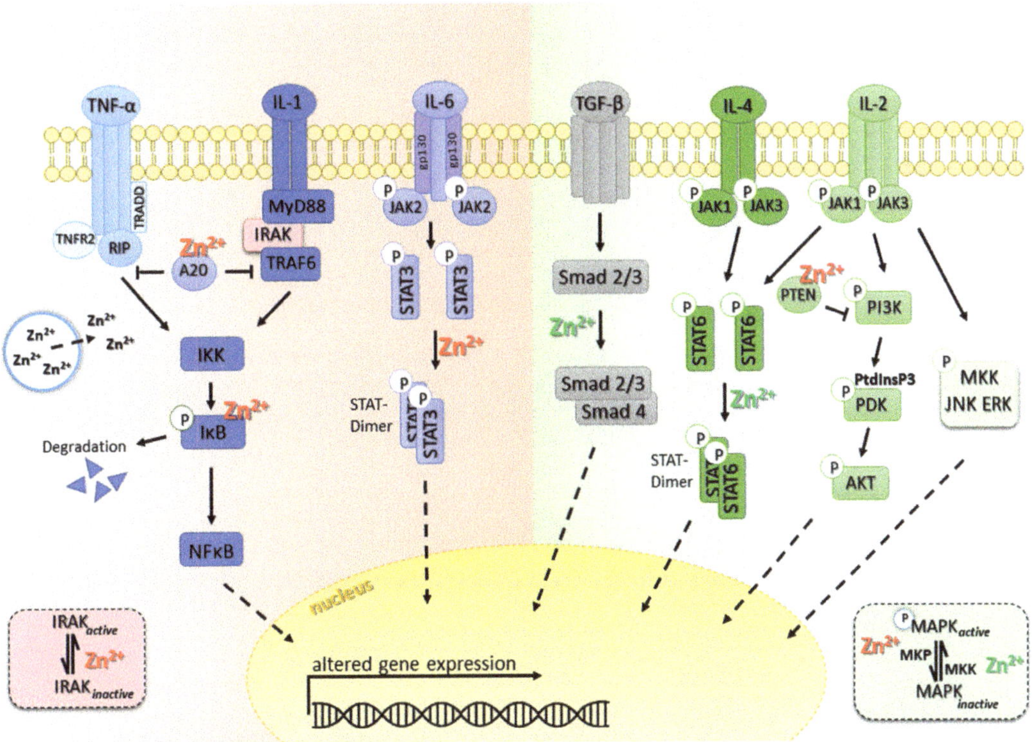

Figure 1. Inhibitory (red) and activating (green) effects of zinc supply on human pro- and anti-inflammatory signaling pathways.

Over the last few decades, contradictory studies have reported zinc to have activating and inhibiting properties on this process, respectively. This is a great and complex example to illustrate how differently zinc is able to fine-tune cell signaling depending on which model is considered. Thus, it is obvious that the influence of zinc cannot be interpreted using a unilateral approach [232].

In vitro studies revealed that increased intracellular zinc concentrations are able to inhibit dephosphorylation of mitogen-activated protein kinase (MAPK) and to support phosphorylation of IKKα/β. Hence, NFκB can translocate into the nucleus and activate expression of proinflammatory cytokines [233]. Using chelating agents to prevent zinc signals, such as TPEN (N,N,N0,N0-Tetrakis(2-pyridylmethyl)ethylenediamine), NFκB is not able to translocate to the nucleus anymore. On the one hand, this might be due to the inhibition of kinases ERK1/2, IKKβ, MKK3/6 and IκB, which might be associated with the effect of zinc on PTP and PTK activities as described earlier [234,235]. On the other hand, IRAK1 destruction might be avoided, causing its accumulation in the cytosol. This might be due to the inhibition of zinc-dependent MMPs, which are probably responsible for IRAK1 destruction during adequate zinc conditions. IRAK1 is responsible for the degradation of NFκB inhibitors by IKK.

In contrast, a growing body of literature supports the role of zinc as a negative regulator of NFκB signaling:

For instance, zinc supplementation is described to influence the zinc-finger protein A20 expression, which is an anti-inflammatory protein. A20 negatively regulates tumor necrosis factor receptor (TNFR)- and toll-like receptor (TLR)-initiated NFκB pathways. Expression of A20 mRNA and protein expression can be induced by zinc, as shown in premonocytic, endothelial, and cancer cells [236]. A20 deubiquitinates receptor-interacting protein 1 (RIP1), which prevents its interaction with NFκB essential modulator IKK. Additionally, it inhibits TLR signaling by removing polyubiquitin chains from TNFR-associated factor 6 (TRAF6). Moreover, zinc supplementation was able to downregulate pro-inflammatory cytokine production by decreasing the gene expression of IL-1β and TNF-α through upregulation of mRNA- and DNA-specific binding for A20 [237]. In addition to zinc, proinflammatory cytokines are known inducers for A20 in macrophages, possibly representing a feedback mechanism to prevent hyperinflammatory reactions [238].

Proinflammatory pathways: In TNF-α and IL-1 signaling pathways, zinc supply prevents the translocation of nuclear translocation of NFκB and inhibits subsequent inflammation by modulation of the common IκB kinase complex that phosphorylates the NFκB inhibitory protein. Zinc also inhibits the IL-6-mediated activation of STAT3.

Anti-inflammatory pathways: In TGF-β signaling, zinc supply promotes Smad 2/3 nuclear translocation and transcriptional activity. Zinc promotes the IL-2 signaling pathway by blocking MAP kinase phosphatase (MKP) in extracellular signal-regulated kinase (ERK) 1/2 pathways and phosphatase and tensin homologue (PTEN), which opposes the phosphoinositide 3-kinase (PI3K) function in the PI3K/AKT pathway. In IL-4 signaling, zinc induces STAT6 phosphorylation, promotes dimerization and translocation into the nucleus, and thus promotes anti-inflammatory effects.

Elevated cGMP level also might be responsible for preventing NFκB activation and translocation into the nucleus. Zinc acts as an inhibitor of cyclic nucleotide phosphodiesterase (PDE). The cGMP level rises when PDE is inhibited, thus activating protein kinase A (PKA). PKA suppresses Raf-1 by phosphorylation and subsequently inactivates MAPK and NFκB signaling [239]. Similarly, studies on mast cell signaling show that zinc can bind to a zinc-finger-like motif found on protein kinase C (PKC), which inhibits PMA-mediated PKC translocation to the membrane. Thus, NFκB activity is indirectly inhibited [240].

Besides NFκB signaling, IL-6-mediated signaling is also very important for modulation of an adequate proinflammatory immune response. IL-6 is one of the most potent inflammatory cytokines, activating immune cells by binding to the IL-6 receptor alpha chain and tow gp130 molecules, which are consequently activated and function as signal transducers (Figure 1). As a result, MAPK pathways and Janus kinase (JAK)-signal transducer and activator of transcription 3 (STAT3) pathways are activated, leading to STAT3 activation and dimerization. In particular, the phosphorylation of STAT3 is broadly affected by zinc homeostasis. During zinc deficiency, the JAK-STAT3 signaling pathway is activated in an IL-6 dependent manner. IL-6 was found to be highly upregulated while zinc-deficient. These observations were confirmed by DNA methylation profile characterization, showing a progressive demethylation of IL-6 promoter correlating with IL-6 expression [241]. However, zinc reconstitution normalizes the effect, and zinc supplementation was shown to inhibit IL-6/IL-1β-induced STAT3 phosphorylation [135,242]. Moreover, STAT3 phosphorylation is known to be regulated by other kinases, such as the SH2 domain-containing phosphatases (Shp), which belong to a class of non-transmembrane protein tyrosine phosphatases [243]. Furthermore, alteration in the redox state, e.g., via oxidative stress, seems to alter STAT3 phosphorylation [244], illustrating the complexity of regulation. STAT3 seems to be an important mediator in cell proliferation and differentiation, especially during inflammatory diseases such as sepsis [244]. Studies highlight a fundamental role for zinc-dependent STAT3 regulation in the innate as well as in the adaptive immune response. Zinc chloride increased transient STAT3 phosphorylation in hematopoietic stem cells, maintaining pluripotency and cellular self-renewal. STAT3 inhibition abrogates the zinc-related effect, underlining the importance of STAT3 as a central molecule in hematopoiesis. Moreover, B cell proliferation was increased during zinc deficiency, mediated amongst others by increased STAT3 phosphorylation [245]. However, IL-4-induced STAT6 phosphorylation was diminished [245]. Since the IL-4-induced pathway is essential for the immunoglobulin (Ig) class switch to IgE, altered signaling could explain the increased susceptibility towards parasite infections during zinc deficiency. IL-6 is also important for the induction of terminal B cell development into plasma cells, providing an additional reason for the decrease in antibodies during zinc deficiency [246]. For T cells, zinc supplementation was shown to inhibit the IL-6-induced STAT3 signaling cascade that is essential for T helper cell (Th) 17 development [247]. In the elderly population, secretion of proinflammatory IL-6 is pathologically increased while T-cell activation is reduced, as is likewise the response to stimulation or vaccination [248]. Interestingly, all pathologies can be ameliorated by zinc intake, which highlights the importance of zinc for a balanced immune response.

IL-1 promotes cell proliferation but also induces the expression of inflammation-related genes such as IFN-γ. When IL-1 binds to the IL-1 receptor (IL-1R) an induction of an intracellular zinc signal is described. Following the general signal transduction pathway, the IL-1 receptor-associated kinase (IRAK) is activated, which causes NFκB translocation into the nucleus that alters gene activation (Figure 1). While zinc-deficient, the production of IL-1β has been reported to be elevated, as found in in vitro studies of PBMC of zinc-deficient adults [249]. In line with that, epigenetic changes due to long-term zinc deficiency are known to promote changes of the chromatin structures of IL-1 β and TNF-α promoters, enabling the expression of both genes [227]. In contrast, zinc supply was shown to decrease IRAK activity (similarly to the effect on TLR4-signaling), which results in repression of the memory Th17 response, hence favoring the anti-inflammatory immune response [52,250].

An in vivo rodent model showed that zinc supplementation skewed the regulatory T cell (Treg)-Th17 balance towards Treg cells. After a pulmonary fungal challenge with *H. capsulatum*, zinc impaired dendritic cells (DC) to mount a proinflammatory response to fungal infection and stimulation by TLR ligands. Hence, zinc triggers the tolerogenic phenotype of DC and promotes the T-cell differentiation towards Treg cells [251]. In this context, zinc is known to keep unwanted T-cell-mediated immune responses in check by suppressing apoptosis, proliferation of immune cells, and production of proinflammatory cytokines, as well as IL-2 and IL-10 [252,253]. One important trigger for Treg development is the TGF-β-induced Smad 2/3 signaling pathway, which induces the expression of transcription factor forkhead-box-protein p3 (FoxP3). This is an important transcription factor for T-cell lineage development. Zinc supplementation amplifies TGF-β-induced Smad 2/3 signaling, and consequently FoxP3 expression [254]. Additionally, proteasomal FoxP3 degradation is inhibited by zinc due to histone deacetylase Sirtuin-1 (Sirt1) inhibition [255]. Other molecular targets, the interferon regulatory factor (IRF)-1 and Krüppel-like-factor 10 (KLF-10), are able to modulate FoxP3 activity, thereby influencing Treg cell development [256]. IRF-1 is referred to as a negative regulator of FoxP3 expression, as in vivo IRF-1 deficiency results in a selective and marked increase in highly differentiated and activated Treg cells [257]. KLF-10 is an essential transcription factor for proper Treg cell function, since KLF-10 deficiency leads to impaired cell differentiation, skewed cytokine profiles with enhanced Th1, Th2, and Th17 cytokines, and a reduced capacity for suppression of effector cells [258]. Zinc supplementation dampened IRF-1 activity and induced KLF-10 expression, leading to strengthened pro-tolerogenic immune response [256].

During zinc deficiency, studies reported an imbalance between Th1 and Th2 cell differentiation, resulting in a diminished Th1 immune response with lowered IFN-γ and IL-2 production, whereas the Th2 cytokines (IL-4, IL-6, and IL-10) remain unaffected [249,259,260]. In contrast, it was reported that zinc promotes a Th1 immune response by augmenting the gene expression of IFN-γ and IL-2 [261].

IL-2 is known to induce proliferation and development of effector functions in T cells, inducing antiapoptotic and cell-cycle-related genes as well as certain cytokines and lineage decisive factors [262].

When IL-2 binds to the IL-2 receptor, signal transduction via Jak1 and Jak3 activates numerous signaling pathways, including PI3K/Akt and MEK1/2-ERK1/2. Similar to IL-1 binding to its receptor, binding of IL-2 provokes an intracellular zinc level increase. Zinc is released from lysosomes (zincosomes), which is linked to ERK1/2 and Akt activation. In general, Akt activation is mediated by PI3K-produced PI(3,4,5)P3, which is degraded by phosphatase and tensin homologue deleted on chromosome 10 (PTEN). Interestingly, the mean inhibitory constant (IC50) of zinc for PTEN was found to be 0.59 nM, which would allow its blockade by intracellular zinc level, explaining the activation of Akt [263,264]. Similar results were uncovered for healthy men having a restricted zinc diet for 10 weeks, resulting in reduced IL-2R production [265]. In parallel, studies uncovered the transcription factor cAMP-responsive element modulator alpha (CREMα) to be upregulated due to zinc deficiency, leading to dampened IL-2 production [266,267].

The complexity of zinc's effect on cell development and activation is highly dependent on environment, cellular status, and experimental setup, which may provide a possible explanation for different outcomes observed in the above-mentioned studies. In summary, these results show opposing effects of zinc deficiency and zinc supply on inflammatory and anti-inflammatory signaling pathways, underlying the importance of zinc for proper immune function.

7. Membrane Barrier Function and Zinc

One vital mechanism to defend invading pathogens are membrane barriers, such as those of the human skin or mucosa. The latter is particularly important in areas such as the lung and intestine, which are constantly exposed to myriads of pathogens and harmful substances. Zinc deficiency is associated with severe consequences, such as critically impaired intestinal and/or pulmonary health. Those tissues have a high turnover rate and are therefore the first to be affected by zinc deficiency, as indicated by morphological changes [268,269]; reduced self-renewal; severe degeneration of the intestinal epithelium [270,271]; reduction in integrity, resulting in increased membrane permeability [272]; and altered cell function, illustrated by impaired activity of brush border enzymes [273]. Studies even indicate disturbed production of the gastrointestinal mucus layer, which normally covers the whole gastrointestinal tract (GIT) and serves as an additional physical barrier for the underlying epithelium, protecting it against chemical and physical damage and pathogens [274]. Moreover, it serves as habitat for numerous commensal bacteria [275], and is important for nutritional absorption of macro- as well as micronutrients, such as zinc [276]. During zinc deficiency, mucus glycosylation is disordered, making the epithelium more vulnerable against pathogens, leading to overall degeneration of the mucosa that increases the occurrence of intestinal infections such as diarrhea [277]. Hence, zinc supplementation is recommended in acute diarrhea [117].

In general, zinc supply is reported to regenerate the intestinal epithelium, improve the absorption of water and electrolytes, increase levels of brush border enzymes, and allow better clearance of pathogens by modulating the immune system [273,278,279]. In detail, zinc inhibits cyclic adenosine monophosphate (cAMP)-induced chloride-dependent fluid secretion in enterocytes. Increased cAMP or cGMP levels mediate translocation of additional chloride channels to secrete more ions than usual. Due to osmotic reasons, water follows the salt concentration, making the stool more liquid, leading to diarrhea. Zinc supplementation results in a substantial reduction in cholera toxin (CT)-induced ion secretion and cAMP concentration [280]. On the molecular level, studies reveal structural proteins of the epithelium barrier to be influenced by zinc deficiency. Cellular connection of adjacent cells is facilitated by complexes called tight junctions (TJ) and adherens junctions (AJ). Important tight junction proteins and transmembrane proteins, such as ZO-1 and Claudin-1 or E-cadherin, are degraded during zinc deficiency, leading to increased leakage across the intestinal and pulmonal mucosa [272,281,282]. Thus, neutrophil leukocytes can migrate through the disrupted mucosa, causing acute and/or chronic infection that eventually contributes to intestinal and lung diseases.

Zinc supplementation is able to restore membrane function and structure [279], since it acts as a stabilizer of cell membranes and as essential cofactor in transcription factors and enzymes. For instance, zinc gives resistance to epithelial apoptosis through cytoprotection against reactive oxygen species and bacterial toxins, possibly through antioxidant activity of the cysteine-rich metallothioneins [283]. Investigations in patients suffering from *Shigella* infection showed that zinc supplementation improves the intestinal mucosal permeability, alkaline phosphatase activity, and nitrogen absorption [184]. Furthermore, lymphocyte proliferation is augmented in response to phytohemagglutinin activation, and antigen-specific IgG titers are increased [185]. In addition, elevated serum antibody titers, B cell count, and plasma cell counts are observed [186]. These effects indicate an immune modulatory effect of zinc supplementation.

Studies on *Mycobacterium* infections unravel lower severity of erythema nodosum, reduced infiltration and bacterial index of granuloma, earlier sputum conversion, and resolution of X-ray lesion areas in response to zinc supplementation (see Table 2) [170,181]. Effective clearance of mycobacterial infections requires a Th1-mediated activation of infected macrophages by IFN-γ [284]. Zinc improves the imbalance in T-cell subpopulations, as described earlier.

Moreover, zinc supplementation reduced the incidence of respiratory tract infections and pneumonia in children in developing countries and in nursing home residents [285,286]. For instance, the recently upcoming COVID-19 infection predominantly affects the respiratory system, resulting in pneumonia and acute respiratory distress syndrome [287], which often leads to the requirement of mechanical ventilation [288] and/or to long COVID symptoms [289]. Zinc supplementation, especially in combination with zinc ionophore pyrithione, has been shown to inhibit SARS-coronavirus RNA polymerase (RNA dependent RNA polymerase, RdRp) activity by decreasing its replication [290]. Moreover, ciliary length and ciliary beat frequency were improved in bronchial epithelium in vitro [291,292]. Another zinc-dependent effect is the upregulation of antiviral enzymes, such as latent ribonuclease (RNaseL) and protein kinase RNA-activated (PKR), which are involved in viral RNA degradation and inhibition of viral RNA translation [293]. In addition, viral entry into the cell, replication, and viral protein translation are known to be altered by zinc. This was especially shown for Picornaviridae causing common cold [97,294]. The effects of zinc treatment in common cold patients are summarized in Table 2. In respiratory syncytial virus (RSV) infection, viral replication and RSV plaque formation was found to be more than 1000-fold reduced due to zinc treatment, while during zinc deficiency, higher susceptibility to pneumonia was found [295,296]. Regarding COVID-19 infection, numerous clinical trials have been registered for COVID-19 infection [15], and some, but not all, show beneficial effects on hospitalization rates or severity of infection, as listed in Table 2.

Taken together, the above-described effects of zinc supply underscore the fundamental role of the micronutrient zinc in a healthy immune response. There is a great potential to improve the use of zinc as therapeutic agent in various diseases. However, the interactions between micronutrients need further study to develop supplementation measures that target multiple deficiencies to successfully support disease treatment.

Author Contributions: Conceptualization, M.M. and L.R.; data curation, M.M.; writing—original draft preparation, M.M.; writing—review and editing, L.R.; visualization, M.M.; supervision, L.R.; project administration, L.R. All authors have read and agreed to the published version of the manuscript.

Funding: This research received no external funding.

Institutional Review Board Statement: Not applicable.

Informed Consent Statement: Not applicable.

Conflicts of Interest: The authors declare no conflict of interest.

References

1. Maret, W. Zinc in Cellular Regulation: The Nature and Significance of "Zinc Signals". *Int. J. Mol. Sci.* **2017**, *18*, 2285. [CrossRef] [PubMed]
2. Andreini, C.; Banci, L.; Bertini, I.; Rosato, A. Counting the zinc-proteins encoded in the human genome. *J. Proteome Res.* **2006**, *5*, 196–201. [CrossRef] [PubMed]
3. King, J.C. Zinc: An essential but elusive nutrient. *Am. J. Clin. Nutr.* **2011**, *94*, 679S–684S. [CrossRef] [PubMed]
4. Maret, W. The metals in the biological periodic system of the elements: Concepts and conjectures. *Int. J. Mol. Sci.* **2016**, *17*, 66. [CrossRef]

5. Lim, K.H.; Riddell, L.J.; Nowson, C.A.; Booth, A.O.; Szymlek-Gay, E.A. Iron and zinc nutrition in the economically-developed world: A review. *Nutrients* **2013**, *5*, 3184–3211. [CrossRef]
6. Prasad, A.S.; Miale, A., Jr.; Farid, Z.; Sandstead, H.; Schulert, A. Zinc metabolism in patients with the syndrome of iron deficiency anemia, hepatosplenomegaly, dwarfism, and hypogonadism. *J. Lab. Clin. Med.* **1963**, *61*, 537–549.
7. Jones, S.; Farr, G.; Nimmanon, T.; Ziliotto, S.; Gee, J.M.; Taylor, K.M. The importance of targeting signalling mechanisms of the SLC39A family of zinc transporters to inhibit endocrine resistant breast cancer. *Explor. Target. Anti-Tumor Ther.* **2022**, *3*, 224–239.
8. Wan, Y.; Zhang, B. The Impact of Zinc and Zinc Homeostasis on the Intestinal Mucosal Barrier and Intestinal Diseases. *Biomolecules* **2022**, *12*, 900. [CrossRef]
9. Bitirim, C.V. The role of zinc transporter proteins as predictive and prognostic biomarkers of hepatocellular cancer. *PeerJ* **2021**, *9*, e12314. [CrossRef]
10. Fong, L.Y.; Farber, J.; Croce, C.M. Zinc intake, microRNA dysregulation, and esophageal cancer. *Aging* **2016**, *8*, 1161–1162. [CrossRef]
11. MacKenzie, S.; Bergdahl, A. Zinc Homeostasis in Diabetes Mellitus and Vascular Complications. *Biomedicines* **2022**, *10*, 139. [CrossRef]
12. Maret, W. Zinc in pancreatic islet biology, insulin sensitivity, and diabetes. *Prev. Nutr. Food Sci.* **2017**, *22*, 1. [CrossRef] [PubMed]
13. Wessels, I.; Maywald, M.; Rink, L. Zinc as a gatekeeper of immune function. *Nutrients* **2017**, *9*, 1286. [CrossRef] [PubMed]
14. Maret, W. Zinc and human disease. *Met. Ions Life Sci.* **2013**, *13*, 389–414. [CrossRef] [PubMed]
15. Wessels, I.; Rolles, B.; Slusarenko, A.J.; Rink, L. Zinc deficiency as a possible risk factor for increased susceptibility and severe progression of Corona Virus Disease 19. *Br. J. Nutr.* **2022**, *127*, 214–232. [CrossRef]
16. Plum, L.M.; Rink, L.; Haase, H. The essential toxin: Impact of zinc on human health. *Int. J. Environ. Res. Public Health* **2010**, *7*, 1342–1365. [CrossRef]
17. Ross, A.C.; Caballero, B.H.; Cousins, R.J.; Tucker, K.L.; Ziegler, T.R. *Modern Nutrition in Health and Disease*, 11th ed.; Wolters Kluwer Health Adis (ESP): London, UK, 2012.
18. National Institutes of Health. Zinc, Fact Sheet for Health Professionals. Available online: https://ods.od.nih.gov/factsheets/Zinc-HealthProfessional/ (accessed on 6 October 2022).
19. Institute of Medicine (US) Panel on Micronutrients. *Dietary Reference Intakes for Vitamin A, Vitamin K, Arsenic, Boron, Chromium, Copper, Iodine, Iron, Manganese, Molybdenum, Nickel, Silicon, Vanadium, and Zinc*; National Academies Press (US): Washington, DC, USA, 2001.
20. Sandstead, H.H.; William, A. *Handbook on the Toxicology of Metals*; Nordberg, G.F., Fowler, B.A., Nordberg, M., Friberg, L.T., Eds.; Elsevier Inc: Amsterdam, The Netherlands, 2007; pp. 925–947.
21. Hall, A.G.; King, J.C.; McDonald, C.M. Comparison of Serum, Plasma, and Liver Zinc Measurements by AAS, ICP-OES, and ICP-MS in Diverse Laboratory Settings. *Biol. Trace Elem. Res.* **2022**, *200*, 2606–2613. [CrossRef]
22. Maret, W. Optical methods for measuring zinc binding and release, zinc coordination environments in zinc finger proteins, and redox sensitivity and activity of zinc-bound thiols. In *Methods in Enzymology*; Sies, H., Packer, L., Eds.; Academic Press: Cambridge, MA, USA, 2002; Volume 348, pp. 230–237.
23. Carpenter, M.C.; Lo, M.N.; Palmer, A.E. Techniques for measuring cellular zinc. *Arch. Biochem. Biophys.* **2016**, *611*, 20–29. [CrossRef]
24. Rink, L. *Zinc in Human Health*; Ios Press: Amsterdam, The Netherlands, 2011.
25. Jackson, M. Physiology of zinc: General aspects. In *Zinc in Human Biology*; Springer: Berlin/Heidelberg, Germany, 1989; pp. 1–14.
26. Maares, M.; Haase, H. A guide to human zinc absorption: General overview and recent advances of in vitro intestinal models. *Nutrients* **2020**, *12*, 762. [CrossRef]
27. Livingstone, C. Zinc: Physiology, deficiency, and parenteral nutrition. *Nutr. Clin. Pract.* **2015**, *30*, 371–382. [CrossRef]
28. Vallee, B.L.; Falchuk, K.H. The biochemical basis of zinc physiology. *Physiol. Rev.* **1993**, *73*, 79–118. [CrossRef] [PubMed]
29. Beyersmann, D.; Haase, H. Functions of zinc in signaling, proliferation and differentiation of mammalian cells. *Biometals* **2001**, *14*, 331–341. [CrossRef] [PubMed]
30. Vallee, B.L. Metallothionein: Historical review and perspectives. *Experientia Suppl.* **1979**, *34*, 19–39. [PubMed]
31. Colvin, R.A.; Holmes, W.R.; Fontaine, C.P.; Maret, W. Cytosolic zinc buffering and muffling: Their role in intracellularzinc homeostasis. *Metallomics* **2010**, *2*, 306–317. [CrossRef]
32. Haase, H.; Beyersmann, D. Uptake and intracellular distribution of labile and total Zn(II) in C6 rat glioma cells investigated with fluorescent probes and atomic absorption. *Biometals* **1999**, *12*, 247–254. [CrossRef]
33. Blaby-Haas, C.E.; Merchant, S.S. Lysosome-related organelles as mediators of metal homeostasis. *J. Biol. Chem.* **2014**, *289*, 28129–28136. [CrossRef]
34. Krężel, A.; Maret, W. The biological inorganic chemistry of zinc ions. *Arch. Biochem. Biophys.* **2016**, *611*, 3–19. [CrossRef]
35. Thomas, R.C.; Coles, J.A.; Deitmer, J.W. Homeostatic muffling. *Nature* **1991**, *350*, 564. [CrossRef]
36. Haase, H.; Maret, W. The regulatory and signaling functions of zinc ions in human cellular physiology. In *Cellular and Molecular Biology of Metals*; CRC Press: Boca Raton, FL, USA, 2010; pp. 191–222.
37. Rink, L. Zinc and the immune system. *Proc. Nutr. Soc.* **2000**, *59*, 541–552. [CrossRef]
38. Gibson, R.S.; King, J.C.; Lowe, N. A review of dietary zinc recommendations. *Food Nutr. Bull.* **2016**, *37*, 443–460. [CrossRef]
39. Krebs, N.F. Overview of zinc absorption and excretion in the human gastrointestinal tract. *J. Nutr.* **2000**, *130*, 1374S–1377S. [CrossRef] [PubMed]
40. Steinhardt, H.J.; Adibi, S.A. Interaction between transport of zinc and other solutes in human intestine. *Am. J. Physiol.* **1984**, *247*, G176–G182. [CrossRef] [PubMed]

41. Lee, H.H.; Prasad, A.S.; Brewer, G.J.; Owyang, C. Zinc absorption in human small intestine. *Am. J. Physiol.* **1989**, *256*, G87–G91. [CrossRef] [PubMed]
42. Ford, D. Intestinal and placental zinc transport pathways. *Proc. Nutr. Soc.* **2004**, *63*, 21–29. [CrossRef]
43. Trame, S.; Wessels, I.; Haase, H.; Rink, L. A short 18 items food frequency questionnaire biochemically validated to estimate zinc status in humans. *J. Trace Elem. Med. Biol.* **2018**, *49*, 285–295. [CrossRef]
44. Hunt, J.R.; Beiseigel, J.M.; Johnson, L.K. Adaptation in human zinc absorption as influenced by dietary zinc and bioavailability. *Am. J. Clin. Nutr.* **2008**, *87*, 1336–1345. [CrossRef]
45. Lönnerdal, B. Dietary factors influencing zinc absorption. *J. Nutr.* **2000**, *130*, 1378s–1383s. [CrossRef]
46. Medicine, I.o. *Dietary Reference Intakes: The Essential Guide to Nutrient Requirements*; The National Academies Press: Washington, DC, USA, 2006; p. 1344.
47. Huth, P.J.; Fulgoni, V.L.; Keast, D.R.; Park, K.; Auestad, N. Major food sources of calories, added sugars, and saturated fat and their contribution to essential nutrient intakes in the U.S. diet: Data from the National Health and Nutrition Examination Survey (2003–2006). *Nutr. J.* **2013**, *12*, 116. [CrossRef]
48. King, J.C.; Brown, K.H.; Gibson, R.S.; Krebs, N.F.; Lowe, N.M.; Siekmann, J.H.; Raiten, D.J. Biomarkers of Nutrition for Development (BOND)-Zinc Review. *J. Nutr.* **2015**, *146*, 858s–885s. [CrossRef]
49. Deutsche Gesellschaft für Ernährunf e. V. Referenzwerte für die Nährstoffzufuhr. Available online: https://www.dge.de/wissenschaft/referenzwerte/?L=0 (accessed on 6 October 2022).
50. World Health Organization. *Trace Elements in Human Nutrition and Health*; World Health Organization: Geneva, Switzerland, 1996.
51. Agostoni, C.; Canani, R.; Fairweather-Tait, S.; Heinonen, M.; Korhonen, H.; La Vieille, S.; Marchelli, R.; Martin, A.; Naska, A.; Neuhäuser-Berthold, M. EFSA Panel on Dietetic Products, Nutrition and Allergies. *Sci. Opin. Diet. Ref. Values Folate. EFSA J.* **2014**, *12*, 3893.
52. Brieger, A.; Rink, L. Zink und Immunfunktionen. *Ernährung Med.* **2010**, *25*, 156–160. [CrossRef]
53. Ezzati, M.; Lopez, A.D.; Rodgers, A.A.; Murray, C.J.L. *Comparative Quantification of Health Risks: Global and Regional Burden of Disease Attributable to Selected Major Risk Factors*; World Health Organization: Geneva, Switzerland, 2004.
54. Sandstead, H.H. Zinc deficiency. A public health problem? *Am. J. Dis. Child.* **1991**, *145*, 853–859. [CrossRef] [PubMed]
55. Haase, H.; Rink, L. Multiple impacts of zinc on immune function. *Metallomics* **2014**, *6*, 1175–1180. [CrossRef] [PubMed]
56. World Health Organization. *Vitamin and Mineral Requirements in Human Nutrition*; World Health Organization: Geneva, Switzerland, 2005.
57. Baarz, B.R.; Rink, L. Rebalancing the unbalanced aged immune system-a special focus on zinc. *Ageing Res. Rev.* **2021**, *74*, 101541. [CrossRef] [PubMed]
58. Lestienne, I.; Icard-Vernière, C.; Mouquet, C.; Picq, C.; Trèche, S. Effects of soaking whole cereal and legume seeds on iron, zinc and phytate contents. *Food Chem.* **2005**, *89*, 421–425. [CrossRef]
59. Foster, M.; Samman, S. Vegetarian diets across the lifecycle: Impact on zinc intake and status. *Adv. Food Nutr. Res.* **2015**, *74*, 93–131. [CrossRef] [PubMed]
60. Wellenreuther, G.; Cianci, M.; Tucoulou, R.; Meyer-Klaucke, W.; Haase, H. The ligand environment of zinc stored in vesicles. *Biochem. Biophys. Res. Commun.* **2009**, *380*, 198–203. [CrossRef]
61. Kambe, T.; Hashimoto, A.; Fujimoto, S. Current understanding of ZIP and ZnT zinc transporters in human health and diseases. *Cell. Mol. Life Sci.* **2014**, *71*, 3281–3295. [CrossRef]
62. Fukada, T.; Kambe, T. Molecular and genetic features of zinc transporters in physiology and pathogenesis. *Metallomics* **2011**, *3*, 662–674. [CrossRef]
63. Prasad, A.S. Discovery of human zinc deficiency: Its impact on human health and disease. *Adv. Nutr.* **2013**, *4*, 176–190. [CrossRef]
64. Jeong, J.; Eide, D.J. The SLC39 family of zinc transporters. *Mol. Asp. Med.* **2013**, *34*, 612–619. [CrossRef] [PubMed]
65. Tuschl, K.; Meyer, E.; Valdivia, L.E.; Zhao, N.; Dadswell, C.; Abdul-Sada, A.; Hung, C.Y.; Simpson, M.A.; Chong, W.; Jacques, T.S. Mutations in SLC39A14 disrupt manganese homeostasis and cause childhood-onset parkinsonism–dystonia. *Nat. Commun.* **2016**, *7*, 1–16. [CrossRef] [PubMed]
66. Hara, T.; Yoshigai, E.; Ohashi, T.; Fukada, T. Zinc transporters as potential therapeutic targets: An updated review. *J. Pharmacol. Sci.* **2022**, *148*, 221–228. [CrossRef] [PubMed]
67. Huang, L.; Tepaamorndech, S. The SLC30 family of zinc transporters—A review of current understanding of their biological and pathophysiological roles. *Mol. Asp. Med.* **2013**, *34*, 548–560. [CrossRef] [PubMed]
68. Maret, W. Redox biochemistry of mammalian metallothioneins. *J. Biol. Inorg. Chem.* **2011**, *16*, 1079–1086. [CrossRef]
69. Kimura, T.; Kambe, T. The Functions of Metallothionein and ZIP and ZnT Transporters: An Overview and Perspective. *Int. J. Mol. Sci.* **2016**, *17*, 336. [CrossRef]
70. Sutherland, D.E.; Stillman, M.J. The "magic numbers" of metallothionein. *Metallomics* **2011**, *3*, 444–463. [CrossRef]
71. Thirumoorthy, N.; Shyam Sunder, A.; Manisenthil Kumar, K.; Senthil Kumar, M.; Ganesh, G.; Chatterjee, M. A review of metallothionein isoforms and their role in pathophysiology. *World J. Surg. Oncol.* **2011**, *9*, 54. [CrossRef]
72. Vašák, M.; Meloni, G. Mammalian Metallothionein-3: New Functional and Structural Insights. *Int. J. Mol. Sci.* **2017**, *18*, 1117. [CrossRef]
73. Dai, H.; Wang, L.; Li, L.; Huang, Z.; Ye, L. Metallothionein 1: A New Spotlight on Inflammatory Diseases. *Front. Immunol.* **2021**, *12*, 739918. [CrossRef]
74. Yin, X.; Knecht, D.A.; Lynes, M.A. Metallothionein mediates leukocyte chemotaxis. *BMC Immunol.* **2005**, *6*, 21. [CrossRef] [PubMed]

75. Ling, X.B.; Wei, H.W.; Wang, J.; Kong, Y.Q.; Wu, Y.Y.; Guo, J.L.; Li, T.F.; Li, J.K. Mammalian Metallothionein-2A and Oxidative Stress. *Int. J. Mol. Sci.* **2016**, *17*, 1493. [CrossRef]
76. Klaasen, C. *Metallothionein iv*; Birkhäuser: Basel, Switzerland, 2012.
77. Min, K.S.; Kim, H.; Fujii, M.; Tetsuchikawahara, N.; Onosaka, S. Glucocorticoids suppress the inflammation-mediated tolerance to acute toxicity of cadmium in mice. *Toxicol. Appl. Pharmacol.* **2002**, *178*, 1–7. [CrossRef] [PubMed]
78. Sobocinski, P.Z.; Canterbury, W.J., Jr.; Mapes, C.A.; Dinterman, R.E. Involvement of hepatic metallothioneins in hypozincemia associated with bacterial infection. *Am. J. Physiol.* **1978**, *234*, E399–E406. [CrossRef] [PubMed]
79. Emeny, R.T.; Kasten-Jolly, J.; Mondal, T.; Lynes, M.A.; Lawrence, D.A. Metallothionein differentially affects the host response to Listeria infection both with and without an additional stress from cold-restraint. *Cell Stress Chaperones* **2015**, *20*, 1013–1022. [CrossRef] [PubMed]
80. Leclerc, E.; Heizmann, C.W. The importance of Ca^{2+}/Zn^{2+} signaling S100 proteins and RAGE in translational medicine. *Front. Biosci. (School Ed.)* **2011**, *3*, 1232–1262. [CrossRef]
81. Gilston, B.A.; Skaar, E.P.; Chazin, W.J. Binding of transition metals to S100 proteins. *Sci. China Life Sci.* **2016**, *59*, 792–801. [CrossRef]
82. Singh, P.; Ali, S.A. Multifunctional Role of S100 Protein Family in the Immune System: An Update. *Cells* **2022**, *11*, 2274. [CrossRef]
83. Heizmann, C.W.; Cox, J.A. New perspectives on S100 proteins: A multi-functional Ca(2+)-, Zn(2+)- and Cu(2+)-binding protein family. *Biometals* **1998**, *11*, 383–397. [CrossRef]
84. Mocchegiani, E.; Malavolta, M. Zinc dyshomeostasis, ageing and neurodegeneration: Implications of A2M and inflammatory gene polymorphisms. *J. Alzheimer's Dis.* **2007**, *12*, 101–109. [CrossRef]
85. Guryanova, S.V.; Ovchinnikova, T.V. Immunomodulatory and Allergenic Properties of Antimicrobial Peptides. *Int. J. Mol. Sci.* **2022**, *23*, 2499. [CrossRef] [PubMed]
86. Zygiel, E.M.; Nolan, E.M. Transition Metal Sequestration by the Host-Defense Protein Calprotectin. *Annu. Rev. Biochem.* **2018**, *87*, 621–643. [CrossRef] [PubMed]
87. Bozzi, A.T.; Nolan, E.M. Avian MRP126 Restricts Microbial Growth through Ca(II)-Dependent Zn(II) Sequestration. *Biochemistry* **2020**, *59*, 802–817. [CrossRef] [PubMed]
88. Corbin, B.D.; Seeley, E.H.; Raab, A.; Feldmann, J.; Miller, M.R.; Torres, V.J.; Anderson, K.L.; Dattilo, B.M.; Dunman, P.M.; Gerads, R.; et al. Metal chelation and inhibition of bacterial growth in tissue abscesses. *Science* **2008**, *319*, 962–965. [CrossRef] [PubMed]
89. Brinkmann, V.; Reichard, U.; Goosmann, C.; Fauler, B.; Uhlemann, Y.; Weiss, D.S.; Weinrauch, Y.; Zychlinsky, A. Neutrophil extracellular traps kill bacteria. *Science* **2004**, *303*, 1532–1535. [CrossRef] [PubMed]
90. Arcanjo, A.; Logullo, J.; Menezes, C.C.B.; de Souza Carvalho Giangiarulo, T.C.; Dos Reis, M.C.; de Castro, G.M.M.; da Silva Fontes, Y.; Todeschini, A.R.; Freire-de-Lima, L.; Decoté-Ricardo, D.; et al. The emerging role of neutrophil extracellular traps in severe acute respiratory syndrome coronavirus 2 (COVID-19). *Sci. Rep.* **2020**, *10*, 19630. [CrossRef]
91. Liang, C.; Lian, N.; Li, M. The emerging role of neutrophil extracellular traps in fungal infection. *Front. Cell. Infect. Microbiol.* **2022**, *12*, 900895. [CrossRef]
92. Pappas, P.G.; Lionakis, M.S.; Arendrup, M.C.; Ostrosky-Zeichner, L.; Kullberg, B.J. Invasive candidiasis. *Nat. Rev. Dis. Prim.* **2018**, *4*, 18026. [CrossRef]
93. Subramanian Vignesh, K.; Landero Figueroa, J.A.; Porollo, A.; Caruso, J.A.; Deepe, G.S., Jr. Granulocyte macrophage-colony stimulating factor induced Zn sequestration enhances macrophage superoxide and limits intracellular pathogen survival. *Immunity* **2013**, *39*, 697–710. [CrossRef]
94. Kehl-Fie, T.E.; Skaar, E.P. Nutritional immunity beyond iron: A role for manganese and zinc. *Curr. Opin. Chem. Biol.* **2010**, *14*, 218–224. [CrossRef]
95. Botella, H.; Stadthagen, G.; Lugo-Villarino, G.; de Chastellier, C.; Neyrolles, O. Metallobiology of host-pathogen interactions: An intoxicating new insight. *Trends Microbiol.* **2012**, *20*, 106–112. [CrossRef] [PubMed]
96. Branch, A.H.; Stoudenmire, J.L.; Seib, K.L.; Cornelissen, C.N. Acclimation to Nutritional Immunity and Metal Intoxication Requires Zinc, Manganese, and Copper Homeostasis in the Pathogenic Neisseriae. *Front. Cell. Infect. Microbiol.* **2022**, *12*, 909888. [CrossRef] [PubMed]
97. Ishida, T. Review on the role of Zn^{2+} ions in viral pathogenesis and the effect of Zn^{2+} ions for host cell-virus growth inhibition. *Am. J. Biomed. Sci. Res.* **2019**, *2*, 28–37. [CrossRef]
98. Alamir, O.F.; Oladele, R.O.; Ibe, C. Nutritional immunity: Targeting fungal zinc homeostasis. *Heliyon* **2021**, *7*, e07805. [CrossRef] [PubMed]
99. Aydemir, T.B.; Chang, S.M.; Guthrie, G.J.; Maki, A.B.; Ryu, M.S.; Karabiyik, A.; Cousins, R.J. Zinc transporter ZIP14 functions in hepatic zinc, iron and glucose homeostasis during the innate immune response (endotoxemia). *PLoS ONE* **2012**, *7*, e48679. [CrossRef]
100. Gaetke, L.M.; McClain, C.J.; Talwalkar, R.T.; Shedlofsky, S.I. Effects of endotoxin on zinc metabolism in human volunteers. *Am. J. Physiol.* **1997**, *272*, E952–E956. [CrossRef]
101. Lonergan, Z.R.; Skaar, E.P. Nutrient Zinc at the Host-Pathogen Interface. *Trends Biochem. Sci.* **2019**, *44*, 1041–1056. [CrossRef]
102. Lappann, M.; Danhof, S.; Guenther, F.; Olivares-Florez, S.; Mordhorst, I.L.; Vogel, U. In vitro resistance mechanisms of Neisseria meningitidis against neutrophil extracellular traps. *Mol. Microbiol.* **2013**, *89*, 433–449. [CrossRef]
103. Mikhaylina, A.; Ksibe, A.Z.; Scanlan, D.J.; Blindauer, C.A. Bacterial zinc uptake regulator proteins and their regulons. *Biochem. Soc. Trans.* **2018**, *46*, 983–1001. [CrossRef]

104. Maunders, E.A.; Ganio, K.; Hayes, A.J.; Neville, S.L.; Davies, M.R.; Strugnell, R.A.; McDevitt, C.A.; Tan, A. The Role of ZntA in Klebsiella pneumoniae Zinc Homeostasis. *Microbiol. Spectr.* **2022**, *10*, e0177321. [CrossRef]
105. Price, S.L.; Vadyvaloo, V.; DeMarco, J.K.; Brady, A.; Gray, P.A.; Kehl-Fie, T.E.; Garneau-Tsodikova, S.; Perry, R.D.; Lawrenz, M.B. Yersiniabactin contributes to overcoming zinc restriction during Yersinia pestis infection of mammalian and insect hosts. *Proc. Natl. Acad. Sci. USA* **2021**, *118*, e2104073118. [CrossRef]
106. Huang, K.; Wang, D.; Frederiksen, R.F.; Rensing, C.; Olsen, J.E.; Fresno, A.H. Investigation of the Role of Genes Encoding Zinc Exporters zntA, zitB, and fieF during Salmonella Typhimurium Infection. *Front. Microbiol.* **2017**, *8*, 2656. [CrossRef] [PubMed]
107. Overbeck, S.; Rink, L.; Haase, H. Modulating the immune response by oral zinc supplementation: A single approach for multiple diseases. *Arch. Immunol. Et Ther. Exp.* **2008**, *56*, 15–30. [CrossRef] [PubMed]
108. Eby, G.A.; Davis, D.R.; Halcomb, W.W. Reduction in duration of common colds by zinc gluconate lozenges in a double-blind study. *Antimicrob. Agents Chemother.* **1984**, *25*, 20–24. [CrossRef]
109. Godfrey, J.C.; Conant Sloane, B.; Smith, D.S.; Turco, J.H.; Mercer, N.; Godfrey, N.J. Zinc gluconate and the common cold: A controlled clinical study. *J. Int. Med. Res.* **1992**, *20*, 234–246. [CrossRef] [PubMed]
110. Petrus, E.J.; Lawson, K.A.; Bucci, L.R.; Blum, K. Randomized, double-masked, placebo-controlled clinical study of the effectiveness of zinc acetate lozenges on common cold symptoms in allergy-tested subjects. *Curr. Ther. Res. Clin. Exp.* **1998**, *59*, 595–607. [CrossRef]
111. Prasad, A.S.; Fitzgerald, J.T.; Bao, B.; Beck, F.W.; Chandrasekar, P.H. Duration of symptoms and plasma cytokine levels in patients with the common cold treated with zinc acetate. A randomized, double-blind, placebo-controlled trial. *Ann. Intern. Med.* **2000**, *133*, 245–252. [CrossRef]
112. Turner, R.B.; Cetnarowski, W.E. Effect of treatment with zinc gluconate or zinc acetate on experimental and natural colds. *Clin. Infect. Dis.* **2000**, *31*, 1202–1208. [CrossRef]
113. Veverka, D.V.; Wilson, C.; Martinez, M.A.; Wenger, R.; Tamosuinas, A. Use of zinc supplements to reduce upper respiratory infections in United States Air Force Academy cadets. *Complement. Ther. Clin. Pract.* **2009**, *15*, 91–95. [CrossRef]
114. Al-Nakib, W.; Higgins, P.G.; Barrow, I.; Batstone, G.; Tyrrell, D.A. Prophylaxis and treatment of rhinovirus colds with zinc gluconate lozenges. *J. Antimicrob. Chemother.* **1987**, *20*, 893–901. [CrossRef]
115. Brooks, W.A.; Yunus, M.; Santosham, M.; Wahed, M.A.; Nahar, K.; Yeasmin, S.; Black, R.E. Zinc for severe pneumonia in very young children: Double-blind placebo-controlled trial. *Lancet* **2004**, *363*, 1683–1688. [CrossRef]
116. Dhingra, U.; Kisenge, R.; Sudfeld, C.R.; Dhingra, P.; Somji, S.; Dutta, A.; Bakari, M.; Deb, S.; Devi, P.; Liu, E.; et al. Lower-Dose Zinc for Childhood Diarrhea—A Randomized, Multicenter Trial. *N. Engl. J. Med.* **2020**, *383*, 1231–1241. [CrossRef] [PubMed]
117. World Health Organization. Diarrhoeal Disease. Available online: https://www.who.int/news-room/fact-sheets/detail/diarrhoeal-disease (accessed on 6 October 2022).
118. Nowak, J.E.; Harmon, K.; Caldwell, C.C.; Wong, H.R. Prophylactic zinc supplementation reduces bacterial load and improves survival in a murine model of sepsis. *Pediatr. Crit. Care Med.* **2012**, *13*, e323–e329. [CrossRef] [PubMed]
119. Weismann, K.; Jakobsen, J.P.; Weismann, J.E.; Hammer, U.M.; Nyholm, S.M.; Hansen, B.; Lomholt, K.E.; Schmidt, K. Zinc gluconate lozenges for common cold. A double-blind clinical trial. *Dan. Med. Bull.* **1990**, *37*, 279–281.
120. Douglas, R.M.; Miles, H.B.; Moore, B.W.; Ryan, P.; Pinnock, C.B. Failure of effervescent zinc acetate lozenges to alter the course of upper respiratory tract infections in Australian adults. *Antimicrob. Agents Chemother.* **1987**, *31*, 1263–1265. [CrossRef] [PubMed]
121. Walker, C.L.F.; Bhutta, Z.A.; Bhandari, N.; Teka, T.; Shahid, F.; Taneja, S.; Black, R.E.; Group, Z.S. Zinc during and in convalescence from diarrhea has no demonstrable effect on subsequent morbidity and anthropometric status among infants <6 mo of age. *Am. J. Clin. Nutr.* **2007**, *85*, 887–894.
122. Abbasinazari, M.; Alavian, S.M.; Behnava, B.; Asgharinia, M.; Salimi, S.; Keshvari, M.; Mehrnoush, L.; Karim, P. Effect of zinc supplementation on viral response in patients with chronic hepatitis C and Beta thalassemia major, a pilot study. *J. Clin. Diagn. Res. JCDR* **2014**, *8*, HC16. [CrossRef]
123. Natarajan, S.; Anbarasi, C.; Sathiyarajeswaran, P.; Manickam, P.; Geetha, S.; Kathiravan, R.; Prathiba, P.; Pitchiahkumar, M.; Parthiban, P.; Kanakavalli, K.; et al. Kabasura Kudineer (KSK), a poly-herbal Siddha medicine, reduced SARS-CoV-2 viral load in asymptomatic COVID-19 individuals as compared to vitamin C and zinc supplementation: Findings from a prospective, exploratory, open-labeled, comparative, randomized controlled trial, Tamil Nadu, India. *Trials* **2021**, *22*, 623. [CrossRef]
124. Berendsen, A.A.M.; Kramer, C.S.; de Groot, L. The Newly Developed Elderly Nutrient-Rich Food Score Is a Useful Tool to Assess Nutrient Density in European Older Adults. *Front. Nutr.* **2019**, *6*, 119. [CrossRef]
125. Mezzetti, A.; Pierdomenico, S.D.; Costantini, F.; Romano, F.; De Cesare, D.; Cuccurullo, F.; Imbastaro, T.; Riario-Sforza, G.; Di Giacomo, F.; Zuliani, G. Copper/zinc ratio and systemic oxidant load: Effect of aging and aging-related degenerative diseases. *Free. Radic. Biol. Med.* **1998**, *25*, 676–681. [CrossRef]
126. Kawahara, M.; Tanaka, K.I.; Kato-Negishi, M. Crosstalk of copper and zinc in the pathogenesis of vascular dementia. *J. Clin. Biochem. Nutr.* **2022**, *71*, 7–15. [CrossRef] [PubMed]
127. Fang, A.; Zhu, H. Serum Copper and Zinc Levels and Risk of Hepatocellular Carcinoma: A 1: 1 Matched Case-Control Study. *Curr. Dev. Nutr.* **2020**, *4*, 1397. [CrossRef]
128. Lin, Y.; Wu, C.; Yan, W.; Guo, S.; Liu, B. Five serum trace elements associated with risk of cardia and noncardia gastric cancer in a matched case–control study. *Cancer Manag. Res.* **2020**, *12*, 4441. [CrossRef] [PubMed]
129. Emokpae, M.A.; Fatimehin, E.B.; Obazelu, P.A. Serum levels of copper, zinc and disease severity scores in sickle cell disease patients in Benin City, Nigeria. *Afr. Health Sci.* **2019**, *19*, 2798–2805. [CrossRef] [PubMed]

130. Pourfallah, F.; Javadian, S.; Zamani, Z.; Saghiri, R.; Sadeghi, S.; Zarea, B.; Faiaz, S.; Mirkhani, F.; Fatemi, N. Evaluation of serum levels of zinc, copper, iron, and zinc/copper ratio in cutaneous leishmaniasis. *Iran. J. Arthropod-Borne Dis.* **2009**, *3*, 7.
131. Tuncay, M.E.; Neselioglu, S.; Kalkan, E.A.; Inan, O.; Akkus, M.S.; Ates, I.; Erel, O. Modified Proline Metabolism and Prolidase Enzyme in COVID-19. *Lab. Med.* **2022**, *53*, 453–458. [CrossRef]
132. Farr, B.M.; Conner, E.M.; Betts, R.F.; Oleske, J.; Minnefor, A.; Gwaltney, J.M., Jr. Two randomized controlled trials of zinc gluconate lozenge therapy of experimentally induced rhinovirus colds. *Antimicrob Agents Chemother* **1987**, *31*, 1183–1187. [CrossRef]
133. Mossad, S.B.; Macknin, M.L.; Medendorp, S.V.; Mason, P. Zinc gluconate lozenges for treating the common cold. A randomized, double-blind, placebo-controlled study. *Ann. Intern. Med.* **1996**, *125*, 81–88. [CrossRef]
134. Kurugöl, Z.; Akilli, M.; Bayram, N.; Koturoglu, G. The prophylactic and therapeutic effectiveness of zinc sulphate on common cold in children. *Acta Paediatr.* **2006**, *95*, 1175–1181. [CrossRef]
135. Kahmann, L.; Uciechowski, P.; Warmuth, S.; Plümäkers, B.; Gressner, A.M.; Malavolta, M.; Mocchegiani, E.; Rink, L. Zinc supplementation in the elderly reduces spontaneous inflammatory cytokine release and restores T cell functions. *Rejuvenation Res.* **2008**, *11*, 227–237. [CrossRef]
136. Zazzo, J.F.; Rouveix, B.; Rajagopalon, P.; Levacher, M.; Girard, P.M. Effect of zinc on the immune status of zinc-depleted AIDS related complex patients. *Clin. Nutr.* **1989**, *8*, 259–261. [CrossRef] [PubMed]
137. Bobat, R.; Coovadia, H.; Stephen, C.; Naidoo, K.L.; McKerrow, N.; Black, R.E.; Moss, W.J. Safety and efficacy of zinc supplementation for children with HIV-1 infection in South Africa: A randomised double-blind placebo-controlled trial. *Lancet* **2005**, *366*, 1862–1867. [CrossRef] [PubMed]
138. Green, J.A.; Lewin, S.R.; Wightman, F.; Lee, M.; Ravindran, T.S.; Paton, N.I. A randomised controlled trial of oral zinc on the immune response to tuberculosis in HIV-infected patients. *Int. J. Tuberc. Lung Dis.* **2005**, *9*, 1378–1384. [PubMed]
139. Fawzi, W.W.; Villamor, E.; Msamanga, G.I.; Antelman, G.; Aboud, S.; Urassa, W.; Hunter, D. Trial of zinc supplements in relation to pregnancy outcomes, hematologic indicators, and T cell counts among HIV-1-infected women in Tanzania. *Am. J. Clin. Nutr.* **2005**, *81*, 161–167. [CrossRef]
140. Villamor, E.; Aboud, S.; Koulinska, I.N.; Kupka, R.; Urassa, W.; Chaplin, B.; Msamanga, G.; Fawzi, W.W. Zinc supplementation to HIV-1-infected pregnant women: Effects on maternal anthropometry, viral load, and early mother-to-child transmission. *Eur. J. Clin. Nutr.* **2006**, *60*, 862–869. [CrossRef]
141. Deloria-Knoll, M.; Steinhoff, M.; Semba, R.D.; Nelson, K.; Vlahov, D.; Meinert, C.L. Effect of zinc and vitamin A supplementation on antibody responses to a pneumococcal conjugate vaccine in HIV-positive injection drug users: A randomized trial. *Vaccine* **2006**, *24*, 1670–1679. [CrossRef]
142. Baum, M.K.; Lai, S.; Sales, S.; Page, J.B.; Campa, A. Randomized, controlled clinical trial of zinc supplementation to prevent immunological failure in HIV-infected adults. *Clin. Infect. Dis.* **2010**, *50*, 1653–1660. [CrossRef]
143. Asdamongkol, N.; Phanachet, P.; Sungkanuparph, S. Low plasma zinc levels and immunological responses to zinc supplementation in HIV-infected patients with immunological discordance after antiretroviral therapy. *Jpn. J. Infect. Dis.* **2013**, *66*, 469–474. [CrossRef]
144. Lodi, S.; Freiberg, M.; Gnatienko, N.; Blokhina, E.; Yaroslavtseva, T.; Krupitsky, E.; Murray, E.; Samet, J.H.; Cheng, D.M. Per-protocol analysis of the ZINC trial for HIV disease among alcohol users. *Trials* **2021**, *22*, 226. [CrossRef]
145. Silva, M.; Montes, C.G.; Canals, A.; Mackenna, M.J.; Wolff, M. Role and effects of zinc supplementation in HIV-infected patients with immunovirological discordance: A randomized, double blind, case control study. *PLoS ONE* **2021**, *16*, e0244823. [CrossRef]
146. Freiberg, M.S.; Cheng, D.M.; Gnatienko, N.; Blokhina, E.; Coleman, S.M.; Doyle, M.F.; Yaroslavtseva, T.; Bridden, C.; So-Armah, K.; Tracy, R.; et al. Effect of Zinc Supplementation vs Placebo on Mortality Risk and HIV Disease Progression Among HIV-Positive Adults with Heavy Alcohol Use: A Randomized Clinical Trial. *JAMA Netw. Open* **2020**, *3*, e204330. [CrossRef]
147. Dirajlal-Fargo, S.; Yu, J.; Kulkarni, M.; Sattar, A.; Funderburg, N.; Barkoukis, H.; McComsey, G.A. Brief Report: Zinc Supplementation and Inflammation in Treated HIV. *J. Acquir. Immune. Defic. Syndr.* **2019**, *82*, 275–280. [CrossRef] [PubMed]
148. Lodha, R.; Shah, N.; Mohari, N.; Mukherjee, A.; Vajpayee, M.; Singh, R.; Singla, M.; Saini, S.; Bhatnagar, S.; Kabra, S.K. Immunologic effect of zinc supplementation in HIV-infected children receiving highly active antiretroviral therapy: A randomized, double-blind, placebo-controlled trial. *J. Acquir. Immune. Defic. Syndr.* **2014**, *66*, 386–392. [CrossRef]
149. Takagi, H.; Nagamine, T.; Abe, T.; Takayama, H.; Sato, K.; Otsuka, T.; Kakizaki, S.; Hashimoto, Y.; Matsumoto, T.; Kojima, A.; et al. Zinc supplementation enhances the response to interferon therapy in patients with chronic hepatitis C. *J. Viral Hepat.* **2001**, *8*, 367–371. [CrossRef] [PubMed]
150. Nagamine, T.; Takagi, H.; Takayama, H.; Kojima, A.; Kakizaki, S.; Mori, M.; Nakajima, K. Preliminary study of combination therapy with interferon-alpha and zinc in chronic hepatitis C patients with genotype 1b. *Biol. Trace Elem. Res.* **2000**, *75*, 53–63. [CrossRef] [PubMed]
151. Ko, W.S.; Guo, C.H.; Hsu, G.S.; Chiou, Y.L.; Yeh, M.S.; Yaun, S.R. The effect of zinc supplementation on the treatment of chronic hepatitis C patients with interferon and ribavirin. *Clin. Biochem.* **2005**, *38*, 614–620. [CrossRef]
152. Kim, K.I.; Kim, S.R.; Sasase, N.; Akimoto, Y.; Shikata, M.; Ohtani, A.; Hirooka, T.; Tanaka, K. Blood cell, liver function, and response changes by PEG-interferon-alpha2b plus ribavirin with polaprezinc therapy in patients with chronic hepatitis C. *Hepatol. Int.* **2008**, *2*, 111–115. [CrossRef]
153. Suzuki, H.; Takagi, H.; Sohara, N.; Kanda, D.; Kakizaki, S.; Sato, K.; Mori, M. Triple therapy of interferon and ribavirin with zinc supplementation for patients with chronic hepatitis C: A randomized controlled clinical trial. *World J. Gastroenterol. WJG* **2006**, *12*, 1265. [CrossRef]

154. Himoto, T.; Hosomi, N.; Nakai, S.; Deguchi, A.; Kinekawa, F.; Matsuki, M.; Yachida, M.; Masaki, T.; Kurokochi, K.; Watanabe, S. Efficacy of zinc administration in patients with hepatitis C virus-related chronic liver disease. *Scand. J. Gastroenterol.* **2007**, *42*, 1078–1087. [CrossRef]
155. Matsumura, H.; Nirei, K.; Nakamura, H.; Arakawa, Y.; Higuchi, T.; Hayashi, J.; Yamagami, H.; Matsuoka, S.; Ogawa, M.; Nakajima, N. Zinc supplementation therapy improves the outcome of patients with chronic hepatitis C. *J. Clin. Biochem. Nutr.* **2012**, *51*, 178–184. [CrossRef]
156. Murakami, Y.; Koyabu, T.; Kawashima, A.; Kakibuchi, N.; Kawakami, T.; Takaguchi, K.; Kita, K.; Okita, M. Zinc supplementation prevents the increase of transaminase in chronic hepatitis C patients during combination therapy with pegylated interferon α-2b and ribavirin. *J. Nutr. Sci. Vitaminol.* **2007**, *53*, 213–218. [CrossRef] [PubMed]
157. Rerksuppaphol, S.; Rerksuppaphol, L. A randomized controlled trial of zinc supplementation as adjuvant therapy for dengue viral infection in Thai children. *Int. J. Prev. Med.* **2018**, *9*, 88. [CrossRef] [PubMed]
158. Carlucci, P.M.; Ahuja, T.; Petrilli, C.; Rajagopalan, H.; Jones, S.; Rahimian, J. Zinc sulfate in combination with a zinc ionophore may improve outcomes in hospitalized COVID-19 patients. *J. Med. Microbiol.* **2020**, *69*, 1228. [CrossRef]
159. Frontera, J.A.; Rahimian, J.O.; Yaghi, S.; Liu, M.; Lewis, A.; Mainali, S.; Huang, J.; Scher, E.; Wisniewski, T.; Troxel, A.B. Treatment with Zinc is Associated with Reduced In-Hospital Mortality among COVID-19 Patients: A Multi-Center Cohort Study. *Res. Sq.* **2020**, rs-3–rs-94509. [CrossRef]
160. Derwand, R.; Scholz, M.; Zelenko, V. COVID-19 outpatients: Early risk-stratified treatment with zinc plus low-dose hydroxychloroquine and azithromycin: A retrospective case series study. *Int. J. Antimicrob. Agents* **2020**, *56*, 106214. [CrossRef] [PubMed]
161. Finzi, E. Treatment of SARS-CoV-2 with high dose oral zinc salts: A report on four patients. *Int. J. Infect. Dis.* **2020**, *99*, 307–309. [CrossRef]
162. Patel, O.; Chinni, V.; El-Khoury, J.; Perera, M.; Neto, A.S.; McDonald, C.; See, E.; Jones, D.; Bolton, D.; Bellomo, R. A pilot double-blind safety and feasibility randomized controlled trial of high-dose intravenous zinc in hospitalized COVID-19 patients. *J. Med. Virol.* **2021**, *93*, 3261–3267. [CrossRef]
163. Abd-Elsalam, S.; Soliman, S.; Esmail, E.S.; Khalaf, M.; Mostafa, E.F.; Medhat, M.A.; Ahmed, O.A.; El Ghafar, M.S.A.; Alboraie, M.; Hassany, S.M. Do zinc supplements enhance the clinical efficacy of hydroxychloroquine?: A randomized, multicenter trial. *Biol. Trace Elem. Res.* **2021**, *199*, 3642–3646. [CrossRef]
164. Stambouli, N.; Driss, A.; Gargouri, F.; Bahrini, K.; Arfaoui, B.; Abid, R.; Taamallah, K.; Hannachi, S.; Boughariou, S.; Rebai, A. COVID-19 prophylaxis with doxycycline and zinc in health care workers: A prospective, randomized, double-blind clinical trial. *Int. J. Infect. Dis.* **2022**, *122*, 553–558. [CrossRef]
165. Malik, A.; Taneja, D.K.; Devasenapathy, N.; Rajeshwari, K. Zinc supplementation for prevention of acute respiratory infections in infants: A randomized controlled trial. *Indian Pediatr.* **2014**, *51*, 780–784. [CrossRef]
166. Shah, U.H.; Abu-Shaheen, A.K.; Malik, M.A.; Alam, S.; Riaz, M.; Al-Tannir, M.A. The efficacy of zinc supplementation in young children with acute lower respiratory infections: A randomized double-blind controlled trial. *Clin. Nutr.* **2013**, *32*, 193–199. [CrossRef]
167. Martinez-Estevez, N.; Alvarez-Guevara, A.; Rodriguez-Martinez, C. Effects of zinc supplementation in the prevention of respiratory tract infections and diarrheal disease in Colombian children: A 12-month randomised controlled trial. *Allergol. Et Immunopathol.* **2016**, *44*, 368–375. [CrossRef] [PubMed]
168. Sazawal, S.; Black, R.E.; Jalla, S.; Mazumdar, S.; Sinha, A.; Bhan, M.K. Zinc supplementation reduces the incidence of acute lower respiratory infections in infants and preschool children: A double-blind, controlled trial. *Pediatrics* **1998**, *102*, 1–5. [CrossRef] [PubMed]
169. Mahalanabis, D.; Lahiri, M.; Paul, D.; Gupta, S.; Gupta, A.; Wahed, M.A.; Khaled, M.A. Randomized, double-blind, placebo-controlled clinical trial of the efficacy of treatment with zinc or vitamin A in infants and young children with severe acute lower respiratory infection. *Am. J. Clin. Nutr.* **2004**, *79*, 430–436. [CrossRef]
170. Karyadi, E.; West, C.E.; Schultink, W.; Nelwan, R.H.; Gross, R.; Amin, Z.; Dolmans, W.M.; Schlebusch, H.; van der Meer, J.W. A double-blind, placebo-controlled study of vitamin A and zinc supplementation in persons with tuberculosis in Indonesia: Effects on clinical response and nutritional status. *Am. J. Clin. Nutr.* **2002**, *75*, 720–727. [CrossRef] [PubMed]
171. Baruah, A.; Saikia, H. Effect of Zinc Supplementation in Children with Severe Pneumonia: A Randomised Controlled Study. *J. Clin. Diagn. Res.* **2018**, *12*, 8–11. [CrossRef]
172. Basnet, S.; Shrestha, P.S.; Sharma, A.; Mathisen, M.; Prasai, R.; Bhandari, N.; Adhikari, R.K.; Sommerfelt, H.; Valentiner-Branth, P.; Strand, T.A. A randomized controlled trial of zinc as adjuvant therapy for severe pneumonia in young children. *Pediatrics* **2012**, *129*, 701–708. [CrossRef]
173. Bose, A.; Coles, C.L.; John, H.; Moses, P.; Raghupathy, P.; Kirubakaran, C.; Black, R.E.; Brooks, W.A.; Santosham, M. Efficacy of zinc in the treatment of severe pneumonia in hospitalized children <2 y old. *Am. J. Clin. Nutr.* **2006**, *83*, 1089–1096.
174. Fataki, M.R.; Kisenge, R.R.; Sudfeld, C.R.; Aboud, S.; Okuma, J.; Mehta, S.; Spiegelman, D.; Fawzi, W.W. Effect of zinc supplementation on duration of hospitalization in Tanzanian children presenting with acute pneumonia. *J. Trop. Pediatr.* **2014**, *60*, 104–111. [CrossRef]
175. Howie, S.; Bottomley, C.; Chimah, O.; Ideh, R.; Ebruke, B.; Okomo, U.; Onyeama, C.; Donkor, S.; Rodrigues, O.; Tapgun, M. Zinc as an adjunct therapy in the management of severe pneumonia among Gambian children: Randomized controlled trial. *J. Glob. Health* **2018**, *8*, 010418. [CrossRef]

176. Sempértegui, F.; Estrella, B.; Rodríguez, O.; Gómez, D.; Cabezas, M.; Salgado, G.; Sabin, L.L.; Hamer, D.H. Zinc as an adjunct to the treatment of severe pneumonia in Ecuadorian children: A randomized controlled trial. *Am. J. Clin. Nutr.* **2014**, *99*, 497–505. [CrossRef] [PubMed]
177. Shah, G.S.; Dutta, A.K.; Shah, D.; Mishra, O.P. Role of zinc in severe pneumonia: A randomized double bind placebo controlled study. *Ital. J. Pediatr.* **2012**, *38*, 1–5. [CrossRef] [PubMed]
178. Srinivasan, M.G.; Ndeezi, G.; Mboijana, C.K.; Kiguli, S.; Bimenya, G.S.; Nankabirwa, V.; Tumwine, J.K. Zinc adjunct therapy reduces case fatality in severe childhood pneumonia: A randomized double blind placebo-controlled trial. *BMC Med.* **2012**, *10*, 14. [CrossRef]
179. Shehzad, N.; Anwar, M.I.; Muqaddas, T. Zinc supplementation for the treatment of severe pneumonia in hospitalized children: A randomized controlled trial. *Sudan. J. Paediatr.* **2015**, *15*, 37. [PubMed]
180. Qasemzadeh, M.J.; Fathi, M.; Tashvighi, M.; Gharehbeglou, M.; Yadollah-Damavandi, S.; Parsa, Y.; Rahimi, E. The effect of adjuvant zinc therapy on recovery from pneumonia in hospitalized children: A double-blind randomized controlled trial. *Scientifica* **2014**, *2014*, 694193. [CrossRef] [PubMed]
181. Mathur, N.; Bumb, R.; Mangal, H. Oral zinc in recurrent Erythema Nodosum Leprosum reaction. *Lepr. India* **1983**, *55*, 547–552.
182. Pakasi, T.A.; Karyadi, E.; Suratih, N.M.; Salean, M.; Darmawidjaja, N.; Bor, H.; van der Velden, K.; Dolmans, W.M.; van der Meer, J.W. Zinc and vitamin A supplementation fails to reduce sputum conversion time in severely malnourished pulmonary tuberculosis patients in Indonesia. *Nutr. J.* **2010**, *9*, 41. [CrossRef]
183. Zolfaghari, B.; Ghanbari, M.; Musavi, H.; Bavandpour Baghshahi, P.; Taghikhani, M.; Pourfallah, F. Investigation of Zinc Supplement Impact on the Serum Biochemical Parameters in Pulmonary Tuberculosis: A Double Blinded Placebo Control Trial. *Rep. Biochem. Mol. Biol.* **2021**, *10*, 173–182. [CrossRef] [PubMed]
184. Alam, A.N.; Sarker, S.A.; Wahed, M.A.; Khatun, M.; Rahaman, M.M. Enteric protein loss and intestinal permeability changes in children during acute shigellosis and after recovery: Effect of zinc supplementation. *Gut* **1994**, *35*, 1707–1711. [CrossRef]
185. Raqib, R.; Roy, S.K.; Rahman, M.J.; Azim, T.; Ameer, S.S.; Chisti, J.; Andersson, J. Effect of zinc supplementation on immune and inflammatory responses in pediatric patients with shigellosis. *Am. J. Clin. Nutr.* **2004**, *79*, 444–450. [CrossRef]
186. Rahman, M.J.; Sarker, P.; Roy, S.K.; Ahmad, S.M.; Chisti, J.; Azim, T.; Mathan, M.; Sack, D.; Andersson, J.; Raqib, R. Effects of zinc supplementation as adjunct therapy on the systemic immune responses in shigellosis. *Am. J. Clin. Nutr.* **2005**, *81*, 495–502. [CrossRef]
187. Roy, S.K.; Raqib, R.; Khatun, W.; Azim, T.; Chowdhury, R.; Fuchs, G.J.; Sack, D.A. Zinc supplementation in the management of shigellosis in malnourished children in Bangladesh. *Eur. J. Clin. Nutr.* **2008**, *62*, 849–855. [CrossRef]
188. Kashimura, H.; Suzuki, K.; Hassan, M.; Ikezawa, K.; Sawahata, T.; Watanabe, T.; Nakahara, A.; Mutoh, H.; Tanaka, N. Polaprezinc, a mucosal protective agent, in combination with lansoprazole, amoxycillin and clarithromycin increases the cure rate of Helicobacter pylori infection. *Aliment. Pharmacol. Ther.* **1999**, *13*, 483–487. [CrossRef]
189. Abd El-Ghaffar, Y.S.; Shouman, A.E.; Hakim, S.A.; El Gendy, Y.G.A.; Wahdan, M.M.M. Effect of Zinc Supplementation in Children Less than 5 Years on Diarrhea Attacks: A Randomized Controlled Trial. *Glob. Pediatr. Health* **2022**, *9*, 2333794X221099266. [CrossRef] [PubMed]
190. Hoque, K.M.; Binder, H.J. Zinc in the treatment of acute diarrhea: Current status and assessment. *Gastroenterology* **2006**, *130*, 2201–2205. [CrossRef]
191. Lamberti, L.M.; Fischer Walker, C.L.; Chan, K.Y.; Jian, W.-Y.; Black, R.E. Oral zinc supplementation for the treatment of acute diarrhea in children: A systematic review and meta-analysis. *Nutrients* **2013**, *5*, 4715–4740. [CrossRef]
192. Laghari, G.S.; Hussain, Z.; Shahzad, H. Effect of zinc supplementation on the frequency and consistency of stool in children with acute diarrhea. *Cureus* **2019**, *11*, e4217. [CrossRef] [PubMed]
193. Ahmadipour, S.; Mohsenzadeh, A.; Alimadadi, H.; Salehnia, M.; Fallahi, A. Treating viral diarrhea in children by probiotic and zinc supplements. *Pediatr. Gastroenterol. Hepatol. Nutr.* **2019**, *22*, 162–170. [CrossRef]
194. Barffour, M.A.; Hinnouho, G.M.; Wessells, K.R.; Kounnavong, S.; Ratsavong, K.; Sitthideth, D.; Bounheuang, B.; Sengnam, K.; Chanhthavong, B.; Arnold, C.D.; et al. Effects of therapeutic zinc supplementation for diarrhea and two preventive zinc supplementation regimens on the incidence and duration of diarrhea and acute respiratory tract infections in rural Laotian children: A randomized controlled trial. *J. Glob. Health* **2020**, *10*, 010424. [CrossRef] [PubMed]
195. Negruț, N. Zinc therapy for acute diarrhea in children under three years of age. *BMC Infect. Dis.* **2014**, *14*, P80. [CrossRef]
196. Zeba, A.N.; Sorgho, H.; Rouamba, N.; Zongo, I.; Rouamba, J.; Guiguemdé, R.T.; Hamer, D.H.; Mokhtar, N.; Ouedraogo, J.B. Major reduction of malaria morbidity with combined vitamin A and zinc supplementation in young children in Burkina Faso: A randomized double blind trial. *Nutr. J.* **2008**, *7*, 7. [CrossRef] [PubMed]
197. Shankar, A.H.; Genton, B.; Baisor, M.; Paino, J.; Tamja, S.; Adiguma, T.; Wu, L.; Rare, L.; Bannon, D.; Tielsch, J.M.; et al. The influence of zinc supplementation on morbidity due to Plasmodium falciparum: A randomized trial in preschool children in Papua New Guinea. *Am. J. Trop. Med. Hyg.* **2000**, *62*, 663–669. [CrossRef] [PubMed]
198. Bates, C.J.; Evans, P.H.; Dardenne, M.; Prentice, A.; Lunn, P.G.; Northrop-Clewes, C.A.; Hoare, S.; Cole, T.J.; Horan, S.J.; Longman, S.C.; et al. A trial of zinc supplementation in young rural Gambian children. *Br. J. Nutr.* **1993**, *69*, 243–255. [CrossRef] [PubMed]
199. Müller, O.; Becher, H.; van Zweeden, A.B.; Ye, Y.; Diallo, D.A.; Konate, A.T.; Gbangou, A.; Kouyate, B.; Garenne, M. Effect of zinc supplementation on malaria and other causes of morbidity in west African children: Randomised double blind placebo controlled trial. *BMJ* **2001**, *322*, 1567. [CrossRef]

200. Zinc Against Plasmodium Study Group. Effect of zinc on the treatment of Plasmodium falciparum malaria in children: A randomized controlled trial. *Am. J. Clin. Nutr.* **2002**, *76*, 805–812. [CrossRef]
201. Richard, S.A.; Zavaleta, N.; Caulfield, L.E.; Black, R.E.; Witzig, R.S.; Shankar, A.H. Zinc and iron supplementation and malaria, diarrhea, and respiratory infections in children in the Peruvian Amazon. *Am. J. Trop. Med. Hyg.* **2006**, *75*, 126–132. [CrossRef]
202. Darling, A.M.; Mugusi, F.M.; Etheredge, A.J.; Gunaratna, N.S.; Abioye, A.I.; Aboud, S.; Duggan, C.; Mongi, R.; Spiegelman, D.; Roberts, D.; et al. Vitamin A and Zinc Supplementation among Pregnant Women to Prevent Placental Malaria: A Randomized, Double-Blind, Placebo-Controlled Trial in Tanzania. *Am. J. Trop. Med. Hyg.* **2017**, *96*, 826–834. [CrossRef]
203. Somé, J.W.; Abbeddou, S.; Yakes Jimenez, E.; Hess, S.Y.; Ouédraogo, Z.P.; Guissou, R.M.; Vosti, S.A.; Ouédraogo, J.B.; Brown, K.H. Effect of zinc added to a daily small-quantity lipid-based nutrient supplement on diarrhoea, malaria, fever and respiratory infections in young children in rural Burkina Faso: A cluster-randomised trial. *BMJ Open* **2015**, *5*, e007828. [CrossRef]
204. Owusu-Agyei, S.; Newton, S.; Mahama, E.; Febir, L.G.; Ali, M.; Adjei, K.; Tchum, K.; Alhassan, L.; Moleah, T.; Tanumihardjo, S.A. Impact of vitamin A with zinc supplementation on malaria morbidity in Ghana. *Nutr. J.* **2013**, *12*, 131. [CrossRef]
205. Veenemans, J.; Milligan, P.; Prentice, A.M.; Schouten, L.R.; Inja, N.; van der Heijden, A.C.; de Boer, L.C.; Jansen, E.J.; Koopmans, A.E.; Enthoven, W.T.; et al. Effect of supplementation with zinc and other micronutrients on malaria in Tanzanian children: A randomised trial. *PLoS Med.* **2011**, *8*, e1001125. [CrossRef]
206. Sharquie, K.E.; Najim, R.A.; Farjou, I.B.; Al-Timimi, D.J. Oral zinc sulphate in the treatment of acute cutaneous leishmaniasis. *Clin. Exp. Dermatol.* **2001**, *26*, 21–26. [CrossRef] [PubMed]
207. Guzman-Rivero, M.; Rojas, E.; Verduguez-Orellana, A.; Pardo, H.; Torrico, M.C.; Cloetens, L.; Akesson, B.; Sejas, E. Nutritional status in patients with cutaneous leishmaniasis and a study of the effects of zinc supplementation together with antimony treatment. *Food Nutr. Res.* **2014**, *58*, 23353. [CrossRef] [PubMed]
208. Farajzadeh, S.; Ahmadi, R.; Mohammadi, S.; Pardakhty, A.; Khalili, M.; Aflatoonian, M. Evaluation of the efficacy of intralesional Glucantime plus niosomal zinc sulphate in comparison with intralesional Glucantime plus cryotherapy in the treatment of acute cutaneous leishmaniasis, a randomized clinical trial. *J. Parasit. Dis.* **2018**, *42*, 616–620. [CrossRef]
209. Carbone, D.C.B.; Zanoni, L.Z.G.; Cônsolo, F.Z.; Sanches, S.C.; Reis, V.Q.D.; Muller, K.T.C.; Carvalho, C.M.E.; Silva, M.C. Potential role of zinc in the visceromegaly regression and recovery of hematological parameters during treatment of visceral leishmaniasis in children from an endemic area. *Rev. Inst. Med. Trop. Sao Paulo* **2018**, *60*, e50. [CrossRef] [PubMed]
210. Maleki, M.; Karimi, G.; Tafaghodi, M.; Raftari, S.; Nahidi, Y. Comparison of intralesional two percent zinc sulfate and glucantime injection in treatment of acute cutaneous leishmaniasis. *Indian J. Dermatol.* **2012**, *57*, 118–122. [CrossRef] [PubMed]
211. Iraji, F.; Vali, A.; Asilian, A.; Shahtalebi, M.A.; Momeni, A.Z. Comparison of intralesionally injected zinc sulfate with meglumine antimoniate in the treatment of acute cutaneous leishmaniasis. *Dermatology* **2004**, *209*, 46–49. [CrossRef] [PubMed]
212. Xie, J.; Zhu, L.; Zhu, T.; Jian, Y.; Ding, Y.; Zhou, M.; Feng, X. Zinc supplementation reduces Candida infections in pediatric intensive care unit: A randomized placebo-controlled clinical trial. *J. Clin. Biochem. Nutr.* **2019**, *64*, 170–173. [CrossRef]
213. Nathan, C. Nonresolving inflammation redux. *Immunity* **2022**, *55*, 592–605. [CrossRef]
214. Gudernatsch, V.; Stefańczyk, S.A.; Mirakaj, V. Novel Resolution Mediators of Severe Systemic Inflammation. *ImmunoTargets Ther.* **2020**, *9*, 31–41. [CrossRef]
215. Gammoh, N.Z.; Rink, L. Zinc in Infection and Inflammation. *Nutrients* **2017**, *9*, 624. [CrossRef]
216. de Oliveira, S.; Rosowski, E.E.; Huttenlocher, A. Neutrophil migration in infection and wound repair: Going forward in reverse. *Nat. Rev. Immunol.* **2016**, *16*, 378–391. [CrossRef] [PubMed]
217. Takeuchi, O.; Akira, S. Pattern recognition receptors and inflammation. *Cell* **2010**, *140*, 805–820. [CrossRef]
218. Hasan, R.; Rink, L.; Haase, H. Zinc signals in neutrophil granulocytes are required for the formation of neutrophil extracellular traps. *Innate Immun.* **2013**, *19*, 253–264. [CrossRef] [PubMed]
219. Hasan, R.; Rink, L.; Haase, H. Chelation of Free Zn^{2+} Impairs Chemotaxis, Phagocytosis, Oxidative Burst, Degranulation, and Cytokine Production by Neutrophil Granulocytes. *Biol. Trace Elem. Res.* **2016**, *171*, 79–88. [CrossRef] [PubMed]
220. Businco, L.; Menghi, A.M.; Rossi, P.; D'Amelio, R.; Galli, E. Zinc-dependent chemotactic defect in an infant with acrodermatitis. *Arch. Dis. Child.* **1980**, *55*, 966–968. [CrossRef]
221. Bellomo, E.; Birla Singh, K.; Massarotti, A.; Hogstrand, C.; Maret, W. The metal face of protein tyrosine phosphatase 1B. *Coord. Chem. Rev.* **2016**, *327–328*, 70–83. [CrossRef] [PubMed]
222. Mustelin, T.; Vang, T.; Bottini, N. Protein tyrosine phosphatases and the immune response. *Nat. Rev. Immunol.* **2005**, *5*, 43–57. [CrossRef]
223. Hasegawa, H.; Suzuki, K.; Suzuki, K.; Nakaji, S.; Sugawara, K. Effects of zinc on the reactive oxygen species generating capacity of human neutrophils and on the serum opsonic activity in vitro. *Luminescence* **2000**, *15*, 321–327. [CrossRef] [PubMed]
224. Sheline, C.T.; Behrens, M.M.; Choi, D.W. Zinc-induced cortical neuronal death: Contribution of energy failure attributable to loss of NAD(+) and inhibition of glycolysis. *J. Neurosci.* **2000**, *20*, 3139–3146. [CrossRef]
225. Huang, J.; Hong, W.; Wan, M.; Zheng, L. Molecular mechanisms and therapeutic target of NETosis in diseases. *MedComm* **2022**, *3*, e162. [CrossRef]
226. Wessels, I.; Jansen, J.; Rink, L.; Uciechowski, P. Immunosenescence of polymorphonuclear neutrophils. *ScientificWorldJournal* **2010**, *10*, 145–160. [CrossRef] [PubMed]

227. Wessels, I.; Haase, H.; Engelhardt, G.; Rink, L.; Uciechowski, P. Zinc deficiency induces production of the proinflammatory cytokines IL-1β and TNFα in promyeloid cells via epigenetic and redox-dependent mechanisms. *J. Nutr. Biochem.* **2013**, *24*, 289–297. [CrossRef]
228. Wessels, I.; Cousins, R.J. Zinc dyshomeostasis during polymicrobial sepsis in mice involves zinc transporter Zip14 and can be overcome by zinc supplementation. *Am. J. Physiol. Gastrointest. Liver Physiol.* **2015**, *309*, G768–G778. [CrossRef] [PubMed]
229. Besecker, B.Y.; Exline, M.C.; Hollyfield, J.; Phillips, G.; Disilvestro, R.A.; Wewers, M.D.; Knoell, D.L. A comparison of zinc metabolism, inflammation, and disease severity in critically ill infected and noninfected adults early after intensive care unit admission. *Am. J. Clin. Nutr.* **2011**, *93*, 1356–1364. [CrossRef] [PubMed]
230. Jarosz, M.; Olbert, M.; Wyszogrodzka, G.; Młyniec, K.; Librowski, T. Antioxidant and anti-inflammatory effects of zinc. Zinc-dependent NF-κB signaling. *Inflammopharmacology* **2017**, *25*, 11–24. [CrossRef] [PubMed]
231. Perkins, N.D. Integrating cell-signalling pathways with NF-kappaB and IKK function. *Nat. Rev. Mol. Cell Biol.* **2007**, *8*, 49–62. [CrossRef] [PubMed]
232. Foster, M.; Samman, S. Zinc and regulation of inflammatory cytokines: Implications for cardiometabolic disease. *Nutrients* **2012**, *4*, 676–694. [CrossRef]
233. Haase, H.; Ober-Blöbaum, J.L.; Engelhardt, G.; Hebel, S.; Heit, A.; Heine, H.; Rink, L. Zinc signals are essential for lipopolysaccharide-induced signal transduction in monocytes. *J. Immunol.* **2008**, *181*, 6491–6502. [CrossRef]
234. Cho, J.; Tsichlis, P.N. Phosphorylation at Thr-290 regulates Tpl2 binding to NF-kappaB1/p105 and Tpl2 activation and degradation by lipopolysaccharide. *Proc. Natl. Acad. Sci. USA* **2005**, *102*, 2350–2355. [CrossRef]
235. Wan, Y.; Petris, M.J.; Peck, S.C. Separation of zinc-dependent and zinc-independent events during early LPS-stimulated TLR4 signaling in macrophage cells. *FEBS Lett.* **2014**, *588*, 2928–2935. [CrossRef]
236. Prasad, A.S.; Bao, B.; Beck, F.W.; Sarkar, F.H. Zinc-suppressed inflammatory cytokines by induction of A20-mediated inhibition of nuclear factor-κB. *Nutrition* **2011**, *27*, 816–823. [CrossRef] [PubMed]
237. Prasad, A.S.; Bao, B.; Beck, F.W.; Kucuk, O.; Sarkar, F.H. Antioxidant effect of zinc in humans. *Free. Radic. Biol. Med.* **2004**, *37*, 1182–1190. [CrossRef] [PubMed]
238. Boone, D.L.; Turer, E.E.; Lee, E.G.; Ahmad, R.C.; Wheeler, M.T.; Tsui, C.; Hurley, P.; Chien, M.; Chai, S.; Hitotsumatsu, O.; et al. The ubiquitin-modifying enzyme A20 is required for termination of Toll-like receptor responses. *Nat. Immunol.* **2004**, *5*, 1052–1060. [CrossRef] [PubMed]
239. von Bülow, V.; Dubben, S.; Engelhardt, G.; Hebel, S.; Plümäkers, B.; Heine, H.; Rink, L.; Haase, H. Zinc-dependent suppression of TNF-alpha production is mediated by protein kinase A-induced inhibition of Raf-1, I kappa B kinase beta, and NF-kappa B. *J. Immunol.* **2007**, *179*, 4180–4186. [CrossRef]
240. Brieger, A.; Rink, L.; Haase, H. Differential regulation of TLR-dependent MyD88 and TRIF signaling pathways by free zinc ions. *J. Immunol.* **2013**, *191*, 1808–1817. [CrossRef]
241. Wong, C.P.; Rinaldi, N.A.; Ho, E. Zinc deficiency enhanced inflammatory response by increasing immune cell activation and inducing IL6 promoter demethylation. *Mol. Nutr. Food Res.* **2015**, *59*, 991–999. [CrossRef] [PubMed]
242. Hu, J.; Yang, Z.; Wang, J.; Yu, J.; Guo, J.; Liu, S.; Qian, C.; Song, L.; Wu, Y.; Cheng, J. Zinc Chloride Transiently Maintains Mouse Embryonic Stem Cell Pluripotency by Activating Stat3 Signaling. *PLoS ONE* **2016**, *11*, e0148994. [CrossRef]
243. Fraker, P.; King, L. Changes in regulation of lymphopoiesis and myelopoiesis in the zinc-deficient mouse. *Nutr. Rev.* **1998**, *56*, S65–S69. [CrossRef]
244. Supasai, S.; Aimo, L.; Adamo, A.M.; Mackenzie, G.G.; Oteiza, P.I. Zinc deficiency affects the STAT1/3 signaling pathways in part through redox-mediated mechanisms. *Redox Biol.* **2017**, *11*, 469–481. [CrossRef]
245. Gruber, K.; Maywald, M.; Rosenkranz, E.; Haase, H.; Plumakers, B.; Rink, L. Zinc deficiency adversely influences interleukin-4 and interleukin-6 signaling. *J. Biol. Regul. Homeost. Agents* **2013**, *27*, 661–671.
246. Lue, C.; Kiyono, H.; McGhee, J.R.; Fujihashi, K.; Kishimoto, T.; Hirano, T.; Mestecky, J. Recombinant human interleukin 6 (rhIL-6) promotes the terminal differentiation of in vivo-activated human B cells into antibody-secreting cells. *Cell. Immunol.* **1991**, *132*, 423–432. [CrossRef] [PubMed]
247. Kitabayashi, C.; Fukada, T.; Kanamoto, M.; Ohashi, W.; Hojyo, S.; Atsumi, T.; Ueda, N.; Azuma, I.; Hirota, H.; Murakami, M.; et al. Zinc suppresses Th17 development via inhibition of STAT3 activation. *Int. Immunol.* **2010**, *22*, 375–386. [CrossRef] [PubMed]
248. Gustafson, C.E.; Kim, C.; Weyand, C.M.; Goronzy, J.J. Influence of immune aging on vaccine responses. *J. Allergy Clin. Immunol.* **2020**, *145*, 1309–1321. [CrossRef] [PubMed]
249. Beck, F.W.; Prasad, A.S.; Kaplan, J.; Fitzgerald, J.T.; Brewer, G.J. Changes in cytokine production and T cell subpopulations in experimentally induced zinc-deficient humans. *Am. J. Physiol.* **1997**, *272*, E1002–E1007. [CrossRef] [PubMed]
250. Lee, H.; Kim, B.; Choi, Y.H.; Hwang, Y.; Kim, D.H.; Cho, S.; Hong, S.J.; Lee, W.W. Inhibition of interleukin-1β-mediated interleukin-1 receptor-associated kinase 4 phosphorylation by zinc leads to repression of memory T helper type 17 response in humans. *Immunology* **2015**, *146*, 645–656. [CrossRef] [PubMed]
251. George, M.M.; Subramanian Vignesh, K.; Landero Figueroa, J.A.; Caruso, J.A.; Deepe, G.S., Jr. Zinc Induces Dendritic Cell Tolerogenic Phenotype and Skews Regulatory T Cell-Th17 Balance. *J. Immunol.* **2016**, *197*, 1864–1876. [CrossRef] [PubMed]
252. Faber, C.; Gabriel, P.; Ibs, K.H.; Rink, L. Zinc in pharmacological doses suppresses allogeneic reaction without affecting the antigenic response. *Bone Marrow Transpl.* **2004**, *33*, 1241–1246. [CrossRef]

253. Schubert, C.; Guttek, K.; Grüngreiff, K.; Thielitz, A.; Bühling, F.; Reinhold, A.; Brocke, S.; Reinhold, D. Oral zinc aspartate treats experimental autoimmune encephalomyelitis. *Biometals* **2014**, *27*, 1249–1262. [CrossRef]
254. Maywald, M.; Meurer, S.K.; Weiskirchen, R.; Rink, L. Zinc supplementation augments TGF-β1-dependent regulatory T cell induction. *Mol. Nutr. Food Res.* **2017**, *61*, 201600493. [CrossRef]
255. Rosenkranz, E.; Metz, C.H.; Maywald, M.; Hilgers, R.D.; Weßels, I.; Senff, T.; Haase, H.; Jäger, M.; Ott, M.; Aspinall, R.; et al. Zinc supplementation induces regulatory T cells by inhibition of Sirt-1 deacetylase in mixed lymphocyte cultures. *Mol. Nutr. Food Res.* **2016**, *60*, 661–671. [CrossRef]
256. Maywald, M.; Rink, L. Zinc supplementation induces CD4(+)CD25(+)Foxp3(+) antigen-specific regulatory T cells and suppresses IFN-γ production by upregulation of Foxp3 and KLF-10 and downregulation of IRF-1. *Eur. J. Nutr.* **2017**, *56*, 1859–1869. [CrossRef] [PubMed]
257. Fragale, A.; Gabriele, L.; Stellacci, E.; Borghi, P.; Perrotti, E.; Ilari, R.; Lanciotti, A.; Remoli, A.L.; Venditti, M.; Belardelli, F.; et al. IFN regulatory factor-1 negatively regulates CD4+ CD25+ regulatory T cell differentiation by repressing Foxp3 expression. *J. Immunol.* **2008**, *181*, 1673–1682. [CrossRef] [PubMed]
258. Cao, Z.; Wara, A.K.; Icli, B.; Sun, X.; Packard, R.R.; Esen, F.; Stapleton, C.J.; Subramaniam, M.; Kretschmer, K.; Apostolou, I.; et al. Kruppel-like factor KLF10 targets transforming growth factor-beta1 to regulate CD4(+)CD25(-) T cells and T regulatory cells. *J. Biol. Chem.* **2009**, *284*, 24914–24924. [CrossRef]
259. Cakman, I.; Rohwer, J.; Schütz, R.M.; Kirchner, H.; Rink, L. Dysregulation between TH1 and TH2 T cell subpopulations in the elderly. *Mech. Ageing Dev.* **1996**, *87*, 197–209. [CrossRef] [PubMed]
260. Prasad, A.S. Effects of zinc deficiency on Th1 and Th2 cytokine shifts. *J. Infect. Dis.* **2000**, *182* (Suppl. S1), S62–S68. [CrossRef] [PubMed]
261. Bao, B.; Prasad, A.S.; Beck, F.W.; Godmere, M. Zinc modulates mRNA levels of cytokines. *Am. J. Physiol. Endocrinol. Metab.* **2003**, *285*, E1095–E1102. [CrossRef] [PubMed]
262. Chiang, Y.J.; Hodes, R.J. T-cell development is regulated by the coordinated function of proximal and distal Lck promoters active at different developmental stages. *Eur. J. Immunol.* **2016**, *46*, 2401–2408. [CrossRef]
263. Kaltenberg, J.; Plum, L.M.; Ober-Blöbaum, J.L.; Hönscheid, A.; Rink, L.; Haase, H. Zinc signals promote IL-2-dependent proliferation of T cells. *Eur. J. Immunol.* **2010**, *40*, 1496–1503. [CrossRef]
264. Plum, L.M.; Brieger, A.; Engelhardt, G.; Hebel, S.; Nessel, A.; Arlt, M.; Kaltenberg, J.; Schwaneberg, U.; Huber, M.; Rink, L.; et al. PTEN-inhibition by zinc ions augments interleukin-2-mediated Akt phosphorylation. *Metallomics* **2014**, *6*, 1277–1287. [CrossRef]
265. Pinna, K.; Kelley, D.S.; Taylor, P.C.; King, J.C. Immune functions are maintained in healthy men with low zinc intake. *J. Nutr.* **2002**, *132*, 2033–2036. [CrossRef]
266. Kloubert, V.; Wessels, I.; Wolf, J.; Blaabjerg, K.; Janssens, V.; Hapala, J.; Wagner, W.; Rink, L. Zinc deficiency leads to reduced interleukin-2 production by active gene silencing due to enhanced CREMα expression in T cells. *Clin. Nutr.* **2021**, *40*, 3263–3278. [CrossRef] [PubMed]
267. Baarz, B.R.; Laurentius, T.; Wolf, J.; Wessels, I.; Bollheimer, L.C.; Rink, L. Short-term zinc supplementation of zinc-deficient seniors counteracts CREMα—Mediated IL-2 suppression. *Immun. Ageing* **2022**, *19*, 40. [CrossRef] [PubMed]
268. Southon, S.; Gee, J.M.; Johnson, I.T. Hexose transport and mucosal morphology in the small intestine of the zinc-deficient rat. *Br. J. Nutr.* **1984**, *52*, 371–380. [CrossRef]
269. Southon, S.; Gee, J.M.; Bayliss, C.E.; Wyatt, G.M.; Horn, N.; Johnson, I.T. Intestinal microflora, morphology and enzyme activity in zinc-deficient and Zn-supplemented rats. *Br. J. Nutr.* **1986**, *55*, 603–611. [CrossRef]
270. Elmes, M.E.; Jones, J.G. Ultrastructural changes in the small intestine of zinc deficient rats. *J. Pathol.* **1980**, *130*, 37–43. [CrossRef] [PubMed]
271. Southon, S.; Livesey, G.; Gee, J.M.; Johnson, I.T. Intestinal cellular proliferation and protein synthesis in zinc-deficient rats. *Br. J. Nutr.* **1985**, *53*, 595–603. [CrossRef]
272. Finamore, A.; Massimi, M.; Conti Devirgiliis, L.; Mengheri, E. Zinc deficiency induces membrane barrier damage and increases neutrophil transmigration in Caco-2 cells. *J. Nutr.* **2008**, *138*, 1664–1670. [CrossRef]
273. Park, J.H.; Grandjean, C.J.; Antonson, D.L.; Vanderhoof, J.A. Effects of short-term isolated zinc deficiency on intestinal growth and activities of several brush border enzymes in weaning rats. *Pediatr. Res.* **1985**, *19*, 1333–1336. [CrossRef]
274. Leal, J.; Smyth, H.D.C.; Ghosh, D. Physicochemical properties of mucus and their impact on transmucosal drug delivery. *Int. J. Pharm.* **2017**, *532*, 555–572. [CrossRef]
275. Johansson, M.E.; Hansson, G.C. Immunological aspects of intestinal mucus and mucins. *Nat. Rev. Immunol.* **2016**, *16*, 639–649. [CrossRef]
276. Maares, M.; Keil, C.; Koza, J.; Straubing, S.; Schwerdtle, T.; Haase, H. In Vitro Studies on Zinc Binding and Buffering by Intestinal Mucins. *Int. J. Mol. Sci.* **2018**, *19*, 2662. [CrossRef] [PubMed]
277. McGuckin, M.A.; Lindén, S.K.; Sutton, P.; Florin, T.H. Mucin dynamics and enteric pathogens. *Nat. Rev. Microbiol.* **2011**, *9*, 265–278. [CrossRef] [PubMed]
278. Shankar, A.H.; Prasad, A.S. Zinc and immune function: The biological basis of altered resistance to infection. *Am. J. Clin. Nutr.* **1998**, *68*, 447s–463s. [CrossRef] [PubMed]
279. Hassan, A.; Sada, K.K.; Ketheeswaran, S.; Dubey, A.K.; Bhat, M.S. Role of Zinc in Mucosal Health and Disease: A Review of Physiological, Biochemical, and Molecular Processes. *Cureus* **2020**, *12*, e8197. [CrossRef]
280. Canani, R.B.; Cirillo, P.; Buccigrossi, V.; Ruotolo, S.; Passariello, A.; De Luca, P.; Porcaro, F.; De Marco, G.; Guarino, A. Zinc inhibits cholera toxin-induced, but not Escherichia coli heat-stable enterotoxin-induced, ion secretion in human enterocytes. *J. Infect. Dis.* **2005**, *191*, 1072–1077. [CrossRef]

281. Bao, S.; Knoell, D.L. Zinc modulates cytokine-induced lung epithelial cell barrier permeability. *Am. J. Physiol. Lung Cell. Mol. Physiol.* **2006**, *291*, L1132–L1141. [CrossRef]
282. Roscioli, E.; Jersmann, H.P.; Lester, S.; Badiei, A.; Fon, A.; Zalewski, P.; Hodge, S. Zinc deficiency as a codeterminant for airway epithelial barrier dysfunction in an ex vivo model of COPD. *Int. J. Chron. Obstruct. Pulmon. Dis.* **2017**, *12*, 3503–3510. [CrossRef]
283. Lansdown, A.B.; Mirastschijski, U.; Stubbs, N.; Scanlon, E.; Agren, M.S. Zinc in wound healing: Theoretical, experimental, and clinical aspects. *Wound Repair Regen.* **2007**, *15*, 2–16. [CrossRef]
284. Flynn, J.L.; Chan, J.; Triebold, K.J.; Dalton, D.K.; Stewart, T.A.; Bloom, B.R. An essential role for interferon gamma in resistance to Mycobacterium tuberculosis infection. *J. Exp. Med.* **1993**, *178*, 2249–2254. [CrossRef]
285. Meydani, S.N.; Barnett, J.B.; Dallal, G.E.; Fine, B.C.; Jacques, P.F.; Leka, L.S.; Hamer, D.H. Serum zinc and pneumonia in nursing home elderly. *Am. J. Clin. Nutr.* **2007**, *86*, 1167–1173. [CrossRef]
286. Lassi, Z.S.; Kurji, J.; Oliveira, C.S.; Moin, A.; Bhutta, Z.A. Zinc supplementation for the promotion of growth and prevention of infections in infants less than six months of age. *Cochrane Database Syst. Rev.* **2020**, *4*, Cd010205. [CrossRef] [PubMed]
287. Rothan, H.A.; Byrareddy, S.N. The epidemiology and pathogenesis of coronavirus disease (COVID-19) outbreak. *J. Autoimmun.* **2020**, *109*, 102433. [CrossRef] [PubMed]
288. Ñamendys-Silva, S.A. Respiratory support for patients with COVID-19 infection. *Lancet Respir. Med.* **2020**, *8*, e18. [CrossRef]
289. Mantovani, A.; Morrone, M.C.; Patrono, C.; Santoro, M.G.; Schiaffino, S.; Remuzzi, G.; Bussolati, G. Long Covid: Where we stand and challenges ahead. *Cell Death Differ.* **2022**, *29*, 1891–1900. [CrossRef] [PubMed]
290. Te Velthuis, A.J.; van den Worm, S.H.; Sims, A.C.; Baric, R.S.; Snijder, E.J.; van Hemert, M.J. Zn^{2+} inhibits coronavirus and arterivirus RNA polymerase activity in vitro and zinc ionophores block the replication of these viruses in cell culture. *PLoS Pathog.* **2010**, *6*, e1001176. [CrossRef] [PubMed]
291. Woodworth, B.A.; Zhang, S.; Tamashiro, E.; Bhargave, G.; Palmer, J.N.; Cohen, N.A. Zinc increases ciliary beat frequency in a calcium-dependent manner. *Am. J. Rhinol. Allergy* **2010**, *24*, 6–10. [CrossRef]
292. Darma, A.; Ranuh, I.G.M.R.G.; Merbawani, W.; Setyoningrum, R.A.; Hidajat, B.; Hidayati, S.N.; Endaryanto, A.; Sudarmo, S.M. Zinc supplementation effect on the bronchial cilia length, the number of cilia, and the number of intact bronchial cell in zinc deficiency rats. *Indones. Biomed. J.* **2020**, *12*, 78–84. [CrossRef]
293. Lin, F.-c.; Young, H.A. Interferons: Success in anti-viral immunotherapy. *Cytokine Growth Factor Rev.* **2014**, *25*, 369–376. [CrossRef]
294. Read, S.A.; Obeid, S.; Ahlenstiel, C.; Ahlenstiel, G. The Role of Zinc in Antiviral Immunity. *Adv. Nutr.* **2019**, *10*, 696–710. [CrossRef]
295. Suara, R.O.; Crowe, J.E., Jr. Effect of zinc salts on respiratory syncytial virus replication. *Antimicrob. Agents Chemother.* **2004**, *48*, 783–790. [CrossRef]
296. Che, Z.; Sun, J. Investigation on relationship between whole blood zinc and Fe elements with children pneumonia caused by respiratory syncytial virus. *Int. J. Lab. Med.* **2016**, *2401–2402*, 2405.

biomolecules

MDPI

Review

Zinc's Association with the CmPn/CmP Signaling Network in Breast Cancer Tumorigenesis

Mellisa Renteria, Ofek Belkin, Justin Aickareth, David Jang, Majd Hawwar and Jun Zhang *

Department of Molecular and Translational Medicine (MTM), Texas Tech University Health Science Center El Paso, El Paso, TX 79905, USA
* Correspondence: jun.zhang2000@gmail.com or jh.zhang@ttuhsc.edu

Abstract: It is well-known that serum and cellular concentrations of zinc are altered in breast cancer patients. Specifically, there are notable zinc hyper-aggregates in breast tumor cells when compared to normal mammary epithelial cells. However, the mechanisms responsible for zinc accumulation and the consequences of zinc dysregulation are poorly understood. In this review, we detailed cellular zinc regulation/dysregulation under the influence of varying levels of sex steroids and breast cancer tumorigenesis to try to better understand the intricate relationship between these factors based on our current understanding of the CmPn/CmP signaling network. We also made some efforts to propose a relationship between zinc signaling and the CmPn/CmP signaling network.

Keywords: Zinc; CmPn signaling network; CmP signaling network; CCM signaling complex (CSC); progesterone (PRG); classic nuclear progesterone receptors (nPRs); non-classic membrane progesterone receptors (mPRs/PAQRs)

1. Introduction

Despite recent scientific advancements in cancer research, breast cancer remains one of the leading causes of cancer-related deaths globally and is expected to increase in parallel with the upward projection of the elderly population [1–3]. It is estimated that the majority of breast cancer subtypes are responsive to sex steroid hormones (particularly estrogen and progesterone) due to their expression of classic estrogen receptors (ERs), classic progesterone receptors (nPRs), or both, and are therefore termed hormone-responsive cancers [4,5]. Additional steroid hormone receptors, such as non-classic ERs/PRs, have been discovered in recent advances in the field [6–9]. In comparison, triple-negative breast cancer (TNBC) is a distinct form of breast cancer in which cells lack expression of ERs, nPRs, and epidermal growth factor receptor 2 (HER2) [10]. As a result, TNBC is considered a non-responsive breast cancer due to lack of sex steroid hormone-receptor expression and this property of TNBCs is largely responsible for the disease's aggressive nature, high metastatic potential and worse prognosis, as well as the limited therapeutic options [11,12]. Various properties that contribute to efficacious metastasis include cell motility, invasion, plasticity, and microenvironmental modulation [13].

Zinc, one of the essential micronutrients, is involved in a variety of cellular processes such as structural or catalytic components in an assortment of proteins and has been noted to be involved in these metastatic processes as well [13,14]. Zinc is required for the action of more than 300 metalloenzymes, 3000 transcription factors, and thousands of other cellular proteins, which play crucial roles in physiological processes, including antioxidant, anti-inflammatory and immune responses, as well as apoptosis [14,15]. Furthermore, zinc plays an important role in the polymeric organization of macromolecules such as DNA and RNA, protein synthesis, and cell division [16]. In addition, an estimated 10% of human proteins are known or predicted to bind zinc ions [17]. Aberrant zinc levels may lead to various human conditions, including cancers [18]. However, zinc supplementation could ameliorate the inflammation in cancer [18]. Unfortunately, research on the regulatory effect and influence of zinc deficiency during tumorigenesis remains elusive. As zinc supplementation has gradually been applied in nutritional support therapy for cancer patients, further exploration of the regulation of enzyme activity by zinc deficiency in cancer may provide comprehensive guidance for the application of zinc therapy [19]. Previous studies demonstrated that zinc concentration is especially important in the microenvironment modulation of tumor cells' acquisition of tumorigenic potential [13]. Additionally, there is evidence that zinc dyshomeostasis caused by a dysfunctional zinc transporter can contribute to the initiation of various cancers such as prostate, breast, and pancreatic cancers [20]. Serum and intracellular zinc concentrations are altered in breast cancer patients, and it remains unclear if this redistribution is a direct consequence of zinc nutrient alteration or a tissue-specific response [21]. Zinc signaling directly controls mammary gland development and aberrant signaling has been observed in malignant cells during breast cancer progression, accumulation, or the redistribution of zinc in the mammary gland [22].

2. Cellular Zinc Level Is Influenced by Dietary Supplementation

2.1. Cellular Zinc Concentrations Depend on Dietary Consumption

Minerals and other trace elements are essential micronutrients for normal physiological functioning and well-being [3]. The daily intake and homeostasis of these micronutrients are largely dependent on dietary habits [3]. Although zinc is the most abundant intracellular micronutrient, the human body is incapable of storing sufficient amounts of zinc; therefore, an inadequate diet can rapidly lead to zinc deficiencies [15,23]. It is estimated that nearly 50% of United States adults over the age of 50 are consuming suboptimal amounts of zinc and nearly two billion people worldwide may be zinc-deficient [15]. In rodents, nutritional zinc deficiencies have shown to predispose subjects to a reversible carcinogenesis, in which the subsequent replenishment of dietary zinc led to a reduction in cellular multiplicity and subsequent progression of the malignancy, likely by altering cellular proliferation and gene expression [24,25]. Epidemiological studies have shown that an inverse relationship exists between dietary zinc consumption and the development of breast cancer [23,26]. Specifically, evidence shows that biopsies of breast tumor cells contained significantly higher intracellular zinc concentrations and increased zinc transmembrane protein expression compared with normal tissue cells [26]. Therefore, the aberrant expression and homeostasis of zinc in breast tumors correlates with malignancy and could contribute to the severity of this cancer subtype [26].

2.2. Zinc Cellular Specific Actions

Zinc has shown to modulate immune system functioning, as well as the regulation of various metabolic, genetic, and cell-signaling pathways [14,15]. Studies have shown that zinc plays a protective role in tumor initiation and development by reducing oxidative stress and protecting DNA from reactive oxygen species (ROS) and the subsequent development of oncogenic mutations [27]. Specifically, zinc's function as an antioxidant provides genomic stability by decreasing oxidative DNA damage [26]. However, other studies suggest that cytotoxic levels of zinc are also known to cause DNA damage, oxidative stress, and the formation of ROS [20]. Several findings suggest that zinc's function is largely concentration- and cell-specific and this property of zinc may result in both pro-apoptotic and anti-apoptotic properties of the micronutrient [14]. While zinc has been shown to play a role in numerous malignancies, as a result of the complex nature of zinc homeostasis, a delineated relationship between zinc and tumorigenesis has yet to be established [14].

3. Zinc Plays a Significant Role in Tumorigenesis

3.1. Function of Zinc Contributes to the Progression of Cell Tumorigenesis

The biological effects of zinc can be exerted through the intra- and extracellular zinc regulatory functions and its interactions with proteins [28]. Zinc can act as either extracellular stimuli or intracellular messengers. Therefore, a precise working model of zinc regulatory mechanisms is needed to obtain a better understanding of homeostatic control for transients, subcellular distribution and trafficking, organellar homeostasis, and vesicular storage and exocytosis of zinc ions [29]. The vast majority (95%) of zinc is located intracellularly so that the extracellular concentration available is low. Intracellular zinc homeostatic molecules include cytosolic zinc-binding proteins, transporters localized to cytoplasmic and organellar membranes, and sensors of cytoplasmic free zinc ions [30]. Circulatory zinc is mainly bound to albumin, transferrin, and α2-macroglobulin but remains accessible to zinc transporters to control the cellular zinc balance [31]. Intracellular levels of zinc are largely coordinated by zinc transport channels, which are capable of both zinc influx and efflux [15]. There is evidence suggesting that zinc accumulates in elevated levels in breast cancer cells and other malignant cell lines in comparison to normal mammary epithelial cells [32]. Specifically, breast cancer cells showed a 72% increase in intracellular zinc concentrations in comparison to other, non-malignant breast cells, while serum levels of zinc were decreased from baseline [33]. One meta-analysis of 36 studies containing more than 5700 patients found significantly decreased serum zinc levels in breast cancer patients in comparison to controls and patients with benign breast diseases [34]. The findings of increased intracellular zinc concentrations were also observed at the histological level, suggesting that cellular zinc concentrations may be clinically useful in determining malignancy grading, as well as serving as predictive biomarkers for breast cancers [35]. However, the exact mechanisms that are responsible for the accumulation and dysregulation of zinc in breast tumors are not well understood and it remains unclear as to whether intracellular zinc accumulation causes the disease or is a consequence of this disease [15,32].

3.2. Zinc Transport Protein in Breast Cancer Cells

While the exact mechanisms of zinc's role in tumorigenesis is not well-understood, there are many speculations. Zinc acts as a signaling molecule, and both its intracellular and extracellular concentrations must be tightly regulated for proper physiological functioning [36]. Therefore, there is a complex regulatory system for the precise homeostatic control of cellular zinc transport, distribution, trafficking, organelle homeostasis, vesicular storage, and exocytosis of zinc ions [29,37]. There are several regulators of free intracellular zinc, such as zinc transporters, inhibitory factors, and sensors. Among the zinc transporters, Zrt-/Irt-like proteins (*ZIPs*) are most frequently studied [38–40]. Cellular zinc levels are strictly controlled by two families of transport proteins: ZIP channels (SLC39A) and ZnT transporters (SLC30A). ZIP channels increase cytosolic zinc levels by importing zinc into cells or releasing zinc from endoplasmic reticulum (ER) [41–43]. One subfamily of ZIP, estrogen-regulated LIV-1 (SLC39A6) has been implicated in breast cancer [42,44]. For example, the gene expression levels of LIV-1, a membranous zinc transporter, have been shown to increase four-fold under exposure to estrogen treatment [23,24]. Similarly, increased expression levels of LIV-1 were also observed under progesterone (PRG) treatment [23]. LIV-1 is one of the few zinc transporters identified to contain a metalloproteinase motif, which is responsible for breaking down the basement membrane and allowing for the metastasis of breast cancer cells [25]. Likewise, studies have shown that the expression levels of another zinc transporter, ZIP10, were significantly increased in metastatic breast cancer cell lines (such as MDA-MB-231 and MDA-MB-435S) when compared to less invasive cell lines (such as T47D, MCF7, ZR75-1, and ZR75-30). In addition, attenuating ZIP10 or intracellular zinc chelation in MDA-MB-231 cell lines lead to the inhibition of malignant cell migration [18,26]. Furthermore, ZIP7 was found to be able to release zinc from the ER, which leads to zinc-mediated tyrosine kinase signaling to activate cell migration and growth. This result suggests that ZIP7 might be a novel therapeutic target for breast cancer [45]. Contradicting results for the roles of LIV1 in breast cancer tumorigenesis have been reported: some studies demonstrate that ER-positive(+) breast cancer cells, which, according to the aforementioned mechanism above, have increased LIV-1 and intracellular zinc levels, are associated with better outcomes [46], as they are responsive to anti-estrogenic therapies such as Tamoxifen and aromatase inhibitors [27]. Similarly, ZIP6 deficiency disturbs intracellular Zn(2+) homeostasis, leading to increased cell survival [44]. However, breast samples from patients showed significant increases in both ZIP7 and ZIP6 in tumors, and the Kaplan–Meier curve revealed that high ZIP7 levels are correlated with decreased overall survival of patients [47]. These contradicting results can be explained by the tumor-specific hormonal response for different members of ZIP. Another cellular zinc transport receptor, ZnT2, has been shown to function in zinc sequestration, protecting cells from the cytotoxic effects of excess intracellular zinc, ROS formation, and subsequent apoptosis [28]. This study demonstrated that increased expression levels of ZnT2 transporters in malignant breast cancer cells protects these cells from apoptosis and that, conversely, tumor cells with decreased expression levels of ZnT2 transporters were less viable [28]. As a result, attenuating intracellular zinc-sequestering mechanisms may be a viable strategy for treating malignant breast cancers [28]. One study displayed a correlation between zinc concentration and histological and molecular grading and subtypes, showing elevated zinc levels in TNBC [29]. This study also displayed that increased intracellular levels of zinc were correlated with increased aggressiveness of breast cancers, with the highest zinc concentrations being present in HER2-positive breast cancers and TNBCs [29]. Additional zinc cellular regulatory factors are zinc inhibitory factor (ZIF) and zinc-sensing G-protein coupled receptor (ZnR/GPR39). ZIF reduces free intracellular zinc by inhibiting zinc transport in the oocyte before ovulation [48]. As a G-protein coupled receptor, ZnR/GPR39 triggers intracellular Ca^{2+} release and subsequently activates downstream MAPK or PI3K/AKT pathways controlling cell proliferation [49]. ZnR/GPR39 activity has been found to be enhanced in breast cancer [50–52].

4. Zinc and the CmPn/CmP Signaling Network

4.1. Zinc Is a Critical Nutrient in Mammalian Female Reproductive System

Although zinc has been recognized as a critical nutrient in mammalian female reproduction, its specific role in the female reproductive system was only recently recognized [53–55]. Accumulated data indicates that zinc depletion causes multiple defects in the female reproductive system in mice [56]. Zinc deficiency in females can lead to series of reproductive problems, including impaired synthesis and secretion of follicle stimulating hormone (FSH) and luteinizing hormone (LH), abnormal ovarian development, disruption of the estrous cycle, increased risk of abortion, prolonged gestational periods, teratogenicity, stillbirths, difficulty in parturition, pre-eclampsia, toxemia, and low fetal birth weights [57]. In addition, it is well known that zinc deficiency during pregnancy in experimental animals results in fetal anomalies [58–60]. Furthermore, additional data indicated that maternal supplementation of zinc drastically improved female pregnancies in sheep [61]. However, this conclusion has not yet been fully validated in humans [62].

4.2. Zinc Supplementation for Cancer Prevention

Due to its crucial functions and effects on multiple processes in the tumorigenesis and progression of cancers, the potential impact of zinc supplementation on cancer prevention and intervention as an anticancer and antitoxicity agent cannot be ignored [63–65]. To date, some notable progress has been made in this area [66]. As an essential micronutrient in foods and mineral supplements, zinc has been examined in many clinical trials for the possible prevention of gynecological cancers [64], like breast [26] and ovarian cancers [65], gastrointestinal and hepatic cancers [67–70], and lung [71] and oral cancers [72], in addition to cardiovascular disease (CVD) [73,74]. However, if breast cancers sequester and use more zinc than normal cells, therapeutic use of an excessive amount of zinc needs to be examined in animal models before any attempt at clinical trials for cancer treatment.

4.3. Zinc Is an Essential Nutrient in PRG Biogenesis and PRG-Mediated Signaling

Zinc has been shown to function in the production of FSH and LH [75], which play a role in ovulation and the subsequent production of PRG. As a result, it has been determined that dietary zinc deficiencies have a negative impact on the developmental potential of oocytes [76,77], which could affect *corpus luteum* (CL) development and PRG production [78]. Likewise, serum zinc levels have been found to be significantly lower in PRG-associated reproductive disorders in sheep [79], suggesting that there is a positive correlation between zinc levels and PRG production. However, relevant reports on the relationship between zinc and PRG are not without controversy. One report determined that two weeks' supplementation with low-dose zinc had no significant effect on serum estradiol and PRG concentration in postmenopausal women [80]. Another report suggested that zinc may play an inhibitory role in the onset and maintenance of PRG production in a mouse study. However, the increase in PRG under zinc-depleted conditions could be due to the removal of inhibitory pathways of zinc-mediated signaling [78], as signaling through the zinc-binding the SMAD (mothers against decapentaplegic homolog) transcriptional pathway is known to inhibit PRG production [81–83]. Therefore, the accumulated data indicate that the potential impact of zinc on the female reproductive system should not be ignored [75].

As mentioned prior, zinc is a key modulator of mammary gland development and maintenance [84]. Similarly, PRG and nPRs are responsible for the growth in alveolar epithelial structures, milk production, and secretion [85]. Clinical data indicate that PRG is a risk factor for breast cancer and that alterations in PRG-nPR signaling pathways contribute to early-stage human breast cancer progression [86]. Furthermore, it has been demonstrated that zinc can change the binding properties of PRG to its nPRs in the human endometrial cytosol [87]. Membrane progesterone receptors/progestin and adipoQ receptors (mPRs; mPRα-ε is identical to PAQR5-9) share structural similarities with other progestin and adipoQ receptors (PAQRs) that have a binding pocket for free fatty acid [88]. Protein structural data indicated that PAQR1 has an arginine residue (positive charge) in the binding pocket that is occupied with an oleic acid. PAQR1 was only able to bind PRG under the concentration of 100 μM zinc. Similarly, since mPRα (PAQR7) has the same binding pocket, the addition of high concentrations of zinc would form a salt with the fatty acid in the binding pocket, allowing for PRG to bind mPRα [89]. Therefore, sufficient data indicate that zinc can modulate the binding of PRG to either nPRs and/or mPRs [9,87], three key components of the CmPn (CSC-mPRs-PRG-nPRs) signaling network.

4.4. Zinc Associated with the CSC

Since the CSC (CCM signaling complex) [90–93] is the first component in newly defined CmPn (CSC-mPRs-PRG-nPRs)/CmP (CSC-mPRs-PRG) signaling networks in nPR (+/−) breast cancer cells [94–96], its relationship with zinc will be explored before other key components of the CmPn/CmP signaling networks. It has been reported that acute and subacute hemorrhage were found in head and neck cancers [97,98]. Furthermore, adenocarcinomas have shown to exhibit a depletion of zinc due to the down-regulated gene expression of hZIP1, a zinc transporter [99]. This indicates that local zinc deficiency in cancer lesions is a critical early event in tumorigenesis and can indirectly lead to unexpected brain hemorrhage. Therefore, a similar dietary supplementation strategy has been proposed for the prevention of CVDs, similarly to the proposed supplementation for cancer prevention [74]. However, contradicting in vitro and in vivo data demonstrated that both zinc deficiency and over-supplementation were associated with disintegration of the blood–brain barrier (BBB) leading to hemorrhagic stroke [100–103], implicating zinc as an independent risk factor for hemorrhagic stroke [103]. This once again emphasizes the importance of the tight regulation of intra- or extracellular zinc concentrations in proper homeostatic function [28,29]. The core components of the CSC are composed of three CCM proteins (CCM1, 2, 3) [93]. Deficiencies in any one of these will compromise microvascular integrity [92,93,104] and lead to hemorrhagic stroke in familial CCMs [90], suggesting a possible link between the CSC and zinc. Zinc localization and transporters were reported to be altered in Ccm1 zebrafish models and, in the same study, transcriptome analyses showed that solute carrier family 39 member 6, or LIV1 with 9 ZIPs, 4–8, 10, 12–14 (SLC39) zinc transporter genes [42] are misregulated in brain microvascular endothelial cells (BMECs) isolated from *Ccm1* null mice. These results indicate a perturbed transcription of SLC39 zinc transporter genes in the depletion of *CCM1* [105]. Subsequently, SLC39 zinc transporters have been linked to the CSC signaling cascades, and been defined as the direct downstream target of hyper-activated MEKK3–MEK5–ERK5 kinase cascade [105,106]. SLC39 is currently recognized and utilized as a key factor in the CSC signaling in various reports [107–116]. Ironically, all these reports relied on a single report that showed a "conserved CCM complex" (Ccm1/2 in the CSC) modulates ERK-5/MAPK-KLF-3 signaling factor to facilitate the expression of the SLC39 ortholog (or paralog, 24–30% identity in amino acid sequence between human and *C. elegans/Drosophila*), zipt-2.3, in *C. elegans*. A deficiency of Ccm1/2 will lead to abnormal zinc transporter (SLC39) expression and zinc storage in the intestinal granules [105]. Recently, comparative genomics in *vertebrate and invertebrate CCM* models have demonstrated that, despite the fact that CCM3 is evolutionarily conserved as an ancient gene (40–48% identity in amino acid sequence between human and its invertebrate counterparts, *C. elegans/D. melanogaster*), in contrast to a previous report [105], CCM1 and CCM2 orthologs have not been identified in *invertebrates* [117–119]. This suggests one of two possibilities: either the CCM1/2 gene-pair may be the latest evolutionary newcomers in the CSC or, more likely, this gene pair form the CSC first and later recruits CCM3 into this signaling complex that specializes in angiogenic endothelial cell (EC) maintenance in the closed circulatory system of *vertebrates* [117,119]. Unlike *vertebrates*, most organisms in invertebrates, such as arthropods (*D. melanogaster*), have an open circulation system; however, nematodes (*C. elegans*) have *no* circulatory system at all. This fact calls the existence of only an invertebrate ortholog of CCM1 deposited in the genebank into question (in fact, a short appearance of a drosophila ortholog of CCM1 was quickly removed from the genebank, and an in *vertebrate* CCM2 ortholog was never published), which could be an CCM1-irrelevant gene involving the signaling cascades affecting *C. elegans* lifespan [120,121]. Therefore, excessive caution should be applied when dealing with these kri-1/ SLC39 associated signaling pathways/cascades in the CSC mediated pathogenesis of CCMs [107–113,115].

Interestingly, more data demonstrate the functional association of zinc with the CSC signaling in an indirect fashion, such as zinc-containing superoxide dismutase (SODs) and related reactive oxygen species (ROS) [122–125], a downstream target of SODs, matrix metalloproteinases (MMPs) [126] and its counterpart, a tissue inhibitor of metalloproteinase-1 (TIMP-1) [127–134].

4.5. Involvement of Zinc within the CmPn/CmP Signaling Network

It is clinically well-known that a variety of breast cancer subtypes are hormone-responsive and that these hormones have potentially debilitating effects on the growth, metastatic properties, and subsequent prognosis of breast cancer subtypes. As a matter of fact, it is estimated that nearly 70% of breast cancers express ERs, nPRs, or both [4,5]. The exact intracellular physiological mechanisms by which these hormones function remains largely unstudied. In a previous study, we provided evidence suggesting that the CCM signaling complex (CSC) functions as a bridge for crosstalk among nPRs, mPRs, and their ligands, to form what we proposed to be the CSC-mPRs-PRG-nPR (CmPn)/CmP signaling network [91,92,94–96,135–137]. Studies have shown variable expression patterns of mediators of this pathway across various breast cancers, suggesting that this pathway is largely involved in breast tumorigenesis [94–96,136]. Some breast cancer subtypes expressing mPRs have partially contributed to the discovery of mPR-specific PRG actions of the CmP signaling network (CSC-mPRs-PRG). Furthermore, the findings of elevated zinc levels in TNBCs, zinc's potential for modulating effects of PRG and nPRs, and the nucleocytoplasmic shuttling and localization properties of mPRs under the influence of zinc all call for further investigation into zinc and its function within the CmPn/CmP signaling network in the breast cancer tumorigenesis of TNBCs [138].

In this review, based on the presented evidence that zinc affects all key components of the CmPn/CmP signaling networks, we propose an intricate mechanism by which zinc functions within the CmPn/CmP signaling networks in modulating breast cancer tumorigenesis (Figure 1). We suggest that altered expression levels of CmPn/CmP mediators, as well as altered expression levels of ZIP channels (SLC39A) and ZnT transporters (SLC30A), and serum and intracellular zinc levels, may serve as diagnostic and prognostic biomarkers due to their shared involvement in the malignant transformation of breast cells [94–96,136]. In this review, we provided published evidence of a critical relationship between zinc and CmPn/CmP signaling networks, which may be responsible for the increased risks with breast cancers, necessitating further investigation into the physiologic dynamic between zinc and the CmPn/CmP signaling network (Table 1).

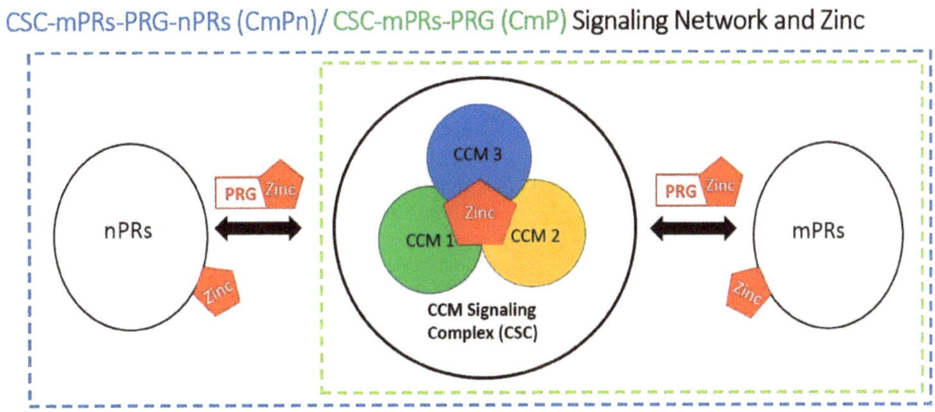

Figure 1. Signaling network bridge "crosstalk" among key players within CmPn signaling network. The diagram displays the relationship among PRG, nPR, mPR, CSC, and zinc (Zn). We show that the zinc interacts with all major players within the CmPn signaling network and involved in CmPn-modulating signaling during breast cancer tumorigenesis.

Table 1. Detailed information regarding the CmPn signaling network. A list of recent key publications describing CmPn and CmP signaling networks among nPR(+) and nPR(-) cancers and vascular ECs.

Main Point	Key Findings	Pubmed	References
The CSC is linked to tumorigenesis with mPRs in various cancers.	Differential expression of CCM and mPR correlated with various types and grades of major human cancers, especially breast and live cancers.	**PMID:** 32186778	[136,138]
Establishing the CmPn signaling network in nPR(+) breast cancers	The CSC role in coupling classic, non-classic, or combined PRG signaling pathways via the effects of an intricate homeostatic concentration of progesterone to form the CmPn signaling network	**PMID:** 35971177	[96,138]
Establishing the CmP signaling network in nPR(-) breast cancers	Through establishingCmP signaling network in nPR(-) breast cancers, we discovered novel biomarker signature panels for Triple-Negative Breast Cancers (TNBCs) between African and Caucasian Women	**PMID:** 35431232; 35481969	[94,95,138]
Establishing the CmP signaling network in nPR(-) vasular ECs	Deficiency of any CCM genes, in combination with mPR-specific PRG actions, leads to perturbed CmP signaling network in nPR(-) ECs both in vitro and in vivo, result in compromising blood brain barrier integrity and increases the risk of hemorrhage	**PMID:** 36077089; 35098046	[135,137,138]
Exploring molecuular signaling within the CmPn/CmP signaling networks with multiomics	molecuular signaling within the CmPn/CmP signaling networks were investigated with multiomics, such as RNAseq and proteomics. Major molecular pathways were presented with pathway analysis and visualization	**PMID:** 36077089; 35098046	[91,94–96,135,136]

5. Conclusions

The relationship between serum and cellular concentrations of zinc and tumorigenesis is rather intricate, and the exact mechanism by which zinc promotes tumorigenesis is not well understood. Furthermore, many studies that have previously studied the relationship between zinc and oncogenesis failed to elaborate on whether their findings relate to serum or intracellular concentrations of zinc and the association with the potential initiation of tumorigenesis. This has led to many contradicting results. In this review, we attempted to summarize the critical associations between dynamic cellular levels of zinc during breast tumorigenesis in an attempt to elaborate on the relationship between the sex steroid hormones, estrogen and PRG, their corresponding receptors, and zinc. We summarize our review using our proposed model, which depicts the mechanisms by which the CmPn/CmP signaling complex and zinc modulate breast cell tumorigenesis.

Author Contributions: J.Z.: Conceptualization, Methodology, Writing—Original draft preparation, Writing—Reviewing and Editing; M.R.: Writing—Original draft preparation, Writing—Reviewing and Editing; O.B.: Writing—Original draft preparation, Writing—Reviewing and Editing; J.A.: Writing—Original draft preparation, Writing—Reviewing and Editing; D.J.: Writing—Original draft preparation, Writing—Reviewing and Editing; M.H.: Writing—Original draft preparation, Writing—Reviewing and Editing. All authors have read and agreed to the published version of the manuscript.

Funding: This research received no external funding.

Institutional Review Board Statement: Not applicable.

Informed Consent Statement: Not applicable.

Data Availability Statement: Not applicable.

Acknowledgments: We wish to thank Johnathan Abou-Fadel for providing at Texas Tech University Health Science Center El Paso (TTUHSCEP) for his technical help during the experiments.

Conflicts of Interest: The authors declare no conflict of interest.

Abbreviations

Breast Cancer (Breast cancer); Cerebral cavernous malformations (CCMs); CCM signaling complex (CSC); CCM proteins (CCM1, 2, 3); progesterone (PRG); triple-negative breast cancer (TNBC); membrane progesterone receptors (mPRs); progestin and adipoQ receptors (PAQRs); nuclear Progesterone Receptors (nPRs); classic estrogen receptors (ERs); classic progesterone receptors (nPRs); CSC-mPRs-PRG-nPRs (CmPn); CSC-mPRs-PRG (CmPn); epidermal growth factor receptor 2 (HER2); reactive oxygen species (ROS); ER-positive [ER(+)]; nPR-positive [nPR(+)]; Zrt-/Irt-like protein (*ZIP*); zinc inhibitory factor (ZIF); zinc-sensing G-protein coupled receptor (ZnR/GPR39); secretion of follicle stimulating hormone (FSH); luteinizing hormone (LH); *corpus luteum* (CL); SMAD (mothers against decapentaplegic homolog); cardiovascular disorders (CVD); blood–brain barrier (BBB); superoxide dismutases (SODs); reactive oxygen species (ROS); matrix metalloproteinases (MMPs); thrombospondin (TSP); tissue inhibitor of metalloproteinases (TIMPs); tight junction (TJ); Genome-wide association studies (GWAS); knockout (KO); endothelial cells (ECs); solute carrier family 39 member 6 (SLC39); single nucleotide polymorphism (SNP).

References

1. Bauer, K.R.; Brown, M.; Cress, R.D.; Parise, C.A.; Caggiano, V. Descriptive analysis of estrogen receptor (ER)-negative, progesterone receptor (PR)-negative, and HER2-negative invasive breast cancer, the so-called triple-negative phenotype: A population-based study from the California cancer Registry. *Cancer* **2007**, *109*, 1721–1728. [CrossRef] [PubMed]
2. Baquet, C.R.; Mishra, S.I.; Commiskey, P.; Ellison, G.L.; DeShields, M. Breast cancer epidemiology in blacks and whites: Disparities in incidence, mortality, survival rates and histology. *J. Natl. Med. Assoc.* **2008**, *100*, 480–488. [CrossRef]
3. Venturelli, S.; Leischner, C.; Helling, T.; Renner, O.; Burkard, M.; Marongiu, L. Minerals and Cancer: Overview of the Possible Diagnostic Value. *Cancers* **2022**, *14*, 1256. [CrossRef] [PubMed]

4. Gonzalez-Angulo, A.M.; Morales-Vasquez, F.; Hortobagyi, G.N. Overview of resistance to systemic therapy in patients with breast cancer. *Adv. Exp. Med. Biol.* **2007**, *608*, 1–22. [CrossRef] [PubMed]
5. Kohler, B.A.; Sherman, R.L.; Howlader, N.; Jemal, A.; Ryerson, A.B.; Henry, K.A.; Boscoe, F.P.; Cronin, K.A.; Lake, A.; Noone, A.M.; et al. Annual Report to the Nation on the Status of Cancer, 1975–2011, Featuring Incidence of Breast Cancer Subtypes by Race/Ethnicity, Poverty, and State. *J. Natl. Cancer Inst.* **2015**, *107*, djv048. [CrossRef] [PubMed]
6. Fuentes, N.; Silveyra, P. Estrogen receptor signaling mechanisms. *Adv. Protein Chem. Struct. Biol.* **2019**, *116*, 135–170. [CrossRef]
7. Vail, G.; Roepke, T.A. Membrane-initiated estrogen signaling via Gq-coupled GPCR in the central nervous system. *Steroids* **2019**, *142*, 77–83. [CrossRef]
8. Azeez, J.M.; Susmi, T.R.; Remadevi, V.; Ravindran, V.; Sasikumar Sujatha, A.; Ayswarya, R.N.S.; Sreeja, S. New insights into the functions of progesterone receptor (PR) isoforms and progesterone signaling. *Am. J. Cancer Res.* **2021**, *11*, 5214–5232.
9. Thomas, P. Membrane Progesterone Receptors (mPRs, PAQRs): Review of Structural and Signaling Characteristics. *Cells* **2022**, *11*, 1785. [CrossRef]
10. Carey, L.A.; Perou, C.M.; Livasy, C.A.; Dressler, L.G.; Cowan, D.; Conway, K.; Karaca, G.; Troester, M.A.; Tse, C.K.; Edmiston, S.; et al. Race, breast cancer subtypes, and survival in the Carolina Breast Cancer Study. *JAMA* **2006**, *295*, 2492–2502. [CrossRef]
11. Perou, C.M.; Sorlie, T.; Eisen, M.B.; van de Rijn, M.; Jeffrey, S.S.; Rees, C.A.; Pollack, J.R.; Ross, D.T.; Johnsen, H.; Akslen, L.A.; et al. Molecular portraits of human breast tumours. *Nature* **2000**, *406*, 747–752. [CrossRef] [PubMed]
12. Goldhirsch, A.; Wood, W.C.; Coates, A.S.; Gelber, R.D.; Thurlimann, B.; Senn, H.J.; Panel, M. Strategies for subtypes—Dealing with the diversity of breast cancer: Highlights of the St. Gallen International Expert Consensus on the Primary Therapy of Early Breast Cancer 2011. *Ann. Oncol.* **2011**, *22*, 1736–1747. [CrossRef] [PubMed]
13. Vogel-Gonzalez, M.; Musa-Afaneh, D.; Rivera Gil, P.; Vicente, R. Zinc Favors Triple-Negative Breast Cancer's Microenvironment Modulation and Cell Plasticity. *Int. J. Mol. Sci.* **2021**, *22*, 9188. [CrossRef]
14. Franklin, R.B.; Costello, L.C. The Important Role of the Apoptotic Effects of Zinc in the Development of Cancers. *J. Cell Biochem.* **2009**, *106*, 750–757. [CrossRef] [PubMed]
15. Grattan, B.J.; Freake, H.C. Zinc and Cancer: Implications for LIV-1 in Breast Cancer. *Nutrients* **2012**, *4*, 648–675. [CrossRef] [PubMed]
16. McCall, K.A.; Huang, C.; Fierke, C.A. Function and mechanism of zinc metalloenzymes. *J. Nutr.* **2000**, *130*, 1437S–1446S. [CrossRef]
17. Andreini, C.; Banci, L.; Bertini, I.; Rosato, A. Counting the zinc-proteins encoded in the human genome. *J. Proteome Res.* **2006**, *5*, 196–201. [CrossRef]
18. Taccioli, C.; Chen, H.; Jiang, Y.; Liu, X.P.; Huang, K.; Smalley, K.J.; Farber, J.L.; Croce, C.M.; Fong, L.Y. Dietary zinc deficiency fuels esophageal cancer development by inducing a distinct inflammatory signature. *Oncogene* **2012**, *31*, 4550–4558. [CrossRef]
19. Cheng, Y.; Chen, H. Aberrance of Zinc Metalloenzymes-Induced Human Diseases and Its Potential Mechanisms. *Nutrients* **2021**, *13*, 4456. [CrossRef]
20. Wang, J.; Zhao, H.; Xu, Z.; Cheng, X. Zinc dysregulation in cancers and its potential as a therapeutic target. *Cancer Biol. Med.* **2020**, *17*, 612–625. [CrossRef]
21. Woo, W.; Xu, Z.M. Body zinc distribution profile during N-methyl-N-nitrosourea-induced mammary tumorigenesis in rats at various levels of dietary zinc intake. *Biol. Trace Elem Res.* **2002**, *87*, 157–169. [PubMed]
22. Chakraborty, M.; Hershfinkel, M. Zinc Signaling in the Mammary Gland: For Better and for Worse. *Biomedicines* **2021**, *9*, 1204. [CrossRef] [PubMed]
23. Skrajnowska, D.; Bobrowska-Korczak, B. Role of Zinc in Immune System and Anti-Cancer Defense Mechanisms. *Nutrients* **2019**, *11*, 2273. [CrossRef]
24. Fong, L.Y.Y.; Nguyen, V.T.; Farber, J.L. Esophageal cancer prevention in zinc-deficient rats: Rapid induction of apoptosis by replenishing zinc. *J. Natl. Cancer Inst.* **2001**, *93*, 1525–1533. [CrossRef] [PubMed]
25. Fong, L.Y.Y.; Jiang, Y.; Riley, M.; Liu, X.; Smalley, K.J.; Guttridge, D.C.; Farber, J.L. Prevention of upper aerodigestive tract cancer in zinc-deficient rodents: Inefficacy of genetic or pharmacological disruption of COX-2. *Int. J. Cancer* **2008**, *122*, 978–989. [CrossRef]
26. Alam, S.; Kelleher, S.L. Cellular mechanisms of zinc dysregulation: A perspective on zinc homeostasis as an etiological factor in the development and progression of breast cancer. *Nutrients* **2012**, *4*, 875–903. [CrossRef]
27. Yildiz, A.; Kaya, Y.; Tanriverdi, O. Effect of the Interaction Between Selenium and Zinc on DNA Repair in Association With Cancer Prevention. *J. Cancer Prev.* **2019**, *24*, 146–154. [CrossRef]
28. Maret, W. Regulation of Cellular Zinc Ions and Their Signaling Functions. In *Zinc Signaling*; Fukada, T., Kambe, T., Eds.; Springer: Singapore, 2019; pp. 5–22.
29. Maret, W. Zinc in Cellular Regulation: The Nature and Significance of "Zinc Signals". *Int. J. Mol. Sci.* **2017**, *18*, 2285. [CrossRef]
30. Colvin, R.A.; Holmes, W.R.; Fontaine, C.P.; Maret, W. Cytosolic zinc buffering and muffling: Their role in intracellular zinc homeostasis. *Metallomics* **2010**, *2*, 306–317. [CrossRef]
31. Lonergan, Z.R.; Skaar, E.P. Nutrient Zinc at the Host-Pathogen Interface. *Trends Biochem. Sci.* **2019**, *44*, 1041–1056. [CrossRef]
32. Chandler, P.; Kochupurakkal, B.S.; Alam, S.; Richardson, A.L.; Soybel, D.I.; Kelleher, S.L. Subtype-specific accumulation of intracellular zinc pools is associated with the malignant phenotype in breast cancer. *Mol. Cancer* **2016**, *15*, 2. [CrossRef] [PubMed]
33. Margalioth, E.J.; Schenker, J.G.; Chevion, M. Copper and zinc levels in normal and malignant tissues. *Cancer* **1983**, *52*, 868–872. [CrossRef]

34. Feng, Y.; Zeng, J.W.; Ma, Q.; Zhang, S.; Tang, J.; Feng, J.F. Serum copper and zinc levels in breast cancer: A meta-analysis. *J. Trace Elem. Med. Biol.* **2020**, *62*, 126629. [CrossRef] [PubMed]
35. Riesop, D.; Hirner, A.V.; Rusch, P.; Bankfalvi, A. Zinc distribution within breast cancer tissue: A possible marker for histological grading? *J. Cancer Res. Clin. Oncol.* **2015**, *141*, 1321–1331. [CrossRef]
36. Yamasaki, S.; Sakata-Sogawa, K.; Hasegawa, A.; Suzuki, T.; Kabu, K.; Sato, E.; Kurosaki, T.; Yamashita, S.; Tokunaga, M.; Nishida, K.; et al. Zinc is a novel intracellular second messenger. *J. Cell Biol.* **2007**, *177*, 637–645. [CrossRef]
37. Nagamatsu, S.; Nishito, Y.; Yuasa, H.; Yamamoto, N.; Komori, T.; Suzuki, T.; Yasui, H.; Kambe, T. Sophisticated expression responses of ZNT1 and MT in response to changes in the expression of ZIPs. *Sci. Rep.* **2022**, *12*, 7334. [CrossRef]
38. Eide, D. Molecular biology of iron and zinc uptake in eukaryotes. *Curr. Opin. Cell Biol.* **1997**, *9*, 573–577. [CrossRef]
39. Guerinot, M.L. The ZIP family of metal transporters. *Biochim. Biophys. Acta* **2000**, *1465*, 190–198. [CrossRef]
40. Cuajungco, M.P.; Ramirez, M.S.; Tolmasky, M.E. Zinc: Multidimensional Effects on Living Organisms. *Biomedicines* **2021**, *9*, 208. [CrossRef]
41. Taylor, K.M.; Morgan, H.E.; Johnson, A.; Nicholson, R.I. Structure-function analysis of HKE4, a member of the new LIV-1 subfamily of zinc transporters. *Biochem. J.* **2004**, *377*, 131–139. [CrossRef]
42. Taylor, K.M.; Morgan, H.E.; Smart, K.; Zahari, N.M.; Pumford, S.; Ellis, I.O.; Robertson, J.F.; Nicholson, R.I. The emerging role of the LIV-1 subfamily of zinc transporters in breast cancer. *Mol. Med.* **2007**, *13*, 396–406. [CrossRef] [PubMed]
43. Taylor, K.M. A distinct role in breast cancer for two LIV-1 family zinc transporters. *Biochem. Soc. Trans* **2008**, *36*, 1247–1251. [CrossRef] [PubMed]
44. Matsui, C.; Takatani-Nakase, T.; Hatano, Y.; Kawahara, S.; Nakase, I.; Takahashi, K. Zinc and its transporter ZIP6 are key mediators of breast cancer cell survival under high glucose conditions. *FEBS Lett.* **2017**, *591*, 3348–3359. [CrossRef] [PubMed]
45. Hogstrand, C.; Kille, P.; Nicholson, R.I.; Taylor, K.M. Zinc transporters and cancer: A potential role for ZIP7 as a hub for tyrosine kinase activation. *Trends Mol. Med.* **2009**, *15*, 101–111. [CrossRef] [PubMed]
46. Kasper, G.; Weiser, A.A.; Rump, A.; Sparbier, K.; Dahl, E.; Hartmann, A.; Wild, P.; Schwidetzky, U.; Castanos-Velez, E.; Lehmann, K. Expression levels of the putative zinc transporter LIV-1 are associated with a better outcome of breast cancer patients. *Int. J. Cancer* **2005**, *117*, 961–973. [CrossRef]
47. Jones, S.; Farr, G.; Nimmanon, T.; Ziliotto, S.; Gee, J.M.W.; Taylor, K.M. The importance of targeting signalling mechanisms of the SLC39A family of zinc transporters to inhibit endocrine resistant breast cancer. *Explor. Target Antitumor Ther.* **2022**, *3*, 224–239. [CrossRef]
48. Lisle, R.S.; Anthony, K.; Randall, M.A.; Diaz, F.J. Oocyte-cumulus cell interactions regulate free intracellular zinc in mouse oocytes. *Reproduction* **2013**, *145*, 381–390. [CrossRef]
49. Sharir, H.; Zinger, A.; Nevo, A.; Sekler, I.; Hershfinkel, M. Zinc released from injured cells is acting via the Zn2+-sensing receptor, ZnR, to trigger signaling leading to epithelial repair. *J. Biol. Chem.* **2010**, *285*, 26097–26106. [CrossRef]
50. Hershfinkel, M. The Zinc Sensing Receptor, ZnR/GPR39, in Health and Disease. *Int. J. Mol. Sci.* **2018**, *19*, 439. [CrossRef]
51. Ventura-Bixenshpaner, H.; Asraf, H.; Chakraborty, M.; Elkabets, M.; Sekler, I.; Taylor, K.M.; Hershfinkel, M. Enhanced ZnR/GPR39 Activity in Breast Cancer, an Alternative Trigger of Signaling Leading to Cell Growth. *Sci. Rep.* **2018**, *8*, 8119. [CrossRef]
52. Mero, M.; Asraf, H.; Sekler, I.; Taylor, K.M.; Hershfinkel, M. ZnR/GPR39 upregulation of K(+)/Cl(-)-cotransporter 3 in tamoxifen resistant breast cancer cells. *Cell Calcium* **2019**, *81*, 12–20. [CrossRef] [PubMed]
53. Kim, A.M.; Vogt, S.; O'Halloran, T.V.; Woodruff, T.K. Zinc availability regulates exit from meiosis in maturing mammalian oocytes. *Nat. Chem. Biol.* **2010**, *6*, 674–681. [CrossRef] [PubMed]
54. Bernhardt, M.L.; Kong, B.Y.; Kim, A.M.; O'Halloran, T.V.; Woodruff, T.K. A zinc-dependent mechanism regulates meiotic progression in mammalian oocytes. *Biol. Reprod.* **2012**, *86*, 114. [CrossRef] [PubMed]
55. Que, E.L.; Duncan, F.E.; Bayer, A.R.; Philips, S.J.; Roth, E.W.; Bleher, R.; Gleber, S.C.; Vogt, S.; Woodruff, T.K.; O'Halloran, T.V. Zinc sparks induce physiochemical changes in the egg zona pellucida that prevent polyspermy. *Integr. Biol.* **2017**, *9*, 135–144. [CrossRef] [PubMed]
56. Tian, X.; Diaz, F.J. Zinc depletion causes multiple defects in ovarian function during the periovulatory period in mice. *Endocrinology* **2012**, *153*, 873–886. [CrossRef]
57. Bedwal, R.S.; Bahuguna, A. Zinc, copper and selenium in reproduction. *Experientia* **1994**, *50*, 626–640. [CrossRef]
58. Sandstead, H.H.; Fosmire, G.J.; McKenzie, J.M.; Halas, E.S. Zinc deficiency and brain development in the rat. *Fed. Proc.* **1975**, *34*, 86–88.
59. Keen, C.L.; Hurley, L.S. Effects of zinc deficiency on prenatal and postnatal development. *Neurotoxicology* **1987**, *8*, 379–387.
60. Uriu-Adams, J.Y.; Keen, C.L. Zinc and reproduction: Effects of zinc deficiency on prenatal and early postnatal development. *Birth Defects Res. B Dev. Reprod. Toxicol.* **2010**, *89*, 313–325. [CrossRef]
61. Aliarabi, H.; Fadayifar, A.; Alimohamady, R.; Dezfoulian, A.H. The Effect of Maternal Supplementation of Zinc, Selenium, and Cobalt as Slow-Release Ruminal Bolus in Late Pregnancy on Some Blood Metabolites and Performance of Ewes and Their Lambs. *Biol. Trace Elem. Res.* **2019**, *187*, 403–410. [CrossRef]
62. Tamura, T.; Goldenberg, R.L.; Johnston, K.E.; DuBard, M. Maternal plasma zinc concentrations and pregnancy outcome. *Am. J. Clin. Nutr.* **2000**, *71*, 109–113. [CrossRef] [PubMed]
63. Grober, U.; Holzhauer, P.; Kisters, K.; Holick, M.F.; Adamietz, I.A. Micronutrients in Oncological Intervention. *Nutrients* **2016**, *8*, 163. [CrossRef] [PubMed]

64. Michalczyk, K.; Cymbaluk-Ploska, A. The Role of Zinc and Copper in Gynecological Malignancies. *Nutrients* **2020**, *12*, 3732. [CrossRef]
65. Zhang, R.; Zhao, G.; Shi, H.; Zhao, X.; Wang, B.; Dong, P.; Watari, H.; Pfeffer, L.M.; Yue, J. Zinc regulates primary ovarian tumor growth and metastasis through the epithelial to mesenchymal transition. *Free Radic. Biol. Med.* **2020**, *160*, 775–783. [CrossRef]
66. Huang, H.Y.; Caballero, B.; Chang, S.; Alberg, A.J.; Semba, R.D.; Schneyer, C.R.; Wilson, R.F.; Cheng, T.Y.; Vassy, J.; Prokopowicz, G.; et al. The efficacy and safety of multivitamin and mineral supplement use to prevent cancer and chronic disease in adults: A systematic review for a National Institutes of Health state-of-the-science conference. *Ann. Intern. Med.* **2006**, *145*, 372–385. [CrossRef] [PubMed]
67. Harvie, M. Nutritional supplements and cancer: Potential benefits and proven harms. *Am. Soc. Clin. Oncol. Educ. Book* **2014**, *34*, e478–e486. [CrossRef] [PubMed]
68. Alam, W.; Ullah, H.; Santarcangelo, C.; Di Minno, A.; Khan, H.; Daglia, M.; Arciola, C.R. Micronutrient Food Supplements in Patients with Gastro-Intestinal and Hepatic Cancers. *Int. J. Mol. Sci.* **2021**, *22*, 8014. [CrossRef]
69. Grungreiff, K.; Gottstein, T.; Reinhold, D.; Blindauer, C.A. Albumin Substitution in Decompensated Liver Cirrhosis: Don't Forget Zinc. *Nutrients* **2021**, *13*, 4011. [CrossRef]
70. Lewandowska, A.; Religioni, U.; Czerw, A.; Deptala, A.; Karakiewicz, B.; Partyka, O.; Pajewska, M.; Sygit, K.; Cipora, E.; Kmiec, K.; et al. Nutritional Treatment of Patients with Colorectal Cancer. *Int. J. Environ. Res. Public Health* **2022**, *19*, 6881. [CrossRef]
71. Kasprzyk, A.; Bilmin, K.; Chmielewska-Ignatowicz, T.; Pawlikowski, J.; Religioni, U.; Merks, P. The Role of Nutritional Support in Malnourished Patients With Lung Cancer. *Vivo* **2021**, *35*, 53–60. [CrossRef]
72. Nett, H.; Steegmann, J.; Tollkuhn-Prott, B.; Holzle, F.; Modabber, A. A prospective randomized comparative trial evaluating postoperative nutritional intervention in patients with oral cancer. *Sci. Rep.* **2022**, *12*, 14213. [CrossRef] [PubMed]
73. Fortmann, S.P.; Burda, B.U.; Senger, C.A.; Lin, J.S.; Whitlock, E.P. Vitamin and mineral supplements in the primary prevention of cardiovascular disease and cancer: An updated systematic evidence review for the U.S. Preventive Services Task Force. *Ann. Intern. Med.* **2013**, *159*, 824–834. [CrossRef] [PubMed]
74. Schwingshackl, L.; Boeing, H.; Stelmach-Mardas, M.; Gottschald, M.; Dietrich, S.; Hoffmann, G.; Chaimani, A. Dietary Supplements and Risk of Cause-Specific Death, Cardiovascular Disease, and Cancer: A Systematic Review and Meta-Analysis of Primary Prevention Trials. *Adv. Nutr.* **2017**, *8*, 27–39. [CrossRef] [PubMed]
75. Nasiadek, M.; Stragierowicz, J.; Klimczak, M.; Kilanowicz, A. The Role of Zinc in Selected Female Reproductive System Disorders. *Nutrients* **2020**, *12*, 2464. [CrossRef] [PubMed]
76. Hurley, L.S.; Shrader, R.E. Abnormal development of preimplantation rat eggs after three days of maternal dietary zinc deficiency. *Nature* **1975**, *254*, 427–429. [CrossRef]
77. Peters, J.M.; Wiley, L.M.; Zidenberg-Cherr, S.; Keen, C.L. Influence of short-term maternal zinc deficiency on the in vitro development of preimplantation mouse embryos. *Proc. Soc. Exp. Biol. Med.* **1991**, *198*, 561–568. [CrossRef]
78. Garner, T.B.; Hester, J.M.; Carothers, A.; Diaz, F.J. Role of zinc in female reproduction. *Biol. Reprod.* **2021**, *104*, 976–994. [CrossRef]
79. Derar, D.; Ali, A.; Almundarij, T.; Abd-Elmoniem, E.; Alhassun, T.; Zeitoun, M. Association between Serum Trace Elements Levels, Steroid Concentrations, and Reproductive Disorders in Ewes and Does. *Vet. Med. Int.* **2022**, *2022*, 8525089. [CrossRef]
80. Sunar, F.; Gormus, Z.I.; Baltaci, A.K.; Mogulkoc, R. The effect of low dose zinc supplementation to serum estrogen and progesterone levels in post-menopausal women. *Biol. Trace Elem. Res.* **2008**, *126* (Suppl. S1), S11–S14. [CrossRef]
81. Miro, F.; Smyth, C.D.; Hillier, S.G. Development-related effects of recombinant activin on steroid synthesis in rat granulosa cells. *Endocrinology* **1991**, *129*, 3388–3394. [CrossRef]
82. Chang, H.M.; Cheng, J.C.; Klausen, C.; Leung, P.C. BMP15 suppresses progesterone production by down-regulating StAR via ALK3 in human granulosa cells. *Mol. Endocrinol.* **2013**, *27*, 2093–2104. [CrossRef] [PubMed]
83. Fang, L.; Chang, H.M.; Cheng, J.C.; Leung, P.C.; Sun, Y.P. TGF-beta1 downregulates StAR expression and decreases progesterone production through Smad3 and ERK1/2 signaling pathways in human granulosa cells. *J. Clin. Endocrinol. Metab.* **2014**, *99*, E2234–E2243. [CrossRef] [PubMed]
84. Lee, S.; Kelleher, S.L. Molecular regulation of lactation: The complex and requisite roles for zinc. *Arch. Biochem. Biophys.* **2016**, *611*, 86–92. [CrossRef] [PubMed]
85. Brisken, C.; Park, S.; Vass, T.; Lydon, J.P.; O'Malley, B.W.; Weinberg, R.A. A paracrine role for the epithelial progesterone receptor in mammary gland development. *Proc. Natl. Acad. Sci. USA* **1998**, *95*, 5076–5081. [CrossRef] [PubMed]
86. Obr, A.E.; Edwards, D.P. The biology of progesterone receptor in the normal mammary gland and in breast cancer. *Mol. Cell Endocrinol.* **2012**, *357*, 4–17. [CrossRef]
87. Habib, F.K.; Maddy, S.Q.; Stitch, S.R. Zinc induced changes in the progesterone binding properties of the human endometrium. *Eur. J. Endocrinol.* **1980**, *94*, 99–106. [CrossRef]
88. Vasiliauskaite-Brooks, I.; Sounier, R.; Rochaix, P.; Bellot, G.; Fortier, M.; Hoh, F.; De Colibus, L.; Bechara, C.; Saied, E.M.; Arenz, C.; et al. Structural insights into adiponectin receptors suggest ceramidase activity. *Nature* **2017**, *544*, 120–123. [CrossRef]
89. Kelder, J.; Pang, Y.; Dong, J.; Schaftenaar, G.; Thomas, P. Molecular modeling, mutational analysis and steroid specificity of the ligand binding pocket of mPRalpha (PAQR7): Shared ligand binding with AdipoR1 and its structural basis. *J. Steroid. Biochem. Mol. Biol.* **2022**, *219*, 106082. [CrossRef]
90. Padarti, A.; Zhang, J. Recent advances in cerebral cavernous malformation research. *Vessel Plus* **2018**, *2*, 21. [CrossRef]

91. Abou-Fadel, J.; Vasquez, M.; Grajeda, B.; Ellis, C.; Zhang, J. Systems-wide analysis unravels the new roles of CCM signal complex (CSC). *Heliyon* **2019**, *5*, e02899. [CrossRef]
92. Abou-Fadel, J.; Smith, M.; Falahati, K.; Zhang, J. Comparative omics of CCM signaling complex (CSC). *Chin. Neurosurg. J.* **2020**, *6*, 4. [CrossRef] [PubMed]
93. Abou-Fadel, J.; Zhang, J. Systems Wide Analysis of CCM Signaling Complex Alterations in CCM-Deficient Models Using Omics Approaches. *Methods Mol. Biol.* **2020**, *2152*, 325–344. [CrossRef] [PubMed]
94. Abou-Fadel, J.; Bhalli, M.; Grajeda, B.; Zhang, J. CmP Signaling Network Leads to Identification of Prognostic Biomarkers for Triple-Negative Breast Cancer in Caucasian Women. *Genet Test Mol. Biomark.* **2022**, *26*, 198–219. [CrossRef]
95. Abou-Fadel, J.; Grajeda, B.; Jiang, X.; Cailing-De La, O.A.; Flores, E.; Padarti, A.; Bhalli, M.; Le, A.; Zhang, J. CmP signaling network unveils novel biomarkers for triple negative breast cancer in African American women. *Cancer Biomark* **2022**, Preprint. [CrossRef]
96. Abou-Fadel, J.; Jiang, X.; Grajeda, B.; Padarti, A.; Ellis, C.C.; Flores, E.; Cailing-De La, O.A.; Zhang, J. CCM signaling complex (CSC) couples both classic and non-classic Progesterone receptor signaling. *Cell Commun. Signal* **2022**, *20*, 120. [CrossRef] [PubMed]
97. Sesterhenn, A.M.; Iwinska-Zelder, J.; Dalchow, C.V.; Bien, S.; Werner, J.A. Acute haemorrhage in patients with advanced head and neck cancer: Value of endovascular therapy as palliative treatment option. *J. Laryngol. Otol.* **2006**, *120*, 117–124. [CrossRef] [PubMed]
98. Zhao, L.B.; Shi, H.B.; Park, S.; Lee, D.G.; Shim, J.H.; Lee, D.H.; Suh, D.C. Acute bleeding in the head and neck: Angiographic findings and endovascular management. *Am. J. Neuroradiol.* **2014**, *35*, 360–366. [CrossRef]
99. Franklin, R.B.; Feng, P.; Milon, B.; Desouki, M.M.; Singh, K.K.; Kajdacsy-Balla, A.; Bagasra, O.; Costello, L.C. hZIP1 zinc uptake transporter down regulation and zinc depletion in prostate cancer. *Mol. Cancer* **2005**, *4*, 32–45. [CrossRef]
100. Meerarani, P.; Ramadass, P.; Toborek, M.; Bauer, H.-C.; Bauer, H.; Hennig, B. Zinc protects against apoptosis of endothelial cells induced by linoleic acid and tumor necrosis factor α. *Am. J. Clin. Nutr.* **2000**, *71*, 81–87. [CrossRef]
101. Karadas, S.; Sayin, R.; Aslan, M.; Gonullu, H.; Kati, C.; Dursun, R.; Duran, L.; Gonullu, E.; Demir, H. Serum Levels of Trace Elements and Heavy Metals in Patients with Acute Hemorrhagic Stroke. *J. Membrane Biol.* **2014**, *247*, 175–180. [CrossRef]
102. Mammadova-Bach, E.; Braun, A. Zinc Homeostasis in Platelet-Related Diseases. *Int. J. Mol. Sci.* **2019**, *20*, 5258. [CrossRef] [PubMed]
103. Grungreiff, K.; Gottstein, T.; Reinhold, D. Zinc Deficiency-An Independent Risk Factor in the Pathogenesis of Haemorrhagic Stroke? *Nutrients* **2020**, *12*, 3548. [CrossRef] [PubMed]
104. Liu, H.; Rigamonti, D.; Badr, A.; Zhang, J. Ccm1 regulates microvascular morphogenesis during angiogenesis. *J. Vasc. Res.* **2011**, *48*, 130–140. [CrossRef]
105. Chapman, E.M.; Lant, B.; Ohashi, Y.; Yu, B.; Schertzberg, M.; Go, C.; Dogra, D.; Koskimaki, J.; Girard, R.; Li, Y.; et al. A conserved CCM complex promotes apoptosis non-autonomously by regulating zinc homeostasis. *Nat. Commun.* **2019**, *10*, 1791. [CrossRef]
106. Ito, S.; Greiss, S.; Gartner, A.; Derry, W.B. Cell-nonautonomous regulation of C. elegans germ cell death by kri-1. *Curr. Biol.* **2010**, *20*, 333–338. [CrossRef]
107. Maddaluno, L.; Rudini, N.; Cuttano, R.; Bravi, L.; Giampietro, C.; Corada, M.; Ferrarini, L.; Orsenigo, F.; Papa, E.; Boulday, G.; et al. EndMT contributes to the onset and progression of cerebral cavernous malformations. *Nature* **2013**, *498*, 492–496. [CrossRef]
108. Zhou, Z.; Rawnsley, D.R.; Goddard, L.M.; Pan, W.; Cao, X.-J.; Jakus, Z.; Zheng, H.; Yang, J.; Arthur, J.; Simon, C.; et al. The Cerebral Cavernous Malformation Pathway Controls Cardiac Development via Regulation of Endocardial MEKK3 Signaling and KLF Expression. *Developmental. Cell* **2015**, *32*, 168–180. [CrossRef]
109. Zhou, Z.; Tang, A.T.; Wong, W.-Y.; Bamezai, S.; Goddard, L.M.; Shenkar, R.; Zhou, S.; Yang, J.; Wright, A.C.; Foley, M.; et al. Cerebral cavernous malformations arise from endothelial gain of MEKK3–KLF2/4 signalling. *Nature* **2016**, *532*, 122–126. [CrossRef]
110. .Cunha, S.I.; Magnusson, P.U.; Dejana, E.; Lampugnani, M.G. Deregulated TGF-β/BMP Signaling in Vascular Malformations. *Circ. Res.* **2017**, *121*, 981–999. [CrossRef]
111. Lopez-Ramirez, M.A.; Fonseca, G.; Zeineddine, H.A.; Girard, R.; Moore, T.; Pham, A.; Cao, Y.; Shenkar, R.; de Kreuk, B.-J.; Lagarrigue, F.; et al. Thrombospondin1 (TSP1) replacement prevents cerebral cavernous malformations. *J. Exp. Med.* **2017**, *214*, 3331–3346. [CrossRef]
112. Choi, J.P.; Wang, R.; Yang, X.; Wang, X.; Wang, L.; Ting, K.K.; Foley, M.; Cogger, V.; Yang, Z.; Liu, F.; et al. Ponatinib (AP24534) inhibits MEKK3-KLF signaling and prevents formation and progression of cerebral cavernous malformations. *Sci. Adv.* **2018**, *4*, eaau0731. [CrossRef]
113. Lopez-Ramirez, M.A.; Pham, A.; Girard, R.; Wyseure, T.; Hale, P.; Yamashita, A.; Koskimaki, J.; Polster, S.; Saadat, L.; Romero, I.A.; et al. Cerebral cavernous malformations form an anticoagulant vascular domain in humans and mice. *Blood* **2019**, *133*, 193–204. [CrossRef]
114. Su, V.L.; Calderwood, D.A. Signalling through cerebral cavernous malformation protein networks. *Open Biol.* **2020**, *10*, 200263. [CrossRef]
115. Wei, S.; Li, Y.; Polster, S.P.; Weber, C.R.; Awad, I.A.; Shen, L. Cerebral Cavernous Malformation Proteins in Barrier Maintenance and Regulation. *Int. J. Mol. Sci.* **2020**, *21*, 675. [CrossRef]

116. Lopez-Ramirez, M.A.; Lai, C.C.; Soliman, S.I.; Hale, P.; Pham, A.; Estrada, E.J.; McCurdy, S.; Girard, R.; Verma, R.; Moore, T.; et al. Astrocytes propel neurovascular dysfunction during cerebral cavernous malformation lesion formation. *J. Clin. Investig.* **2021**, *131*. [CrossRef]
117. Jiang, X.; Padarti, A.; Qu, Y.; Sheng, S.; Abou-Fadel, J.; Badr, A.; Zhang, J. Alternatively spliced isoforms reveal a novel type of PTB domain in CCM2 protein. *Sci. Rep.* **2019**, *9*, 15808–15819. [CrossRef]
118. Antwi-Adjei, E.; Burguete, A.S.; Ghabrial, A.S. Furry is a component of the CCM3-GCKIII signaling pathway. *Vessel Plus* **2021**, *5*, 35. [CrossRef]
119. Zhang, J. Learn from the past, review the present, and look towards the future. *Vessel Plus* **2022**, *6*, 20. [CrossRef]
120. Berman, J.R.; Kenyon, C. Germ-cell loss extends C. elegans life span through regulation of DAF-16 by kri-1 and lipophilic-hormone signaling. *Cell* **2006**, *124*, 1055–1068. [CrossRef]
121. Yamawaki, T.M.; Arantes-Oliveira, N.; Berman, J.R.; Zhang, P.; Kenyon, C. Distinct activities of the germline and somatic reproductive tissues in the regulation of Caenorhabditis elegans' longevity. *Genetics* **2008**, *178*, 513–526. [CrossRef]
122. Baumbach, G.L.; Didion, S.P.; Faraci, F.M. Hypertrophy of cerebral arterioles in mice deficient in expression of the gene for CuZn superoxide dismutase. *Stroke* **2006**, *37*, 1850–1855. [CrossRef] [PubMed]
123. Didion, S.P.; Kinzenbaw, D.A.; Schrader, L.I.; Faraci, F.M. Heterozygous CuZn superoxide dismutase deficiency produces a vascular phenotype with aging. *Hypertension* **2006**, *48*, 1072–1079. [CrossRef] [PubMed]
124. Didion, S.P.; Kinzenbaw, D.A.; Faraci, F.M. Critical role for CuZn-superoxide dismutase in preventing angiotensin II-induced endothelial dysfunction. *Hypertension* **2005**, *46*, 1147–1153. [CrossRef]
125. Didion, S.P.; Ryan, M.J.; Didion, L.A.; Fegan, P.E.; Sigmund, C.D.; Faraci, F.M. Increased superoxide and vascular dysfunction in CuZnSOD-deficient mice. *Circ. Res.* **2002**, *91*, 938–944. [CrossRef] [PubMed]
126. Morita-Fujimura, Y.; Fujimura, M.; Gasche, Y.; Copin, J.C.; Chan, P.H. Overexpression of copper and zinc superoxide dismutase in transgenic mice prevents the induction and activation of matrix metalloproteinases after cold injury-induced brain trauma. *J Cerebr. Blood Flow Metab.* **2000**, *20*, 130–138. [CrossRef]
127. Goitre, L.; Balzac, F.; Degani, S.; Degan, P.; Marchi, S.; Pinton, P.; Retta, S.F. KRIT1 regulates the homeostasis of intracellular reactive oxygen species. *PLoS ONE* **2010**, *5*, e11786. [CrossRef]
128. Goitre, L.; De Luca, E.; Braggion, S.; Trapani, E.; Guglielmotto, M.; Biasi, F.; Forni, M.; Moglia, A.; Trabalzini, L.; Retta, S.F. KRIT1 loss of function causes a ROS-dependent upregulation of c-Jun. *Free Radic. Biol. Med.* **2014**, *68*, 134–147. [CrossRef]
129. Jung, K.H.; Han, D.M.; Jeong, S.G.; Choi, M.R.; Chai, Y.G.; Cho, G.W. Proteomic analysis reveals KRIT1 as a modulator for the antioxidant effects of valproic acid in human bone-marrow mesenchymal stromal cells. *Drug Chem. Toxicol.* **2015**, *38*, 286–292. [CrossRef]
130. Trapani, E.; Retta, S.F. Cerebral cavernous malformation (CCM) disease: From monogenic forms to genetic susceptibility factors. *J. Neurosurg Sci.* **2015**, *59*, 201–209.
131. Retta, S.F.; Glading, A.J. Oxidative stress and inflammation in cerebral cavernous malformation disease pathogenesis: Two sides of the same coin. *Int. J. Biochem. Cell Biol.* **2016**, *81*, 254–270. [CrossRef]
132. Fontes-Dantas, F.L.; da Fontoura Galvao, G.; Veloso da Silva, E.; Alves-Leon, S.; Cecilia da Silva Rego, C.; Garcia, D.G.; Marques, S.A.; Blanco Martinez, A.M.; Reis da Silva, M.; Marcondes de Souza, J. Novel CCM1 (KRIT1) Mutation Detection in Brazilian Familial Cerebral Cavernous Malformation: Different Genetic Variants in Inflammation, Oxidative Stress, and Drug Metabolism Genes Affect Disease Aggressiveness. *World Neurosurg.* **2020**, *138*, 535–540. [CrossRef] [PubMed]
133. Perrelli, A.; Retta, S.F. Fluorescence Analysis of Reactive Oxygen Species (ROS) in Cellular Models of Cerebral Cavernous Malformation Disease. *Methods Mol. Biol.* **2020**, *2152*, 451–465. [CrossRef] [PubMed]
134. Swamy, H.; Glading, A.J. Contribution of protein-protein interactions to the endothelial-barrier-stabilizing function of KRIT1. *J. Cell Sci.* **2022**, *135*, jcs258816. [CrossRef] [PubMed]
135. Abou-Fadel, J.; Jiang, X.; Padarti, A.; Goswami, D.; Smith, M.; Grajeda, B.; Walker, W.; Zhang, J. CCM signaling complex (CSC) is a master regulator governing homeostasis of progestins and their mediated signaling cascades. *bioRxiv* **2020**. [CrossRef]
136. Abou-Fadel, J.; Qu, Y.; Gonzalez, E.; Smith, M.; Zhang, J. Emerging roles of CCM genes during tumorigenesis with potential application as novel biomarkers across major types of cancers. *Oncol. Rep.* **2020**, *43*, 1945–1963. [CrossRef]
137. Abou-Fadel, J.; Jiang, X.; Padarti, A.; Goswami, D.G.; Smith, M.; Grajeda, B.; Bhalli, M.; Le, A.; Walker, W.E.; Zhang, J. mPR-Specific Actions Influence Maintenance of the Blood–Brain Barrier (BBB). *Int. J. Mol. Sci.* **2022**, *23*, 9684.
138. Renteria, M.; Belkin, O.; Jang, D.; Aickareth, J.; Bhalli, M.; Zhang, J. CmPn signaling networks in the tumorigenesis of breast cancer. *Front. Endocrinol.* **2022**, *13*, 1013892. [CrossRef]

biomolecules

MDPI

Review

Investigating the Role of Zinc in Atherosclerosis: A Review

Tong Shen [1,†], Qing Zhao [2,†], Yumin Luo [1,3,*] and Tao Wang [4,*]

1 Department of Neurology, Institute of Cerebrovascular Diseases Research, Xuanwu Hospital of Capital Medical University, No. 45 Changchun Street, Beijing 100053, China
2 Peking Union Medical College, No. 9 Dongdansantiao Street, Beijing 100730, China
3 Beijing Institute for Brain Disorders, No. 10 Xitoutiao, You An Men, Beijing 100069, China
4 Department of Neurosurgery, China International Neuroscience Institute, Xuanwu Hospital of Capital Medical University, National Center for Neurological Disorders, No. 45 Changchun Street, Beijing 100053, China
* Correspondence: yumin111@ccmu.edu.cn (Y.L.); wangtao_dr@xwhosp.org (T.W.); Tel.: +86-01083198129 (Y.L.); +86-10-83199060 (T.W.)
† These authors contributed equally to this work.

Abstract: Zinc, an indispensable micronutrient for human health, might play an important role in the development of atherosclerosis. Zinc could be involved in the atherogenic process through interaction with atherogenic cells, such as endothelial cells (ECs), vascular smooth muscle cells (VSMCs), and immune cells. In addition, zinc also exerts important positive or negative functions in various atherosclerosis-related risk factors, including lipid metabolism, glucose metabolism, and blood pressure. Currently, evidence focusing on the relationship between zinc status and atherogenic risk factors has been well established, while the direct interaction between zinc and atherosclerosis has not been fully understood. In this review, we aimed to summarize the association between zinc and atherosclerosis and explore current findings on how zinc and zinc homeostasis-associated proteins act in the atherogenic processes.

Keywords: zinc; atherosclerosis; zinc homeostasis-associated proteins; endothelial cells; vascular smooth muscle cell; immune cell; cardiovascular risk factors; animal study; human study

1. Introduction

Zinc, an abundant micronutrient, plays a vital role in numerous biochemical pathways in cells and is indispensable to all creatures. Following iron, zinc is the second most abundant transition metal ion in all living organisms [1]. The human adult body contains approximately 1.4–2.3 g of zinc [2]. Zinc is also a broadly distributed metal ion in the human body, with approximately 85% of zinc concentration found in muscle and bones, 11% in skin and liver, and the remaining 4% in other tissues [3]. The distribution and homeostasis of zinc in the body are highly regulated by various zinc homeostasis-associated proteins, including Zn transporters (ZnT), Zrt- and Irt-like proteins (ZIP), and metallothioneins (MTs). Zinc conducts multiple crucial biological functions in organisms. Firstly, zinc is a redox-neutral metal ion and can stably bind to practically 10% of the total human proteome and still maintain its structural integrity [4]. Secondly, zinc is required in the catalysis and co-catalysis of more than 300 enzymes, which control various metabolic cellular processes [5]. Thirdly, zinc, being "the calcium of the 21st century", functions as a signaling mediator and targets multiple enzymes involved in cellular signaling, such as kinases and phosphatases [6,7].

Considering the critical biological roles that zinc plays in human health, even a minor zinc deficiency could be disastrous. Multiple abnormalities can cause zinc deficiency, including inadequate intake, decreased absorption, and increased loss. Zinc deficiency is estimated to affect nearly two billion people worldwide, with the estimated prevalence of suboptimal zinc status ranging from 4% to 73% in different countries [8]. Severe zinc deficiency is associated with profound consequences, such as growth retardation and death [7]. Studies estimate that zinc deficiency may be associated with a 4% morbidity and mortality rate of children in developing countries. Even marginal zinc deficiency can cause a variety of pathological conditions, especially age-related pathologies, such as atherosclerosis [7].

Atherosclerosis is a pathological condition characterized by atherosclerotic plaque formation in the vascular intima of large and medium-sized arteries. Given the popularity of the western lifestyle, the worldwide prevalence of atherosclerosis is gradually increasing and might develop into an epidemic in the future [9]. Atherosclerosis is the principal cause of cardiovascular diseases (CVDs), including coronary artery disease (CAD), stroke, and peripheral arterial disease. Among adults aged $\geq$ 20 years in the US, 37.4% of men and 35.9% of women have some form of CVDs, and the estimated prevalence of CVDs in the US adult population could reach 43.9% by 2030 [10]. The identification of risk factors for atherosclerosis and CVDs is crucial for its prevention and treatment. Recent evidence suggests that zinc deficiency plays a vital role in the development of atherosclerosis [11].

In this review, we aim to summarize the current findings on the association between zinc and atherosclerosis and investigate the potential effect of zinc and zinc homeostasis-associated proteins in atherogenic processes. Since zinc homeostasis-associated proteins are critical for the physiology and homeostasis of zinc, we first provide a summary of zinc homeostasis-associated proteins and their functions.

2. Zinc-Related Proteins

As previously mentioned, zinc plays a crucial role as a structural, catalytic, and regulatory component of many metabolic processes; therefore, it is necessary to clarify the mechanism of zinc homeostasis. In the human body, serum zinc takes up only 0.1% of total zinc, the majority of which is bound to proteins, such as albumin and α2-macroglobulin, with only a small fraction existing as free ions [12,13]. The distribution of cellular zinc is the cytoplasm (50%), nucleus (30–40%), and cell membrane (10%) [13]. Cellular zinc is stored in compartmentalized regions, such as organelles and vesicles, or bound to metalloproteins/metalloenzymes and MTs. Consequently, the concentration of cytosolic free zinc is likewise considerably low, ranging between picomolar and nanomolar [14,15]. Through the release or combination of zinc within proteins and organelles, organisms can cope with transient changes in zinc fluctuation and maintain zinc concentration at a stable level. The four pools of cellular zinc [16] are as follows: (1) zinc that is tightly conjugated to metalloproteins/metalloenzymes as a structural component or cofactor; (2) zinc that binds to MTs with low affinity, which executes the functions as both receptor and donor of zinc in case of zinc perturbation; (3) zinc that is stored in organelles and vesicles and serves as zinc provider through zinc transporters; (4) cytosolic-free zinc that participates in signaling transduction [17]. The last three pools act as labile zinc pools and play a key role when zinc status fluctuates.

The underlying mechanisms of cellular zinc homeostasis maintenance are known as "zinc buffering" and "zinc muffling" [18]. Buffering is realized by cytosolic zinc-binding proteins to obtain consistent levels of free zinc ions in a way that is similar to proton buffering [19,20]. In parallel, the muffling mechanism is instrumental in the instance of transient changes with higher zinc loads in the cytosol and functions by mobilizing the zinc-binding proteins and zinc transporters to move zinc ions into sequestrated subcellular compartments or out of cells [18]. When cellular zinc concentration fluctuates, metal response element-binding transcription factor-1 (MTF-1) can regulate the transcription of MTs and several ZnT transporters [21,22]. The mutual interaction between zinc concentration and transcription factors contributes to the maintenance of cellular zinc homeostasis (see Figure 1). MTs and the zinc transporter family play a pivotal role in not only controlling cellular zinc homeostasis but also the progression of cardiovascular diseases.

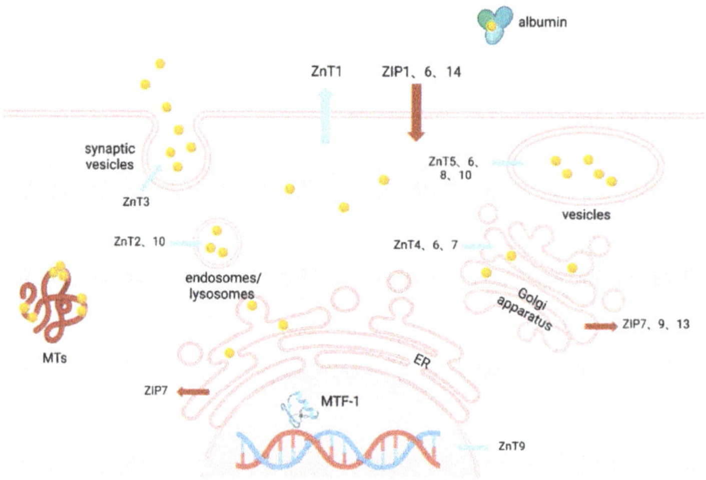

Figure 1. The localization of zinc transporters Zrt-, Irt-like proteins (ZIP, red arrow), Zn transporters (ZnT, blue arrow), metallothioneins (MTs), and metal response element-binding transcription factor-1 (MTF-1) in the cell.

2.1. ZnT/ZIP

Since the concentration of labile zinc ion is exceptionally low, subtle changes in labile zinc can cause tremendous alternation in cell function. The distribution of zinc in eukaryotic cells is characterized by compartmentalization and is rigorously controlled by membrane zinc transporters, which restricts the availability of zinc ions and provides better buffering ability toward zinc loads [19]. Distinct from iron, more than 30 proteins are involved in the storage and transportation of zinc rather than ferritin alone, which attributes to the deficiency in humoral mediators of zinc status [13]. In humans, 10 members of the ZnT/solute carrier 30A (SLC30A) family were recognized to function in zinc efflux from the cytosol to extracellular space or intracellular organelles, whereas 14 members of the ZIP/solute carrier 39A (SLC39A) family function in mobilizing zinc in a converse way [23]. ZnTs and ZIPs are responsible for the development of atherosclerosis.

ZnT transporters are zinc-proton antiporters residing in the inner membrane of subcellular compartments, except for ZnT-1, which is located in the plasma membrane [24]. When zinc ion conjugates with the metal-binding sites of ZnT protein, it triggers structural changes in the protein and causes conformational transformation, extruding zinc out of the cytosol [13]. ZnT-1 and ZnT-4 have a potential association with CVDs due to their participation in vascular physiology [25]. ZIP transporters, in contrast, are mainly localized to the plasma membrane [26,27] and possibly work as a co-transporter with protons as well [28]. Twenty-four ZIP transporters and twelve MTs are recognized within human heart muscle tissues [29,30], indicating the connection between ZIP proteins and CVDs. A sophisticated interaction exists between ZnTs, ZIPs, and MTs by regulating gene expression reciprocally in response to zinc homeostasis [31]. The function of zinc transporters may be regulated by protein phosphorylation and may thus be linked to many diseases [32].

2.2. MTs

MTs are low molecular metal-binding proteins that have selective binding capacity to zinc, cadmium, and copper ions [33]. Human MTs have 11 functional isoforms, which are divided into four subgroups, MT-1 to MT-4 [16]. MT-1 and MT-2 are ubiquitously expressed in many organs and tissues and are engaged in zinc homeostasis, while MT-3 and MT-4 are cell-specific. Each molecule of MTs has up to seven ion-binding sites, attributing to its distinct affinity to zinc and other divalent metals [34].

MTs function is crucial for stabilizing the fluctuation of intracellular zinc concentration. Thioneins are translated in the ribosome and chelate with cytosolicfree zinc ion, synthesizing metallothioneins to lower zinc concentration. Conversely, when cellular zinc level is insufficient to maintain MT protein, it rapidly proteolyzes and releases conjugated zinc to keep intracellular zinc concentration stable [35,36]. MT is a potent electrophilic scavenger and cytoprotective agent [37,38] and plays a critical role in immune protection and signal transduction [39,40]. Experiments show that cytokines can either up or down-regulate the expression of MTs, and MTs, in turn, affect the production of cytokines [41–43]. In addition, MTs overexpression exerts a positive effect on alleviating diabetes and diabetic cardiovascular complications through attenuating oxidative stress [11,44,45], whereas another study indicated a negative correlation between MT-1 overexpression and insulin secretion [46].

2.3. *Zinc Finger Proteins*

Zinc finger proteins (ZFPs) are the most abundant proteins in human genome [47], which, as the name suggests, are characterized by their zinc finger motifs. Many ZFPs function as transcription factors by binding to specific DNA sequences. The most famous ZFPs in atherosclerosis might be the Krüppel-like factor (KLF) family. KLF2 and KLF4 are assumed to play a central role in maintaining endothelial cell homeostasis and vascular health. Multiple studies have confirmed the atheroprotective effect of KLF2 and KLF4 through inhibition of inflammatory response [48–52]. The activator of KLF2 presented anti-inflammatory effects and attenuated the adhesion of monocytes to endothelial cells, which could be a promising therapeutic strategy for atherosclerosis [53]. Other ZFPs, such as KLF14, zinc finger E-box binding homeobox 1 (ZEB1), and early growth response gene-1 (Egr-1), also participate in the regulation of atherogenesis [54–56].

3. Zinc and Atherogenic Cells

Zinc plays a direct role in atherosclerosis through the interaction with atherogenic cells, such as endothelial cells (ECs), vascular smooth muscle cells (VSMCs), and immune cells. This process is primarily realized by the regulation of cellular functional molecules. Here, we elaborate on the direct interaction between zinc and proatherogenic cells.

3.1. *Zinc and Vascular Cells*

Atherosclerosis is the most common cause of CVDs, which are characterized by lesions to the arterial wall, subsequent subendothelial plaque formation, and vessel occlusion. It is an artery-centered disease that eventually results in multiple organ damage. Vascular cells, including ECs and VSMCs, are major contributors to the initiation, development, and progression of atherogenic process.

3.1.1. Zinc and ECs

ECs are the innermost vascular layer and function as a barrier between the bloodstream and the vascular walls. Endothelial cell dysfunction (ECD) is a crucial link in the atherogenic process. Early speculation on the role of ECD in atherosclerosis was referred to as the "Response-to-Injury Hypothesis", which focused on the loss of integrity in the intima of the arterial wall. However, careful examination of the fatty streak lesions in diet-induced animal models indicated no sign of intimal injury or platelet adhesion [57]. With the expansion of understanding of ECD, accumulating evidence suggests that the changes of hemodynamic forces in athero-susceptible regions induce differential gene expression, alternation in cell types, and focal ECD [58]. Unidirectional, high laminar shear stress can upregulate the expression of anti-inflammatory, antioxidative, and antithrombotic genes, such as KLF2, KLF4, and nuclear factor (erythroid-derived 2)–like 2 (NRF2), while in contrast, low shear stress exerts proatherogenic functions by initiating multiple vascular events, including focal permeation, ROS generation, NF-κB activity, and expression of receptors of adhesion molecules, and cytokines that recruit leukocytes [57,59,60]. A former study indicated the upregulation of MTs and ZnT under low shear stress and an increase in free zinc under high shear stress in human umbilical vein endothelial cells (HUVECs) [61]. The dysfunction of ECs results in focal permeation under an intact endothelium and the trapping of circulating lipoproteins in the subendothelial space, thereby initiating the complex atherogenic process [62]. ECD also plays a role in the complications of atherosclerosis, such as vascular occlusion and plaque rapture, through the chronic lesion towards the endothelium [63]. Recent studies suggested that zinc is essential for the proper functioning of ECs through different pathways, as mentioned below [64].

Firstly, zinc regulates the synthesis of nitric oxide (NO) in ECs. NO, an endothelium-derived vasodilator, is synthesized via endothelial NO synthase (eNOS). The eNOS-derived NO critically influences the function of ECs and exerts multiple anti-atherosclerotic effects, such as vasodilation, suppression of VSMCs proliferation, and inhibition of leucocyte adhesion, platelet adhesion, and platelet aggregation [65]. The decreased availability of NO, resulting from the reduced expression or activity of eNOS, is actively involved in the atherogenic process [66]. The eNOS is catalytically inactive in its monomeric form and only exerts catalytic function as dimeric eNOS. The dimerization of eNOS is mediated by zinc, located at the dimer interface [25]. N,N,N,N-Tetrakis (2-pyridylmethyl)-ethylenediamine (TPEN), a zinc chelator, was shown to convert eNOS dimers into monomers in ECs [67]. This finding indirectly suggests that zinc is critical for the dimerization of eNOS. Studies found that dietary zinc deficiency during the fetal life, lactation, and postnatal life of rats reduced the expression and activity of Enos [68]. Zhuang et al. [69] further found that zinc supplementation effectively enhances intracellular NO production by elevating the expression and enzymatic activity of eNOS. NO participates in regulating the homeostasis of intracellular zinc in ECs. MTs, proteins functioning as buffer for intracellular-free zinc, release zinc when stimulated by NO, thereby increasing labile zinc level in ECs [25]. This may create a positive feedback loop since the released zinc could then promote the dimerization and activation of eNOS and amplify the anti-atherosclerotic effect of NO. Except for being synthesized by eNOS, NO in ECs can also be produced by inducible NOS (iNOS). Unlike eNOS, iNOS is only expressed under pro-inflammatory conditions. Evidence suggests that iNOS is detrimental in diseases because of its capability of producing excessive NO in response to inflammatory signals [70,71]. Miyoshi et al. [72] indicated that genetic deficiency of iNOS in ApoE-deficient mice reduced the formation of atherosclerotic lesions. Cortese-Krott et al. [73] have shown that zinc down-regulates the expression and activity of iNOS in ECs, and, therefore, limits the undesirable influence of the high levels of iNOS-derived NO.

Secondly, atherosclerosis is considered a chronic inflammatory disorder, while zinc exhibits anti-inflammatory effects in ECs. Faced with varieties of stimuli, ECs express multiple cytokines, chemokines, and cell adhesion molecules, recruiting leukocytes and facilitating their migration to the neointimal space of the vascular wall. Activated ECs could trigger chemokine secretion and recruit the circulating monocytes to the vascular wall, known as monocyte-EC interaction, which is a key event in the formation of atherosclerotic lesions [74]. ZEB is a key member of the zinc finger-homeodomain family. Studies indicated that the overexpression of ZEB1 could decrease the expression of C-X-C motif chemokine ligand 1 (CXCL1), which was a chemoattractant that promoted the monocyte–ECs interaction [75]. These results suggested that ZEB1 inhibits the monocyte–ECs interaction, thereby preventing the aggravation of atherosclerosis. Vascular cell adhesion molecule-1 (VCAM-1), an important cell adhesion molecule, triggers monocytes' attachment to ECs and promotes the development of atherosclerotic plaques. Zinc deficiency increases the production of VCAM-1 and intercellular adhesion molecules (ICAM-1), thereby promoting inflammation [76]. Zinc additionally affects the expression of multiple pro-inflammatory cytokines in ECs. Zinc deficiency could exacerbate chronic inflammation in ECs by permitting the expression of inflammatory genes [73]. A recent systematic review indicated that zinc status was inversely associated with the development of endothelial inflammation or atherosclerosis [76]. Multiple signaling pathways of various inflammatory cytokines converge with the release of nuclear transcription factor kappa B (NF-κB) [77]. Atherosclerosis is a chronic inflammatory disease of the arterial wall, and the activation of NF-κB has been shown to be involved in all stages of the development of atherosclerosis through mediating downstream canonical and non-canonical signaling pathways [78,79]. Previous studies indicated that zinc deficiency upregulates the activity of NF-κB [80], while zinc supplementation suppresses its activity in ECs [69]. KLF family is another important category of zinc finger transcription factors. Studies indicated that KLF2, KLF4, and KLF11 all conferred important anti-inflammatory actions in ECs by inhibiting the activation of NF-κB [81]. The transactivation activity of peroxisome proliferator-activated receptor (PPAR), the inhibitor of NF-κB signaling, also requires adequate zinc. Zinc deficiency down-regulates the expression of PPARα and PPARγ in cultured ECs [82,83], which then upregulates the DNA binding activity of NF-κB. A20, a protein act as a general inhibitor of NF-κB activation. A positive association between zinc levels and A20 protein concentration was reported in human ECs [84]. This finding suggests that zinc could also suppress the generation of NF-κB-regulated inflammatory cytokines by inducing A20.

Thirdly, oxidative stress contributes to the development and progression of atherosclerosis, while zinc exhibits anti-oxidative actions in ECs. Oxidative stress is the result of the imbalance between reactive oxygen species (ROS) generation and the antioxidant system of the body [85]. In cultured ECs, zinc deficiency increases oxidative stress, while subsequent zinc supplementation blocks this effect [80,82]. Further studies indicated that zinc deficiency could significantly aggravate oxidative stress by increasing the level of ROS and decreasing antioxidants in the lung and kidneys of mice [86,87]. A recent meta-analysis indicated that zinc supplementation significantly decreases the total antioxidant capacity, as well as malondialdehyde, which is an important metabolite of lipid peroxide [88,89]. Zinc is not redox-active under physiological conditions. Thus, the anti-oxidative action of zinc is realized through indirect mechanisms, such as participation in the synthesis of antioxidative enzymes or working as their catalyzers [90]. Zago et al. [91] found that the effect of zinc in protecting the cell membrane from lipid oxidation may be realized by competition with other redox-active metals for negative charges in the lipid bilayer.

Finally, ECs apoptosis impaired the barrier function of the endothelium and enables the infiltration of circulating leukocytes and lipoproteins in the vascular wall, initiating the atherogenic process [92]. Tricot et al. [93] noted that the turnover and apoptosis of ECs increased in the endothelium of human atherosclerotic plaques. These observations highlighted the important role of the apoptotic cell death of ECs in the development of atherosclerosis. Therefore, suppressing the apoptosis of ECs may be an attractive approach for realizing anti-atherosclerotic effects. ZEB1 could inhibit the apoptosis and senescence of ECs via reducing p21 protein, protecting ECs from oxidative stress-induced dysfunction [94]. Thambiayya et al. [95] found that the decrease in zinc level may contribute to the lipopolysaccharide (LPS)-induced apoptosis. These authors further discovered that the increase in intracellular labile zinc conducts essential functions in the resistance to LPS-induced apoptosis mediated by ZIP-14 or NO [96]. Caspases are pivotal enzymes in the apoptotic pathways, and studies indicate that zinc inhibits caspases [97]. Compared with control cultures, ECs with zinc-deficient cultures were associated with the upregulation of caspase-3 and increased cytokine- or lipid-induced apoptotic cell death [98]. These authors further clarified that subsequent zinc supplementation significantly blocks the activation of caspase-3. Likewise, Fanzo et al. [99] found that the activity of caspase-3 in human aortic ECs was inversely associated with the zinc level in the culture media. Zinc can also inhibit other apoptotic regulators beyond caspase-3, such as caspase-6, caspase-9, and calcium-magnesium-dependent endonuclease [100,101].

3.1.2. Zinc and VSMCs

VSMC is a type of smooth muscle cell that is in the media layer of elastic arteries and is responsible for arterial contraction and the production of extracellular matrix (ECM) proteins. VSMCs are the main contributor to atherosclerotic plaque composition. Studies suggest that 30–70% of the plaque cells express VSMC-specific molecules in murine models of atherosclerosis [102–104]. The historical view of VSMCs in atherosclerosis typically assumes that VSMCs play an entirely beneficial role in advanced plaques by generating ECM to form the fibrous cap, which stabilizes the vulnerable plaques, preventing them from rupturing. However, genetic lineage tracing studies reveal a more comprehensive and integral perspective on VSMCs function in atherogenesis [102–105]. During the progression of atherosclerosis, VSMCs undergo de-dedifferentiation and "phenotypic switching" [104,106], which, specifically, refers to the transformation from a contractile phenotype to a proliferative phenotype, which is considered an essential part of the development of atherosclerosis, as well as the development of endothelial damage and lipid deposition [107]. VSMCs in the plaque exhibit a co-expression of both VSMC- and macrophage-specific biomarkers, and the concomitant macrophage properties, such as secreting pro-inflammatory molecules and inducing the migration of inflammatory cells [104,106]. These findings indicate a potential dual role of VSMCs in atherogenesis, and zinc might influence the progress by regulating the function of transcription factors containing zinc-finger structures, such as KLF4 and ZEB1 [55,81,108].

Zinc is associated with VSMC apoptosis and proliferation according to studies conducted by Keith Allen-Redpath et al. [109] and Ethel H. Alcantara et al. [110]. VSMCs from marginally zinc deficient diet rats present with an increased apoptotic effect in the carotid artery and aorta, which is not observed in the endothelial cell layer. The underlying mechanism of interchange to the apoptotic phenotype could be related to increased expression and dephosphorylation of Bcl-2-associated death promoter protein (BAD) (a member of the pro-apoptotic Bcl-2 family), and the decreased activation of the pro-survival extracellular signal-regulated kinase 1/2 (ERK1/2) pathway, which maintains the phosphorylation of BAD. An in vitro study indicates that the apoptotic effect induced by zinc deficiency is potentially mediated by a type of humoral factor since the VSMCs incubated by plasma from zinc-deficient rats also show increased apoptosis [109]. Ou et al. [111] confirmed this hypothesis and demonstrated that zinc not only regulated gene expression directly but also affected the bioactivity of a low-molecular-weight humoral factor. This factor is potentially associated with immune functionality due to the responsive actions of cytokine pathways, but the derivation of this factor has not yet been confirmed. Another study indicates that chronic zinc deprivation can accelerate the proliferation of VSMCs via decreased mitogen-activated protein kinase/c-Jun N-terminal kinase (MAPK/JNK) signaling pathway (apoptotic) activation [110]. Zinc may also play a role in preventing VSMC calcification [112]. Kuzan et al. [113] used the Inductively Coupled Plasma Optical Emission Spectrometer (ICP-OES) to determine the content of microelements in the human aorta, however, the correlation between serum zinc and the severity of atherosclerosis was not observed. Zinc could play a role in vasodilation, via cofunction with calcium. In conclusion, further research is required to establish a correlation between zinc, VSMCs, and atherogenesis.

3.2. *Zinc and Immune Cells*

Atherosclerosis is known as a chronic inflammatory condition of the arterial walls initiated by the deposition of subendothelial lipids. The immune reaction is recognized as a pivotal part of the atherosclerotic mechanism [114]. During the atherogenic process, oxidized low-density lipoprotein (oxLDL) accumulates in the intima and is engulfed by macrophages which later transform into an inflammatory phenotype and turn into foam cells [115–117]. The initiation of innate immunity has an early onset and depends mostly on macrophages with the upregulation of adhesive molecules and proinflammatory cytokines, whereas adaptive immunity has a later onset and a dual effect on atherogenesis [118]. T lymphocytes facilitate the progression of atherosclerosis, but T helper 2 cells (Th2) and regulatory T cells (Treg) can dampen the process through the production of transforming growth factor (TGF)-β and interleukin (IL)-10 [114,119]. B lymphocytes have conflicting functions, since B1 cells perform an atheroprotective effect, while B2 cells are pro-atherogenic [116,119]. Zinc, an important trace element, is crucial to the function of

immune cells. In the acute phase of inflammation, zinc undergoes a transient redistribution from serum to organs, especially the liver, which is mediated by the increased expression of ZIP-14 and up-regulation of MT in liver cells. This transient redistribution leads to temporary hypozincemia and upregulation of IL-1β, tumor necrosis factor (TNF)-α, S100 calcium-binding protein A8 (S100A8), and matrix metalloproteinase-8(MMP-8) [120]. Long before plasma zinc concentration changes, immune cells have the ability to respond to zinc deficiency early on [121]. The potential mechanism of how zinc influences the immune cells and the progress of atherosclerosis is explained below.

3.2.1. Zinc, Macrophage, and Cytokines

OxLDL is characterized by the properties of damage-associated molecule patterns (DAMPs) and can interact with scavenger receptors or toll-like receptors, triggering an inflammatory response [116,117,122]. Circulating myeloid-derived monocytes migrate and relocate in the intima to differentiate into macrophages with a pro-inflammatory phenotype. Zinc affects the phagocytosis and inflammatory function of macrophages through a series of regulatory actions [123] and acts as an intracellular second messenger in signaling pathways [124]. Dubben et al. [125] found that zinc has a specific negative impact on monocyte differentiation, which is not observed in granulocytes. Zinc chelator TPEN can promote monocyte differentiation through augmentation of cyclic adenosine monophosphate (cAMP) production, which is formerly decreased by an inhibitive function of zinc ion on the cAMP-synthesizing enzyme adenylate cyclase. Another study indicates that zinc is the intermediary between monocyte adhesion to the vascular endothelium [126]. The excessive accumulation of cholesterol in macrophages activates the inflammasome NOD-like receptor thermal protein domain associated protein 3 (NLRP3), a protein that cleaves IL-1β precursor into its active form and enhances the expression of cytokines [116]. Zinc supplementation can reduce NLRP3 inflammasome activation [127], whereas TPEN inversely activates NLRP3 and improves the levels of pro-inflammatory cytokines, such as IL-1β and IL-6 in macrophages [128].

Inflammatory macrophages facilitate the progression of atherosclerosis and cause expansion of the lesion, leading to plaque morphological changes, eventually resulting in plaque rupture and acute lumenal thrombosis via the release of pro-inflammatory cytokines [122]. Zinc deficiency influences the generation of cytokines, and cytokines, in turn, affect zinc distribution. IL-6 and IL-1β are the major pro-inflammatory cytokines that innate immune cells secrete [124]. A meta-analysis conducted by Faghfouri et al. [129] indicated that zinc supplementation can decrease IL-6 levels, and another former study shows that IL-6 induces low zinc ion availability by modulating MT gene expression [130].

3.2.2. Zinc and Lymphocytes

Along with macrophage and dendritic cell accumulation, adaptive immunity joins to exert a deleterious influence on plaque formation. Studies confirmed that the absence of T cells and B cells, which were induced by recombination-activating genes (RAGs) deficit, demonstrated alleviated atherosclerosis in *Apoe*$^{-/-}$ *or Ldlr*$^{-/-}$ mice [131]. T cells play a predominant role in disease progression and constitute 10% of total plaque cells, 70% of which are CD4+ T cells, with B cells only occasionally found in plaque [119]. Atherosclerosis is known as a Th1 disease, as Th1 cells are the major producer of proatherogenic cytokines, such as interferon (IFN)-γ, IL-2, IL-3, and TNF. Zinc deficiency in the experimental human model results in imbalanced Th1/Th2 function. The products of Th1 cells, such as IL-2 and IFN-γ, decline under zinc deficit while Th2 cell products (such as IL-4, IL-6, and IL-10) remain unaffected [132,133].

Zinc contributes to T cell regulation in two ways: the signaling pathway initiated by T cell receptor (TCR) or by cytokine stimulation [124]. In the TCR activation-mediated signaling pathway, the CD3 subunits of the TCR complex are phosphorylated by lymphocyte-specific protein-tyrosine kinase (LCK) and promote the assembly of the immunological synapse, leading to the activation of the downstream signaling cascade [134]. Zinc facilitates the process by regulating the activity of LCK in multiple ways and plays a vital role in T cell development and activation [135,136]. After the TCR trigger, certain zinc transporters, such as ZIP-6 and ZIP-8, are up-regulated and induce a rise in cellular zinc concentration, which is directly involved in T cell activation [137,138]. On the other hand, the cytokine-mediated signaling pathway is regulated by IL-2, and according to Faghfouri et al., zinc can increase IL-2 levels [129]. IL-2 stimulation can induce T cell proliferation and development through three different signaling pathways: Janus kinase 1/3 (JAK1/3)-signal transducer and activator of transcription 5 (STAT5), ERK1/2, and phosphoinositide 3-kinase (PI3K)/protein kinase B (Akt) [124]. The interaction between zinc and immune cells in atherogenesis has not been directly confirmed through animal model studies. Further research should be conducted in this regard.

4. Role of Zinc in Risk Factors of Atherosclerosis

In addition to the direct role of zinc in atherosclerosis, zinc furthermore exerts important positive or negative functions via various risk factors of the atherogenic process, such as lipid metabolism, glucose metabolism, and blood pressure.

4.1. Role of Zinc in Lipid Metabolism

In humans, adipose tissue is the main lipid repository and the interaction between zinc and adipose tissue is critical in lipid metabolism. Olechnowicz et al. [90] previously proposed that zinc may regulate lipid metabolism by interfering with leptin production. However, other studies found no significant association between zinc supplementation and overall serum leptin [139,140]. A recent study proposed that zinc could induce lipolysis by activating the zinc/MTF-1/PPARα and calcium/calmodulin-dependent protein kinase kinase-β (CaMKKβ)/adenosine 5′-monophosphate-act protein kinase (AMPK) pathway, which further induce autophagy-mediated lipophagy [141]. Zinc-α2-glycoprotein (ZAG), an adipokine, plays a critical role in lipid metabolism, including reducing fatty acid synthesis, mobilizing lipids, and stimulating lipolysis [142]. The expression of ZAG was significantly reduced in the liver and adipose tissue of the obese population [143]. A zinc-binding site exists in the structure of ZAG [144], which suggests that the effect of zinc on lipid metabolism may be mediated by ZAG. In addition, zinc prevents the oxidation of LDL, stopping the main mechanism of atherogenesis [145].

High zinc levels increase hepatocyte activity and improve hepatic lipid metabolism in rats, while zinc deficiency impairs hepatic lipid metabolism [146]. Zinc deficiency in the prenatal and postnatal life of Wistar rats was shown to increase hypertriglyceridemia and adipocyte hypertrophy and adequate dietary zinc in postweaning growth could partially reverse this change [147]. Zinc supplementation could reverse abnormal lipid metabolism and elevated liver enzymes (such as aspartate aminotransferase and alanine transaminase) found in high-fat diet (HFD)-fed rabbits [148]. In lipid profiles, dietary zinc supplementation significantly reduced cholesterol and cholesterol oxidation products in rat serum [149]. In humans, obesity and being overweight were associated with decreased serum zinc levels [150]. Zinc deficiency or marginal zinc deficiency was associated with lower lipid profiles in healthy individuals [151]. However, the effects of zinc on specific lipid-conjugated molecules remain contradictory. A cross-sectional study on 7597 United States adults found that zinc intake was inversely associated with the ratio of total cholesterol (TC) to high-density lipoprotein (HDL) cholesterol [152]. However, Yary et al. [153] noticed an inverse relationship between serum zinc level and HDL cholesterol in middle-aged and older eastern Finnish men. In a meta-analysis, Khazdouz et al. [154] indicated that zinc supplementation could significantly decrease the plasma level of triglyceride, TC, and very-low-density lipoprotein (VLDL). Recently, Saykally et al. [155] noticed that the reduced amount of ZEB1 could increase adiposity during early weight gain in female TCF8+/- mice fed with HFD. Sterol regulatory element-binding protein 2 (SREBP-2), a master transcriptional regulator of cholesterol biosynthetic genes and LDL receptor genes, is highly involved in foam cell formation and atherogenesis [156]. Overexpression of ZEB1 could inhibit the endogenous expression of SREBP-2, thereby regulating the metabolism of cellular cholesterol [157]. Evidence also suggested that ZEB1 can regulate the expression of ATP-binding cassette transporter A1 (ABCA1) to reduce lipid accumulation in macrophages [55]. KLF2 is another important zinc finger transcription factor involved in macrophage form cell formation. Atkins et al. [52] noted that hemizygous deficiency of KLF2 increased diet-induced atherosclerosis in ApoE-deficient mice, indicating its atheroprotective role.

4.2. Role of Zinc in Glucose Metabolism

Zinc performs critical functions in insulin secretion and regulation of glucose metabolism. Zinc deficiency was associated with decreased insulin secretion from pancreatic β cells [158]. Filios et al. [159] indicated that knockdown of ZEB1 could significantly increase the apoptosis of pancreatic β cells. Zinc is believed to modulate the effects of insulin through different mechanisms [158,160]. Furthermore, serum zinc level was inversely associated with insulin resistance [161]. Several zinc homeostasis-associated proteins expressed in adipocytes, including ZnT-7, ZIP-13, and ZIP-14, can influence adipogenesis and the function of adipocytes, thereby participating in insulin resistance [158,162]. Apart from the indirect regulation of glucose metabolism via insulin regulation, zinc also directly regulates glucose metabolism. Studies have shown that zinc plays a vital role in the inhibition of gluconeogenesis, as well as the stimulation of glucose uptake and glycogen synthesis [163]. Moreover, it should be noted that a complex correlation exists between zinc, insulin, and diabetes. Diabetes can affect zinc homeostasis, by increasing the urinary loss of zinc and decreasing the total bodily zinc level [164].

Sisnande et al. [165] found that zinc restriction in male Swiss mice was associated with smaller pancreatic islets, mediated by increased apoptosis. Zinc supplementation could enhance the function of pancreatic β cells in HFD-fed mice, thereby improving glucose homeostasis [166]. Zinc supplementation in mice with HFD-induced liver injury could reduce glucose production and lipid deposition and promote glucose absorption [167]. These findings suggested that high zinc levels may alleviate the abnormal metabolism of glucose and lipid induced by HFD. In prediabetic populations, zinc supplementation significantly improved pancreatic β-cell function and decreased fasting blood glucose (FBG), 2 h postprandial glucose levels in oral glucose tolerance test, and insulin resistance [168]. A recent meta-analysis of randomized controlled trials (RCTs) focusing on the effects of zinc supplementation on cardiometabolic risk factors also indicated that zinc supplementation significantly decreases FBG and hemoglobin A1c (HbA1c) levels [154]. However, the association between serum zinc and type 2 diabetes mellitus (T2DM) remains controversial, with an inverse association [169], positive association [170,171], and no association [172] found in different studies. Fernández-Cao et al. [173] carried out a meta-analysis and found that slightly elevated dietary zinc intake may prevent the development of T2DM, while the elevation of serum zinc level might increase the risk of T2DM. Future studies focusing on the role of dietary or serum zinc levels in T2DM pathophysiology are warranted.

4.3. Role of Zinc in the Regulation of Blood Pressure

Zinc may regulate blood pressure (BP) through different mechanisms. Firstly, copper-zinc superoxide dismutase (Cu/Zn-SOD) conducts an important function in regulating hypertension by influencing oxidative stress. Studies have demonstrated that both excessive and insufficient zinc intake inhibits the activity of Cu/Zn-SOD [174,175]. Secondly, zinc may regulate BP by regulating the renin-angiotensin-aldosterone system (RAAS). Zinc is an active component of the angiotensin-converting enzyme (ACE), and its deficiency decreases serum ACE activity [176]. During critical development periods of Wistar rats, zinc deficiency was associated with morphological renal alterations and long-term/permanent changes in the renal RAAS system [177]. Thirdly, renal Na+-Cl- cotransporter (NCC) is a zinc-regulated transporter, and zinc deficiency could upregulate NCC expression and renal Na^+ transport, contributing to an increase in BP [178]. In addition, NO is a key molecule involved in the tonic relaxation of systemic vasculature and BP regulation [179].

As previously described, zinc deficiency decreases the expression and activity of NOS in the vascular and renal system, leading to a decrease in NO activity [180]. Finally, zinc may also regulate BP by influencing intracellular calcium homeostasis [181].

Zinc may regulate BP in a bidirectional way. Some animal studies found that zinc restriction was associated with higher systolic BP (SBP) [147], while others indicated that excessive zinc intake also significantly elevated SBP, presumably through superoxide-induced oxidative stress [182]. The results from human studies focused on the relationship between zinc and hypertension are also inconsistent. A recent meta-analysis on 9 RCTs with 544 patients found that zinc supplementation significantly reduces the SBP but has no significant effect on diastolic BP (DBP) [183]. Several clinical trials and population-based studies found that zinc deficiency is associated with prehypertension in apparently healthy subjects [184], and zinc supplementation is beneficial for BP regulation [168,185]. Other studies found no association between zinc supplementation or serum zinc level and SBP [186–189]. All these findings suggested that both a lack and excess of zinc intake may have an adverse effect on elevated BP, indicating a U-type or J-type relationship between zinc and hypertension. Additional high-quality clinical trials focusing on the effects of zinc intake on blood pressure are warranted for further evidence.

5. Zinc and Atherosclerosis: Animal and Human Studies

5.1. Animal Studies

Animal studies found that zinc deficiency was associated with increased plasma concentration of triacyl glycerides and cholesterol and a higher lipoprotein-cholesterol distribution toward the non-HDL fraction [190,191]. ApoE knock-out (AEKO) mice with suboptimal dietary zinc intake were associated with raised proatherogenic lipoprotein levels and developed more aortic plaques than mice consuming adequate zinc [192]. In atherogenic mouse models, zinc supplementation suppressed the abnormal plasma lipid profile and the proinflammatory events in atherosclerosis [190]. Studies also found that dietary zinc supplementation significantly reduced total cholesterol accumulation in the aorta and inhibit the development of aortic atherogenesis in HFD-fed New Zealand rabbits [193–196]. A recent study conducted by Cheng et al. [197] suggests the role of zinc influx in monocyte adhesion and recruitment. Ex vivo and AEKO mice model experiments confirmed ZIP8 overexpression and increased zinc influx in monocyte adhesion to aortas and nascent atherosclerosis lesions, which potentially promote the inflammatory response towards arterial walls. Ren et al. [194] further found that zinc levels in aortic atheromatous lesions of zinc-supplied rabbits were not significantly different from that in control rabbits. However, zinc supplementation was associated with significantly lower iron lesion levels. Based on these findings, they proposed that the antiatherogenic effect of zinc may be mediated by inhibiting iron-catalyzed free radical reactions. Several studies indicate the anti-apoptosis function of zinc in VSMCs [109] and the detrimental effect of zinc deficiency on vascular health [198] (see Table 1). A variety of zinc finger proteins also exhibit significant influence in the pathogenesis of atherosclerosis by regulating the transcription process. Sayin et al. [199] found that zinc finger protein 148 (ZFP148) deficiency exerted an atheroprotective effect by increasing the activity of p53 and inhibiting the proliferation of macrophages in AEKO mice. KLF14, like other transcription factors in the KLF family, shows an anti-atherosclerotic effect in $ApoE^{-/-}$ mice by promoting cholesterol efflux and inhibiting inflammatory response [200]. In addition, POZ/BTB and AT-hook-containing zinc finger protein 1 (PATZ1) and Juxtaposed with another zinc finger gene 1 (JAZF1) also play a protective role in atherosclerosis [201,202].

Table 1. Animal and human studies that report the effects of zin on atherosclerosis.

Author, Year	Subjects	N	Study Design	Intervention	Outcome
Animal studies					
Reiterer et al. 2005 [190]	LDL receptor knock-out (LDL-R-/-) mice, 5 wk	11 (ZD) 11 (Z) 11 (ZS)	/	ZD: AIN-93 diets with 0 μmol Zn/g diet for 4 wk; Z: AIN-93 diets with 0.45 μmol Zn/g diet for 4 wk; ZS: AIN-93 diets with 1.529 μmol Zn/g diet for 4 wk	**Zinc deficiency was associated with:** - Higher mRNA expression of the antioxidant enzyme glutathione reductase, VCAM-1, and PPARγ; - Higher DNA binding activity of NF-κB in liver tissues; - Lower DNA binding activity of PPARγ. **Zinc supplementation was associated with:** - Reduced TG, total plasma cholesterol, and VLDL and HDL fractions of free cholesterol.
Shen et al. 2007 [191]	LDL receptor knock-out (LDL-R-/-) male mice, 5 wk	2 (ZD) 2 (Z)	/	ZD: low-fat (LF) diet with 0.4 mg/kg of zinc for 21 d, and then high-fat (HF) diet with 0.4 mg/kg of zinc for 1 wk; Z: low-fat (LF) diet with 33.1 mg/kg of zinc for 21 d, and then high-fat (HF) diet with 33.1 mg/kg of zinc for 1 wk	**Zinc deficiency was associated with:** - Higher total plasma cholesterol, total plasma fatty acids, and non-HDL fraction (VLDL, IDL, LDL); - Inhibited PPARγ transactivation activity induced by rosiglitazone.
Beattie et al. 2012 [192]	ApoE knock-out (AEKO) mice, 5 wk	24 (ZD) 12 (Z)	/	ZD: high saturated fat (21% w/w) and high cholesterol (0.15%) semi-synthetic diets containing suboptimal zinc (3 or 8 mg/kg) for 25 wk; Z: high saturated fat (21% *w*/*w*) and high cholesterol (0.15%) semi-synthetic diets containing adequate zinc (35 mg/kg) for 25 wk	**Zinc deficiency was associated with:** - Higher plaque area; - Elevated cytokine IL-1β, IL-6 and sVCAM-1; - Elevated total protein and cholesterol level; - Increased intramural calcium and phosphorous deposits (atheromatous plaque).
Alissa et al. 2004 [193]	Male New Zealand White rabbits, 6–10 wk	8 (ZS) 8 (C)	/	ZS: diet containing 0.25–1% (*w*/*w*) cholesterol plus zinc as its sulfate at 0.5% (*w*/*w*) for 12 wk; C: diet containing 0.25–1% (*w*/*w*) cholesterol for 12 wk	**Zinc supplementation was associated with:** - No significant difference in weight gain, integrated serum cholesterol levels, erythrocyte superoxide dismutase, hepatic enzyme activity; - Reduction in aortic atherogenesis lesion.
Ren et al. 2006 [194]	Male New Zealand white rabbits,/	6 (ZS) 6 (C)	/	ZS: zinc-supplemented diet SF03-017 (modified guinea pig and rabbit + 1% cholesterol + 1000 ppm (1 g/kg) zinc as zinc carbonate) for 8 wk; C: high-cholesterol diet (HCD) SF00- 221 (modified guinea pig and rabbit + 1% cholesterol) for 8 wk	**Zinc supplementation was associated with:** - No significant difference in total cholesterol, triglyceride, and LDL; - Lower HDL; - Decreased lesion areas from sections of aorta; - Decreased Fe lesion levels.
Abdelhalim et al. 2013 [195]	Male New Zealand white rabbits, 12 wk	5 (ZS) 5 (C)	/	ZS: zinc-supplemented high-cholesterol diet: NOR Purina Certified Rabbit Chow no. 5321 with 1.0% cholesterol plus 1.0% olive oil (100 g/day), with 350 ppm zinc for 12 wk; C: high-cholesterol diet: NOR Purina Certified Rabbit Chow no. 5321 with 1.0% cholesterol plus 1.0% olive oil (100 g/day) for 12 wk.	**Zinc supplementation was associated with:** - Delayed or retarded progression of atherosclerosis.
Jenner et al. 2007 [196]	Male New Zealand white rabbits,/	5 (ZS) 6 (C)	/	ZS: zinc-supplemented high cholesterol diet [GPR + 1% cholesterol + 1000 ppm (1 g/kg of diet) zinc as zinc carbonate] for 8 wk; C: High cholesterol diet [GPR + 1% cholesterol] for 8 wk.	**Zinc supplementation was associated with:** - No significant effect on levels of LDL or plasma triglyceride; - Lower total arterial wall cholesterol levels; - Decreased average lesion cross-sectional areas; - Decreased total F2-isoprostanes in plasma and arterial wall atherosclerosis; - Decreased cholesterol oxidation products in complete arterial wall.
Beattie et al. 2008 [198]	Sprague–Dawley rats, 100 g, 4 wk	Study 1: 9(MZD) 9(MZA) Study 2: 10(AZD) 10(APF) 10(AZA)	/	AIN-76 diet that contained 35, 6 or, 1 mg Zn/kg; MZD: 6 mg zinc/kg for 43 days; MZA: 35 mg zinc/kg for 43 days; AZD: 1 mg zinc/kg for 39 days; APF: 35 mg zinc/kg for 39 days, pair-fed with AZD group; AZA: 35 mg zinc/kg for 39 days.	**Zinc deficiency was associated with:** - Aorta cell structure, fatty acid metabolism, and carbohydrate metabolism; - Changes in smooth muscle contractility, insulin resistance, and perturbation of key Zn-dependent transcription factors; - Detriment to the maintenance of vascular health.

Table 1. *Cont.*

Author, Year	Subjects	N	Study Design	Intervention	Outcome
Human studies					
Yang et al. 2010 [203]	Middle-aged and elderly Korean populations, 40–89 y	4564	Cross-sectional analyses	/	- Zinc intake was inversely related to subclinical atherosclerosis.
Masley et al. 2015 [204]	Population attending an executive evaluation program, 23–65 y	592	Prospective cross-sectional analysis	/	- Lower intake of zinc was associated with higher mean carotid intima media thickness.
Chen et al. 2020 [205]	Noninstitutionalized civilian US populations, ≥40 y	1764 (AAC) 771 (noAAC)	Cross-sectional analyses	/	- Patients with severe AAC had a lower intake of dietary zinc; - A higher intake of dietary zinc was associated with lower odds of having severe AAC; - A higher serum zinc level was associated with higher supplemental zinc intake but not with dietary or total zinc intake; - Those with severe AAC had a higher intake of supplemental and total zinc.
Dziedzic et al. 2022 [206]	Patients with a history of previous MI who were treated with coronary Angioplasty, 37–95 y	133	Cross-sectional analyses	/	- No significant statistical differences in zinc levels between groups with different severity of CAD; - A negative correlation was identified between zinc content in the hair and serum TG concentration.
Giannoglou et al. 2010 [207]	Underwent diagnostic coronary angiography for evaluating chest pain. CAD: 66 ± 43 y; control: 61 ± 51 y	40 (CAD) 32 (control)	A pilot study	/	- No significant difference in serum zinc level; - CAD was significantly associated with higher urinary zinc and lower serum zinc/24-h urine zinc ratio; - Angiographic indices were positively associated with higher urinary zinc and negatively associated with serum zinc/24-h urine zinc ratio.
Milton et al. 2018 [208]	Mid-age Australian women, 50–61 y	9264	Large longitudinal study	/	- Dietary zinc intake was associated with a higher CVD incidence in women aged ≥ 50 years.
Mattern et al. 2021 [209]	Black and White participants, ≥45 y	2944	Population-based prospective longitudinal study	/	- Serum zinc levels were inversely associated with the incidence of ischemic stroke after adjustment for potential confounders.
Gao et al. 2021 [210]	Multi-ethnic participants, 45–84 y	6814	Prospective population-based observational cohort study	/	- Higher dietary zinc intake was significantly associated with a lower risk of CAC progression for women and men; - This association was found in higher dietary zinc intake from other sources, but not from red meat.
de Oliveira Otto et al. 2012 [211]	Adults free of clinical CVD from six U.S. communities, 45–84 y	6814	Prospective population-based study	/	- Zinc derived from red meat sources was associated with greater risk of incident metabolic syndrome; - Zinc from red meat, but not zinc from other sources, was associated with the risk of CVD.
Jung et al. 2013 [212]	Populations from multi-rural Korean communities, ≥40 y	5532	Cross-sectional analysis	/	- Phytate: zinc molar ratio was positively related to subclinical atherosclerosis risk in men, but not in women.
Alissa et al. 2006 [213]	Saudi male subjects with established CVD, 55.6 ± 12.1 y age-matched controls, 55.0 ± 11.6 y	130 (CVD) 130 (control)	Population-based study	/	- Urinary zinc was significantly lower in patients with CVD; - Serum zinc was negatively associated with atherosclerosis.
Qazmooz et al. 2021 [214]	Patients with atherosclerosis and age, sex-matched healthy controls	120 (atherosclerosis) 58 (control)	Machine learning studies	/	- Serum zinc decreased from controls to patients with atherosclerosis, then to atherosclerotic patients with unstable angina.

Table 1. *Cont.*

Author, Year	Subjects	N	Study Design	Intervention	Outcome
Lee et al. 2005 [215]	Postmenopausal women, 55–69 y	34,492	Population-based study	/	- In the total sample, dietary zinc intake was not associated with the risk of CVD mortality; - Among alcohol drinkers who consumed ≥10 g alcohol per day, dietary zinc showed an inverse association with CVD mortality.
Eshak et al. 2018 [216]	Middle-aged residents in 45 Japanese communities, ≥40 y	58,646	Population-based prospective cohort study	/	- Populations with higher zinc intake were less likely to be hypertensives but more likely to be diabetics; - Dietary zinc intake was inversely associated with mortality from CHD in men but not women.
Shi et al. 2018 [217]	Based on a subsample of the Chinese national nutrition and health survey representing Jiangsu province, ≥20 y	2832	Population-based prospective cohort study	/	- Zinc intake is positively related to all-cause mortality and there was no significant association between relative zinc intake and CVD mortality.
Bao et al. 2010 [218]	40 healthy elderly, 56–83 y	20(ZS) 20(C)	RCT	ZS: 45 mg zinc/d as gluconate for 6 mo; C: placebo for 6 mo	- Zinc supplementation increased plasma zinc concentration and decreased the concentrations of plasma hsCRP, IL-6, MCP-1, VCAM-1, sPLA, and MDA+HAE; - Zinc supplementation decreased the generation of TNF-α, IL-1β, VCAM-1, MDA+HAE, activation of NF-κB, and increased anti-inflammatory proteins A20 and PPAR-α in human monocytic leukemia THP-1 cells and human aortic endothelial cells.
Costarelli et al. 2010 [219]	Overweight/obese individuals (BMI ≥25 kg/m^2), 43 ± 5 y	100 (Group 1) 123 (Group 2)	Clinical trial	Group 1: low-zinc dietary intake (<7 mg/day for females and <9.5 for males); Group 2: normal-zinc dietary intake (≥7 mg/day for females and ≥9.5 for males)	- Zinc deficiency group presents higher inflammatory markers (A2M and CRP) and - lipid assets (total and LDL cholesterol, triglycerides); - Zinc deficiency group shows decreased zinc homeostasis genes (SLC30A1, MT-1A, and MTF-1) and up-regulated inflammatory genes (IL6, IL1-α, IL1-β, and A2M).
Seet et al. 2011 [220]	Male T2DM patients, ≥21 y	20(ZS) 20(C)	RCT	ZS: 240 mg zinc/day as gluconate for 3 mo; C: 2 tablets of placebo (99% microcrystalline cellulose, 1% magnesium stearate) for 3 mo	- Oral zinc supplementation did not exert change in markers of oxidative damage and vascular function.

Abbreviations: A2M: alpha 2-macroglobulin; AAC: abdominal aortic calcification; APF: pair-fed control group for acute zinc deficiency; AZA: zinc adequate control group for acute zinc deficiency; AZD: acute zinc deficiency; C: without zinc supplementation; CAC: coronary artery calcium; CAD: coronary artery disease; CHD: coronary heart disease; CVD: cardiovascular diseases; HDL: high-density lipoprotein; hsCRP: high-sensitivity C-reactive protein; IDL: intermediate density lipoprotein; IL: interleukin; MCP-1: macrophage chemo-attractant protein 1; MDA+HAE: malondialdehyde and hydroxyalkenals; MT: metallothionein; MTF-1: metal response element-binding transcription factor-1; MZA: zinc adequate control group for marginal zinc deficiency; MZD: marginal zinc deficiency; NF-κB: nuclear transcription factor kappa B; PF: pair-fed control group; PPAR: peroxisome proliferator-activated receptor; TG: triacylglycerides; TNF-α: tumor necrosis factor α; SLC30A: solute carrier 30A; sPLA: secretory phospholipase A2; VCAM-1: vascular cell adhesion molecule-1; VLDL: very low-density lipoprotein; Z: adequate zinc; ZD: zinc deficiency; ZS: zinc supplementation.

5.2. Human Studies

Epidemiological studies of zinc deficiency and atherosclerosis are scarce and most human studies are cohort studies and RCTs. Several studies focusing on zinc and vascular physiology indicated an inverse relationship between dietary zinc intake and carotid intima-media thickness [203,204] or aortic calcification [205]. A recent study conducted by Dziedzic et al. [206] measured the zinc level of hair samples from 133 CAD patients that were evaluated by angiography. A lower level of zinc was detected in the hair of patients with T2DM and high serum TG concentration. However, there are no significant statistical differences among groups with varying severity of CAD, which is in contrast with many other studies.

A pilot study found an association between the low ratio of serum zinc to 24 h urine zinc and severe coronary atherosclerosis in angiography [207]. The incidence of cardiovascular disease-related events was reported to be influenced by the level of both dietary zinc intake [208] and serum zinc [209]. A systematic review found that a lower incidence of CVDs was associated with higher serum zinc, but not higher dietary zinc intake [221]. The difference in the conclusions of these studies might be partially explained by the difference in sources of dietary zinc. Higher dietary zinc intake could elevate the level of serum zinc, however, a poor correlation exists between the level of serum zinc and zinc intake, since zinc intake from various sources varies greatly in its bioavailability and metabolism [222,223]. Coronary artery calcium progression is considered an important marker of subclinical coronary atherosclerosis. In clinical CVDs-free populations, Gao et al. [210] found an association between a lower risk of coronary artery calcium progression and higher dietary zinc intake from non-red meat sources. De Oliveira Otto et al. [211] reported a positive association between higher dietary zinc intake from red meat sources and the risk of CVDs. Considering the inconsistency between dietary zinc intake and serum zinc, it may be beneficial to investigate how zinc bioavailability is associated with atherosclerosis. Phytate/zinc molar ratio is inversely associated with zinc bioavailability. Jung et al. [212] found that the molar ratio of phytate to zinc was positively associated with the risk of atherosclerosis in women aged ≤ 65. This finding might suggest an inverse relationship between zinc bioavailability and the risk of subclinical atherosclerosis.

Zinc deficiency has been recognized to function negatively in atherogenesis. Compared with healthy controls, patients with established atherosclerosis were associated with a significantly lower serum zinc level, which further decreases in patients with unstable angina [213,214]. Zinc supplementation was shown to be a promising treatment in patients with severely symptomatic and inoperable atherosclerotic disease [224]. Lee et al. [215] found that among alcohol drinkers who consumed ≥ 10 g of alcohol per day, CVD-associated mortality was inversely associated with dietary zinc intake. A prospective study on healthy Japanese men and women without CVDs found that higher dietary zinc intake was inversely associated with CVD-associated mortality in men but not women [216]. However, a ten-year follow-up study in Jiangsu province reported no significant relationship between relative zinc intake and CVD-associated mortality [217]. RCTs have been conducted to verify the positive effects of zinc. Bao et al. [218] found that increased plasma zinc level is associated with decreased plasma high-sensitivity C-reactive protein (hsCRP), IL-6, VCAM-1, secretory phospholipase A2 (sPLA), malondialdehyde, and hydroxyalkenals (MDA+HAE). Furthermore, zinc supplementation decreased the production of TNF-α, IL-1β, VCAM-1, and MDA+HAE, as well as the activation of NF-κB, while an increase in anti-inflammatory proteins A20 and PPAR-α were found in human monocytic leukemia THP-1 cells and human aortic endothelial cells. Costarelli et al. [219] confirmed these findings about inflammatory molecules via a zinc deficit clinical trial on overweight/obese adults. Results indicated that the zinc deficit group presented higher inflammatory markers, such as α 2-macroglobulin (A2M) and CRP, and lipid assets, such as TC, LDL, and triglycerides, in parallel with decreased zinc homeostasis genes (SLC30A-1, MT-1A, and MTF-1) and up-regulated inflammatory genes (IL-6, IL-1α, IL-1β, and A2M). However, a later study by Seet et al. [220] indicated that oral zinc supplementation did not exert change in markers of oxidative damage and vascular function. More RCTs are required for further understanding and instruction of zinc supplementation and patient benefits (see Table 1).

6. Limitations and Future Perspectives

Zinc is an indispensable micronutrient involved in multiple biological processes in organisms. Due to the lack of functional reserve for zinc, human zinc homeostasis primarily relies on daily dietary intake. The recommended dose of zinc intake is 11 mg/d and 8 mg/d for men and women, respectively, and should not exceed 40 mg/d [225]. Zinc deficiency is more commonly found in the elderly [226] and vegetarians [227] and usually progresses in a chronic and latent way, of which the manifestation is hard to connect with zinc homeostasis directly, much different from that of iron and iodine. The absence of a reliable and specific biomarker for the indication of in vivo zinc levels also restricts further study of zinc in clinical trials. The most adopted measurements for the estimation of adult zinc status at present are dietary zinc intake and plasma/serum zinc concentration [228]. Other indicators include hair, urinary, and blood cell zinc. However, none of them are able to represent an overall and long-term status of body zinc and zinc-binding proteins. Since zinc has an extensive and profound influence on cardiovascular physiology and cell metabolism, it is necessary to seek novel zinc status biomarkers that are less influenced by individual variability or develop an innovative technology to measure cellular zinc levels in clinical trials.

All in all, the animal studies and epidemical research have indicated the potential negative role of zinc deficiency in atherosclerosis. However, specific pathways or interaction patterns between zinc and plaque formation or atherosclerotic progression remain unclear. In this review, we aim to investigate the underlying mechanism of how zinc affects atherogenesis progress through a review of the role that zinc plays in proatherogenic cells, such as ECs, VSMCs, and immune cells, as well as the risk factors, such as lipid metabolism, glucose metabolism, and blood pressure. Most evidence prompts a positive role of zinc in multiple physiologic processes, such as NO generation, anti-inflammation, anti-oxidation, and anti-apoptosis, while paradoxical effects are also indicated in immune cells. More research is required in this field, especially those that directly reveal the effect of zinc on atherosclerosis models, such as in vitro studies with atherosclerosis-specific stimuli, assays on plaque composition and morphological changes in animal models, or evaluation of plaque and stenosis degree in atherosclerosis patients. Therefore, further studies are needed to better elucidate the role of zinc in atherosclerosis. Taken together, zinc participates in the regulation of multiple cellular processes and may exert a pivotal role in atherogenesis and disease progression. Zinc supplementation is a potential adjunctive therapy for atherosclerosis and its complications. Further efforts should be expended on a comprehensive understanding of zinc and atherosclerosis and optimal prophylactic administration for the targeted populations.

Author Contributions: Writing—original draft preparation, T.S. and Q.Z.; writing—review and editing, Y.L. and T.W.; supervision, Y.L.; funding acquisition, Y.L. All authors have read and agreed to the published version of the manuscript.

Funding: This research was funded by National Natural Science Foundation of China, grant number 82171303 & 82171301, and Capital Funds for Health Improvement and Research, grant number 2020-2-1032.

Institutional Review Board Statement: Not applicable.

Informed Consent Statement: Not applicable.

Data Availability Statement: Not applicable.

Conflicts of Interest: The authors declare no conflict of interest.

References

1. Chasapis, C.T.; Ntoupa, P.-S.A.; Spiliopoulou, C.A.; Stefanidou, M.E. Recent aspects of the effects of zinc on human health. *Arch. Toxicol.* **2020**, *94*, 1443–1460. [CrossRef] [PubMed]
2. Stefanidou, M.; Maravelias, C.; Dona, A.; Spiliopoulou, C. Zinc: A multipurpose trace element. *Arch. Toxicol.* **2005**, *80*, 1–9. [CrossRef]
3. King, J.C.; Shames, D.M.; Woodhouse, L.R. Zinc Homeostasis in Humans. *J. Nutr.* **2000**, *130*, 1360S–1366S. [CrossRef]
4. Valko, M.; Jomova, K.; Rhodes, C.J.; Kuča, K.; Musílek, K. Redox- and non-redox-metal-induced formation of free radicals and their role in human disease. *Arch. Toxicol.* **2016**, *90*, 1–37. [CrossRef]
5. Mocchegiani, E.; Muzzioli, M.; Giacconi, R. Zinc and immunoresistance to infection in aging: New biological tools. *Trends Pharmacol. Sci.* **2000**, *21*, 205–208. [CrossRef]

6. Choi, S.; Cui, C.; Luo, Y.; Kim, S.; Ko, J.; Huo, X.; Ma, J.; Fu, L.; Souza, R.F.; Korichneva, I.; et al. Selective inhibitory effects of zinc on cell proliferation in esophageal squamous cell carcinoma through Orai1. *FASEB J.* **2017**, *32*, 404–416. [CrossRef]
7. Pan, Z.; Choi, S.; Ouadid-Ahidouch, H.; Yang, J.-M.; Beattie, J.H.; Korichneva, I. Zinc transporters and dysregulated channels in cancers. *Front. Biosci.* **2017**, *22*, 623–643. [CrossRef]
8. Prasad, A.S. Discovery of human zinc deficiency: 50 years later. *J. Trace Elements Med. Biol.* **2012**, *26*, 66–69. [CrossRef]
9. Bonow, R.O.; Smaha, L.A.; Smith, S.C.; Mensah, G.; Lenfant, C. World Heart Day. *Circulation* **2002**, *106*, 1602–1605. [CrossRef]
10. Benjamin, E.J.; Blaha, M.J.; Chiuve, S.E.; Cushman, M.; Das, S.R.; Deo, R.; de Ferranti, S.D.; Floyd, J.; Fornage, M.; Gillespie, C. Heart Disease and Stroke Statistics-2017 Update: A Report From the American Heart Association. *Circulation* **2017**, *135*, e146–e603. [CrossRef]
11. Choi, S.; Liu, X.; Pan, Z. Zinc deficiency and cellular oxidative stress: Prognostic implications in cardiovascular diseases. *Acta Pharmacol. Sin.* **2018**, *39*, 1120–1132. [CrossRef] [PubMed]
12. Lu, J.; Stewart, A.J.; Sadler, P.J.; Pinheiro, T.J.; Blindauer, C.A. Albumin as a zinc carrier: Properties of its high-affinity zinc-binding site. *Biochem. Soc. Trans.* **2008**, *36*, 1317–1321. [CrossRef] [PubMed]
13. Kambe, T.; Tsuji, T.; Hashimoto, A.; Itsumura, N. The Physiological, Biochemical, and Molecular Roles of Zinc Transporters in Zinc Homeostasis and Metabolism. *Physiol. Rev.* **2015**, *95*, 749–784. [CrossRef] [PubMed]
14. Outten, C.E.; O'Halloran, T.V. Femtomolar sensitivity of metalloregulatory proteins controlling zinc homeostasis. *Science* **2001**, *292*, 2488–2492. [CrossRef] [PubMed]
15. Vinkenborg, J.L.; Nicolson, T.J.; Bellomo, E.A.; Koay, M.S.; Rutter, G.A.; Merkx, M. Genetically encoded FRET sensors to monitor intracellular Zn^{2+} homeostasis. *Nat. Methods* **2009**, *6*, 737–740. [CrossRef]
16. Kimura, T.; Kambe, T. The Functions of Metallothionein and ZIP and ZnT Transporters: An Overview and Perspective. *Int. J. Mol. Sci.* **2016**, *17*, 336. [CrossRef]
17. Kambe, T.; Taylor, K.M.; Fu, D. Zinc transporters and their functional integration in mammalian cells. *J. Biol. Chem.* **2021**, *296*, 100320. [CrossRef]
18. Colvin, R.A.; Holmes, W.R.; Fontaine, C.P.; Maret, W. Cytosolic zinc buffering and muffling: Their role in intracellular zinc homeostasis. *Metallomics* **2010**, *2*, 306–317. [CrossRef]
19. Maret, W. Zinc Biochemistry: From a Single Zinc Enzyme to a Key Element of Life. *Adv. Nutr. Int. Rev. J.* **2013**, *4*, 82–91. [CrossRef]
20. Krezel, A.; Maret, W. Zinc-buffering capacity of a eukaryotic cell at physiological pZn. *JBIC J. Biol. Inorg. Chem.* **2006**, *11*, 1049–1062. [CrossRef]
21. Laity, J.H.; Andrews, G.K. Understanding the mechanisms of zinc-sensing by metal-response element binding transcription factor-1 (MTF-1). *Arch. Biochem. Biophys.* **2007**, *463*, 201–210. [CrossRef] [PubMed]
22. Jia, R.; Song, Z.; Lin, J.; Li, Z.; Shan, G.; Huang, C. Gawky modulates MTF-1-mediated transcription activation and metal discrimination. *Nucleic Acids Res.* **2021**, *49*, 6296–6314. [CrossRef] [PubMed]
23. Baltaci, A.K.; Yuce, K. Zinc Transporter Proteins. *Neurochem. Res.* **2017**, *43*, 517–530. [CrossRef]
24. Huang, L.; Tepaamorndech, S. The SLC30 family of zinc transporters—A review of current understanding of their biological and pathophysiological roles. *Mol. Asp. Med.* **2013**, *34*, 548–560. [CrossRef]
25. Zalewski, P.D.; Beltrame, J.F.; Wawer, A.A.; Abdo, A.I.; Murgia, C. Roles for endothelial zinc homeostasis in vascular physiology and coronary artery disease. *Crit. Rev. Food Sci. Nutr.* **2018**, *59*, 3511–3525. [CrossRef] [PubMed]
26. Takagishi, T.; Hara, T.; Fukada, T. Recent Advances in the Role of SLC39A/ZIP Zinc Transporters In Vivo. *Int. J. Mol. Sci.* **2017**, *18*, 2708. [CrossRef]
27. Bowers, K.; Srai, S.K.S. The trafficking of metal ion transporters of the Zrt- and Irt-like protein family. *Traffic* **2018**, *19*, 813–822. [CrossRef]
28. Hoch, E.; Levy, M.; Hershfinkel, M.; Sekler, I. Elucidating the H^+ Coupled Zn^{2+} Transport Mechanism of ZIP4; Implications in Acrodermatitis Enteropathica. *Int. J. Mol. Sci.* **2020**, *21*, 734. [CrossRef]
29. Fagerberg, L.; Hallström, B.M.; Oksvold, P.; Kampf, C.; Djureinovic, D.; Odeberg, J.; Habuka, M.; Tahmasebpoor, S.; Danielsson, A.; Edlund, K.; et al. Analysis of the human tissue-specific expression by genome-wide integration of transcriptomics and antibody-based proteomics. *Mol. Cell. Proteom.* **2014**, *13*, 397–406. [CrossRef]
30. Thul, P.J.; Åkesson, L.; Wiking, M.; Mahdessian, D.; Geladaki, A.; Ait Blal, H.; Alm, T.; Asplund, A.; Björk, L.; Breckels, L.M.; et al. A subcellular map of the human proteome. *Science* **2017**, *356*, eaal3321. [CrossRef]
31. Nagamatsu, S.; Nishito, Y.; Yuasa, H.; Yamamoto, N.; Komori, T.; Suzuki, T.; Yasui, H.; Kambe, T. Sophisticated expression responses of ZNT1 and MT in response to changes in the expression of ZIPs. *Sci. Rep.* **2022**, *12*, 7334. [CrossRef] [PubMed]
32. Thingholm, T.E.; Rönnstrand, L.; Rosenberg, P.A. Why and how to investigate the role of protein phosphorylation in ZIP and ZnT zinc transporter activity and regulation. *Experientia* **2020**, *77*, 3085–3102. [CrossRef] [PubMed]
33. Karin, M. Metallothioneins: Proteins in search of function. *Cell* **1985**, *41*, 9–10. [CrossRef]
34. Krężel, A.; Maret, W. The Bioinorganic Chemistry of Mammalian Metallothioneins. *Chem. Rev.* **2021**, *121*, 14594–14648. [CrossRef] [PubMed]
35. Krezel, A.; Maret, W. Different redox states of metallothionein/thionein in biological tissue. *Biochem. J.* **2007**, *402*, 551–558. [CrossRef]
36. Krężel, A.; Maret, W. The Functions of Metamorphic Metallothioneins in Zinc and Copper Metabolism. *Int. J. Mol. Sci.* **2017**, *18*, 1237. [CrossRef]

37. Jarosz, M.; Olbert, M.; Wyszogrodzka, G.; Mlyniec, K.; Librowski, T. Antioxidant and anti-inflammatory effects of zinc. Zinc-dependent NF-κB signaling. *Inflammopharmacology* **2017**, *25*, 11–24. [CrossRef]
38. Hübner, C.; Haase, H. Interactions of zinc- and redox-signaling pathways. *Redox Biol.* **2021**, *41*, 101916. [CrossRef]
39. Subramanian Vignesh, K.; Deepe, G.S., Jr. Metallothioneins: Emerging Modulators in Immunity and Infection. *Int. J. Mol. Sci.* **2017**, *18*, 2197. [CrossRef]
40. Rahman, M.T.; Karim, M.M. Metallothionein: A Potential Link in the Regulation of Zinc in Nutritional Immunity. *Biol. Trace Element Res.* **2017**, *182*, 1–13. [CrossRef]
41. Cousins, R.J.; Leinart, A.S. Tissue-specific regulation of zinc metabolism and metallothionein genes by interleukin 1. *FASEB J.* **1988**, *2*, 2884–2890. [CrossRef] [PubMed]
42. Ugajin, T.; Nishida, K.; Yamasaki, S.; Suzuki, J.; Mita, M.; Kubo, M.; Yokozeki, H.; Hirano, T. Zinc-binding metallothioneins are key modulators of IL-4 production by basophils. *Mol. Immunol.* **2015**, *66*, 180–188. [CrossRef] [PubMed]
43. Dai, H.; Wang, L.; Li, L.; Huang, Z.; Ye, L. Metallothionein 1: A New Spotlight on Inflammatory Diseases. *Front. Immunol.* **2021**, *12*, 739918. [CrossRef] [PubMed]
44. Park, Y.; Zhang, J.; Cai, L. Reappraisal of metallothionein: Clinical implications for patients with diabetes mellitus. *J. Diabetes* **2017**, *10*, 213–231. [CrossRef]
45. Huang, S.; Wang, J.; Men, H.; Tan, Y.; Lin, Q.; Gozal, E.; Zheng, Y.; Cai, L. Cardiac metallothionein overexpression rescues diabetic cardiomyopathy in Akt2-knockout mice. *J. Cell. Mol. Med.* **2021**, *25*, 6828–6840. [CrossRef]
46. Bensellam, M.; Shi, Y.-C.; Chan, J.Y.; Laybutt, D.R.; Chae, H.; Abou-Samra, M.; Pappas, E.G.; Thomas, H.E.; Gilon, P.; Jonas, J.-C. Metallothionein 1 negatively regulates glucose-stimulated insulin secretion and is differentially expressed in conditions of beta cell compensation and failure in mice and humans. *Diabetologia* **2019**, *62*, 2273–2286. [CrossRef]
47. Laity, J.H.; Lee, B.M.; E Wright, P. Zinc finger proteins: New insights into structural and functional diversity. *Curr. Opin. Struct. Biol.* **2001**, *11*, 39–46. [CrossRef]
48. Jain, M.K.; Sangwung, P.; Hamik, A. Regulation of an inflammatory disease: Krüppel-like factors and atherosclerosis. *Arter. Thromb. Vasc. Biol.* **2014**, *34*, 499–508. [CrossRef]
49. Homeister, J.W.; Patterson, C. Zinc Fingers in the pizza pie aorta. *Circ. Res.* **2008**, *103*, 687–689. [CrossRef]
50. Nayak, L.; Lin, Z.; Jain, M.K. "Go with the flow": How krüppel-like factor 2 regulates the vasoprotective effects of shear stress. *Antioxidants Redox Signal.* **2011**, *15*, 1449–1461. [CrossRef]
51. Wittner, J.; Schuh, W. Krüppel-like factor 2 (KLF2) in immune cell migration. *Vaccines* **2021**, *9*, 1171. [CrossRef] [PubMed]
52. Atkins, G.B.; Wang, Y.; Mahabeleshwar, G.H.; Shi, H.; Gao, H.; Kawanami, D.; Natesan, V.; Lin, Z.; Simon, D.I.; Jain, M.K. Hemizygous Deficiency of Krüppel-Like Factor 2 Augments Experimental Atherosclerosis. *Circ. Res.* **2008**, *103*, 690–693. [CrossRef] [PubMed]
53. Xu, Y.; Xu, S.; Liu, P.; Koroleva, M.; Zhang, S.; Si, S.; Jin, Z.G. Suberanilohydroxamic Acid as a Pharmacological Kruppel-Like Factor 2 Activator That Represses Vascular Inflammation and Atherosclerosis. *J. Am. Heart Assoc.* **2017**, *6*, e007134. [CrossRef] [PubMed]
54. Xie, W.; Li, L.; Zheng, X.-L.; Yin, W.-D.; Tang, C.-K. The role of Krüppel-like factor 14 in the pathogenesis of atherosclerosis. *Atherosclerosis* **2017**, *263*, 352–360. [CrossRef] [PubMed]
55. Li, H.; Zou, J.; Yu, X.; Ou, X.; Tang, C. Zinc finger E-box binding homeobox 1 and atherosclerosis: New insights and therapeutic potential. *J. Cell. Physiol.* **2020**, *236*, 4216–4230. [CrossRef] [PubMed]
56. McCaffrey, T.A.; Fu, C.; Du, B.; Eksinar, S.; Kent, K.C.; Bush, H.; Kreiger, K.; Rosengart, T.; Cybulsky, M.I.; Silverman, E.S.; et al. High-level expression of Egr-1 and Egr-1–inducible genes in mouse and human atherosclerosis. *J. Clin. Investig.* **2000**, *105*, 653–662. [CrossRef]
57. Gimbrone, M.A., Jr.; García-Cardeña, G. Endothelial Cell Dysfunction and the Pathobiology of Atherosclerosis. *Circ. Res.* **2016**, *118*, 620–636. [CrossRef] [PubMed]
58. Zhou, J.; Li, Y.-S.; Chien, S. Shear stress–initiated signaling and its regulation of endothelial function. *Arter. Thromb. Vasc. Biol.* **2014**, *34*, 2191–2198. [CrossRef]
59. Baeyens, N.; Bandyopadhyay, C.; Coon, B.G.; Yun, S.; Schwartz, M.A. Endothelial fluid shear stress sensing in vascular health and disease. *J. Clin. Investig.* **2016**, *126*, 821–828. [CrossRef]
60. Munjal, A.; Khandia, R. Atherosclerosis: Orchestrating cells and biomolecules involved in its activation and inhibition. *Adv. Protein. Chem. Struct. Biol.* **2020**, *120*, 85–122. [CrossRef]
61. Conway, D.E.; Lee, S.; Eskin, S.G.; Shah, A.K.; Jo, H.; McIntire, L.V. Endothelial metallothionein expression and intracellular free zinc levels are regulated by shear stress. *Am. J. Physiol. Physiol.* **2010**, *299*, C1461–C1467. [CrossRef] [PubMed]
62. Falk, E. Pathogenesis of atherosclerosis. *J. Am. Coll. Cardiol.* **2006**, *47*, C7–C12. [CrossRef] [PubMed]
63. Higashi, Y.; Noma, K.; Yoshizumi, M.; Kihara, Y. Endothelial function and oxidative stress in cardiovascular diseases. *Circ. J.* **2009**, *73*, 411–418. [CrossRef] [PubMed]
64. Ozyildirim, S.; Baltaci, S.B. Cardiovascular Diseases and Zinc. *Biol. Trace Element Res.* **2022**, 1–12. [CrossRef] [PubMed]
65. Maruhashi, T.; Kihara, Y.; Higashi, Y. Bilirubin and Endothelial Function. *J. Atheroscler. Thromb.* **2019**, *26*, 688–696. [CrossRef] [PubMed]
66. Förstermann, U.; Xia, N.; Li, H. Roles of Vascular Oxidative Stress and Nitric Oxide in the Pathogenesis of Atherosclerosis. *Circ. Res.* **2017**, *120*, 713–735. [CrossRef]

67. Zou, M.-H.; Shi, C.; Cohen, R.A. Oxidation of the zinc-thiolate complex and uncoupling of endothelial nitric oxide synthase by peroxynitrite. *J. Clin. Investig.* **2002**, *109*, 817–826. [CrossRef]
68. Abregú, F.M.G.; Gobetto, M.N.; Juriol, L.V.; Caniffi, C.; Elesgaray, R.; Tomat, A.L.; Arranz, C. Developmental programming of vascular dysfunction by prenatal and postnatal zinc deficiency in male and female rats. *J. Nutr. Biochem.* **2018**, *56*, 89–98. [CrossRef]
69. Zhuang, X.; Pang, X.; Zhang, W.; Wu, W.; Zhao, J.; Yang, H.; Qu, W. Effects of zinc and manganese on advanced glycation end products (AGEs) formation and AGEs-mediated endothelial cell dysfunction. *Life Sci.* **2012**, *90*, 131–139. [CrossRef]
70. Lind, M.; Hayes, A.; Caprnda, M.; Petrovic, D.; Rodrigo, L.; Kruzliak, P.; Zulli, A. Inducible nitric oxide synthase: Good or bad? *Biomed. Pharmacother.* **2017**, *93*, 370–375. [CrossRef]
71. Yuniwati, Y.; Syaban, M.; Anoraga, S.; Sabila, F. Molecular Docking Approach of Bryophyllum Pinnatum Compounds as Atherosclerosis Therapy by Targeting Adenosine Monophosphate-Activated Protein Kinase and Inducible Nitric Oxide Synthase. *Acta Inform. Medica.* **2022**, *30*, 91–95. [CrossRef] [PubMed]
72. Miyoshi, T.; Li, Y.; Shih, D.M.; Wang, X.; Laubach, V.E.; Matsumoto, A.H.; Helm, G.A.; Lusis, A.J.; Shi, W. Deficiency of inducible NO synthase reduces advanced but not early atherosclerosis in apolipoprotein E-deficient mice. *Life Sci.* **2006**, *79*, 525–531. [CrossRef] [PubMed]
73. Cortese-Krott, M.M.; Kulakov, L.; Opländer, C.; Kolb-Bachofen, V.; Kröncke, K.-D.; Suschek, C.V. Zinc regulates iNOS-derived nitric oxide formation in endothelial cells. *Redox Biol.* **2014**, *2*, 945–954. [CrossRef] [PubMed]
74. Mestas, J.; Ley, K. Monocyte-endothelial cell interactions in the development of atherosclerosis. *Trends Cardiovasc. Med.* **2008**, *18*, 228–232. [CrossRef] [PubMed]
75. Li, D.; Lang, W.; Zhou, C.; Wu, C.; Zhang, F.; Liu, Q.; Yang, S.; Hao, J. Upregulation of Microglial ZEB1 Ameliorates Brain Damage after Acute Ischemic Stroke. *Cell Rep.* **2018**, *22*, 3574–3586. [CrossRef]
76. Gholizadeh, M.; Saeedy, S.A.G.; Roodi, P.B.; Saedisomeolia, A. The association between zinc and endothelial adhesion molecules ICAMs and VCAM-1 and nuclear receptors PPAR-α and PPAR-γ: A systematic review on cell culture, animal and human studies. *Microvasc. Res.* **2021**, *138*, 104217. [CrossRef]
77. Verstrepen, L.; Bekaert, T.; Chau, T.-L.; Tavernier, J.; Chariot, A.; Beyaert, R. TLR-4, IL-1R and TNF-R signaling to NF-κB: Variations on a common theme. *Exp.* **2008**, *65*, 2964–2978. [CrossRef]
78. Cheng, W.; Cui, C.; Liu, G.; Ye, C.; Shao, F.; Bagchi, A.K.; Mehta, J.L.; Wang, X. NF-κB, A Potential Therapeutic Target in Cardiovascular Diseases. *Cardiovasc. Drugs Ther.* **2022**, 1–14. [CrossRef]
79. Libby, P.; Ridker, P.M.; Hansson, G.K. Progress and challenges in translating the biology of atherosclerosis. *Nature* **2011**, *473*, 317–325. [CrossRef]
80. Hennig, B.; Meerarani, P.; Toborek, M.; McClain, C.J. Antioxidant-like properties of zinc in activated endothelial cells. *J. Am. Coll. Nutr.* **1999**, *18*, 152–158. [CrossRef]
81. Fan, Y.; Lu, H.; Liang, W.; Hu, W.; Zhang, J.; Chen, Y.E. Krüppel-like factors and vascular wall homeostasis. *J. Mol. Cell Biol.* **2017**, *9*, 352–363. [CrossRef] [PubMed]
82. Shen, H.; Oesterling, E.; Stromberg, A.; Toborek, M.; Macdonald, R.; Hennig, B. Zinc deficiency induces vascular pro-inflammatory parameters associated with NF-κB and PPAR signaling. *J. Am. Coll. Nutr.* **2008**, *27*, 577–587. [CrossRef] [PubMed]
83. Meerarani, P.; Reiterer, G.; Toborek, M.; Hennig, B. Zinc Modulates PPARγ signaling and activation of porcine endothelial Cells. *J. Nutr.* **2003**, *133*, 3058–3064. [CrossRef] [PubMed]
84. Prasad, A.S.; Bao, B.; Beck, F.W.; Sarkar, F.H. Zinc-suppressed inflammatory cytokines by induction of A20-mediated inhibition of nuclear factor-κB. *Nutrition* **2011**, *27*, 816–823. [CrossRef] [PubMed]
85. Kattoor, A.J.; Pothineni, N.V.K.; Palagiri, D.; Mehta, J.L. Oxidative Stress in Atherosclerosis. *Curr. Atheroscler. Rep.* **2017**, *19*, 42. [CrossRef] [PubMed]
86. Zhang, Q.; Xue, Y.; Fu, Y.; Bao, B.; Guo, M.-Y. Zinc Deficiency Aggravates Oxidative Stress Leading to Inflammation and Fibrosis in Lung of Mice. *Biol. Trace Element Res.* **2021**, *200*, 4045–4057. [CrossRef]
87. Xu, Y.; Li, A.; Li, X.; Deng, X.; Gao, X.-J. Zinc Deficiency Induces Inflammation and Apoptosis via Oxidative Stress in the Kidneys of Mice. *Biol. Trace Element Res.* 2022, *online ahead of print*. [CrossRef]
88. Mohammadi, H.; Talebi, S.; Ghavami, A.; Rafiei, M.; Sharifi, S.; Faghihimani, Z.; Ranjbar, G.; Miraghajani, M.; Askari, G. Effects of zinc supplementation on inflammatory biomarkers and oxidative stress in adults: A systematic review and meta-analysis of randomized controlled trials. *J. Trace Elements Med. Biol.* **2021**, *68*, 126857. [CrossRef]
89. Hosseini, R.; Ferns, G.A.; Sahebkar, A.; Mirshekar, M.A.; Jalali, M. Zinc supplementation is associated with a reduction in serum markers of inflammation and oxidative stress in adults: A systematic review and meta-analysis of randomized controlled trials. *Cytokine* **2020**, *138*, 155396. [CrossRef]
90. Olechnowicz, J.; Tinkov, A.; Skalny, A.; Suliburska, J. Zinc status is associated with inflammation, oxidative stress, lipid, and glucose metabolism. *J. Physiol. Sci.* **2017**, *68*, 19–31. [CrossRef]
91. Zago, M.; I Oteiza, P. The antioxidant properties of zinc: Interactions with iron and antioxidants. *Free Radic. Biol. Med.* **2001**, *31*, 266–274. [CrossRef]
92. Libby, P.; Okamoto, Y.; Rocha, V.Z.; Folco, E. Inflammation in Atherosclerosis: Transition From Theory to Practice. *Circ. J.* **2010**, *74*, 213–220. [CrossRef] [PubMed]

93. Tricot, O.; Mallat, Z.; Heymes, C.; Belmin, J.; Lesèche, G.; Tedgui, A. Relation between endothelial cell apoptosis and blood flow direction in human atherosclerotic plaques. *Circulation* **2000**, *101*, 2450–2453. [CrossRef]
94. Magenta, A.; Cencioni, C.; Fasanaro, P.; Zaccagnini, G.; Greco, S.; Sarra-Ferraris, G.; Antonini, A.; Martelli, F.; Capogrossi, M.C. miR-200c is upregulated by oxidative stress and induces endothelial cell apoptosis and senescence via ZEB1 inhibition. *Cell Death Differ.* **2011**, *18*, 1628–1639. [CrossRef] [PubMed]
95. Thambiayya, K.; Wasserloos, K.J.; Huang, Z.; Kagan, V.E.; Croix, C.M.S.; Pitt, B.R. LPS-induced decrease in intracellular labile zinc, [Zn]i, contributes to apoptosis in cultured sheep pulmonary artery endothelial cells. *Am. J. Physiol. Cell. Mol. Physiol.* **2011**, *300*, L624–L632. [CrossRef] [PubMed]
96. Thambiayya, K.; Wasserloos, K.; Kagan, V.E.; Stoyanovsky, D.; Pitt, B.R. A critical role for increased labile zinc in reducing sensitivity of cultured sheep pulmonary artery endothelial cells to LPS-induced apoptosis. *Am. J. Physiol. Cell. Mol. Physiol.* **2012**, *302*, L1287–L1295. [CrossRef] [PubMed]
97. Perry, D.K.; Smyth, M.J.; Stennicke, H.R.; Salvesen, G.S.; Duriez, P.; Poirier, G.G.; Hannun, Y.A. Zinc is a potent inhibitor of the apoptotic protease, caspase-3. *J. Biol. Chem.* **1997**, *272*, 18530–18533. [CrossRef]
98. Meerarani, P.; Ramadass, P.; Toborek, M.; Bauer, H.-C.; Bauer, H.; Hennig, B. Zinc protects against apoptosis of endothelial cells induced by linoleic acid and tumor necrosis factor α. *Am. J. Clin. Nutr.* **2000**, *71*, 81–87. [CrossRef]
99. Fanzo, J.C.; Reaves, S.K.; Cui, L.; Zhu, L.; Lei, K.Y. p53 protein and p21 mRNA levels and caspase-3 activity are altered by zinc status in aortic endothelial cells. *Am. J. Physiol. Physiol.* **2002**, *283*, C631–C638. [CrossRef]
100. Juriol, L.V.; Gobetto, M.N.; Abregú, F.M.G.; Dasso, M.E.; Pineda, G.; Güttlein, L.; Carranza, A.; Podhajcer, O.; Toblli, J.E.; Elesgaray, R.; et al. Cardiac changes in apoptosis, inflammation, oxidative stress, and nitric oxide system induced by prenatal and postnatal zinc deficiency in male and female rats. *Eur. J. Nutr.* **2016**, *57*, 569–583. [CrossRef]
101. Truong-Tran, A.Q.; Carter, J.; Ruffin, R.E.; Zalewski, P.D. The role of zinc in caspase activation and apoptotic cell death. *BioMetals* **2001**, *14*, 315–330. [CrossRef] [PubMed]
102. Albarrán-Juárez, J.; Kaur, H.; Grimm, M.; Offermanns, S.; Wettschureck, N. Lineage tracing of cells involved in atherosclerosis. *Atherosclerosis* **2016**, *251*, 445–453. [CrossRef] [PubMed]
103. Chappell, J.; Harman, J.L.; Narasimhan, V.M.; Yu, H.; Foote, K.; Simons, B.D.; Bennett, M.R.; Jørgensen, H.F. Extensive Proliferation of a Subset of Differentiated, yet Plastic, Medial Vascular Smooth Muscle Cells Contributes to Neointimal Formation in Mouse Injury and Atherosclerosis Models. *Circ. Res.* **2016**, *119*, 1313–1323. [CrossRef] [PubMed]
104. Basatemur, G.L.; Jørgensen, H.F.; Clarke, M.C.H.; Bennett, M.R.; Mallat, Z. Vascular smooth muscle cells in atherosclerosis. *Nat. Rev. Cardiol.* **2019**, *16*, 727–744. [CrossRef]
105. Clarke, M.C.; Littlewood, T.D.; Figg, N.; Maguire, J.J.; Davenport, A.P.; Goddard, M.; Bennett, M.R. Chronic apoptosis of vascular smooth muscle cells accelerates atherosclerosis and promotes calcification and medial degeneration. *Circ. Res.* **2008**, *102*, 1529–1538. [CrossRef]
106. Bennett, M.R.; Sinha, S.; Owens, G.K. Vascular Smooth Muscle Cells in Atherosclerosis. *Circ. Res.* **2016**, *118*, 692–702. [CrossRef]
107. Ross, R.; Glomset, J.A. Atherosclerosis and the arterial smooth muscle cell: Proliferation of smooth muscle is a key event in the genesis of the lesions of atherosclerosis. *Science* **1973**, *180*, 1332–1339. [CrossRef]
108. Shankman, L.S.; Gomez, D.; A Cherepanova, O.; Salmon, M.; Alencar, G.F.; Haskins, R.M.; Swiatlowska, P.; Newman, A.A.C.; Greene, E.S.; Straub, A.C.; et al. KLF4-dependent phenotypic modulation of smooth muscle cells has a key role in atherosclerotic plaque pathogenesis. *Nat. Med.* **2015**, *21*, 628–637. [CrossRef]
109. Allen-Redpath, K.; Ou, O.; Beattie, J.H.; Kwun, I.-S.; Feldmann, J.; Nixon, G.F.; Feldmann, J. Marginal dietary zinc deficiency in vivo induces vascular smooth muscle cell apoptosis in large arteries. *Cardiovasc. Res.* **2013**, *99*, 525–534. [CrossRef]
110. Alcantara, E.H.; Shin, M.Y.; Feldmann, J.; Nixon, G.F.; Beattie, J.H.; Kwun, I.S. Long-term zinc deprivation accelerates rat vascular smooth muscle cell proliferation involving the down-regulation of JNK1/2 expression in MAPK signaling. *Atherosclerosis* **2013**, *228*, 46–52. [CrossRef]
111. Ou, O.; Allen-Redpath, K.; Urgast, D.; Gordon, M.; Campbell, G.; Feldmann, J.; Nixon, G.F.; Mayer, C.; Kwun, I.; Beattie, J.H. Plasma zinc's alter ego is a low-molecular-weight humoral factor. *FASEB J.* **2013**, *27*, 3672–3682. [CrossRef] [PubMed]
112. Shin, M.-Y.; Kwun, I.-S. Zinc Restored the Decreased Vascular Smooth Muscle Cell Viability under Atherosclerotic Calcification Conditions. *Prev. Nutr. Food Sci.* **2014**, *19*, 363–366. [CrossRef] [PubMed]
113. Kuzan, A.; Wujczyk, M.; Wiglusz, R. The Study of the Aorta Metallomics in the Context of Atherosclerosis. *Biomolecules* **2021**, *11*, 946. [CrossRef]
114. Libby, P.; Hansson, G.K. From Focal Lipid Storage to Systemic Inflammation. *J. Am. Coll. Cardiol.* **2019**, *74*, 1594–1607. [CrossRef] [PubMed]
115. Frostegård, J. Immunity, atherosclerosis and cardiovascular disease. *BMC Med.* **2013**, *11*, 117. [CrossRef]
116. Wolf, D.; Ley, K. Immunity and Inflammation in Atherosclerosis. *Circ. Res.* **2019**, *124*, 315–327. [CrossRef]
117. Hansson, G.K.; Hermansson, A. The immune system in atherosclerosis. *Nat. Immunol.* **2011**, *12*, 204–212. [CrossRef]
118. Libby, P. The changing landscape of atherosclerosis. *Nature* **2021**, *592*, 524–533. [CrossRef]
119. Ketelhuth, D.F.; Hansson, G.K. Adaptive Response of T and B Cells in Atherosclerosis. *Circ. Res.* **2016**, *118*, 668–678. [CrossRef]
120. Wessels, I.; Cousins, R.J. Zinc dyshomeostasis during polymicrobial sepsis in mice involves zinc transporter Zip14 and can be overcome by zinc supplementation. *Am. J. Physiol. Liver Physiol.* **2015**, *309*, G768–G778. [CrossRef]

121. Westin, G.; Schaffner, W. A zinc-responsive factor interacts with a metal-regulated enhancer element (MRE) of the mouse metallothionein-I gene. *EMBO J.* **1988**, *7*, 3763–3770. [CrossRef] [PubMed]
122. Tabas, I.; Lichtman, A.H. Monocyte-Macrophages and T Cells in Atherosclerosis. *Immunity* **2017**, *47*, 621–634. [CrossRef] [PubMed]
123. Gao, H.; Dai, W.; Zhao, L.; Min, J.; Wang, F. The Role of Zinc and Zinc Homeostasis in Macrophage Function. *J. Immunol. Res.* **2018**, *2018*, 6872621. [CrossRef]
124. Kim, B. Regulatory Role of Zinc in Immune Cell Signaling. *Mol. Cells* **2021**, *44*, 335–341. [CrossRef] [PubMed]
125. Dubben, S.; Hönscheid, A.; Winkler, K.; Rink, L.; Haase, H. Cellular zinc homeostasis is a regulator in monocyte differentiation of HL-60 cells by 1α,25-dihydroxyvitamin D_3. *J. Leukoc. Biol.* **2010**, *87*, 833–844. [CrossRef] [PubMed]
126. Kojima, C.; Kawakami, A.; Takei, T.; Nitta, K.; Yoshida, M. Angiotensin-converting enzyme inhibitor attenuates monocyte adhesion to vascular endothelium through modulation of intracellular zinc. *J. Pharmacol. Exp. Ther.* **2007**, *323*, 855–860. [CrossRef]
127. Fan, Y.; Zhang, X.; Yang, L.; Wang, J.; Hu, Y.; Bian, A.; Liu, J.; Ma, J. Zinc inhibits high glucose-induced NLRP3 inflammasome activation in human peritoneal mesothelial cells. *Mol. Med. Rep.* **2017**, *16*, 5195–5202. [CrossRef]
128. Summersgill, H.; England, H.; Lopez-Castejon, G.; Lawrence, C.; Luheshi, N.M.; Pahle, J.; Mendes, P.; E Brough, D. Zinc depletion regulates the processing and secretion of IL-1β. *Cell Death Dis.* **2014**, *5*, e1040. [CrossRef]
129. Faghfouri, A.H.; Baradaran, B.; Khabbazi, A.; Bishak, Y.K.; Zarezadeh, M.; Tavakoli-Rouzbehani, O.M.; Faghfuri, E.; Payahoo, L.; Alipour, M.; Alipour, B. Profiling inflammatory cytokines following zinc supplementation: A systematic review and meta-analysis of controlled trials. *Br. J. Nutr.* **2021**, *126*, 1441–1450. [CrossRef]
130. Mocchegiani, E.; Bertoni-Freddari, C.; Marcellini, F.; Malavolta, M. Brain, aging and neurodegeneration: Role of zinc ion availability. *Prog. Neurobiol.* **2005**, *75*, 367–390. [CrossRef]
131. Zhou, X.; Nicoletti, A.; Elhage, R.; Hansson, G.K. Transfer of CD4 $^+$ T cells aggravates atherosclerosis in immunodeficient apolipoprotein E knockout mice. *Circulation* **2000**, *102*, 2919–2922. [CrossRef] [PubMed]
132. Prasad, A.S. Effects of Zinc Deficiency on Th1 and Th2 Cytokine Shifts. *J. Infect. Dis.* **2000**, *182*, S62–S68. [CrossRef] [PubMed]
133. Prasad, A.S. Zinc: Mechanisms of Host Defense. *J. Nutr.* **2007**, *137*, 1345–1349. [CrossRef]
134. Hwang, J.-R.; Byeon, Y.; Kim, D.; Park, S.-G. Recent insights of T cell receptor-mediated signaling pathways for T cell activation and development. *Exp. Mol. Med.* **2020**, *52*, 750–761. [CrossRef]
135. Lin, R.S.; Rodriguez, C.; Veillette, A.; Lodish, H.F. Zinc is essential for binding of p56(lck) to CD4 and CD8alpha. *J. Biol. Chem.* **1998**, *273*, 32878–32882. [CrossRef]
136. Romir, J.; Lilie, H.; Egerer-Sieber, C.; Bauer, F.; Sticht, H.; Muller, Y.A. Crystal structure analysis and solution studies of human Lck-SH3; zinc-induced homodimerization competes with the binding of proline-rich motifs. *J. Mol. Biol.* **2007**, *365*, 1417–1428. [CrossRef]
137. Aydemir, T.B.; Liuzzi, J.P.; McClellan, S.; Cousins, R.J. Zinc transporter ZIP8 (SLC39A8) and zinc influence IFN-γ expression in activated human T cells. *J. Leukoc. Biol.* **2009**, *86*, 337–348. [CrossRef]
138. Liu, M.-J.; Bao, S.; Gálvez-Peralta, M.; Pyle, C.J.; Rudawsky, A.C.; Pavlovicz, R.E.; Killilea, D.W.; Li, C.; Nebert, D.W.; Wewers, M.D.; et al. ZIP8 Regulates Host Defense through Zinc-Mediated Inhibition of NF-κB. *Cell Rep.* **2013**, *3*, 386–400. [CrossRef]
139. Khorshidi, M.; Zarezadeh, M.; Sadeghi, A.; Teymouri, A.; Emami, M.R.; Kord-Varkaneh, H.; Aryaeian, N.; Rahmani, J.; Mousavi, S.M. The Effect of Zinc Supplementation on Serum Leptin Levels: A Systematic Review and Meta-Analysis of Randomized Controlled Trials. *Horm. Metab. Res.* **2019**, *51*, 503–510. [CrossRef]
140. Tabatabaie, M.; Soltani, S.; Mozaffari-Khosravi, H.; Salehi-Abargouei, A. Zinc Supplementation Might Not Affect Serum Leptin and Adiponectin Levels in Adults: A Systematic Review and Meta-Analysis of Randomized Controlled Clinical Trials. *Iran. J. Public Health* **2021**, *50*, 245–256. [CrossRef]
141. Wei, C.; Luo, Z.; Hogstrand, C.; Xu, Y.; Wu, L.; Chen, G.; Pan, Y.; Song, Y. Zinc reduces hepatic lipid deposition and activates lipophagy via Zn^{2+}/MTF-1/PPARα and Ca^{2+}/CaMKKβ/AMPK pathways. *FASEB J.* **2018**, *32*, 6666–6680. [CrossRef] [PubMed]
142. Severo, J.S.; Morais, J.B.S.; Beserra, J.B.; dos Santos, L.R.; de Sousa Melo, S.R.; de Sousa, G.S.; de Matos Neto, E.M.; Henriques, G.S.; do Nascimento Marreiro, D. Role of Zinc in Zinc-α2-Glycoprotein Metabolism in Obesity: A Review of Literature. *Biol. Trace Elem. Res.* **2020**, *193*, 81–88. [CrossRef] [PubMed]
143. Balaz, M.; Vician, M.; Janakova, Z.; Kurdiova, T.; Surova, M.; Imrich, R.; Majercikova, Z.; Penesova, A.; Vlcek, M.; Kiss, A.; et al. Subcutaneous adipose tissue zinc-α2-glycoprotein is associated with adipose tissue and whole-body insulin sensitivity. *Obesity* **2014**, *22*, 1821–1829. [CrossRef] [PubMed]
144. Wei, X.; Liu, X.; Tan, C.; Mo, L.; Wang, H.; Peng, X.; Deng, F.; Chen, L. Expression and Function of Zinc-α2-Glycoprotein. *Neurosci. Bull.* **2019**, *35*, 540–550. [CrossRef] [PubMed]
145. Vasto, S.; Mocchegiani, E.; Candore, G.; Listì, F.; Colonna-Romano, G.; Lio, D.; Malavolta, M.; Giacconi, R.; Cipriano, C.; Caruso, C. Inflammation, genes and zinc in ageing and age-related diseases. *Biogerontology* **2006**, *7*, 315–327. [CrossRef]
146. Li, X.; Guan, Y.; Shi, X.; Ding, H.; Song, Y.; Li, C.; Liu, R.; Liu, G. Effects of High Zinc Levels on the Lipid Synthesis in Rat Hepatocytes. *Biol. Trace Element Res.* **2013**, *154*, 97–102. [CrossRef]
147. Abregú, F.M.G.; Gobetto, M.N.; Castañón, A.; Lucero, D.; Caniffi, C.; Elesgaray, R.; Schreier, L.; Arranz, C.; Tomat, A.L. Fetal and postnatal zinc restriction: Sex differences in metabolic alterations in adult rats. *Nutrition* **2019**, *65*, 18–26. [CrossRef]
148. Xu, C.; Huang, Z.; Liu, L.; Luo, C.; Lu, G.; Li, Q.; Gao, X. Zinc Regulates Lipid Metabolism and MMPs Expression in Lipid Disturbance Rabbits. *Biol. Trace Element Res.* **2015**, *168*, 411–420. [CrossRef]

149. Stawarska, A.; Czerwonka, M.; Wyrębiak, R.; Wrzesień, R.; Bobrowska-Korczak, B. Zinc Affects Cholesterol Oxidation Products and Fatty Acids Composition in Rats' Serum. *Nutrients* **2021**, *13*, 1563. [CrossRef]
150. Rios-Lugo, M.J.; Madrigal-Arellano, C.; Gaytán-Hernández, D.; Hernández-Mendoza, H.; Romero-Guzmán, E.T. Association of Serum Zinc Levels in Overweight and Obesity. *Biol. Trace Element Res.* **2020**, *198*, 51–57. [CrossRef]
151. Yokokawa, H.; Fukuda, H.; Saita, M.; Miyagami, T.; Takahashi, Y.; Hisaoka, T.; Naito, T. Serum zinc concentrations and characteristics of zinc deficiency/marginal deficiency among Japanese subjects. *J. Gen. Fam. Med.* **2020**, *21*, 248–255. [CrossRef] [PubMed]
152. Ma, X.; Jiang, S.; Yan, S.; Li, M.; Wang, C.; Pan, Y.; Sun, C.; Jin, L.; Yao, Y.; Li, B. Association Between Copper, Zinc, Iron, and Selenium Intakes and TC/HDL-C Ratio in US Adults. *Biol. Trace Element Res.* **2019**, *197*, 43–51. [CrossRef] [PubMed]
153. Yary, T.; Virtanen, J.; Ruusunen, A.; Tuomainen, T.-P.; Voutilainen, S. Association between serum zinc and later development of metabolic syndrome in middle aged and older men: The Kuopio Ischaemic Heart Disease Risk Factor Study. *Nutrition* **2017**, *37*, 43–47. [CrossRef] [PubMed]
154. Khazdouz, M.; Djalalinia, S.; Zadeh, S.S.; Hasani, M.; Shidfar, F.; Ataie-Jafari, A.; Asayesh, H.; Zarei, M.; Gorabi, A.M.; Noroozi, M.; et al. Effects of Zinc Supplementation on Cardiometabolic Risk Factors: A Systematic Review and Meta-analysis of Randomized Controlled Trials. *Biol. Trace Element Res.* **2019**, *195*, 373–398. [CrossRef] [PubMed]
155. Saykally, J.N.; Dogan, S.; Cleary, M.P.; Sanders, M.M. The ZEB1 transcription factor is a novel repressor of adiposity in female mice. *PLoS ONE* **2009**, *4*, e8460. [CrossRef]
156. Gopoju, R.; Panangipalli, S.; Kotamraju, S. Metformin treatment prevents SREBP2-mediated cholesterol uptake and improves lipid homeostasis during oxidative stress-induced atherosclerosis. *Free Radic. Biol. Med.* **2018**, *118*, 85–97. [CrossRef]
157. Zhao, Z.; Hao, D.; Wang, L.; Li, J.; Meng, Y.; Li, P.; Wang, Y.; Zhang, C.; Zhou, H.; Gardner, K.; et al. CtBP promotes metastasis of breast cancer through repressing cholesterol and activating TGF-β signaling. *Oncogene* **2018**, *38*, 2076–2091. [CrossRef]
158. Tamura, Y. The Role of Zinc Homeostasis in the Prevention of Diabetes Mellitus and Cardiovascular Diseases. *J. Atheroscler. Thromb.* **2021**, *28*, 1109–1122. [CrossRef]
159. Filios, S.R.; Xu, G.; Chen, J.; Hong, K.; Jing, G.; Shalev, A. MicroRNA-200 is induced by thioredoxin-interacting protein and regulates Zeb1 protein signaling and beta cell apoptosis. *J. Biol. Chem.* **2014**, *289*, 36275–36283. [CrossRef]
160. Cruz, K.J.C.; de Oliveira, A.R.S.; Morais, J.B.S.; Severo, J.S.; Mendes, P.M.V.; Melo, S.R.D.S.; de Sousa, G.S.; Marreiro, D.D.N. Zinc and Insulin Resistance: Biochemical and Molecular Aspects. *Biol. Trace Element Res.* **2018**, *186*, 407–412. [CrossRef]
161. Ahn, B.-I.; Kim, M.J.; Koo, H.S.; Seo, N.; Joo, N.-S.; Kim, Y.-S. Serum zinc concentration is inversely associated with insulin resistance but not related with metabolic syndrome in nondiabetic korean adults. *Biol. Trace Element Res.* **2014**, *160*, 169–175. [CrossRef] [PubMed]
162. Fukunaka, A.; Fukada, T.; Bhin, J.; Suzuki, L.; Tsuzuki, T.; Takamine, Y.; Bin, B.-H.; Yoshihara, T.; Ichinoseki-Sekine, N.; Naito, H.; et al. Zinc transporter ZIP13 suppresses beige adipocyte biogenesis and energy expenditure by regulating C/EBP-β expression. *PLoS Genet.* **2017**, *13*, e1006950. [CrossRef] [PubMed]
163. Ranasinghe, P.; Pigera, S.; Galappatthy, P.; Katulanda, P.; Constantine, G.R. Zinc and diabetes mellitus: Understanding molecular mechanisms and clinical implications. *DARU J. Pharm. Sci.* **2015**, *23*, 44. [CrossRef]
164. Bandeira, V.D.S.; Pires, L.V.; Hashimoto, L.L.; de Alencar, L.L.; Almondes, K.G.S.; Lottenberg, S.A.; Cozzolino, S. Association of reduced zinc status with poor glycemic control in individuals with type 2 diabetes mellitus. *J. Trace Elements Med. Biol.* **2017**, *44*, 132–136. [CrossRef] [PubMed]
165. Sisnande, T.; Lima, C.K.; da Silva, D.C.; Beninatto, T.M.; Alves, N.L.; Amaral, M.J.; Miranda-Alves, L.; Lima, L.M.T. Dietary zinc restriction promotes degeneration of the endocrine pancreas in mice. *Biochim. Biophys. Acta. Mol. Basis. Dis.* **2020**, *1866*, 165675. [CrossRef]
166. Cooper-Capetini, V.; De Vasconcelos, D.A.A.; Martins, A.R.; Hirabara, S.M.; Donato, J., Jr.; Carpinelli, A.R.; Abdulkader, F.; Donato, J. Zinc Supplementation Improves Glucose Homeostasis in High Fat-Fed Mice by Enhancing Pancreatic β-Cell Function. *Nutrients* **2017**, *9*, 1150. [CrossRef]
167. Qi, Y.; Zhang, Z.; Liu, S.; Aluo, Z.; Zhang, L.; Yu, L.; Li, Y.; Song, Z.; Zhou, L. Zinc Supplementation Alleviates Lipid and Glucose Metabolic Disorders Induced by a High-Fat Diet. *J. Agric. Food Chem.* **2020**, *68*, 5189–5200. [CrossRef]
168. Ranasinghe, P.; Wathurapatha, W.S.; Galappatthy, P.; Katulanda, P.; Jayawardena, R.; Constantine, G.R. Zinc supplementation in prediabetes: A randomized double-blind placebo-controlled clinical trial. *J. Diabetes* **2018**, *10*, 386–397. [CrossRef]
169. Skalnaya, M.G.; Skalny, A.V.; Yurasov, V.V.; Demidov, V.A.; Grabeklis, A.R.; Radysh, I.V.; Tinkov, A.A. Serum Trace Elements and Electrolytes Are Associated with Fasting Plasma Glucose and HbA1c in Postmenopausal Women with Type 2 Diabetes Mellitus. *Biol. Trace Element Res.* **2016**, *177*, 25–32. [CrossRef]
170. Hansen, A.F.; Simić, A.; Åsvold, B.O.; Romundstad, P.R.; Midthjell, K.; Syversen, T.; Flaten, T.P. Trace elements in early phase type 2 diabetes mellitus—A population-based study. The HUNT study in Norway. *J. Trace Elem. Med. Biol.* **2017**, *40*, 46–53. [CrossRef]
171. Zhang, J.; Hu, J.; Zhao, J.; Li, J.; Cai, X. Serum zinc concentrations and prediabetes and diabetes in the general population. *Biol. Trace Element Res.* **2021**, *200*, 1071–1077. [CrossRef] [PubMed]
172. Simić, A.; Hansen, A.F.; Åsvold, B.O.; Romundstad, P.R.; Midthjell, K.; Syversen, T.; Flaten, T.P. Trace element status in patients with type 2 diabetes in Norway: The HUNT3 Survey. *J. Trace Elements Med. Biol.* **2017**, *41*, 91–98. [CrossRef] [PubMed]
173. Fernández-Cao, J.C.; Warthon-Medina, M.; Moran, V.H.; Arija, V.; Doepking, C.; Serra-Majem, L.; Lowe, N.M. Zinc Intake and Status and Risk of Type 2 Diabetes Mellitus: A Systematic Review and Meta-Analysis. *Nutrients* **2019**, *11*, 1027. [CrossRef]

174. Dimitrova, A.A.; Strashimirov, D.; Betova, T.; Russeva, A.; Alexandrova, M. Zinc content in the diet affects the activity of Cu/ZnSOD, lipid peroxidation and lipid profile of spontaneously hypertensive rats. *Acta Biol. Hung.* **2008**, *59*, 305–314. [CrossRef] [PubMed]
175. Yanagisawa, H.; Miyazaki, T.; Nodera, M.; Miyajima, Y.; Suzuki, T.; Kido, T.; Suka, M. Zinc-Excess Intake Causes the Deterioration of Renal Function Accompanied by an Elevation in Systemic Blood Pressure Primarily Through Superoxide Radical-Induced Oxidative Stress. *Int. J. Toxicol.* **2014**, *33*, 288–296. [CrossRef]
176. Tomat, A.L.; Weisstaub, A.R.; Jauregui, A.; Piñeiro, A.; Balaszczuk, A.M.; A Costa, M.; Arranz, C.T. Moderate zinc deficiency influences arterial blood pressure and vascular nitric oxide pathway in growing rats. *Pediatr. Res.* **2005**, *58*, 672–676. [CrossRef]
177. Gobetto, M.N.; Abregú, F.M.G.; Caniffi, C.; Veiras, L.; Elesgaray, R.; Gironacci, M.; Tomat, A.L.; Arranz, C. Fetal and postnatal zinc restriction: Sex differences in the renal renin-angiotensin system of newborn and adult Wistar rats. *J. Nutr. Biochem.* **2020**, *81*, 108385. [CrossRef]
178. Williams, C.R.; Mistry, M.; Cheriyan, A.M.; Williams, J.M.; Naraine, M.K.; Ellis, C.L.; Mallick, R.; Mistry, A.C.; Gooch, J.L.; Ko, B.; et al. Zinc deficiency induces hypertension by promoting renal Na^+ reabsorption. *Am. J. Physiol. Physiol.* **2019**, *316*, F646–F653. [CrossRef]
179. Navarro, J.; Sanchez, A.; Saiz, J.; Ruilope, L.M.; García-Estañ, J.; Romero, J.C.; Moncada, S.; Lahera, V. Hormonal, renal, and metabolic alterations during hypertension induced by chronic inhibition of NO in rats. *Am. J. Physiol. Integr. Comp. Physiol.* **1994**, *267*, R1516–R1521. [CrossRef]
180. Tomat, A.; Elesgaray, R.; Zago, V.; Fasoli, H.; Fellet, A.; Balaszczuk, A.M.; Schreier, L.; Costa, M.; Arranz, C. Exposure to zinc deficiency in fetal and postnatal life determines nitric oxide system activity and arterial blood pressure levels in adult rats. *Br. J. Nutr.* **2010**, *104*, 382–389. [CrossRef]
181. Xiao, H.; Yan, Y.; Gu, Y.; Zhang, Y. Strategy for sodium-salt substitution: On the relationship between hypertension and dietary intake of cations. *Food Res. Int.* **2021**, *156*, 110822. [CrossRef] [PubMed]
182. Yanagisawa, H.; Sato, M.; Nodera, M.; Wada, O. Excessive zinc intake elevates systemic blood pressure levels in normotensive rats ??? potential role of superoxide-induced oxidative stress. *J. Hypertens.* **2004**, *22*, 543–550. [CrossRef] [PubMed]
183. Mousavi, S.M.; Mofrad, M.D.; Nascimentoe, I.J.B.D.; Milajerdi, A.; Mokhtari, T.; Esmaillzadeh, A. The effect of zinc supplementation on blood pressure: A systematic review and dose–response meta-analysis of randomized-controlled trials. *Eur. J. Nutr.* **2020**, *59*, 1815–1827. [CrossRef] [PubMed]
184. Nevárez-López, S.C.; Simental-Mendía, L.E.; Guerrero-Romero, F.; Burciaga-Nava, J.A. Zinc deficiency is an independent risk factor for prehypertension in healthy subjects. *Int. J. Vitam. Nutr. Res.* **2021**, *91*, 25–30. [CrossRef] [PubMed]
185. Naghizadeh, S.; Kheirouri, S.; Ojaghi, H.; Kaffash, A.J. Zinc supplementation attenuate diabetic indices in patients with diabetic retinopathy. *Prog. Nutr.* **2018**, *20*, 263–269. [CrossRef]
186. Yao, J.; Hu, P.; Zhang, D. Associations Between Copper and Zinc and Risk of Hypertension in US Adults. *Biol. Trace Element Res.* **2018**, *186*, 346–353. [CrossRef]
187. Darroudi, S.; Saberi-Karimian, M.; Tayefi, M.; Tayefi, B.; Khashyarmanesh, Z.; Fereydouni, N.; Haghighi, H.M.; Mahmoudi, A.A.; Kharazmi-Khorassani, J.; Gonoodi, K.; et al. Association Between Hypertension in Healthy Participants and Zinc and Copper Status: A Population-Based Study. *Biol. Trace Element Res.* **2018**, *190*, 38–44. [CrossRef]
188. Bastola, M.M.; Locatis, C.; Maisiak, R.; Fontelo, P. Selenium, copper, zinc and hypertension: An analysis of the National Health and Nutrition Examination Survey (2011–2016). *BMC Cardiovasc. Disord.* **2020**, *20*, 45. [CrossRef]
189. Suliburska, J.; Skrypnik, K.; Szulińska, M.; Kupsz, J.; Bogdanski, P. Effect of hypotensive therapy combined with modified diet or zinc supplementation on biochemical parameters and mineral status in hypertensive patients. *J. Trace Elements Med. Biol.* **2018**, *47*, 140–148. [CrossRef]
190. Reiterer, G.; Macdonald, R.; Browning, J.D.; Morrow, J.; Matveev, S.V.; Daugherty, A.; Smart, E.; Toborek, M.; Hennig, B. Zinc Deficiency increases plasma lipids and atherosclerotic markers in LDL-receptor–deficient mice. *J. Nutr.* **2005**, *135*, 2114–2118. [CrossRef]
191. Shen, H.; MacDonald, R.; Bruemmer, D.; Stromberg, A.; Daugherty, A.; Li, X.-A.; Toborek, M.; Hennig, B. Zinc deficiency alters lipid metabolism in LDL receptor–deficient mice treated with rosiglitazone. *J. Nutr.* **2007**, *137*, 2339–2345. [CrossRef] [PubMed]
192. Beattie, J.H.; Gordon, M.-J.; Duthie, S.J.; McNeil, C.J.; Horgan, G.W.; Nixon, G.F.; Feldmann, J.; Kwun, I.-S. Suboptimal dietary zinc intake promotes vascular inflammation and atherogenesis in a mouse model of atherosclerosis. *Mol. Nutr. Food Res.* **2012**, *56*, 1097–1105. [CrossRef] [PubMed]
193. Alissa, E.M.; Bahijri, S.M.; Lamb, D.J.; Ferns, G.A.A. The effects of coadministration of dietary copper and zinc supplements on atherosclerosis, antioxidant enzymes and indices of lipid peroxidation in the cholesterol-fed rabbit. *Int. J. Exp. Pathol.* **2004**, *85*, 265–275. [CrossRef] [PubMed]
194. Ren, M.; Rajendran, R.; Ning, P.; Huat, B.T.K.; Nam, O.C.; Watt, F.; Jenner, A.; Halliwell, B. Zinc supplementation decreases the development of atherosclerosis in rabbits. *Free Radic. Biol. Med.* **2006**, *41*, 222–225. [CrossRef] [PubMed]
195. Abdelhalim, M.A.K.; Moussa, S.A.A.; Al-Mohy, Y.H. Ultraviolet-visible and fluorescence spectroscopy techniques are important diagnostic tools during the progression of atherosclerosis: Diet zinc supplementation retarded or delayed atherosclerosis. *BioMed Res. Int.* **2013**, *2013*, 652604. [CrossRef]
196. Jenner, A.; Ren, M.; Rajendran, R.; Ning, P.; Huat, B.; Watt, F.; Halliwell, B. Zinc supplementation inhibits lipid peroxidation and the development of atherosclerosis in rabbits fed a high cholesterol diet. *Free Radic. Biol. Med.* **2007**, *42*, 559–566. [CrossRef]

197. Cheng, G.; Chang, F.-J.; You, P.-H.; Lin, J.; Huang, X.-Y.; Wu, H.-Y.; Yan, L.; Deng, J.-Z.; You, H.-J.; Sun, C.-F. ZIP8 induces monocyte adhesion to the aortas ex-vivo by regulating zinc influx. *Int. Immunopharmacol.* **2018**, *62*, 203–211. [CrossRef]
198. Beattie, J.H.; Gordon, M.-J.; Rucklidge, G.J.; Reid, M.D.; Duncan, G.J.; Horgan, G.W.; Cho, Y.-E.; Kwun, I.-S. Aorta protein networks in marginal and acute zinc deficiency. *Proteomics* **2008**, *8*, 2126–2135. [CrossRef]
199. Sayin, V.I.; Khan, O.M.; Pehlivanoglu, L.E.; Staffas, A.; Ibrahim, M.X.; Asplund, A.; Ågren, P.; Nilton, A.; Bergström, G.; Bergo, M.O.; et al. Loss of one copy of Zfp148 reduces lesional macrophage proliferation and atherosclerosis in mice by activating p53. *Circ. Res.* **2014**, *115*, 781–789. [CrossRef]
200. Wang, H.; Guo, Y.; Lu, H.; Luo, Y.; Hu, W.; Liang, W.; Garcia-Barrio, M.T.; Chang, L.; Schwendeman, A.; Zhang, J.; et al. Krüppel-like factor 14 deletion in myeloid cells accelerates atherosclerotic lesion development. *Cardiovasc. Res.* **2021**, *118*, 475–488. [CrossRef]
201. Li, X.; Yang, M.; Wang, H.; Jia, Y.; Yan, P.; Boden, G.; Yang, G.; Li, L. Overexpression of JAZF1 protected ApoE-deficient mice from atherosclerosis by inhibiting hepatic cholesterol synthesis via CREB-dependent mechanisms. *Int. J. Cardiol.* **2014**, *177*, 100–110. [CrossRef] [PubMed]
202. Cho, J.H.; Kim, M.J.; Kim, K.J.; Kim, J.-R. POZ/BTB and AT-hook-containing zinc finger protein 1 (PATZ1) inhibits endothelial cell senescence through a p53 dependent pathway. *Cell Death Differ.* **2011**, *19*, 703–712. [CrossRef] [PubMed]
203. Yang, Y.J.; Choi, B.Y.; Chun, B.-Y.; Kweon, S.-S.; Lee, Y.-H.; Park, P.S.; Kim, M.K. Dietary zinc intake is inversely related to subclinical atherosclerosis measured by carotid intima-media thickness. *Br. J. Nutr.* **2010**, *104*, 1202–1211. [CrossRef]
204. Masley, S.C.; Roetzheim, R.; Masley, L.V.; McNamara, T.; Schocken, U.D. Emerging Risk Factors as Markers for Carotid Intima Media Thickness Scores. *J. Am. Coll. Nutr.* **2015**, *34*, 100–107. [CrossRef]
205. Chen, W.; Eisenberg, R.; Mowrey, W.B.; Wylie-Rosett, J.; Abramowitz, M.K.; A Bushinsky, D.; Melamed, M.L. Association between dietary zinc intake and abdominal aortic calcification in US adults. *Nephrol. Dial. Transplant.* **2019**, *35*, 1171–1178. [CrossRef]
206. Dziedzic, E.A.; Gąsior, J.S.; Tuzimek, A.; Paleczny, J.; Kwaśny, M.; Dąbrowski, M.; Jankowski, P. No Association of Hair Zinc Concentration with Coronary Artery Disease Severity and No Relation with Acute Coronary Syndromes. *Biomolecules* **2022**, *12*, 862. [CrossRef] [PubMed]
207. Giannoglou, G.D.; Konstantinou, D.; Kovatsi, L.; Chatzizisis, Y.S.; Mikhailidis, D.P. Association of reduced zinc status with angiographically severe coronary atherosclerosis: A pilot study. *Angiology* **2010**, *61*, 449–455. [CrossRef] [PubMed]
208. Milton, A.H.; Vashum, K.P.; McEvoy, M.; Hussain, S.; McElduff, P.; Byles, J.; Attia, J. Prospective Study of Dietary Zinc Intake and Risk of Cardiovascular Disease in Women. *Nutrients* **2018**, *10*, 38. [CrossRef]
209. Mattern, L.; Chen, C.; McClure, L.A.; Brockman, J.; Cushman, M.; Judd, S.; Kahe, K. Serum Zinc Levels and Incidence of Ischemic Stroke: The Reasons for Geographic and Racial Differences in Stroke Study. *Stroke* **2021**, *52*, 3953–3960. [CrossRef]
210. Gao, J.-W.; Zhang, S.-L.; Hao, Q.-Y.; Huang, F.-F.; Liu, Z.-Y.; Zhang, H.-F.; Yan, L.; Wang, J.-F.; Liu, P.-M. Association of dietary zinc intake with coronary artery calcium progression: The Multi-Ethnic Study of Atherosclerosis (MESA). *Eur. J. Nutr.* **2021**, *60*, 2759–2767. [CrossRef]
211. de Oliveira Otto, M.C.; Alonso, A.; Lee, D.-H.; Delclos, G.L.; Bertoni, A.G.; Jiang, R.; Lima, J.A.; Symanski, E.; Jacobs, D.R.; Nettleton, J.A. Dietary intakes of zinc and heme iron from red meat, but not from other sources, are associated with greater risk of metabolic syndrome and cardiovascular disease. *J. Nutr.* **2012**, *142*, 526–533. [CrossRef] [PubMed]
212. Jung, S.K.; Kim, M.-K.; Lee, Y.-H.; Shin, D.H.; Shin, M.-H.; Chun, B.-Y.; Choi, B.Y. Lower zinc bioavailability may be related to higher risk of subclinical atherosclerosis in Korean adults. *PLoS ONE* **2013**, *8*, e80115. [CrossRef] [PubMed]
213. Alissa, E.M.; Bahjri, S.M.; Ahmed, W.H.; Al-Ama, N.; Ferns, G.A. Trace element status in Saudi patients with established atherosclerosis. *J. Trace Elements Med. Biol.* **2006**, *20*, 105–114. [CrossRef]
214. Qazmooz, H.A.; Smeism, H.N.; Mousa, R.F.; Al-Hakeim, H.K.; Maes, M. Trace element, immune and opioid biomarkers of unstable angina, increased atherogenicity and insulin resistance: Results of machine learning. *J. Trace Elements Med. Biol.* **2020**, *64*, 126703. [CrossRef]
215. Lee, D.-H.; Folsom, A.R.; Jacobs, D.R. Iron, zinc, and alcohol consumption and mortality from cardiovascular diseases: The Iowa Women's Health Study. *Am. J. Clin. Nutr.* **2005**, *81*, 787–791. [CrossRef] [PubMed]
216. Eshak, E.S.; Iso, H.; Yamagishi, K.; Maruyama, K.; Umesawa, M.; Tamakoshi, A. Associations between copper and zinc intakes from diet and mortality from cardiovascular disease in a large population-based prospective cohort study. *J. Nutr. Biochem.* **2018**, *56*, 126–132. [CrossRef] [PubMed]
217. Shi, Z.; Chu, A.; Zhen, S.; Taylor, A.W.; Dai, Y.; Riley, M.; Samman, S. Association between dietary zinc intake and mortality among Chinese adults: Findings from 10-year follow-up in the Jiangsu Nutrition Study. *Eur. J. Nutr.* **2017**, *57*, 2839–2846. [CrossRef]
218. Bao, B.; Prasad, A.S.; Beck, F.W.; Fitzgerald, J.T.; Snell, D.; Bao, G.W.; Singh, T.; Cardozo, L.J. Zinc decreases C-reactive protein, lipid peroxidation, and inflammatory cytokines in elderly subjects: A potential implication of zinc as an atheroprotective agent. *Am. J. Clin. Nutr.* **2010**, *91*, 1634–1641. [CrossRef]
219. Costarelli, L.; Muti, E.; Malavolta, M.; Cipriano, C.; Giacconi, R.; Tesei, S.; Piacenza, F.; Pierpaoli, S.; Gasparini, N.; Faloia, E.; et al. Distinctive modulation of inflammatory and metabolic parameters in relation to zinc nutritional status in adult overweight/obese subjects. *J. Nutr. Biochem.* **2010**, *21*, 432–437. [CrossRef]
220. Seet, R.C.; Lee, C.-Y.J.; Lim, E.C.; Quek, A.M.; Huang, H.; Huang, S.H.; Looi, W.F.; Long, L.H.; Halliwell, B. Oral zinc supplementation does not improve oxidative stress or vascular function in patients with type 2 diabetes with normal zinc levels. *Atherosclerosis* **2011**, *219*, 231–239. [CrossRef]

221. Chu, A.; Foster, M.; Samman, S. Zinc Status and Risk of Cardiovascular Diseases and Type 2 Diabetes Mellitus—A Systematic Review of Prospective Cohort Studies. *Nutrients* **2016**, *8*, 707. [CrossRef] [PubMed]
222. Chu, A.; Holdaway, C.; Varma, T.; Petocz, P.; Samman, S. Lower Serum Zinc Concentration Despite Higher Dietary Zinc Intake in Athletes: A Systematic Review and Meta-analysis. *Sports Med.* **2017**, *48*, 327–336. [CrossRef] [PubMed]
223. Gonoodi, K.; Moslem, A.; Darroudi, S.; Ahmadnezhad, M.; Mazloum, Z.; Tayefi, M.; Zadeh, S.A.T.; Eslami, S.; Shafiee, M.; Khashayarmanesh, Z.; et al. Serum and dietary zinc and copper in Iranian girls. *Clin. Biochem.* **2018**, *54*, 25–31. [CrossRef]
224. Eby, G.A.; Halcomb, W.W. High-dose zinc to terminate angina pectoris: A review and hypothesis for action by ICAM inhibition. *Med. Hypotheses* **2006**, *66*, 169–172. [CrossRef] [PubMed]
225. Ruz, M.; Carrasco, F.; Rojas, P.; Basfi-Fer, K.; Hernández, M.C.; Pérez, A. Nutritional Effects of Zinc on Metabolic Syndrome and Type 2 Diabetes: Mechanisms and Main Findings in Human Studies. *Biol. Trace Element Res.* **2019**, *188*, 177–188. [CrossRef] [PubMed]
226. Haase, H.; Mocchegiani, E.; Rink, L. Correlation between zinc status and immune function in the elderly. *Biogerontology* **2006**, *7*, 421–428. [CrossRef] [PubMed]
227. Lestienne, I.; Verniere, C.I.; Mouquet-Rivier, C.; Picq, C.; Trèche, S. Effects of soaking whole cereal and legume seeds on iron, zinc and phytate contents. *Food Chem.* **2005**, *89*, 421–425. [CrossRef]
228. King, J.C.; Brown, K.H.; Gibson, R.S.; Krebs, N.F.; Lowe, N.M.; Siekmann, J.H.; Raiten, D.J. Biomarkers of Nutrition for Development (BOND)—Zinc Review. *J. Nutr.* **2015**, *146*, 858S–885S. [CrossRef]

biomolecules

MDPI

Review

Zinc in Cognitive Impairment and Aging

Ruize Sun, Jue Wang, Juan Feng and Bin Cao *

Department of Neurology, Shengjing Hospital of China Medical University, 36 Sanhao St., Shenyang 110004, China; drsunruize@163.com (R.S.); wjw_999@126.com (J.W.); fengjuan99999@hotmail.com (J.F.)
* Correspondence: caob@sj-hospital.org

Abstract: Zinc, an essential micronutrient for life, was first discovered in 1869 and later found to be indispensable for the normal development of plants and for the normal growth of rats and birds. Zinc plays an important role in many physiological and pathological processes in normal mammalian brain development, especially in the development of the central nervous system. Zinc deficiency can lead to neurodegenerative diseases, mental abnormalities, sleep disorders, tumors, vascular diseases, and other pathological conditions, which can cause cognitive impairment and premature aging. This study aimed to review the important effects of zinc and zinc-associated proteins in cognitive impairment and aging, to reveal its molecular mechanism, and to highlight potential interventions for zinc-associated aging and cognitive impairments.

Keywords: zinc-associated protein; zinc; zinc finger protein; aging; cognitive impairments

1. Introduction

Zinc plays an irreplaceable role in growth and development. Studies have shown that zinc supplementation improved serum insulin-like growth factor 1 (IGF-1) levels and promoted growth in infants born at term, with non-organic growth disorders [1]. IGF-1 is associated with infant height, and IGF-1 deficiency may lead to placental and fetal growth restrictions [2]. A randomized double-blind trial on children aged 4–5 years found that vitamin A supplementation along with zinc increased linear growth in height for age and reduced the risk of infection in children. IGF-1 levels were significantly higher in individuals with concomitant supplementation of vitamin A and zinc [3].

The hormone levels and biochemical indicators in the body gradually decline and deteriorate with age. In particular, the effectiveness of antioxidants in the nervous system decreases during the aging process. An increase in reactive oxygen species (ROS) is considered to be the main pathological process of aging. Cellular damage is followed by significant ROS production. The mortality of neuronal cells in a state of oxidative stress increases, which leads to protein degeneration and neurodegenerative lesions. It has been found that substances such as myostatin can reduce aging-induced β-amyloid (Aβ) plaque formation and deposition in brain regions. Furthermore, myostatin improves the level of cognitive function and has an anti-aging effect, although there was no benefit in young rats [4].

Neurons containing zinc ions can be found in various regions of the brain, including the cortex, amygdala, and hippocampus [5]. The average concentration of zinc ions in the brain is approximately 150 μmol/L, which is 10 times higher than that in serum [6]. Unlike iron and copper, zinc is not redox-active under physiological conditions, and is present in the body in three main forms: free zinc, vesicular zinc, and protein-bound zinc [7]. Most zinc in the brain is in the form of free zinc or non-compactly bound zinc. In addition, it is mostly stored in the vesicles of excitatory glutamatergic synaptic neurons and is released into the synaptic gap along with glutamate upon stimulation of neurons [8]. Due to these properties, zinc plays a key structural role, is involved in catalytic and signaling components, and is found in about 10% of proteins [9]. Zinc-containing proteins are fundamental components of more than 2000 enzymes and proteins in the body, which are involved in the regulations of DNA and RNA synthesis and hormone-receptor interactions, and act as second messengers in intracellular signaling, neurotransduction, neurocytogenesis, and neuronal growth [10]. Zinc is a molecular signal for immune cells, and intracellular zinc homeostasis is maintained by 14 zinc-regulated, iron-regulated transporter-like proteins (ZIP), 10 zinc transporters (ZnT), and other zinc-related proteins, such as metallothionein. All of these zinc-related proteins are irreplaceable in zinc homeostasis. Owing to these sophisticated regulatory pathways, the zinc content of the nervous system can be stabilized within an acceptable range to maintain normal brain function when there is a deficiency or excess of zinc in the body circulation, in a short period of time [3].

Cognitive impairment is usually associated with decreased neuronal apoptosis [11], synaptic dysfunction, a reduction in myelinated neurons and brain volume, and a thinner cerebral cortex. With an increase in age, the number of brain neuron stem cells decreases, which stimulates more calcium ions to enter the neurons to protect neurons and enable cognitive functioning [12]. However, at the same time, aging leads to decreases in glucose transporters in neurons and the level of mitochondrial oxidative phosphorylation and an increase in the production of ROS. This results in oxidative stress damage to neurons, decreases their ATP utilization ability [13,14] and mitochondrial division, and causes impairment in autophagy. These factors cause impaired cognitive functions. Zinc is important for the brain to function normally; however, the serum concentration of zinc gradually decreases with age. Zinc deficiency is closely related to cognitive impairment, memory impairment, and other diseases [15]. The purpose of this article is to summarize the role and mechanism of zinc in aging and cognitive impairment, and to identify new ways to intervene in age-related cognitive impairment.

2. Zinc and Neurobiology

2.1. Zinc and Neuronal Development

The role of zinc in growth and development has been widely reported for many years. Experimental studies have confirmed its importance in growth and development. Poor nutrition or poor environmental conditions during early growth and development may have irreversible effects on the nervous system. Zinc deficiency is widespread in developing countries. This review will discuss zinc in relation to neurodevelopment, neurogenesis, and cognitive impairment [16].

Inadequate growth in children is closely associated with abnormal neurodevelopment. Nutritional development, especially brain development, is influenced by many factors such as energy, and levels of protein, fat, carbohydrates, copper, iron, selenium, zinc, iodine, vitamins A, B12, C, D, thiamin, folic acid, choline, and long-chain polyunsaturated fatty acids, which are essential nutrients [17]. Zinc, among other minerals, is involved in the composition of many proteins and enzymes in the body. As early as 1986, zinc deficiency was reported to be linked to abnormal brain development as well as learning and memory deficits in young rats [18]. In mid- to late gestation, administration of maternal lipopolysaccharide (LPS) in mice activates an immune response, causing a redistribution of zinc from the fetus to the mother. The fetus develops a zinc deficiency, and astrocytes proliferate in the fetal brain. Oxidative substances and pro-inflammatory cytokines are released, and apoptosis increases, leading to abnormal neurodevelopment [19]. In recent years, it has been found that zinc supplementation throughout pregnancy in LPS-treated rats increased glutamate decarboxylase 67 (GAD67) expression in male pups but did not affect catechol-O-methyltransferase (COMT), while no changes in either GAD67 or COMT expression levels were found in female pups. This study highlighted the importance of zinc supplementation throughout pregnancy, which may limit the adverse effect of maternal infection on fetal neurodevelopment [20].

Zinc, an essential trace metal element in the body, is involved in many cellular signaling and metabolic pathways. *ZNF292* is a gene that encodes a highly conserved zinc finger protein (ZFP). When *ZNF292* is mutated, a range of neurodevelopmental abnormalities may occur, including intellectual disability (ID), autism spectrum disease (ASD), attention deficit-hyperactivity disorder, etc. [21]. High concentrations of zinc are neurotoxic, so zinc levels must be strictly regulated in the body. Prolonged zinc deficiency can have serious adverse effects on the nervous system. Maintaining zinc homeostasis enhances synaptic plasticity. Zinc supplementation also alleviates synaptic defects caused by ASD-related genetic alterations, particularly in the structure and function of excitatory glutamatergic neuronal synapses [22].

The importance of metal ions and their associated receptors on neurodevelopment has been described. However, given the complexity of neurodevelopment, the underlying mechanisms for these effects remain unclear. Neurodevelopment is regulated by a variety of proteins and metal ions. Zinc is involved in the composition of multiple proteins and enzymes in vivo, and as such, plays a key role in neurodevelopment. Zinc homeostasis is maintained through the regulation of ZFPs. Neurodevelopment may require a stable internal environment. Therefore, moderate amounts of zinc supplementation during periods of active neurodevelopment, i.e., during the neonatal period and pregnancy, may be recommended.

2.2. Zinc and Neurogenesis

Normal brain development requires the involvement of neural stem cells (NSC). NSC can differentiate into different central nervous system (CNS) cells, including neurons, astrocytes, and oligodendrocytes. Neurogenesis is an important process in the development of the CNS. Neurogenesis can be divided into the following steps: 1. proliferation and renewal of neural stem/progenitor cells (NSE), 2. migration of neurons to different regions of the brain, and 3. differentiation of neurons into different neuronal cell types [23].

These processes generally occur during gestation or the neonatal period. Adult neurogenesis differs from fetal or neonatal neurogenesis. Zinc is believed to be a key factor in the regulation of adult brain stem cell proliferation and neurogenesis [21]. Neurogenesis is considered part of the etiopathology of Alzheimer's disease (AD). The hippocampus is one of the most dominant areas of neurogenesis in the brain. Neurogenesis, in turn, influences key factors in the development of AD, such as progerin, APP, and amyloid deposition. Therefore, neurogenesis is considered an important component of AD [24]. It has been reported that a decrease in sonic hedgehog signaling in the hippocampus of patients with AD causes a decrease in NSC proliferation and differentiation, which reduces astrocyte proliferation. It also causes a loss of hippocampal stem cells and decreased neuronal production [25]

It has been shown that CXXC5 is an unmethylated CpG dinucleotide-binding protein with two highly conserved zinc finger structural domains. It is involved in cell proliferation driven by E2-ERα signaling [26]. Cell proliferation is the first step in neurogenesis. Zinc plus cyclo-(His-Pro) (CHP)-treated progenitor cell proliferation was significantly increased in the subgranular zone (SGZ) of the hippocampal dentate gyrus (DG). Zinc deficiency decreases NSC proliferation and inhibits NSC development [27]. However, the specific process of adult neurogenesis refers to the generation of new neurons in the mature brain. It has been shown that adult neurogenesis occurs mainly in two sites: the subventricular zone and the hippocampal DG [28] Vesicular zinc is mainly concentrated in the presynaptic terminals of the DG granule cells. Neurogenesis occurs mainly in the SGZ. New hippocampal neurons are continuously generated. Furthermore, many studies have demonstrated that zinc is important in hippocampal neurogenesis after epilepsy and brain injury [29].

Cognitive impairment may occur after paclitaxel treatment, which is known as chemotherapy-induced cognitive impairment (CICI). A mouse model of CICI showed that vesicular zinc and ZnT-3 expression levels decreased after paclitaxel treatment. This negatively affected the differentiation of progenitor cells but did not affect the value-added effects of progenitor cells. When vesicular zinc stores are disrupted, it may further affect one or more stages of neurogenesis, resulting in cognitive impairment [30].

Notably, because zinc is so important, the question has been raised as to whether zinc supplementation would be beneficial to the organism. One study found that in obese mice, who were fed a high-fat diet, low-dose zinc (15 ppm) supplementation improved lipid oxidation and increased levels of neurogenic markers, such as brain-derived neurotrophic factor (BDNF); conversely, high-dose zinc (60 ppm) supplementation decreased levels of neurogenic markers. Moreover, zinc may affect the neural regeneration of hippocampal neurons [31]. This also suggests that zinc supplementation may not be not beneficial for an organism; thus, zinc homeostasis needs to be maintained. ZnT-3, the classic zinc transporter protein, plays an important role in neurogenesis. The number of mature and immature neurons is reduced in mice with *ZnT-3* gene deletion. ZnT-3 protein was found to co-stain along with axon terminals of Brdu + MAP2+ cells in DG neurons. The expression levels of pERK and pCREB were increased by supplementation with ZINC, given as a mixture of zinc chloride and N-acetylcysteine, a glutathione precursor, in a 1:2 M ratio, further suggesting that zinc may activate CREB activity. CREB increases BDNF mRNA expression through activation of ERK1/2, to increase adult hippocampal neurogenesis and cell proliferation. Zinc supplementation in ZnT-3 genetically deficient mice results in enhanced neurogenesis. Whether extracellular zinc is responsible for these effects is still unclear [32].

Neurogenesis is an important stage in neurodevelopment and is implicated in AD. The importance of zinc in neurogenesis further proves its importance in neurobiology. Whether in newborns, children, or adults, old and new neurons are constantly changing. Neurogenesis occurs throughout the lifespan, so its importance cannot be overstated. Zinc supplementation can increase cell proliferation and neurogenesis through multiple pathways. Furthermore, zinc intervenes in the process of neurogenesis to influence the course of neurodegenerative diseases. Therefore, zinc may be a potential therapeutic target for the treatment of neurodegenerative diseases, especially AD.

3. Zinc and Aging

Zinc plays an irreplaceable role in chronic diseases related to aging. Human aging is a multifactorial process, which is inseparably linked to oxidative stress and chronic inflammation. Serum zinc levels decline with age but decline more rapidly in patients with AD. Zinc deficiency is more severe in patients with AD, when compared to controls of the same age [33].

3.1. Zinc and Hypoxia

Oxidative stress is an underlying factor in many age-related chronic conditions such as atherosclerosis, heart disease, neurodegenerative diseases, and even tumors. Zinc plays a crucial role in antioxidant action during oxidative stress. Zinc inhibits NADPH oxidase and reduces the production of OH^-, O^{2-}, and H_2O_2 (ROS). Zinc is also a cofactor for the enzyme, superoxide dismutase, which catalyzes the conversion of O^{2-} to H_2O_2 [34]. A progressive decline of intracellular ATP levels can be caused by excess zinc, while emodin (6-methyl-1,3,8-trihydroxyanthraquinone) reduces the induction of apoptosis and inhibits cellular ATP production. This could suppress ROS production and endoplasmic reticulum (ER) stress and inactivate the AMPK signaling pathway [35]. Repressor 1 silencing transcription factor (REST), an important transcriptional repressor, was observed in peripheral mononuclear cells among late-onset AD patients. Its minimal promoter region was highly methylated, and a significant reduction in antioxidant enzyme transcription was found. It was demonstrated that REST methylation may induce alterations in AD pathophysiology, which may cause the production of ROS [36].

The relationship between AD, age, and zinc has been studied. Age and sex were found to be risk factors for Aβ accumulation in AD, affecting the regulation of extracellular Zn^{2+} in the brain. Aging-induced brain mitochondrial damage may lead to slow Zn^{2+} reuptake, resulting in approximately four-fold higher extracellular zinc levels in the brain of older animals than in younger adult animals. Zn^{2+}-induced soluble Aβ oligomers may produce transient toxicity. In addition, amyloid bodies in AD are zinc-rich, but not strongly associated with cognitive impairment [37]. In the D-galactose-induced senescence mouse model, considerable damage and deterioration in the hippocampal region was observed.

The hippocampal damage and deterioration were markedly reduced in the H- intracellular zinc polysaccharides (IZPS) group, indicating that IZPS had more antioxidant and anti-aging capacity [38]. With aging, dopaminergic deficiency may also lead to elevated zinc levels, and a lack of dopaminergic regulation leads to impaired memory [39].

3.2. *Zinc and Inflammation*

Aging is characterized by an impaired immune response and systemic chronic inflammation. Zinc deficiency is an important factor in immune aging. When zinc levels in aged mice are restored through dietary zinc supplementation to a level comparable with young mice, various immune marker levels are improved, including reduced MCP1 and increased naive CD4+ T cell markers. Zinc also reduces activation-induced production of IFNc, interleukin-17, and tissue necrosis factor (TNF)-α cytokines in LNT cells. This suggests that improved zinc status can partially reverse the immune dysfunction and reduce chronic inflammation associated with aging [40]. The NLRP3 inflammatory complex accelerates cognitive dysfunction in AD, and disease progression can be mitigated by zinc supplementation. It was found that in mice deficient in the NLRP3 inflammatory complex, cognitive decline due to zinc deficiency was attenuated [39]. Use of the zinc chelator, *N*, *N*, N', N' tetrakis(2-pyridinylmethyl)-1,2-ethanediamine reduces the LPS-induced increase in the formation of NLRP3-level IL-1β, while reducing neuronal death [41]. In addition, nutritional zinc deficiency can increase the levels of the inflammatory factors, CCL2, CXCL2, CXCL5, IL-1, IL-6, and TNF-α in the serum and in colonic mucosa, accelerating the development of colonic tumors [42]. In animal experiments, zinc supplementation of 120 mg/kg improved the growth performance of Pekin ducks; it reduced intestinal damage and increased intestinal enzyme activity. The ability of zinc to inhibit the expression of LPS-induced inflammatory factors and apoptosis-related genes was also demonstrated in in vitro experiments [43].

The prevalence of AD increases with age. The aforementioned studies suggest that AD may be related to oxidative stress, impaired energy metabolism, and chronic inflammatory responses in vivo. Zinc is involved in many of these responses, and zinc plays a key role in a variety of age-related metabolic changes. Future research should investigate how to decrease aging through the modulation of zinc homeostasis or by inhibiting the aforementioned oxidative stress and inflammation pathways.

4. Zinc and AD

The prevalence of neurodegenerative diseases is increasing yearly with the advances in modern medical technology and increased human life expectancy. As a representative disease, AD is not yet treatable with specific drugs, and treatment with zinc has been intensely studied. Zinc can regulate the function and metabolism of APP. This affects the formation of plaques and fibrous tangles. It also affects the progression of AD. The hypothesis of AD pathophysiology are concluded in Figure 1.

4.1. *AD: Background*

AD is the most common form of dementia in the elderly. It is a degenerative disease of the CNS characterized by progressive cognitive dysfunction and behavioral impairment that occurs in old age. Clinical manifestations include memory impairment, aphasia, dysfunction, dyscognition, impairment of visuospatial abilities, impairment of abstract thinking and computation, and personality and behavioral changes.

AD can be divided into familial AD (FAD) and sporadic AD. Cases of sporadic AD, i.e., without a clear family history, are more numerous. Mutations in the genes encoding APP, progerin 1, and progerin 2 in FAD have been widely reported [44]. Despite the differences in mutations, the pathological changes in the AD brain share the same characteristics, including Aβ plaques, neurogenic fiber tangles (NFTs), and extensive neuronal and synaptic losses. APP is cleaved by β- and γ-secretase to produce Aβ peptides that aggregate into plaques or fibrous tangles. As expected, AD develops slowly; decreased glucose metabolism or increased Aβ deposition in the brain can occur as early as 10 or even 15 years before the onset of symptoms [45]. Since clinical trials of drugs targeting Aβ reduction alone have not been proven effective, the treatment of AD is still being studied [46,47]. However, researchers have turned their attention to other topics. The apolipoprotein E gene (*ApoE*) has been identified as a major genetic risk for late-onset, sporadic AD. *ApoE* regulates zinc homeostasis [48]. Alterations in trace metal homeostasis may contribute to the development of AD and other neurodegenerative disorders [49]. One pathophysiological hypothesis of AD has included the cholinergic, amyloid, tau, and metal hypotheses, etc. (Figure 1) [49].

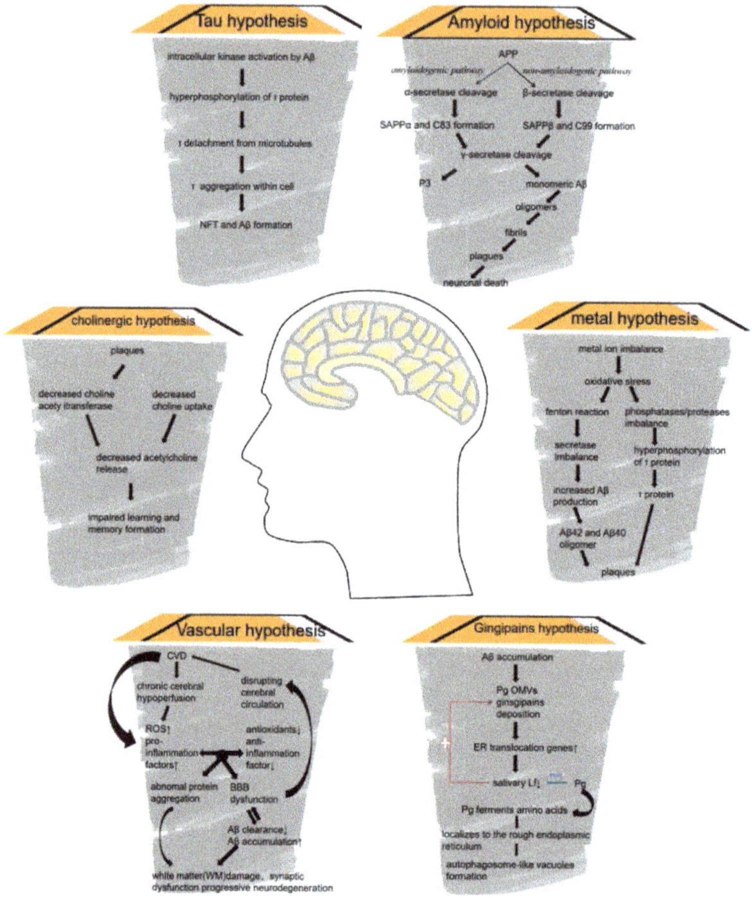

Figure 1. Multiple hypotheses are involved in Alzheimer's disease (AD) pathophysiology.

The infection hypothesis proposes that the brain tissue becomes infected with *Porphyromonas gingivalis*, which produces gingipains, a class of trypsin-like cysteine proteinases, leading to AD. The LPS-PG in *P. gingivalis* is integrated into the outer membrane vesicles (OMV) and is subsequently transported through the OMV to the CNS, regulating the process of vesicle production and formation. Prolonged exposure to LPS from *Porphyromonas* may induce the accumulation of Aβ (Figure 1).

Furthermore, iron imbalances have also been previously reported: Lf, an iron-binding protein, is upregulated in the brain of patients with AD, which blocks Aβ aggregation, spreads the tau lesion, and causes neuronal damage, resulting in neuroprotective effects [50]. The vascular hypothesis proposes that cerebral hypoperfusion causes chronic hypoxia in the whole brain, and an imbalance in oxidative stress homeostasis. Increased pro-inflammatory factors and ROS damage the blood-brain barrier, reducing the clearance of Aβ and eventually causing white matter damage, synaptic dysfunction, and neurodegenerative lesions. Investigators believe this may be one of the important mechanisms in the development of AD (Figure 1) [51]. Although most studies are in the experimental research stage, important progress has been made. A few studies have already entered clinical trials. Therefore, the future treatment of AD may not be limited to a single route of drug therapy.

4.2. APP Processing

The pathological changes in AD are mainly the presence of Aβ plaques and NFT formation, of which Aβ is a key component. Aβ is a product of APP processed by catabolic enzymes. APP is a membrane protein that can be hydrolyzed by two types of enzymes. One involves cleavage by α-secretase in the non-amyloid processing pathway, and the other is cleavage by β-secretase in the amyloid processing pathway. APP is cleaved by α-secretase to form sAPPα and the C-terminal fragment, C83. Conversely, APP is cleaved by β-secretase to form sAPPβ and the C-terminal fragment, C99. The cleaved residues are then cleaved again by γ-secretase to release P3, two major Aβ peptides (Aβ40, Aβ42), and the APP intracellular domain (AICD) (Figure 1). Aβ peptides play important roles in the pathogenesis of AD, with Aβ42 considered harmful [52,53]. Zinc and selenium were found to have neuroprotective effects, probably through their antioxidant properties and inhibitory effects on γ-secretase activity. They reduce the production of Aβ1-40 during APP cleavage processing [54]. Notably, zinc promotes the dimerization and oligomerization of APP-C99 and has an inhibitory effect on γ-secretase. Zinc triggers these effects by binding to substrates at both His-13 and His-14 sites. In contrast, binding of zinc to Lys-28 increases the production of Aβ43 (a precursor of Aβ40) (Figure 2). At different concentrations of zinc, the differences in binding sites lead to different effects. This may also explain the paradoxical role of zinc in AD [55]. Zinc can induce APP trans-dimerization at low concentrations; at low zinc concentrations, C186 and C187 in the E1 structural domain bind to zinc. They have a higher affinity for the appE1 structural domain than the E2 structural domain [56]. The formation of Aβ plaques and NFT after the formation of Aβ peptide aggregation by APP cleavage is the hallmark pathological change of AD. Overexpression of APP is sufficient to cause AD. Recently, APP has been found to be upregulated in astrocytes in an age-dependent fashion. The Alzheimer's disease β-secretase enzyme, BACE1 inhibitors can alter the autonomic function of astrocytes by reducing amyloid production, providing a new target for future treatment of AD [57]. Recently, it has also been shown that scaffold protein, X11L, is a regulator of APP β-site cleavage. X11L forms a tripartite complex with APP and Alca, and regulates metabolism, affecting APP. When Alca is deficient, increased Aβ production leads to enhanced amyloid plaque formation [58]. The combination of fish oil with selenium and zinc inhibits APP overexpression and alleviates learning and memory deficits in a mouse model of aging [59].

In summary, the etiology of AD is difficult to determine and is associated with many factors, including thyroid and cerebrovascular diseases, and family history. The pathogenesis of AD is complex, and there is no fully effective treatment available. The link between zinc and AD has also been heavily investigated in recent years, with the vast majority of studies referring to the maintenance of zinc homeostasis. Although some progress has been made in zinc therapy, there are still shortcomings. Chronic zinc exposure inhibits recognition and memory [60]. Therefore, whether zinc can be used to treat AD still requires extensive experimental and clinical studies. Thus, it becomes very important to continue investigating the pathogenesis of AD to determine appropriate treatments, including zinc therapy.

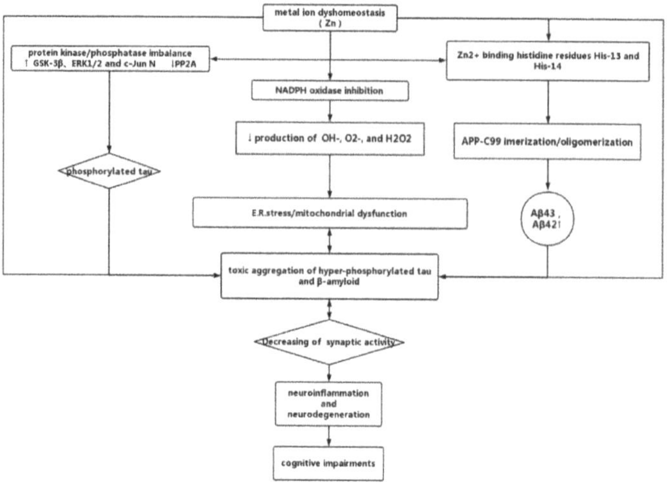

Figure 2. The toxic effects of zinc homeostasis on the pathology associated with AD. Zinc has multiple levels of toxic effects, including oxidative stress and toxic accumulation. Ultimately, cognitive dysfunction is the result.

4.3. Zinc Dyshomeostasis and AD

4.3.1. Zinc in AD

Zinc, copper, iron, calcium, and magnesium were reported to play important roles in AD [61,62]. The zinc homeostasis disorder hypothesis has also been proposed [63]. Zn^{2+} promotes the formation of stable Aβ40 and Aβ42 oligomers. This, in turn, leads to the formation of mature amyloid fibrils. Zn^{2+} is involved in the regulation of synaptic activity, which plays a role in the formation of Aβ oligomers. In turn, synaptic activity plays an important role in the formation of Aβ oligomers [64]. In AD patients, zinc levels are significantly higher in the cortex, hippocampus, and amygdala, and large amounts of Zn^{2+} can be detected in fiber tangles and plaques [65].

Zinc homeostasis is maintained by zinc transporter proteins. The roles of the two families of zinc transport proteins are completely opposed. During the development of AD, there are dynamic changes in the expression of zinc transporter proteins, such as increased ZnT-1 and decreased ZnT-3 expressions. Further control of zinc levels in the brain by modulating expressions of the transporter proteins could be a possible direction for AD treatment [66]. PBT2 acts as a zinc-copper ion carrier. The ability of PBT2 to clear Aβ aggregates in the cortex has been demonstrated in animal studies. Early improvements in cognitive function were observed in phase IIa clinical trials, and the treatment was considered safe [67]. In recent years, it has also been reported that chloroquine (ClioQ) can attenuate the formation of Aβ oligomers in cells [68]. In addition, a protein thought to have similar effects to EDTA, a potent zinc chelator, was found (S100B) to affect Aβ aggregation through two mechanisms; it interacts with fibrotic Aβ42 to decrease amyloid production and may bind competitively to Zn^{2+} to reduce toxic oligomer formation [69].

4.3.2. Zinc Can Be a Double-Edged Sword in AD

It has been shown that 100 μM Zn^{2+} accelerates the protofibrillation of ΔK280 of full-length human tau when compared with 50 μM Zn^{2+}. Pathological concentrations of Zn^{2+} significantly enhance tau aggregation-induced apoptosis and toxicity in Cys-291 and Cys-322-dependent SH-SY5Y cells [70]. Zinc binds directly to tau monomers and subsequently stimulates tau protein phosphorylation by activating GSK-3au monomers. Zinc also induces protein phosphatase 2A inactivation and tau hyperphosphorylation through an Src-dependent pathway, ultimately leading to exacerbation of AD-like tau pathology

(Figure 2) [71]. Zn^{2+} levels are significantly increased in the cortex and hippocampus of AD mice. Selenoprotein P redistributes the levels of intracranial zinc ions. When selenoprotein P expression increases, the level of synaptic zinc and the expression level of ZnT-3 increase simultaneously. The level of membrane channel protein, NMDAR2A, increases and competes with NMDAR2B for extracellular zinc. This can activate the Src-TrkB or/and BDNF pathways to inhibit tau protein pathological processes and Aβ aggregation in AD mouse models. It also preserves the activity of synaptic proteins, thus improving spatial learning and memory impairment [71]. Confocal immunofluorescence microscopy showed that ZnT-3 expression was distributed along the outside of the GFAP-stained astrocyte cytoplasm, and ZnT-3 localized to the cell membrane of astrocytes. Activated astrocytes, with the help of ZnT-3, internalize Zn^{2+} in dystrophic neurons, thus inhibiting the growth of amyloid plaques. The lack of synaptic Zn^{2+} after Aβ aggregation and plaque deposition may cause clinical symptoms of AD [72]. This also confirms that zinc homeostasis is regulated by zinc transport proteins (e.g., ZnT and ZIP) and metallothioneins (MTs). ZIP proteins transport extracellular and vesicular zinc into the cytoplasm to increase cytoplasmic zinc levels. Conversely, ZnT proteins excrete or remove intracellular zinc by vesicular uptake of zinc [73].

In contrast, increasing zinc levels in the brain can decrease AD progression. Nanoparticles (NPs) encapsulated with zinc were applied to the APP23 mouse model as a noninvasive method to increase brain zinc. It was observed that the plaque area was reduced in the experimental group and that IL-6 and IL-18 levels in the experimental group decreased after treatment. Although the level of IL-10 did not significantly correlate, the overall IL-10 level was increased. These results indicated that there was a decrease in the inflammatory response and Aβ deposition after treatment with zinc [74]. Subacute administration of zinc induces increased expression of CCL2 and CCR2. In addition, FGF2 and IGF-1 overexpression may exert antioxidant and anti-inflammatory effects after the hypoxic-ischemic process. It was demonstrated that subacute zinc treatment had a neuroprotective effect and induced an increase in neuronal plasticity [75]. A study of the genetic background revealed that older adults carrying the ZIP2 Leu2 (Arg43Arg) gene showed a marked reduction in the inflammatory factors Mcp-1, TNF-α, and RANTES after zinc supplementation. Not only did this variant prove to be beneficial for maintaining zinc homeostasis, but the study demonstrated a significant anti-inflammatory effect of zinc [76].

Synaptic zinc was found to be involved in cognitive activity. Both deficiency and excess were associated with cognitive decline [77]. Zinc can exert both neuroprotective and neurotoxic effects and promote the formation of amyloid plaques. Different concentrations or forms of zinc do not have the same pathophysiological effects. Zinc deficiency also exacerbates the rate of cognitive decline. Zinc supplementation decreased indicators of inflammation and reduced plaque deposition. Apoptosis was also significantly enhanced with increasing doses of zinc. Whether zinc is beneficial or harmful in patients with AD should be further investigated. The toxic effects of zinc homeostasis on the pathology associated with AD are shown in Figure 2.

5. Zinc and Vascular Dementia

5.1. Vascular Dementia

Among the dementia subtypes, vascular dementia (VD) is the second most common [78]. Post-stroke dementia may occur after stroke, especially in patients with a history of multiple strokes. The manifestations of VD are similar to those of dementia, such as memory impairment, decreased executive function, and decreased social skills. VD is usually progressive, and cognitive impairment may become aggravated over time. It has been shown that ischemia induces APP cleavage and enhances deleterious pathways, such as increased conversion of secreted neurotoxic Aβ and the β-secretase pathways [79]. After the onset of ischemia and reperfusion injury, increased levels of zinc-binding proteins release zinc. Zinc homeostasis in hippocampal neurons is disrupted, and large amounts of zinc accumulate in hippocampal neurons, exhibiting neurotoxicity. The use of zinc chelators can reduce the damage to hippocampal neurons [80], and glutathione can reduce neurotoxicity due to zinc accumulation [81].

5.2. Risk Factors of VD

Obesity may be a risk factor for the accelerated progression of VD. Compared to VD alone, cognitive impairment in VD with obesity is exacerbated and accompanied by disruption of presynaptic membranes and inhibition of BDNF pathways [82]. In addition to obesity, a history of parental dementia, ApoE variants, diabetes, hypertension, and hypercholesterolemia are all strongly associated with VD. Furthermore, studies have revealed a strong association between *rs9923231*, a gene encoding vitamin K epoxide reductase subunit 1, and VD [83]. Atrial fibrillation may also increase the risk of dementia, including not only VD, but also cognitive impairment and AD [84].

5.3. Metal Homeostasis and VD

Metal homeostasis has been found to be important in VD [85]. Following the onset of transient ischemia or stroke, impaired energy metabolism due to abnormal blood supply causes cell membrane depolarization and massive release of glutamate into the synaptic gap. The large amount of glutamate causes overstimulation of its receptors. Currently, known receptors for glutamate include the N-methyl-d-aspartate (NMDA)-type, amino-3-hydroxy-5-methyl-4- isoxazolepropionic acid (AMPA)-type, and erythropoietin-type receptors. Stimulation of the receptors may lead to an imbalance in Ca^{2+} homeostasis in vivo. The entry of Ca^{2+} into glutamate-responsive neurons triggers the death of vulnerable neurons, such as pyramidal neurons in the hippocampus, an area associated with learning and memory [86]. These processes lead to the development of VD. In addition to calcium ions, sublethal concentrations of copper ions (Cu^{2+}) significantly exacerbated Zn^{2+}-induced neurotoxicity. Cu^{2+} and Zn^{2+} co-administration also significantly increases the expression of genes associated with the endoplasmic reticulum stress response, including *CHOP*, *GADD34*, and *ATF4*. Cu^{2+} and Zn^{2+} synergistically promote neuronal death and influence the pathogenesis of VD [87].

Animal models of VD are also being investigated. In these studies, use of a rat 2-VO modified model found that ErbB3 was neuroprotective and had a key role in neural growth. In this rat model, ErbB3 was downregulated, resulting in a diminished role for both ErbB and PI3K-Akt. Both of these are key regulatory components for the protection of the nervous system [88]. In contrast, sublethal doses of zinc increased the expression levels of proteins associated with the phosphatidylinositol 3 kinase (PI3K) pathway, and decreased the expression levels of proteins associated with neuronal cell death, which reduced apoptosis induced by hypoxia in NSCs via its neuroprotective effect [89]. Recently, it has also been found that myostatin (β-alanylhistidine), an endogenous dipeptide, had antioxidant, anti-crosslinking, and anti-glycation effects. It protected neurons from Zn^{2+}-induced neurotoxicity. Myostatin may be a potential drug for the treatment or prevention of VD [90].

As a metal ion, zinc can induce neurotoxicity and induce modulation of the expression of neuroprotection-related channel proteins. The results support the neurotoxic mechanism of zinc, which aggravates VD, AD, and even cognitive impairment. However, some reports suggest the presence of neuroprotective effects of zinc, which can reduce plaque formation [72]. The specific mechanisms and whether zinc is neuroprotective remain unclear. There are few studies on the relationship between zinc and cerebral small vessel disease and the role of zinc in causing VD. Whether zinc can be a therapeutic target or not therefore needs further experimental validation.

6. Sleep Disorders

With the accelerated pace of life in modern society, sleep duration and sleep patterns have changed dramatically, and sleep disorders are an increasing concern. Sleep disorders include insomnia, somnolence, REM sleep behavior disorder, restless legs syndrome, periodic limb movements during sleep (PLMS), and obstructive sleep apnea. The most common sleep disorders in AD patients may manifest as frequent daytime naps, difficulty falling asleep at night, nighttime sleep fragmentation, and early morning awakening. Sleep disorders and AD may be co-morbid [91]. Sleep disturbance is a risk factor for dementia and increases peripheral blood Aβ levels [92]. Chronic sleep restriction (CSR) leads to increased accumulation of Aβ42 in the rat brain, especially in the hippocampus and prefrontal lobes. This is accompanied by increased expression of BACE1 and sAPPβ, suggesting that CSR may increase Aβ accumulation through the BACE1 pathway [93]. The formation and accumulation of Aβ is closely related to zinc levels. Zinc is involved in a variety of catalytic enzymes in the formation of Aβ. Zinc is also closely associated with sleep disorders. Insomnia is probably the most prevalent type of sleep disorder in the elderly. In a study, melatonin, zinc, and magnesium supplementation in 43 screened insomnia patients reduced Pittsburgh Sleep Quality Index scores and improved mood and quality of life [94]. Obesity is also an important factor for sleep disorders; regardless of whether they suffer from sleep apnea or insomnia, obese patients with sleep deprivation have low zinc levels, and sleep apnea patients have even lower blood zinc levels [95]. In addition to Aβ accumulation, sleep-wake rhythmic disturbances are clearly associated with tau-mediated neurodegeneration in a bidirectional manner. Tau protein aggregation leads to synaptic dysfunction, eventually leading to loss of synaptic function [96].

PLMS is a repetitive limb movement caused by repetitive muscle contractions in the legs during sleep. PLMS, a sleep disorder, is thought to be associated with cognitive decline, especially in older men with the periodic limb movement index (PLMI) greater than 30, among whom cognitive dysfunction is more pronounced [97].

Similarly, sleep apnea is associated with decreased cognitive function. Female patients with moderate to severe sleep apnea were reported to be at increased risk of cognitive impairment [98].

Sleep disorders may induce abnormal reactions in the body, such as oxidative stress, inflammatory response, abnormal hypothalamic-pituitary-adrenal (HPA) axis activity, and abnormal dopamine secretion [99]. Cognitive impairment due to sleep disorders has also been frequently reported. Sleep disturbances increase Aβ plaques and tau protein aggregation. Whether zinc plays a role in cognitive impairment, which often co-exists with sleep disorders, must be investigated. Several clinical trials have confirmed the benefit of zinc supplementation in sleep disorders. However, further animal and cytological studies are needed to determine the mechanism and whether zinc is the most important trace metal ion affecting sleep disorders and resulting cognitive changes.

7. Molecular Mechanism of Zinc-Associated Aging and Cognitive Impairment

Zinc is essential for normal brain function, and there is growing evidence that zinc plays a vital role in age-related cognitive impairment (ARCD). Zinc levels in the brain are tightly controlled, as it is essential for various cellular and molecular processes, enzymatic activities, and transcription factor functions [100]. The concentration of zinc in the brain is 10 times higher than that in the body [7]. While most zinc is bound to proteins or amino acids, approximately 10% of zinc is still chelated or free [101]. Under normal conditions, chelated zinc is located in the vesicles of presynaptic glutamatergic neurons and thus plays a crucial role in regulating neuroplasticity and cognition [102]. After depolarization, approximately 300 μM of zinc is released into the synapse along with glutamate [100,102]. Under healthy conditions, zinc levels are maintained by various zinc transporters and other transporters that balance their export/import into discrete cellular compartments. When these transporters are reduced or dysfunctional, it can lead to severe pathology [103]. Current research suggests that two families of zinc-binding proteins play essential roles in age-related cognitive impairment: zinc transporters (ZnTs) and metallothioneins [1,6,104].

7.1. Zinc Transporters

The primary function of ZnTs is to transport zinc out of the cytoplasm or into intracellular organelles [105]. Studies have reported four transcripts of ZnTs in the mammalian brain: ZnT-1, ZnT-3, ZnT-4, and ZnT-6 [105].

7.1.1. Zinc Transporter-1

Zinc transporter-1 (ZnT-1; SLC30A1) is a highly conserved and universally expressed protein [106,107], present on the plasma membrane of neurons and glial cells and is enriched at the postsynaptic density [108–111]. The expression level and activity of ZnT-1 limit the accumulation of postsynaptic Zn^{2+} after translocation from presynaptic release sites [110–112] and are also highly regulated by cytoplasmic Zn^{2+} changes [104,113]; therefore, ZnT-1 plays an essential role in Zn^{2+} homeostasis and protecting neurons from Zn^{2+} toxicity damage. In the presence of acidic extracellular solutions, the cytoplasm acidifies and further accumulates Zn^{2+}. ZnT-1 is expressed in the organelles of acidic cells and acts as a Zn^{2+}/H^{+} exchanger [104,114], inhibiting ZnT-1-dependent Zn^{2+} outflow in neurons, and as such, enabling the ZnT-1 to provide neuroregulation and neuroprotection function. Researchers have found that ZnT-1 in the hippocampus/parahippocampal gyrus (HPG) in preclinical AD patients was significantly decreased [115], and the same was true in mild cognitive impairment (MCI) patients [105]. As the disease progresses, the production of ZnT-1 increases with increases of intracellular Zn^{2+}. Elevated ZnT-1 in late-stage AD (LAD) patients can also lead to increased extracellular zinc, which interacts with Aβ resulting in Aβ aggregation [104,116].

7.1.2. Zinc Transporter-3

Zinc transporter-3 (ZnT-3) is only expressed in the brain and testes, which is the main vesicular zinc transporter in vesicular transporters (VTs). ZnT-3 is located on the membrane of synaptic vesicles (SVs) and transports zinc ions from the cytosol to SVs [117]. Research shows that nearly 36% of SVs carry two VTs, while the SV types containing vesicular glutamate transporter-1 (VGLUT1) and ZnT-3 account for the majority of multi-transporter vesicles (about 34% of all SVs). Co-localization of VGLUT1 and ZnT-3 increases the glutamate content of vesicles, resulting in an enhanced postsynaptic response [118]. The concentration of zinc in SVs is regulated by the activity dependence of glutamate mediated by ZnT-3 [119,120]. Targeted deletion of the ZnT-3 gene also eliminates zinc in SVs, resulting in ARCD and neurodegeneration [121,122]. Zn2þ is a long-term regulator of synaptic plasticity [123]; ZnT-3 mediated synaptic Zn2þ homeostasis is essential for maintaining cognition. In contrast, the loss of ZnT-3 and the resulting synaptic Zn2þ homeostasis disorders may adversely affect memory and cognition. Scientists have found that zinc accumulated in degenerated neurons of ZnT-3-deficient mice [102]. Further research found that after ZnT-3 gene knockout, mice became insensitive to increased Zn^{2+} in the striatum. The electrically induced release of dopamine (DA) was significantly reduced, the clearance rate of DA was higher [124], and ARCD was induced, suggesting that there may be ZnT-3-dependent communication between plasma and neuronal zinc through the cerebrovascular wall [125].

7.1.3. Zinc Transporter-4

ZnT-4 is highly expressed in the brain and mammary gland, while the expression level is low in other tissues. Expression of ZnT-4 decreases significantly with age [126]. ZnT-4 is located on the membrane of intracellular SVs and increases the concentration of zinc in SVs [127], promotes the transport of zinc to endosomes and lysosomes, and protects neurons from zinc poisoning [128,129]. It is reported that the ratio of ZnT-4 mRNA level to neuron-specific enolase mRNA level in the lysosomal system in the brain of post-mortem AD patients increased significantly [130], accelerating the accumulation of Aβ [131–133]. Some investigators found that the levels of ZnT-4 in HPG of patients with AD and in LAD patients increased [134]. ZnT-4 is involved in the zinc homeostasis of the cerebral cortex after hypoxia and ischemia. ZnT-4 expression in cerebral microvessels increases, which further damages the blood–brain barrier, thus damaging the neural cells and glial cells related to cognition [135].

7.1.4. Zinc Transporter-6 (ZnT-6)

ZnT-6 isolates cytoplasmic zinc to the TGN and vesicle compartment during the increase of intracellular zinc, destroys the normal sorting and transportation of essential proteins and lipids through the output function of ZnT-6 on plasma membrane or through the secretory pathway, and leads to neuronal degeneration and cytoplasmic inclusion [114]. ZnT-6 is significantly increased in degenerated neurons, and ZnT-6 immunostaining is increased in neurons containing cytoplasmic inclusions in MCI, AD, and Parkinson's disease (PD). Therefore, it is speculated that ZnT-6 plays a role in the pathogenesis of these lesions [114]. ZnT-6 can promote sphingolipid metabolism by activating smpd1 and converting sphingomyelin into ceramide and choline phosphate [136]. ZnT-6 isolates cytoplasmic zinc to TGN and vesicle compartments during intracellular zinc elevation [137]. ZnT-6 in HPG increased significantly in AD subjects and ZnT-6 increased in patients with MCI, and there was a significant positive correlation between ZnT-6 and Braak score ($p < 0.10$).

7.2. Metallothionein

Metallothionein (MT) is a cysteine-rich low molecular weight protein with high metal content. The reactivity and dynamic coordination of MTs with Zn^{2+} are related to the biological requirements of controlling these intracellular metal ions [138]. It can be used to buffer Zn^{2+} and zinc transporters, exporters, and importers. The transient increase of intracellular (cytosolic) Zn^{2+} concentration plays a role in intracellular signal transduction and participates in long-term potentiation [139], which can prevent the damage caused by ROS [140]. The expression of MT is involved in the regulation and control of zinc-dependent cell signal transduction [141]. MT includes four subfamilies: MT-I, MT-II, MT-III, and MT-IV. MT-I and MT-II exist in most tissues, but the expression of MT-III is limited to the CNS, and the expression of MT-IV is limited to stratified squamous epithelium [142].

7.2.1. Metallothionein-I/II

MT-I and MT-II are multivalent proteins. Their main function is to maintain the cellular homeostasis of essential metals such as zinc. They exist in the spinal cord and brain, mainly in the astrocytes [143]. Relevant studies have shown that the expression of MT-I/II was mainly related to the response of astrocytes in the aging brain to oxidative DNA damage in vivo, and the response to oxidative stress in vitro was induced in human astrocytes [144]. MT-I/II gene expression is regulated by antioxidant response elements. Decreased expression of MT-I/II in cell models or knockout animals may increase the sensitivity to oxidative stress [145,146]. Astrocytes and microglia/macrophages around amyloid plaques have high levels of MT-I and MT-II expression in the cortex and hippocampus [147,148], and MT-I, which is mainly upregulated in active astrocytes, has a neuroprotective effect induced by dual mechanisms in AD, namely through Aβ; neurotoxicity is directly and indirectly attenuated by MT-I inhibition of Aβ-induced microglia activation [149]. MT-I and MT-II-deficient mice show impaired spatial learning, but once learned, they retain information like that of wild-type mice [150]. In a study of a semi-PD model in MT-I and MT-II knockout mice, dopamine-doped cells treated with dopamine and L-dopa injection showed that intrinsic MT prevented dopamine quinone-induced neurotoxicity [151]. Scientists have shown that MTs could affect the production of adult neurons and oligodendrocytes. The data showed that MT-I/II played an important role in cell protection and growth regulation during myelin regeneration activated after toxic demyelination injury. In the injured brain, MT-I + II inhibit macrophages, T-lymphocytes, and their IL and TNF-α, matrix metalloproteinases, and ROS formation. In addition, MT-I + II enhances cell cycle progression, mitosis, and cell survival, while inhibiting neuronal apoptosis. MT-I + II is essential for the recovery of neuropathology and cognitive function [152].

7.2.2. Metallothionein-III

MT-III mainly exists in neurons and astrocytes, especially in neurons with high concentrations of zinc, and contains four divalent metal copper and three or four divalent zinc ions [143]. MT-III is not regulated by metals or glucocorticoids, so it is not inducible. Its neuroanatomical distribution is closely related to the region with a high concentration of zinc [147]. MT-III exists in several brain regions at different concentrations, especially in the presynaptic terminals of zinc-rich neurons in the cerebral cortex and hippocampus [148]. MT-III is related to the pathogenesis of neurodegenerative diseases such as PD and AD. It was reported that MT-III was abundantly expressed in the normal human brain but was greatly reduced in the AD brain; in vitro, MT-III inhibits neurite formation and survival of cortical neurons [153–155]. In patients with AD, MT-III was identified as a growth inhibitory factor (GIF) or neuronal growth inhibitory factor. MT-III upregulation can be observed in glial cells after brain injury [156]. MT-III has anti-Aβ activity, mainly through the elimination of Aβ. The formation of peptide toxic aggregates to antagonize Aβ plays a role in protecting neurons [157]. The hemiplegic rat model induced by 6-hydroxydopamine (6-OHDA) showed that levodopa treatment significantly increased the expression of MT-III mRNA in the contralateral striatum, suggesting that the regulation of MT-III mRNA expression may be related to the progressive degeneration of PD [158]. MT-III can effectively remove ROS; therefore, exogenous MT-III can prevent neurite extension and differentiated cortical neuron death caused by exposure to high oxygen concentration because it scavenges free radicals more effectively than ZnT-7-MT-I/MT-II at the same concentrations [159]. The above studies show that both ZnTs and MTs play a role in maintaining intracellular and extracellular zinc homeostasis, protecting neurons and regulating growth, and have an important impact on ARCD [Table 1]. Whether there are more zinc-associated proteins involved in this regulatory process still needs further study.

Table 1. The molecular mechanism and function of ZnTs and MTs in ARCD.

Zinc Associated Protein		Function	Reference
Zinc transporters	ZnT-1	1. Limit the accumulation of postsynaptic Zn^{2+} after translocation from presynaptic release sites 2. Protect neurons from Zn^{2+} toxicity 3. As a Zn^{2+}/H+ exchanger in acidic cells, it inhibits ZnT-1 dependent Zn^{2+} outflow in neurons 4. ZnT-1 is elevated in AD patients, making Aβ gather	[104,110,112–115]
	ZnT-3	1. Transport Zn^{2+} from cytoplasm to SVs and regulate the concentration of Zn^{2+} in SVs 2. Co-localization with VGLUT1 increased the content of glutamate in VTs and the postsynaptic response 3. Maintain Zn2þ homeostasis in synapses, so as to maintain human cognitive function 4. Plasma and neurons depend on it to exchange through the blood vessel wall	[117–120,123–125]
	ZnT-4	1. Increase the concentration of Zn^{2+} in SVs, promote the transport of Zn^{2+} to endosomes and lysosomes, and protect neurons from zinc poisoning 2. Accelerate the accumulation of Aβ in the brain of patients with AD 3. Further damage the BBB in patients with is, and damage the neural cells and glial cells related to cognition	[127–129,131–133,135]

Table 1. *Cont.*

Zinc Associated Protein		Function	Reference
	ZnT-6	1. It separates Zn^{2+} in the cytoplasm into TGN and VTs in cells with high Zn concentration 2. It plays a transport or secretory function on the plasma membrane, destroys the normal sorting and transport of essential proteins and lipids, and leads to neuronal degeneration and cytoplasmic inclusion 3. Activate smpd1 to promote sphingolipid metabolism and convert sphingomyelin into ceramide and choline phosphate	[114,136,137]
Metallothionein	MT-I/MT-II	1. Maintain intracellular homeostasis of essential metal ions in the human body 2. Induced by oxidative stress in human astrocytes 3. In AD patients, MT-I directly attenuates the neurotoxicity of Aβ and indirectly inhibits Aβ-induced microglia activation and subsequent neurotoxicity 4. Prevention of dopamine quinone induced neurotoxicity in patients with Parkinson's disease 5. It plays the role of cell protection and growth regulation in the process of myelin regeneration activated after toxic demyelination injury 6. Inhibit macrophages, T lymphocytes, and their formation of interleukin and TNF-α, matrix metalloproteinases, and ROS 7. Enhance cell cycle progression, mitosis, and cell survival, and inhibit neuronal apoptosis	[143–146,149,151,152]
	MT-III	1. Inhibition of neurite formation and survival of cortical neurons 2. It has anti-Aβ activity and eliminates the toxic aggregates of Aβ peptide, so as to antagonize the neurotoxic effect of Aβ peptide 3. It can effectively clear ROS and prevent neurite extension and differentiated cortical neuron death caused by exposure to high oxygen concentration	[153–155,159]

8. ZFP and the Cellular Pathways

ZFPs are the largest family of transcription factors in the human genome [160]. They are composed of small domains that bind metal ions (usually zinc) in a tetrahedral geometry using a combination of four cysteine and/or histidine ligands [161–163]. In the absence of zinc, these domains expand; in the presence of zinc, the domains of zinc fingers (ZFs) are folded into three-dimensional structures [161]. After one folding, these domains have the function of specifically recognizing DNA, RNA, or other protein targets [9,161–166]. ZFs can also directly bind to DNA or RNA and participate in many functions, including transcriptional and translational regulation [167]. Different zinc-binding motifs lead to diversity in the structure and function of ZFs in many human tissues, including the brain [168,169]. The main function of ZFs in the brain is to promote the development of different parts of the brain and the differentiation of neural stem cells. Many studies have shown that ZFPs are closely related to the pathogenesis of nervous system diseases [170]. Researchers found that many ZFP DNA-binding domains downregulated the transcriptional activity of nr2f through missense mutations, to regulate cognitive impairment and optic nerve atrophy [171].

8.1. ZFP-1 (CIZ1)

ZFP-1 (CIZ1) is a nuclear protein that interacts with CDKN1A and plays an important role in DNA replication and cell cycle progression [172,173]. Germline or somatic variants in CIZ1 are associated with several neurological and extraneural diseases. Researchers found that after CIZ1 was knocked-down, LPO was upregulated and GSH was downregulated [174], resulting in increased DNA damage, aggravation of neuroinflammation, and aggravation of cell cycle dysfunction [175,176]. Older mice lacking CIZ1 are more likely to acquire age-related cognitive impairment, decreased motor function, and age-related neurological diseases (such as AD). The brain tissue of CIZ1$^{-/-}$ aged mice show significant DNA damage and NF-κB upregulation, oxidative stress, vascular dysfunction, inflammation, and cell death. These harmful effects become more serious in older mice [174]. A variable shear form of CIZ1 was found to be upregulated (CIZ1S) in the hippocampus of AD patients, and the expression of CIZ1S was higher than that of CIZ1 in AD, resulting in the imbalance of the cell cycle [177]. CIZ1 affects defects of the cell cycle and related DNA repair pathways in post-mitotic neurons, which leads to the decline of whole-brain nerve function in the elderly and further aggravates their cognitive impairment [174].

8.2. Unkempt

Unkempt is a highly conserved ZF/ring domain protein, which was first found and expressed in the embryonic nervous system of *Drosophila melanogaster.* Researchers found that unkempt played a role in regulating photoreceptor differentiation in the retina during *Drosophila* development in the downstream pathway of a mammalian target of rapamycin (mTOR) [178,179]. Under nutrient-rich conditions, the reduction of unkempt will not affect cell proliferation, but can change the time of photoreceptor differentiation in *Drosophila* [178,180]. Loss of unkempt leads to premature differentiation and pattern defects of photoreceptor neurons in adult eyes [178]. In addition, unkempt is strongly expressed in the larval nervous system of *Drosophila melanogaster* and plays a role in regulating the cell cycle in neural progenitor cells [181]. Mammalian unkempt has the strongest expression in cell lines with a neuronal origin, and its in vivo expression is the strongest in the developing CNS. The expression of unkempt in the developing brain peaks between embryonic days 12 and 18 and is particularly abundant in neurons expressing tuj-1. In vitro experiments show that mammalian unkempt is a conserved ZFP, which regulates mTOR signaling through its ZF domain binding to mRNA, to affect nerve development and function [182,183]. It was reported that inhibition of mTOR signaling by chronic rapamycin treatment enhanced spatial learning, while chronic dietary restriction inhibited mTORC1 to enhance the memory of young mice [184,185]. Both *Drosophila* and mammalian unkempt protein interact with the mTORC1 component of Raptor [186,187]. In the developing nervous system, the expression of mTORC1 component of Raptor decreases in the mice with specific knockout of unkempt, but it does not affect the proliferation of neural progenitor cells and the overall development of the nervous system [180]. However, studies have shown that unkempt was strongly expressed in the cerebellum and hippocampus of adult mice, and unkempt knockout mice recovered their lost learning ability. Since the loss of unkempt increases cognitive flexibility, researchers have linked it to cognitive disorders such as AD [182,188].

8.3. Zinc Finger Protein 804A (ZFP804A)

ZFP804A belongs to the ZFP family containing the C2H2 domain. ZFP804A is highly expressed in brain tissue and located in neurites and dendritic spines [189–193]. Researchers found that the expression of ZFP804A in the brain increased in the early fetus [194], peaked at 13–24 weeks after conception, decreased after birth, and further gradually decreased in adulthood [193]. Knockdown of ZFP804A in human NPS altered the expression of genes involved in cell adhesion, mitosis, neural differentiation, and the inflammatory response, suggesting that ZFP804A may be involved in neuronal migration, neurite growth, and synapse formation [195]. Cognitive deficits in many neuropsychiatric disorders are thought to be related to dendritic spines or neurite growth [196–199]. In the primary cortical neurons of ZFP804A knockdown rats, the neurite growth and the density of dendritic spines are significantly reduced, and the upregulation and downregulation of ZFP804A significantly damage the dendritic morphology; furthermore, different ZFP804A binding proteins show varying levels of reversal of dendritic and synaptic defects [200]. A recent study showed that ZFP804A was necessary for NPC proliferation and neuronal migration, and the overexpression of the gene encoding neurogranin (an SZ risk gene and the target of ZFP804A) can offset the migration defect caused by ZFP804A knockdown [193]. In addition, abnormal ZFP804A expression levels decrease spinal density or abnormal axonal dendration in the cerebral cortex and hippocampus in vivo and in vitro. Female ZFP804A mutant mice show age-dependent defects in sensorimotor gating. Thinking disorder and distraction occurred only after 6 months of age, while the damage to spatial and fear memory was obvious when they were young [201]. Researchers also found that the deletion of ZFP804A had an impact on the language, reading, and cognitive abilities of the population. Therefore, it is speculated that ZFPs are involved in the regulation of cognitive impairment-related diseases through a variety of cellular pathways [202]. In 2q31.2-2q32.3-deletion syndrome, ZFP804A is in the deletion region [203]. Investigators speculate that the absence of ZFP804A may affect 2q31 2-2q32. Patients with this syndrome exhibit impaired reading and spelling [203]. The ZFP804A-rs1344706 allele may be related to better performance of language-related processes [203]. The impairment of ZFP804A-related cognitive functions (word reading, word spelling, non-word reading, and fast naming) constitutes the characteristic pattern of dyslexia [204]. Different zinc-binding motifs have different binding sites and functions in the brain. CIZ1 binds to the nucleus of neurons through the NF-kB pathway and participates in the regulation of the cell cycle and DNA damage. Unkempt participates in the mTOR pathway at the mRNA level, thereby affecting neural development and function. ZF804A is involved in many aspects of neuronal synaptic growth and development. It also has an important impact on human language, reading, and cognitive ability. In conclusion, ZFPs are closely related to nervous system diseases [Table 2], but further cell, animal, and human studies are still required to understand more about the function of ZFPs in the nervous system, to determine whether cognitive impairment-related diseases can be controlled through in vivo and in vitro regulation of ZFPs.

Table 2. Cellular pathways involved in ZFP regulation in the nervous system.

Zinc Finger Protein	Function	Reference
CIZ1	1. Maintain the balance of LPO and GSH and participate in NF-κB, oxidative stress, vascular dysfunction, inflammation, and cell death 2. Affect the cell cycle and related DNA repair pathways of post-mitotic neurons 3. It is involved in regulating ARCD, motor decline, and age-related neurological diseases	[174–176]
Unkempt	1. Regulation of retinal photoreceptor differentiation in the mTOR downstream pathway 2. Regulate the cell cycle of neural progenitor cells 3. Affect cognitive flexibility and participate in the regulation of cognitive impairment diseases	[178,179,181,188,189]
ZFP804A	1. It affects the expression of genes such as cell adhesion, mitosis, neural differentiation, and inflammatory response, and participates in neuronal migration, axon growth, and synaptic formation 2. Damage dendritic morphology and participate in the regulation of dendritic synaptic defects 3. Affect the spinal cord density or axon dendritic state of the cerebral cortex and hippocampus 4. Regulate people's language, reading, and cognitive abilities, and participate in the regulation of cognitive impairment-related diseases	[193,195,200–202]

9. Zinc and Autophagy

Autophagy is a cellular pathway used to remove and/or recover unnecessary, damaged, and dysfunctional cellular components. Autophagy defects are related to aging-related diseases such as cognitive impairment [205,206]. Zinc plays an indispensable role in apoptosis and cell cycle regulation [207,208]. Convergent evidence from in vivo and in vitro studies shows that zinc is a positive regulator of autophagy [209–212]. As the number of cells increases and an organism grows, the serum concentration of zinc and autophagy decrease [213,214]. Therefore, the effect of zinc on the autophagy pathway may have great potential as a therapeutic target for aging-related and neurodegenerative diseases such as cognitive impairment [215,216].

9.1. Extracellular Signal-Regulated Kinase 1/2 (ERK1/2)

ERK1/2, activated by neurotrophic factors and other chemicals, plays an important role in neuronal differentiation, survival, structural plasticity, long-term enhancement, and memory formation in animal models [217]. The ERK1/2 signaling pathway is associated with many neurodegenerative diseases involving oxidative stress [218]. Abnormal accumulation of activated ERK1/2 occurs in neurons in the brain of patients with AD [219,220]. The mechanism of zinc regulation of autophagy is still unclear, but relevant studies have shown that ERK1/2 plays an indispensable role in regulating autophagy [211,212,221]. Decreased zinc ion concentration in the blood inhibits the activation of the ERK1/2, p38 MAPK, NF-kB, and mTOR pathways [222–224]. Zinc can inactivate mTOR and induce autophagy by regulating ERK1/2 and also activate the Beclin 1-PI3K complex through increased phosphorylation of Bcl-2, the negative regulator of Beclin 1 [225]. In addition, zinc can counteract its ERK1/2-mediated phosphorylation by triggering stronger p53 activation [226].

9.2. MTs

MT, a low molecular weight protein that binds to zinc, is a potential source of zinc for autophagy activation during oxidative stress [209]. Zinc can induce autophagy gene expression by activating metal reactive transcription factor (MTF1) [227]. The MTs proteins can promote autophagy in a cell cycle-dependent manner through E2F transcription factor 4 (E2F4), a newly reported transcriptional activator of cytoprotective autophagy crucial to zinc homeostasis, to degrade MT proteins, improve the distribution of Zn^{2+} in autophagy, and reduce unstable intracellular zinc ions. Under the condition of oxidative stress [228], the expressions of MtnA and other metallothioneins, genes involved in glutathione metabolism, and other genes known to belong to oxidative stress responses or general stress response, increase. MtnA also responds to genes and pathways of additional metabolic processes, autophagy, and apoptosis [229]. Overexpression of MT in cells can protect lysosomes from estrogen-induced damage and maintain viability while inhibiting autophagy [230]. Knockdown of metallothionein-3 inhibits late autophagy in irradiated glioma cells [231].

9.3. Methylation

Gene methylation by N6-methyladenosine (m6A) is the most abundant in mRNA. Zinc can also regulate autophagy by regulating gene methylation [232]. The human n6-methyltransferase complex (MTC) contains a ZF motif sufficient to bind RNA. However, methylation activity requires two ZFPs [233]. YTH domain family protein 3 (YTHDF3) is an important reader in m6A methylation. It plays an important role in regulating the translation of m6A-modified mRNA and affects the decay of methylated mRNA in an m6A methylation-dependent manner [234,235]. Researchers found that ZF E-Box-binding homeobox 1 (ZEB1) is the key downstream target of YTHDF3. YTHDF3 enhances the stability of ZEB1 mRNA in an m6A-dependent manner and positively regulates cell migration, invasion, and the EMT [236]. Arabidopsis HAKAI is necessary for complete m6A methylation, and its knockout leads to a 35% reduction in m6A abundance [237].

ZFPs 1 and 2 (HIZ1 and HIZ2) interact with HAKAI and are members of the m6A writer complex. Their knockdown and knockout cause strong developmental defects [238–240].

9.4. Zinc Homeostasis

Autophagy decreases with age [241]. The dysregulation of zinc metabolism observed in the elderly may be a result of slower autophagy. Similarly, in other age-related cognitive impairment diseases associated with autophagy deficiency, such as AD, autophagy deficiency may be associated with dysregulation of zinc homeostasis [205]. Although zinc is a positive regulator and protector of autophagy, when the zinc concentration exceeds a certain threshold, excessive zinc leads to uncontrolled autophagy and cell death [205,212]. Relevant studies have shown that high concentration of zinc in the culture medium (1000 μM) can stimulate autophagy-driven apoptosis of porcine small intestinal epithelial cells (IPEC-J2) [242]. Zinc ionophores clioquinol and PCI-5002 induce autophagic cell death in mouse astrocytes and non-small cell lung cancer cells, respectively [243,244]. Interestingly, excess zinc is crucial in activating toxic autophagy induced by tamoxifen and H_2O_2 [205,213,245,246]. After spinal cord injury, the serum zinc concentration decreases, the spinal cord zinc concentration increases at the injured site, and the zinc level in infiltrated monocytes increases significantly, which is equivalent to the injured spinal cord, and then zinc infiltrates the lesion area with monocytes [247]. Due to zinc chelation in senile plaques, the local zinc ion dynamics near amyloid aggregates in AD are unbalanced [74]. After long-term injection of zinc nanoparticles into MCAO mice for 2 weeks, the plaque size is significantly reduced, and levels of pro-inflammatory cytokines IL-6 and IL-18 are altered. Whether the autophagy function of nerve cells plays a role is closely related to age-related diseases such as cognitive impairment. Zinc can affect autophagy through ERK1/2, MTs, methylation, zinc homeostasis, and other pathways; however, the specific mechanism of zinc regulating autophagy is still unclear. We still need to further explore the molecular mechanism and cellular pathway of zinc participation in the regulation of autophagy in the nervous system.

10. Interventions for Zinc-Associated Aging and Cognitive Impairment

Zinc affects learning and memory and is an important factor in ARCD because it regulates various cellular and molecular processes [100,248]. For example, during brain depolarization [101,124], zinc is released into synapses together with glutamate to regulate synaptic communication and neural plasticity [249].

10.1. Clioquinol

Chloroiodoquine is a weak zinc chelator (kD $\approx 10^{-7}$). Some researchers have found that chloroiodoquine reduced amyloid protein in APP TG2576 mice (AD model). The weak zinc chelation property of chloroiodohydroquine may be the basis of an anti-amyloid effect. Genetic removal of synaptic zinc in the brain of APP TG2576 mice can greatly reduce the plaque load [250], which is similar to the plaque reduction effect of membrane-permeable metal chelator DP-109 [251]. Chloroiodohydroquine can also act as a zinc ion carrier rather than a zinc chelator in physiological solution [252]. Owing to low intracellular free zinc levels, chloroiodohydroquine transfers zinc downward and creates a concentration gradient [253]. Under physiological conditions, chloroiodohydroquine may carry zinc from the extracellular to the intracellular space. This effect may increase in the brain of AD patients, in which Aβ plaques provide more extracellular zinc stores, which may lead to lower intracellular zinc levels [254]. ZnT-3 can load zinc into presynaptic vesicles of the hippocampus; the expression of ZnT-3 is downregulated in accelerated aging mice [255–257]. Decreased ZnT-3 protein levels and impaired memory were observed in the cortex of ZnT-3 knockout mice, aged mice, and aged humans [125,258]. Researchers found that chloroiodohydroquine acted as a zinc ion carrier under physiological conditions [243], increasing the intracellular zinc level in the cytoplasm and autophagic vesicles in an extracellular zinc-dependent manner, to restore the cognitive ability of ZnT-3

KO mice [125,259]. In addition, PBT2 (prana Biotechnology), an 8-hydroxyquinolone, can improve synaptic plasticity and cognitive function by mediating different cellular pathways (especially involving glutamatergic signal transduction at synapses), improve protein (CamKII, CREB, AMPA, NMDA, and VGLUT1) redistribution of zinc in the brain, and improve the content of total zinc in the hippocampus of elderly subjects [260]. Investigators found that AD-like transgenic mice treated with PBT2 had improved cognitive function and reduced disease pathology [210]. In a short-term phase IIa human clinical trial of AD, it was found that PBT2 played a positive role in improving cognition and restoring normal brain function [260–262].

10.2. Rapamycin

Rapamycin is a well-known mTOR inhibitor, which plays an important role in autophagy and insulin signaling [263,264]. The mTOR is a serine/threonine kinase, which plays a key role in regulating protein synthesis and degradation, age-dependent cognitive decline, and the pathogenesis of AD [265,266]. Abnormal mTOR signaling in the brain can affect several pathways related to metabolism, insulin signaling, protein aggregation, mitochondrial function, and oxidative stress in AD [267]. Increased mTOR expression co-localizes with NFT and mediates tau phosphorylation [265,266,268,269]. Investigators found that intracerebroventricular injection of zinc sulfate induced continuous mTOR (s2448)-P70S6K (t389) phosphorylation in rat hippocampus. Rapamycin can improve zinc-induced tau hyperphosphorylation, oxidative stress injury, and synaptic injury by downregulating mTOR/P70S6K activity, upregulating Nrf2/HO-1 activity, and ameliorating spatial learning defects and age-related cognitive impairments [10].

10.3. Zinc Supplements

Zinc has antioxidant and anti-inflammatory properties. Zinc supplementation within the limits of a normal diet can alleviate cognitive impairment [10]. Melatonin (Mel) is a neurohormone released by the pineal gland. In healthy individuals, Mel levels peak at night. They regulate the circadian rhythm [270], immune system, and blood pressure [271]. They have strong antioxidant properties and neuroprotective effects and act through the promotion of Aβ to alleviate Aβ-induced neurotoxicity [272]. Researchers found that in physical and mental activities, zinc and melatonin protected against AlCl3-induced AD and related liver and kidney injury by activating antioxidant and anti-inflammatory processes and reducing apoptosis. GSK-3β dysregulation is related to the pathogenesis of a variety of diseases affecting the brain, including neuroinflammatory and neurodegenerative diseases [273]. Wnt signaling plays an important role in maintaining synaptic plasticity and memory in the brain [274]. AlCl3-induced neurodegeneration and p-GSK-3β protein expression were significantly increased, followed by a decrease in the level of Wnt and β-catenin. The combination of zinc and Mel therapy can provide additional neuroprotective benefits, because zinc deficiency leads to the downregulation of the Wnt/β-catenin signaling pathway, and Mel activates Wnt/β-catenin signaling to prevent neuronal cell death [275]. In addition, long-term moderate physical activity can regulate the classical Wnt/β-catenin signaling pathway, thereby improving neuroprotection and synaptic plasticity.

Zinc deficiency is closely related to the pathogenesis and process of ARCD. The above-described studies confirm that it can intervene in ARCD by regulating the way zinc participates in the regulation of the nervous system. However, current research is still limited to known cellular pathways, and we still need to explore more methods to treat zinc-related ARCD.

Author Contributions: Conceptualization, R.S. and B.C.; methodology, writing—original draft preparation, J.W. and R.S.; writing—review and editing, J.F.; visualization, B.C.; supervision, J.F.; funding acquisition, J.W. and J.F. All authors have read and agreed to the published version of the manuscript.

Funding: This work was supported by the National Natural Science Foundation of China (Grant No. 81801171; 2018), the National Natural Science Foundation of China (Grant No. 81771271; 2017), and the 345 Talent Project of Shengjing Hospital (Grant No. M0267).

Institutional Review Board Statement: Not applicable.

Informed Consent Statement: Not applicable.

Conflicts of Interest: The authors declare no conflict of interest.

References

1. Min, C.J.; Young, K.J.; Ran, Y.H. Effects of oral zinc supplementation on zinc status and catch-up growth during the first 2 years of life in children with non-organic failure to thrive born preterm and at term. *J. Pediatr. Neonatol.* **2019**, *60*, 201–209.
2. Akram Shahzad, K.; Carlsson-Skwirut, C.; Bhutta Zulfiqar, A.; Söder, O. Placental IGF-I, IGFBP-1, zinc, and iron, and maternal and infant anthropometry at birth. *Acta Paediatr.* **2011**, *100*, 1504–1509. [CrossRef]
3. Merryana, A.; Bambang, W. The effect of adding zinc to vitamin A on IGF-1, bone age and linear growth in stunted children. *J. Trace Elem. Med. Biol.* **2014**, *28*, 431–435.
4. Kanchan, B.; Abdul, M.A.B.; Atish, P. Possible role of metal ionophore against zinc induced cognitive dysfunction in D-galactose senescent mice. *Biometals* **2016**, *29*, 399–409.
5. Seunghyuk, C.; Ki, H.D.; Young, C.B.; Won, S.S. Zinc in the Brain: Friend or Foe? *Int. J. Mol. Sci.* **2020**, *21*, 8941.
6. Mocchegiani, E.; Bertoni-Freddari, C.; Marcellini, F.; Malavolta, M. Brain, aging and neurodegeneration, Role of zinc ion availability. *Prog. Neurobiol.* **2004**, *75*, 367–390. [CrossRef]
7. Portbury, S.D.; Adlard, P.A. Zinc Signal in Brain Diseases. *Int. J. Mol. Sci.* **2017**, *18*, 2506. [CrossRef]
8. Mizuno, D.; Kawahara, M. The molecular mechanisms of zinc neurotoxicity and the pathogenesis of vascular type senile dementia. *Int. J. Mol. Sci.* **2013**, *14*, 22067–22081. [CrossRef]
9. Claudia, A.; Lucia, B.; Ivano, B.; Antonio, R. Counting the zinc-proteins encoded in the human genome. *J. Proteome. Res.* **2006**, *5*, 196–201.
10. Rosanna, S.; Amit, P.; Mario, P.; Abofazl, A.; Mariacarla, V.; Mauro, C.R.; Tjaard, H. Zinc Therapy in Early Alzheimer's Disease: Safety and Potential Therapeutic Efficacy. *Biomolecules* **2020**, *10*, 1164.
11. McCarrey, A.C.; An, Y.; Kitner-Triolo, M.H.; Ferrucci, L.; Resnick, S.M. Sex differences in cognitive trajectories in clinically normal older adults. *Psychol. Aging* **2016**, *31*, 166–175. [CrossRef]
12. Russell, J.K.; Jones, C.K.; Newhouse, P.A. The Role of Estrogen in Brain and Cognitive Aging. *Neurotherapeutics* **2019**, *16*, 649–665. [CrossRef]
13. Yin, F.; Sancheti, H.; Patil, I.; Cadenas, E. Energy metabolism and inflammation in brain aging and Alzheimer's disease. *Free Radic. Biol. Med.* **2016**, *100*, 108–122. [CrossRef]
14. Tönnies, E.; Trushina, E. Oxidative Stress, Synaptic Dysfunction, and Alzheimer's Disease. *J. Alzheimer's Dis. JAD* **2017**, *57*, 1105–1121. [CrossRef]
15. Banks, W.A.; Farr, S.A.; La Scola, M.E.; Morley, J.E. Intravenous human interleukin-1alpha impairs memory processing in mice: Dependence on blood-brain barrier transport into posterior division of the septum. *J. Pharmacol. Exp. Ther.* **2001**, *299*, 536–541.
16. Hossain, Z.; Qasem, W.A.; Friel, J.K.; Omri, A. Effects of Total Enteral Nutrition on Early Growth, Immunity, and Neuronal Development of Preterm Infants. *Nutrients* **2021**, *13*, 2755. [CrossRef]
17. Fenton, T.R.; Barbara, C.; Dena, G.; Roseann, N.; Belal, A.; Misha, E.; Hay, W.W.; Angela, H.; Diane, A.; Frank, B.; et al. "Extrauterine growth restriction" and "postnatal growth failure" are misnomers for preterm infants. *J. Perinatol.* **2020**, *40*, 704–714. [CrossRef]
18. Halas, E.S.; Hunt, C.D.; Eberhardt, M.J. Learning and memory disabilities in young adult rats from mildly zinc deficient dams. *Physiol. Behav.* **1986**, *37*, 451–458. [CrossRef]
19. Chua Joanne, S.C.; Cowley Carina, J.; Manavis, J.; Rofe Allan, M.; Coyle, P. Prenatal exposure to lipopolysaccharide results in neurodevelopmental damage that is ameliorated by zinc in mice. *Brain Behav. Immun.* **2012**, *26*, 326–336. [CrossRef]
20. Alizadeh, F.; Davoodian, N.; Kazemi, H.; Ghasemi-Kasman, M.; Shaerzadeh, F. Prenatal zinc supplementation attenuates lipopolysaccharide-induced behavioral impairments in maternal immune activation model. *Behav. Brain Res.* **2020**, *377*, 112247. [CrossRef]
21. Mirzaa, G.M.; Chong, J.X.; Piton, A.; Popp, B.; Foss, K.; Guo, H.; Harripaul, R.; Xia, K.; Scheck, J.; Aldinger, K.A.; et al. De novo and inherited variants in ZNF292 underlie a neurodevelopmental disorder with features of autism spectrum disorder. *Genet. Med.* **2020**, *22*, 538–546. [CrossRef]
22. Yukti, V.; Yewon, J.; Kevin, L.; Garner, C.C.; Montgomery, J.M. In vitro zinc supplementation alters synaptic deficits caused by autism spectrum disorder-associated Shank2 point mutations in hippocampal neurons. *Mol. Brain* **2021**, *14*, 95.
23. Njoud, A.; Rafah, M.; Tomoshige, K. CH-Type Zinc Finger Proteins in Brain Development, Neurodevelopmental, and Other Neuropsychiatric Disorders: Systematic Literature-Based Analysis. *Front. Neurol.* **2020**, *11*, 32.
24. Maruszak, A.; Pilarski, A.; Murphy, T.; Branch, N.; Thuret, S. Hippocampal neurogenesis in Alzheimer's disease: Is there a role for dietary modulation? *Alzheimers Dis.* **2014**, *38*, 11–38. [CrossRef]

25. Ping, H.; Matthias, S.; Rena, L.; Yong, S. Deficiency of patched 1-induced Gli1 signal transduction results in astrogenesis in Swedish mutated APP transgenic mice. *Hum. Mol. Genet.* **2014**, *23*, 6512–6527.
26. Gamze, A.; Negin, R.; Pelin, Y.; Gizem, K.; Kahraman, D.C.; Saatci, Ö.; Şahin, Ö.; Çetin-Atalay, R.; Muyan, M. CXXC5 as an unmethylated CpG dinucleotide binding protein contributes to estrogen-mediated cellular proliferation. *Sci. Rep.* **2020**, *10*, 5971.
27. Young, C.B.; Yeol, K.I.; Hee, K.J.; Eun, L.B.; Hee, L.S.; Ra, K.A.; Sohn, M.; Won, S.S. Zinc plus cyclo-(His-Pro) promotes hippocampal neurogenesis in rats. *Neuroscience* **2016**, *339*, 634–643.
28. Choi, B.Y.; Kim, I.Y.; Kim, J.H.; Lee, B.E.; Lee, S.H.; Kho, A.R.; Sohn, M.; Suh, S.W. Administration of Zinc plus Cyclo-(His-Pro) Increases Hippocampal Neurogenesis in Rats during the Early Phase of Streptozotocin-Induced Diabetes. *Int. J. Mol. Sci.* **2017**, *18*, 73. [CrossRef]
29. Choi, B.Y.; Kim, J.H.; Kim, H.J.; Lee, B.E.; Kim, I.Y.; Sohn, M.; Suh, S.W. Zinc chelation reduces traumatic brain injury-induced neurogenesis in the subgranular zone of the hippocampal dentate gyrus. *J. Trace Elem. Med. Biol.* **2014**, *28*, 474–481. [CrossRef]
30. Lee, B.E.; Choi, B.Y.; Hong, D.K.; Kim, J.H.; Lee, S.H.; Kho, A.R.; Kim, H.; Choi, H.C.; Suh, S.W. The cancer chemotherapeutic agent paclitaxel (Taxol) reduces hippocampal neurogenesis via down-regulation of vesicular zinc. *Sci. Rep.* **2017**, *7*, 11667. [CrossRef]
31. Nam, S.M.; Kim, J.W.; Kwon, H.J.; Yoo, D.Y.; Jung, H.Y.; Kim, D.W.; Hwang, I.K.; Seong, J.K.; Yoon, Y.S. Differential Effects of Low- and High-dose Zinc Supplementation on Synaptic Plasticity and Neurogenesis in the Hippocampus of Control and High-fat Diet-fed Mice. *Neurochem. Res.* **2017**, *42*, 3149–3159. [CrossRef] [PubMed]
32. Choi, B.Y.; Hong, D.K.; Jeong, J.H.; Lee, B.E.; Koh, J.Y.; Suh, S.W. Zinc transporter 3 modulates cell proliferation and neuronal differentiation in the adult hippocampus. *Stem Cells* **2020**, *38*, 994–1006. [CrossRef] [PubMed]
33. George, J. Copper excess, zinc deficiency, and cognition loss in Alzheimer's disease. *Biofactors* **2012**, *38*, 107–113.
34. Prasad Ananda, S. Zinc: An antioxidant and anti-inflammatory agent: Role of zinc in degenerative disorders of aging. *J. Trace Elem. Med. Biol.* **2014**, *28*, 364–371. [CrossRef] [PubMed]
35. Liu, W.; Fan, Z.; Gao, F.; Ou, L.; Li, M.; Zhou, X.; Luo, W.; Wei, P.; Miao, F. Emodin inhibits zinc-induced neurotoxicity in neuroblastoma SH-SY5Y cells. *Biosci. Rep.* **2019**, *39*, BSR20182378. [CrossRef]
36. González-Mundo, I.; Pérez-Vielma, N.M.; Gómez-López, M.; Fleury, A.; Correa-Basurto, J.; Rosales-Hernández, M.C.; Sixto-López, Y.; Martínez-Godinez, M.L.Á.; Domínguez-López, A.; Miliar-García, A. DNA methylation of the RE-1 silencing transcription factor in peripheral blood mononuclear cells and gene expression of antioxidant enzyme in patients with late-onset Alzheimer disease. *Exp. Gerontol.* **2020**, *136*, 110951. [CrossRef]
37. Datki, Z.; Galik-Olah, Z.; Janosi-Mozes, E.; Szegedi, V.; Kalman, J.; Hunya, Á.G.; Fulop, L.; Tamano, H.; Takeda, A.; Adlard, P.A.; et al. Alzheimer risk factors age and female sex induce cortical Aβ aggregation by raising extracellular zinc. *Mol. Psychiatry* **2020**, *25*, 2728–2741. [CrossRef]
38. Zhang, C.; Gao, Z.; Hu, C.; Zhang, J.; Sun, X.; Rong, C.; Jia, L. Antioxidant, antibacterial and anti-aging activities of intracellular zinc polysaccharides from Grifola frondosa SH-05. *Int. J. Biol. Macromol.* **2017**, *95*, 778–787. [CrossRef]
39. Aydemir, T.B.; Troche, C.; Kim, J.; Kim, M.H.; Teran, O.Y.; Leeuwenburgh, C.; Cousins, R.J. Aging amplifies multiple phenotypic defects in mice with zinc transporter Zip14 (Slc39a14) deletion. *Exp. Gerontol.* **2016**, *85*, 88–94. [CrossRef]
40. Wong, C.P.; Magnusson, K.R.; Sharpton, T.J.; Ho, E. Effects of zinc status on age-related T cell dysfunction and chronic inflammation. *Biometals* **2021**, *34*, 291–301. [CrossRef]
41. Hyun-Seo, P.; Heui, Y.M.; Jae-Young, K. Role of zinc dyshomeostasis in inflammasome formation in cultured cortical cells following lipopolysaccharide or oxygen-glucose deprivation/reperfusion exposure. *Neurobiol. Dis.* **2020**, *137*, 104771.
42. Yin, X.; Zhang, Y.; Wen, Y.; Yang, Y.; Chen, H. Celecoxib alleviates zinc deficiency-promoted colon tumorigenesis through suppressing inflammation. *Aging* **2021**, *13*, 8320–8334. [CrossRef] [PubMed]
43. Xie, Y.; Wen, M.; Zhao, H.; Liu, G.; Chen, X.; Tian, G.; Cai, J.; Jia, G. Effect of zinc supplementation on growth performance, intestinal development, and intestinal barrier function in Pekin ducks with lipopolysaccharide challenge. *Poult. Sci.* **2021**, *100*, 101462. [CrossRef] [PubMed]
44. Thonberg, H.; Chiang, H.H.; Lilius, L.; Forsell, C.; Lindström, A.K.; Johansson, C.; Björkström, J.; Thordardottir, S.; Sleegers, K.; Van Broeckhoven, C.; et al. Identification and description of three families with familial Alzheimer disease that segregate variants in the SORL1 gene. *Acta Neuropathol. Commun.* **2017**, *5*, 43. [CrossRef]
45. Jagust, W. Imaging the evolution and pathophysiology of Alzheimer disease. *Nat. Rev. Neurosci.* **2018**, *19*, 687–700. [CrossRef]
46. Jung, S.J.; Jung, E.S.; Ha, K.C.; Baek, H.I.; Park, Y.K.; Han, S.K.; Chae, S.W.; Lee, S.O.; Chung, Y.C. Efficacy and Safety of Sesame Oil Cake Extract on Memory Function Improvement: A 12-Week, Randomized, Double-Blind, Placebo-Controlled Pilot Study. *Nutrients* **2021**, *13*, 2606. [CrossRef]
47. Tofiq, A.; Zetterberg, H.; Blennow, K.; Basun, H.; Cederholm, T.; Eriksdotter, M.; Faxén-Irving, G.; Hjorth, E.; Jernerén, F.; Schultzberg, M.; et al. Effects of Peroral Omega-3 Fatty Acid Supplementation on Cerebrospinal Fluid Biomarkers in Patients with Alzheimer's Disease: A Randomized Controlled Trial-The OmegAD Study. *J. Alzheimers Dis.* **2021**, *83*, 1291–1301. [CrossRef]
48. Babić, L.M.; Jurasović, J.; Nikolac Perković, M.; Španić, E.; Sekovanić, A.; Orct, T.; Lukinović, Š.V.; Bačić, B.K.; Kiđemet-Piskač, S.; Vogrinc, Ž.; et al. The Association of Essential Metals with APOE Genotype in Alzheimer's Disease. *J. Alzheimers Dis.* **2021**, *82*, 661–672. [CrossRef]

49. Donnelly, P.S.; Caragounis, A.; Du, T.; Laughton, K.M.; Volitakis, I.; Cherny, R.A.; Sharples, R.A.; Hill, A.F.; Li, Q.X.; Masters, C.L.; et al. Selective intracellular release of copper and zinc ions from bis(thiosemicarbazonato) complexes reduces levels of Alzheimer disease amyloid-beta peptide. *J. Biol. Chem.* **2008**, *283*, 4568–4577. [CrossRef]
50. Nara, P.L.; Sindelar, D.; Penn, M.S.; Potempa, J.; Griffin, W.S.T. Porphyromonas gingivalis Outer Membrane Vesicles as the Major Driver of and Explanation for Neuropathogenesis, the Cholinergic Hypothesis, Iron Dyshomeostasis, and Salivary Lactoferrin in Alzheimer's Disease. *J. Alzheimers Dis.* **2021**, *82*, 1417–1450. [CrossRef]
51. Scheffer, S.; Hermkens, D.A.M.; van der Weerd, L.; Vries, H.E.; Daemen, M.J.A.P. Vascular Hypothesis of Alzheimer Disease: Topical Review of Mouse Models. *Arterioscler. Thromb. Vasc. Biol.* **2021**, *41*, 1265–1283. [CrossRef] [PubMed]
52. Tcw, J.; Goate, A.M. Genetics of β-Amyloid Precursor Protein in Alzheimer's Disease. *Cold Spring Harb. Perspect. Med.* **2017**, *7*, a024539. [CrossRef] [PubMed]
53. Bailey, D.K.; Kosman, D.J. Is brain iron trafficking part of the physiology of the amyloid precursor protein? *J. Biol. Inorg. Chem.* **2019**, *24*, 1171–1177. [CrossRef] [PubMed]
54. Feng, C.; Liu, Y.; Yuan, Y.; Cui, W.; Zheng, F.; Ma, Y.; Piao, M. Isoflurane anesthesia exacerbates learning and memory impairment in zinc-deficient APP/PS1 transgenic mice. *Neuropharmacology* **2016**, *111*, 119–129. [CrossRef] [PubMed]
55. Gerber, H.; Wu, F.; Dimitrov, M.; Garcia, O.G.M.; Fraering, P.C. Zinc and Copper Differentially Modulate Amyloid Precursor Protein Processing by γ-Secretase and Amyloid-β Peptide Production. *J. Biol. Chem.* **2017**, *292*, 3751–3767. [CrossRef] [PubMed]
56. August, A.; Schmidt, N.; Klingler, J.; Baumkotter, F.; Lechner, M.; Klement, J.; Eggert, S.; Vargas, C.; Wild, K.; Keller, S.; et al. Copper and zinc ions govern the transdirected dimerization of APP family members in multiple ways. *J. Neurochem.* **2019**, *151*, 1471–4159. [CrossRef]
57. Liang, Y.; Raven, F.; Ward, J.F.; Zhen, S.; Zhang, S.; Sun, H.; Miller, S.J.; Choi, S.H.; Tanzi, R.E.; Zhang, C. Upregulation of Alzheimer's Disease Amyloid-β Protein Precursor in Astrocytes Both in vitro and in vivo. *J. Alzheimers Dis.* **2020**, *76*, 1071–1082. [CrossRef]
58. Gotoh, N.; Saito, Y.; Hata, S.; Saito, H.; Ojima, D.; Murayama, C.; Shigeta, M.; Abe, T.; Konno, D.; Matsuzaki, F.; et al. Amyloidogenic processing of amyloid β protein precursor (APP) is enhanced in the brains of alcadein α-deficient mice. *J. Biol. Chem.* **2020**, *295*, 9650–9662. [CrossRef]
59. Fu, C.X.; Dai, L.; Yuan, X.Y.; Xu, Y.J. Effects of Fish Oil Combined with Selenium and Zinc on Learning and Memory Impairment in Aging Mice and Amyloid Precursor Protein Processing. *Biol. Trace Elem. Res.* **2021**, *199*, 1855–1863. [CrossRef]
60. Yoshida, K.; Gi, M.; Fujioka, M.; Teramoto, I.; Wanibuchi, H. Long-term administration of excess zinc impairs learning and memory in aged mice. *J. Toxicol. Sci.* **2019**, *44*, 681–691. [CrossRef]
61. Boström, F.; Hansson, O.O.; Gerhardsson, L.; Lundh, T.; Minthon, L.; Stomrud, E.; Zetterberg, H.; Londos, E. CSF Mg and Ca as diagnostic markers for dementia with Lewy bodies. *Neurobiol. Aging* **2009**, *30*, 1265–1271. [CrossRef] [PubMed]
62. Sensi, S.L.; Granzotto, A.; Siotto, M.; Squitti, R. Copper and Zinc Dysregulation in Alzheimer's Disease. *Trends Pharmacol. Sci.* **2018**, *39*, 1049–1063. [CrossRef] [PubMed]
63. Adlard, P.A.; Bush, A.I. Metals and Alzheimer's Disease: How Far Have We Come in the Clinic? *J. Alzheimers Dis.* **2018**, *62*, 1369–1379. [CrossRef] [PubMed]
64. Xie, Z.; Wu, H.; Zhao, J. Multifunctional roles of zinc in Alzheimer's disease. *Neurotoxicology* **2020**, *80*, 112–123. [CrossRef] [PubMed]
65. Religa, D.; Strozyk, D.; Cherny, R.A.; Volitakis, I.; Haroutunian, V.; Winblad, B.; Naslund, J.; Bush, A.I. Elevated cortical zinc in Alzheimer disease. *Neurology* **2006**, *67*, 69–75. [CrossRef]
66. Narayanan, S.E.; Rehuman, N.A.; Harilal, S.; Vincent, A.; Rajamma, R.G.; Behl, T.; Uddin, M.S.; Ashraf, G.M.; Mathew, B. Molecular mechanism of zinc neurotoxicity in Alzheimer's disease. *Environ. Sci. Pollut. Res. Int.* **2020**, *27*, 43542–43552. [CrossRef]
67. Chew, E.Y.; Clemons, T.E.; Agrón, E.; Launer Lenore, J.; Grodstein, F.; Bernstein, P.S.; Age-Related Eye Disease Study 2 (AREDS2) Research Group. Effect of Omega-3 Fatty Acids, Lutein/Zeaxanthin, or Other Nutrient Supplementation on Cognitive Function: The AREDS2 Randomized Clinical Trial. *JAMA* **2015**, *314*, 791–801. [CrossRef]
68. Seo, B.R.; Lee, S.J.; Cho, K.S.; Yoon, Y.H.; Koh, J.Y. The zinc ionophore clioquinol reverses autophagy arrest in chloroquine-treated ARPE-19 cells and in APP/mutant presenilin-1-transfected Chinese hamster ovary cells. *Neurobiol. Aging* **2015**, *36*, 3228–3238. [CrossRef]
69. Cristóvão, J.S.; Figueira, A.J.; Carapeto, A.P.; Rodrigues, M.S.; Cardoso, I.; Gomes, C.M. The S100B Alarmin Is a Dual-Function Chaperone Suppressing Amyloid-β Oligomerization through Combined Zinc Chelation and Inhibition of Protein Aggregation. *ACS Chem. Neurosci.* **2020**, *11*, 2753–2760. [CrossRef]
70. Hu, J.Y.; Zhang, D.L.; Liu, X.L.; Li, X.S.; Cheng, X.Q.; Chen, J.; Du, H.N.; Liang, Y. Pathological concentration of zinc dramatically accelerates abnormal aggregation of full-length human Tau and thereby significantly increases Tau toxicity in neuronal cells. *Biochim. Biophys Acta Mol. Basis Dis.* **2017**, *1863*, 414–427. [CrossRef]
71. Yue, C.; Shan, Z.; Tan, Y.; Yao, C.; Liu, Y.; Liu, Q. His-Rich Domain of Selenoprotein P Ameliorates Neuropathology and Cognitive Deficits by Regulating TrkB Pathway and Zinc Homeostasis in an Alzheimer Model of Mice. *ACS Chem. Neurosci.* **2020**, *11*, 4098–4110. [CrossRef] [PubMed]
72. Lee, J.Y.; Cho, E.; Seo, J.W.; Hwang, J.J.; Koh, J.Y. Alteration of the cerebral zinc pool in a mouse model of Alzheimer disease. *J. Neuropathol. Exp. Neurol.* **2012**, *71*, 211–222. [CrossRef] [PubMed]

73. Hara, T.; Yoshigai, E.; Ohashi, T.; Fukada, T. Zinc transporters as potential therapeutic targets: An updated review. *J Pharmacol. Sci.* **2022**, *148*, 221–228. [CrossRef] [PubMed]
74. Vilella, A.; Belletti, D.; Sauer, A.K.; Hagmeyer, S.; Sarowar, T.; Masoni, M.; Stasiak, N.; Mulvihill, J.J.E.; Ruozi, B.; Forni, F.; et al. Reduced plaque size and inflammation in the APP23 mouse model for Alzheimer's disease after chronic application of polymeric nanoparticles for CNS targeted zinc delivery. *J. Trace Elem. Med. Biol.* **2018**, *49*, 210–221. [CrossRef]
75. Blanco-Alvarez, V.M.; Soto-Rodriguez, G.; Gonzalez-Barrios, J.A.; Martinez-Fong, D.; Brambila, E.; Torres-Soto, M.; Aguilar-Peralta, A.K.; Gonzalez-Vazquez, A.; Tomás-Sanchez, C.; Limón, I.D.; et al. Prophylactic Subacute Administration of Zinc Increases CCL2, CCR2, FGF2, and IGF-1 Expression and Prevents the Long-Term Memory Loss in a Rat Model of Cerebral Hypoxia-Ischemia. *Neural. Plast.* **2015**, *2015*, 375391. [CrossRef]
76. Giacconi, R.; Costarelli, L.; Malavolta, M.; Cardelli, M.; Galeazzi, R.; Piacenza, F.; Gasparini, N.; Basso, A.; Mariani, E.; Fulop, T.; et al. Effect of ZIP2 Gln/Arg/Leu (rs2234632) polymorphism on zinc homeostasis and inflammatory response following zinc supplementation. *Biofactors* **2015**, *41*, 414–423. [CrossRef]
77. Takeda, A.; Tamano, H. Cognitive decline due to excess synaptic Zn(2+) signaling in the hippocampus. *Front. Aging Neurosci.* **2014**, *6*, 26. [CrossRef]
78. Cao, Q.; Tan, C.C.; Xu, W.; Hu, H.; Cao, X.P.; Dong, Q.; Tan, L.; Yu, J.T. The Prevalence of Dementia: A Systematic Review and Meta-Analysis. *J. Alzheimers Dis.* **2020**, *73*, 1157–1166. [CrossRef]
79. Babusikova, E.; Dobrota, D.; Turner, A.J.; Hu, H.; Cao, X.P.; Dong, Q.; Tan, L.; Yu, J.T. Effect of Global Brain Ischemia on Amyloid Precursor Protein Metabolism and Expression of Amyloid-Degrading Enzymes in Rat Cortex: Role in Pathogenesis of Alzheimer's Disease. *Biochemistry* **2021**, *86*, 680–692. [CrossRef]
80. Shetty, M.S.; Sharma, M.; Sajikumar, S. Chelation of hippocampal zinc enhances long-term potentiation and synaptic tagging/capture in CA1 pyramidal neurons of aged rats: Implications to aging and memory. *Aging Cell* **2017**, *16*, 136–148. [CrossRef]
81. Higashi, Y.; Aratake, T.; Shimizu, T.; Shimizu, S.; Saito, M. Protective Role of Glutathione in the Hippocampus after Brain Ischemia. *Int. J. Mol. Sci.* **2021**, *22*, 7765. [CrossRef] [PubMed]
82. Kim, Y.; Kim, Y.-J. Effect of Obesity on Cognitive Impairment in Vascular Dementia Rat Model via BDNF-ERK-CREB Pathway. *Biol. Res. Nurs.* **2021**, *23*, 248–257. [CrossRef] [PubMed]
83. Mur, J.; McCartney, D.L.; Chasman, D.I.; Visscher, P.M.; Muniz-Terrera, G.; Cox, S.R.; Russ, T.C.; Marioni, R.E. Variation in VKORC1 Is Associated with Vascular Dementia. *J. Alzheimer's Dis.* **2021**, *80*, 1329–1337. [CrossRef] [PubMed]
84. Papanastasiou, C.A.; Theochari, C.A.; Zareifopoulos, N.; Arfaras-Melainis, A.; Giannakoulas, G.; Karamitsos, T.D.; Palaiodimos, L.; Ntaios, G.; Avgerinos, K.I.; Kapogiannis, D.; et al. Atrial Fibrillation Is Associated with Cognitive Impairment, All-Cause Dementia, Vascular Dementia, and Alzheimer's Disease: A Systematic Review and Meta-Analysis. *J. Gen. Intern. Med.* **2021**, *36*, 3122–3135. [CrossRef]
85. Shams, M.; Martola, J.; Charidimou, A.; Granberg, T.; Ferreira, D.; Westman, E.; Wintermark, M.; Iv, M.; Larvie, M.; Kristoffersen, W.M.; et al. Cerebrospinal Fluid Metals and the Association with Cerebral Small Vessel Disease. *J. Alzheimers Dis.* **2020**, *78*, 1229–1236. [CrossRef]
86. Kawahara, M.; Tanaka, K.-I.; Kato-Negishi, M. Copper as a Collaborative Partner of Zinc-Induced Neurotoxicity in the Pathogenesis of Vascular Dementia. *Int. J. Mol. Sci.* **2021**, *22*, 7242. [CrossRef]
87. Tanaka, K.I.; Kawahara, M. Copper Enhances Zinc-Induced Neurotoxicity and the Endoplasmic Reticulum Stress Response in a Neuronal Model of Vascular Dementia. *Front. Neurosci.* **2017**, *11*, 58. [CrossRef]
88. Hu, Z.; Hu, K.; Wang, R.; Gu, Y.; Ouyang, W.; Zhou, J.; Wen, Y. Differentially expressed genes accompanying neurobehavioral deficits in a modified rat model of vascular dementia. *Neurosci. Lett.* **2021**, *750*, 135774. [CrossRef]
89. Choi, H.; Choi, N.Y.; Park, H.H.; Lee, K.Y.; Lee, Y.J.; Koh, S.H. Sublethal Doses of Zinc Protect Rat Neural Stem Cells Against Hypoxia Through Activation of the PI3K Pathway. *Stem. Cells Dev.* **2019**, *28*, 769–780. [CrossRef]
90. Kawahara, M.; Sadakane, Y.; Mizuno, K.; Kato-Negishi, M.; Tanaka, K.I. Carnosine as a Possible Drug for Zinc-Induced Neurotoxicity and Vascular Dementia. *Int. J. Mol. Sci.* **2020**, *21*, 2570. [CrossRef]
91. Cui, S.Y.; Song, J.Z.; Cui, X.Y.; Hu, X.; Ma, Y.N.; Shi, Y.T.; Luo, Y.; Ge, Y.R.; Ding, H.; Ye, H.; et al. Intracerebroventricular streptozotocin-induced Alzheimer's disease-like sleep disorders in rats: Role of the GABAergic system in the parabrachial complex. *CNS Neurosci. Ther.* **2018**, *24*, 1241–1252. [CrossRef] [PubMed]
92. Gao, Y.; Wei, S.; Gao, F.; Gao, L.; Dang, L.; Shang, S.; Chen, C.; Huo, K.; Wang, J.; Wang, J.; et al. Sleep Disturbance is Associated with Higher Plasma Aβ Levels in Cognitively Normal Adults-A Population-Based Cross-Sectional Study. *Front Aging Neurosci.* **2020**, *12*, 615838. [CrossRef] [PubMed]
93. Zhao, B.; Liu, P.; Wei, M.; Li, Y.; Liu, J.; Ma, L.; Shang, S.; Jiang, Y.; Huo, K.; Wang, J.; et al. Chronic Sleep Restriction Induces Aβ Accumulation by Disrupting the Balance of Aβ Production and Clearance in Rats. *Neurochem. Res.* **2019**, *44*, 859–873. [CrossRef] [PubMed]
94. Rondanelli, M.; Opizzi, A.; Monteferrario, F.; Antoniello, N.; Manni, R.; Klersy, C. The effect of melatonin, magnesium, and zinc on primary insomnia in long-term care facility residents in Italy: A double-blind, placebo-controlled clinical trial. *J. Am. Geriatr. Soc.* **2011**, *59*, 82–90. [CrossRef] [PubMed]
95. Samad, N.; Yasmin, F.; Naheed, S.; Bari, A.Z.; Ayaz, M.M.; Zaman, A.; Akbar, A.; Saeed, S.; Ullah, N. Serum levels of leptin, zinc and tryptophan in obese subjects with sleep deficits. *Pak. J. Pharm. Sci.* **2017**, *30*, 1431–1438.

96. Villa, C.; Ferini-Strambi, L.; Combi, R. The Synergistic Relationship between Alzheimer's Disease and Sleep Disorders: An Update. *J. Alzheimers Dis.* **2015**, *46*, 571–580. [CrossRef]
97. Leng, Y.; Blackwell, T.; Stone, K.L.; Hoang, T.D.; Redline, S.; Yaffe, K. Periodic Limb Movements in Sleep are Associated with Greater Cognitive Decline in Older Men without Dementia. *Sleep* **2016**, *39*, 1807–1810. [CrossRef]
98. Yaffe, K.; Laffan, A.M.; Harrison, S.L.; Redline, S.; Spira, A.P.; Ensrud, K.E.; Ancoli-Israel, S.; Stone, K.L. Sleep-disordered breathing, hypoxia, and risk of mild cognitive impairment and dementia in older women. *JAMA* **2011**, *306*, 613–619. [CrossRef]
99. Chen, T.-I.; Chen, M.; Yu, C. Zinc Is Indispensable in Exercise-Induced Cardioprotection against Intermittent Hypoxia-Induced Left Ventricular Function Impairment in Rats. *PLoS ONE* **2016**, *11*, e0168600. [CrossRef]
100. Levenson, C.W.; Tassabehji, N.M. Role and Regulation of Copper and Zinc Transport Proteins in the Central Nervous System. In *Handbook of Neurochemistry and Molecular Neurobiology*; Springer: New York, NY, USA, 2007; pp. 257–284.
101. Assaf, S.Y.; Shin, H.C. Release of endogenous Zn2+ from brain tissue during activity. *Nature* **1984**, *308*, 734–736. [CrossRef]
102. Frederickson, C.J.; Koh, J.-Y.; Bush, A.I. The neurobiology of zinc in health and disease. *Nat. Rev. Neurosci.* **2005**, *6*, 449–462. [CrossRef] [PubMed]
103. Kambe, T.; Tsuji, T.; Hashimoto, A.; Itsumura, N. The Physiological, Biochemical, and Molecular Roles of Zinc Transporters in Zinc Homeostasis and Metabolism. *Physiol. Rev.* **2014**, *95*, 749–784. [CrossRef] [PubMed]
104. Lovell, M.A.; Smith, J.L.; Xiong, S.; Markesbery, W.R. Alterations in zinc transporter protein-1 (ZnT-1) in the brain of subjects with mild cognitive impairment, early, and late-stage alzheimer's disease. *Neurotox. Res.* **2004**, *7*, 265–271. [CrossRef] [PubMed]
105. Cotrim, C.A.; Russell, J.; Jarrott, J.; Martin, L.; Drew, D. A structural overview of the zinc transporters in the cation diffusion facilitator family. *Acta Crystallogr. Sect. D Struct. Biol.* **2019**, *75*, 357–367. [CrossRef]
106. Andrews, G.K.; Wang, H.; Dey, S.K.; Palmiter, R.D. Mousezinc transporter 1 gene provides an essential function during early embryonic development. *Genesis* **2003**, *40*, 74–81. [CrossRef]
107. Palmiter, R.D.; Findley, S.D. Cloning and functional characterization of a mammalian zinc transporter that confers resistance to zinc. *EMBO J.* **1995**, *14*, 639–649. [CrossRef]
108. Ohana, E.; Sekler, I.; Kaisman, T.; Kahn, N.; Cove, J.; Silverman, W.F.; Amsterdam, A.; Hershfinkel, M. Silencing of ZnT-1 expression enhances heavy metal influx and toxicity. *J. Mol. Med.* **2006**, *84*, 753–763. [CrossRef]
109. Nitzan, Y.B.; Sekler, I.; Hershfinkel, M.; Moran, A.; Silverman, W.F. Postnatal regulation of ZnT-1 expression in the mouse brain. *Dev. Brain Res.* **2002**, *137*, 149–157. [CrossRef]
110. Sekler, I.; Moran, A.; Hershfinkel, M.; Dori, A.; Margulis, A.; Birenzweig, N.; Nitzan, Y.; Silverman, W.F. Distribution of the zinc transporter ZnT-1 in comparison with chelatable zinc in the mouse brain. *J. Comp. Neurol.* **2002**, *447*, 201–209. [CrossRef]
111. Sindreu, C.; Bayés, À.; Altafaj, X.; Pérez-Clausell, J. Zinc transporter-1 concentrates at the postsynaptic density of hippocampal synapses. *Mol. Brain* **2013**, *7*, 16. [CrossRef]
112. Medvedeva, Y.V.; Lin, B.; Shuttleworth, C.W.; Weiss, J.H. Intracellular Zn2+ Accumulation Contributes to Synaptic Failure, Mitochondrial Depolarization, and Cell Death in an Acute Slice Oxygen-Glucose Deprivation Model of Ischemia. *J. Neurosci.* **2008**, *29*, 1105–1114. [CrossRef] [PubMed]
113. Nolte, C.; Gore, A.; Sekler, I.; Kresse, W.; Hershfinkel, M.; Hoffmann, A.; Kettenmann, H.; Moran, A. ZnT-1 expression in astroglial cells protects against zinc toxicity and slows the accumulation of intracellular zinc. *Glia* **2003**, *48*, 145–155. [CrossRef] [PubMed]
114. Lovell, M.A.; Smith, J.L.; Markesbery, W.R. Elevated Zinc Transporter-6 in Mild Cognitive Impairment, Alzheimer Disease, and Pick Disease. *J. Neuropathol. Exp. Neurol.* **2006**, *65*, 489–498. [CrossRef] [PubMed]
115. Baulac, S.; LaVoie, M.J.; Kimberly, W.T.; Strahle, J.; Wolfe, M.S.; Selkoe, D.J.; Xia, W. Functional γ-secretase complex assembly in Golgi/trans-Golgi network: Interactions among presenilin, nicastrin, Aph1, Pen-2, and γ-secretase substrates. *Neurobiol. Dis.* **2002**, *14*, 194–204. [CrossRef]
116. Siman, R.; Velji, J. Localization of presenilin-nicastrin complexes and γ-secretase activity to the trans-Golgi network. *J. Neurochem.* **2002**, *84*, 1143–1153. [CrossRef] [PubMed]
117. Bush, A.; Robert, D.M.; Kari, M.K.; Rudolph, E.T. *Response*: Zinc and Alzheimer's Disease. *Science* **1995**, *268*, 1921–1923. [CrossRef]
118. Upmanyu, N.; Jin, J.; von der Emde, H.; Ganzella, M.; Bösche, L.; Malviya, V.N.; Zhuleku, E.; Politi, A.Z.; Ninov, M.; Silbern, I.; et al. Colocalization of different neurotransmitter transporters on synaptic vesicles is sparse except for VGLUT1 and ZnT3. *Neuron* **2021**, *110*, 1483–1497. [CrossRef]
119. Minckley, T.F.; Zhang, C.; Fudge, D.H.; Dischler, A.M.; LeJeune, K.D.; Xu, H.; Qin, Y. Sub-nanomolar sensitive GZnP3 reveals TRPML1-mediated neuronal Zn2+ signals. *Nat. Commun.* **2019**, *10*, 4806. [CrossRef]
120. Sanford, L.; Carpenter, M.C.; Palmer, A.E. Intracellular Zn2+ transients modulate global gene expression in dissociated rat hippocampal neurons. *Sci. Rep.* **2019**, *9*, 9411. [CrossRef]
121. Kawahara, M.; Midori, K.-N.; Tanaka, K. Amyloids: Regulators of Metal Homeostasis in the Synapse. *Molecules* **2019**, *25*, 1441. [CrossRef]
122. Trist, B.G.; Hare, D.J.; Double, K.L. A Proposed Mechanism for Neurodegeneration in Movement Disorders Characterized by Metal Dyshomeostasis and Oxidative Stress. *Cell Chem. Biol.* **2018**, *25*, 807–816. [CrossRef] [PubMed]
123. Sindreu, C.; Palmiter, R.D.; Storm, D.R. Zinc transporter ZnT-3 regulates pre-synaptic Erk1/2 signaling and hippocampus-dependent memory. *Proc. Natl. Acad. Sci. USA* **2011**, *108*, 3366–3370. [CrossRef] [PubMed]

124. Gomez, J.L.; Bonaventura, J.; Keighron, J.; Wright, K.M.; Marable, D.L.; Rodriguez, L.A.; Lam, S.; Carlton, M.L.; Ellis, R.J.; Jordan, C.J.; et al. Synaptic Zn2+ potentiates the effects of cocaine on striatal dopamine neurotransmission and behavior. *Transl. Psychiatry* **2021**, *11*, 570. [CrossRef] [PubMed]
125. Adlard, P.A.; Parncutt, J.M.; Finkelstein, D.I.; Bush, A.I. Cognitive loss in zinctransporter-3 knock-out mice: A phenocopy for the synaptic and memory deficits of Alzheimer's disease? *J. Neurosci.* **2010**, *30*, 1631–1636. [CrossRef] [PubMed]
126. Olesen, R.H.; Hyde, T.M.; Kleinman, J.E.; Smidt, K.; Rungby, J.; Larsen, A. Obesity and age-related alterations in the gene expression of zinc-transporter proteins in the human brain. *Transl. Psychiatry* **2016**, *6*, e838. [CrossRef] [PubMed]
127. Zhang, L.H.; Wang, X.; Zheng, Z.H.; Ren, H.; Stoltenberg, M.; Danscher, G.; Huang, L.; Rong, M.; Wang, Z.Y. Altered expression and distribution of zinc transporters in APP/PS1 transgenic mouse brain. *Neurobiol. Aging* **2010**, *31*, 74–87. [CrossRef] [PubMed]
128. Colvin, R.A.; Fontaine, C.P.; Laskowski, M.; Thomas, D. Zn^{2+} transporters and Zn^{2+} homeostasis in neurons. *Eur. J. Pharmacol.* **2003**, *479*, 171–185. [CrossRef]
129. Hong, N.; Li, C.; Feng, X.; Cen, J.N. Effects of forced running exercise on cognitive function and its relation to zinc homeostasis-related gene expression in rat hippocampus. *Biol. Trace Elem. Res.* **2011**, *142*, 704–712.
130. Beyer, N.; Coulson, D.T.; Heggarty, S.; Ravid, R.; Hellemans, J.; Irvine, G.B.; Johnston, J.A. Zinc transporter mRNA levels in Alzheimer's disease postmortem brain. *J. Alzheimer's Dis. JAD* **2012**, *29*, 863–873. [CrossRef]
131. Takahashi, R.H.; Milner, T.A.; Li, F.; Nam, E.E.; Edgar, M.A.; Yamaguchi, H.; Beal, M.F.; Xu, H.; Greengard, P.; Gouras, G.K. Intraneuronal Alzheimer abeta42 accumulates in multivesicular bodies and is associated with synaptic pathology. *Am. J. Pathol.* **2002**, *161*, 1869–1879. [CrossRef]
132. Wirths, O.; Multhaup, G.; Czech, C.; Blanchard, V.; Moussaoui, S.; Tremp, G.; Pradier, L.; Beyreuther, K.; Bayer, T.A. Intraneuronal Abeta accumulation precedes plaque formation in beta-amyloid precursor protein and presenilin-1 double-transgenic mice. *Neurosci. Lett.* **2001**, *306*, 116–120. [CrossRef]
133. Shie, F.-S.; Leboeuf, R.C.; Jin, L.-W.; LeBoeur, R.C. Early intraneuronal Abeta deposition in the hippocampus of APP transgenic mice. *Neuroreport* **2003**, *14*, 123–129. [CrossRef] [PubMed]
134. Smith, J.L.; Xiong, S.; Markesbery, W.R.; Lovell, M.A. Altered expression of zinc transporters-4 and -6 in mild cognitive impairment, early and late Alzheimer's disease brain. *Neuroscience* **2006**, *140*, 879–888. [CrossRef]
135. Wang, P.; Pan, R.; Weaver, J.; Jia, M.; Yang, X.; Yang, T.; Liang, J.; Liu, K.J. MicroRNA-30a regulates acute cerebral ischemia-induced blood-brain barrier damage through ZnT4/zinc pathway. *J. Cereb. Blood Flow Metab. Off. J. Int. Soc. Cereb. Blood Flow Metab.* **2021**, *41*, 641–655. [CrossRef] [PubMed]
136. Ueda, S.; Manabe, Y.; Kubo, N.; Morino, N.; Yuasa, H.; Shiotsu, M.; Tsuji, T.; Sugawara, T.; Kambe, T. Early secretory pathway-resident Zn transporter proteins contribute to cellular sphingolipid metabolism through activation of sphingomyelin phosphodiesterase 1. *Am. J. Physiol. Cell Physiol.* **2022**, *322*, C948–C959. [CrossRef]
137. Huang, L.; Kirschke, C.P.; Gitschier, J. Functional characterization of a novel mammalian zinc transporter, ZnT6. *J. Biol. Chem.* **2002**, *277*, 26389–26395. [CrossRef]
138. Maret, W. Molecular aspects of zinc signals. In *Zinc Signals in Cellular Functions and Disorders*; Fukuda, T., Kambe, T., Eds.; Springer: Tokyo, Japan, 2014; pp. 7–26.
139. Tamano, H.; Koike, Y.; Nakada, H.; Shakushi, Y.; Takeda, A. Significance of synaptic Zn2+ signaling in zincergic and non-zincergic synapses in the hippocampus in cognition. *J. Trace. Elem. Med. Biol.* **2016**, *38*, 93–98. [CrossRef]
140. Vasak, M.; Meloni, G. Chemistry and biology of mammalian metallothioneins. *J. Biol. Inorg. Chem.* **2011**, *16*, 1067–1078. [CrossRef]
141. Hardyman, J.E.; Tyson, J.; Jackson, K.A.; Aldridge, C.; Cockell, S.J.; Wakeling, L.A.; Valentine, R.A.; Ford, D. Zinc sensing by metal-responsive transcription factor 1 (MTF1) controls metallothionein and ZnT1 expression to buffer the sensitivity of the transcriptome response to zinc. *Metallomics* **2016**, *8*, 337–343. [CrossRef]
142. Artells, E.; Palacios, O.; Capdevila, M.; Atrian, S. Mammalian MT1 and MT2 metallothioneins differ in their metal binding abilities. *Metallomics* **2013**, *5*, 1397–1410. [CrossRef]
143. Juárez-Rebollar, D.; Rios, C.; Nava-Ruíz, C.; Méndez-Armenta, M. Metallothionein in Brain Disorders. *Oxidative Med. Cell. Longev.* **2017**, *2017*, 5828056. [CrossRef] [PubMed]
144. Waller, R.; Murphy, M.; Garwood, C.J.; Jennings, L.; Heath, P.R.; Chambers, A.; Matthews, F.E.; Brayne, C.; Ince, P.G.; Wharton, S.B.; et al. Cognitive Function and Ageing Neuropathology Study Group. Metallothionein-I/II expression associates with the astrocyte DNA damage response and not Alzheimer-type pathology in the aging brain. *Glia* **2018**, *66*, 2316–2323. [CrossRef] [PubMed]
145. Pitt, B.R.; Schwarz, M.; Woo, E.S.; Yee, E.; Wasserloos, K.; Tran, S.; Lazo, J.S. Overexpression of metallothionein decreases sensitivity of pulmonary endothelial cells to oxidant injury. *Am. J. Physiol.* **1997**, *273*, L856. [CrossRef] [PubMed]
146. Qu, W.; Pi, J.; Waalkes, M.P. Metallothionein blocks oxidative DNA damage in vitro. *Arch. Toxicol.* **2013**, *87*, 311–321. [CrossRef] [PubMed]
147. Jong-Heon, K.; Nam, Y.P.; Jeon, S.M.; Han, H.S.; Suk, K. Amyloid neurotoxicity is attenuated by metallothionein: Dual mechanisms at work. *J. Neurochem.* **2012**, *121*, 751–762.
148. Hidalgo, J.; Penkowa, M.; Espejo, C.; Martínez-Cáceres, E.M.; Carrasco, J.; Quintana, A.; Molinero, A.; Florit, S.; Giralt, M.; Ortega-Aznar, A. Expression of metallothionein-I, -II, and -III in Alzheimer disease and animal models of neuroinflammation. *Exp. Biol. Med.* **2006**, *231*, 1450–1458. [CrossRef]

149. Valko, M.; Leibfritz, D.; Moncol, J.; Cronin, M.T.; Mazur, M.; Telser, J. Free radicals and antioxidants in normal physiological functions and human disease. *Int. J. Biochem. Cell Biol.* **2007**, *39*, 44–84. [CrossRef]
150. McAuliffe, J.J.; Joseph, B.; Hughes, E.; Miles, L.; Vorhees, C.V. Metallothionein I,II deficient mice do not exhibit significantly worse long-term behavioral outcomes following neonatal hypoxia-ischemia: MT-I,II deficient mice have inherent behavioral impairments. *Brain Res.* **2008**, *1190*, 175–185. [CrossRef]
151. Miyazaki, I.; Asanuma, M.; Hozumi, H.; Miyoshi, K.; Sogawa, N. Protective effects of metallothionein against dopamine quinone-induced dopaminergic neurotoxicity. *FEBS Lett.* **2007**, *581*, 5003–5008. [CrossRef]
152. Penkowa, M. Metallothioneins are multipurpose neuroprotectants during brain pathology. *FEBS J.* **2006**, *273*, 1857–1870. [CrossRef]
153. Uchida, Y.; Takio, K.; Titani, K.; Ihara, Y.; Tomonaga, M. The growth inhibitory factor that is deficient in the Alzheimer's disease brain is a 68 amino acid metallothionein-like protein. *Neuron* **1991**, *7*, 337–347. [CrossRef]
154. Tsuji, S.; Kobayashi, H.; Uchida, Y.; Ihara, Y.; Miyatake, T. Molecular cloning of human growth inhibitory factor cDNA and its down-regulation in Alzheimer's disease. *EMBO J.* **1992**, *11*, 4843–4850. [CrossRef] [PubMed]
155. Yu, W.H.; Lukiw, W.J.; Bergeron, C.; Niznik, H.B.; Fraser, P.E. Metallothionein III is reduced in Alzheimer's disease. *Brain Res.* **2001**, *894*, 37–45. [CrossRef]
156. Hidalgo, J. Metallothioneins and brain injury: What transgenic mice tell us. *Environ. Health Prev. Med.* **2004**, *9*, 87–94. [CrossRef]
157. Yoshifumi, I.; Keung, W.M. Anti-amyloid beta activity of metallothionein-III is different from its neuronal growth inhibitory activity: Structure-activity studies. *Brain Res.* **2003**, *960*, 228–234.
158. Miyazaki, I.; Sogawa, C.A.; Asanuma, M.; Higashi, Y.; Tanaka, K.I.; Nakanishi, T.; Ogawa, N. Expression of metallothionein-III mRNA and its regulation by levodopa in the basal ganglia of hemi-parkinsonian rats. *Neurosci. Lett.* **2000**, *293*, 65–68. [CrossRef]
159. Uchida, Y.; Gomi, F.; Masumizu, T.; Miura, Y. Growth inhibitory factor prevents neurite extension and the death of cortical neurons caused by high oxygen exposure through hydroxyl radical scavenging. *J. Biol. Chem.* **2002**, *277*, 32353–32359. [CrossRef]
160. Jen, J.; Wang, Y.C. Zinc finger proteins in cancer progression. *J. Biomed. Sci.* **2016**, *23*, 53. [CrossRef]
161. Wolfe, S.A.; Nekludova, L.; Pabo, C.O. DNA recognition by Cys2His2 zinc finger proteins. *Annu. Rev. Biophys. Biomol. Struct.* **2000**, *29*, 183–212. [CrossRef]
162. Lee, S.J.; Michel, S.L.J. Structural metal sites in nonclassical zinc finger proteins involved in transcriptional and translational regulation. *Acc. Chem. Res.* **2014**, *47*, 2643–2650. [CrossRef]
163. Michalek, J.L.; Besold, A.N.; Michel, S.L. Cysteine and histidine shuffling: Mixing and matching cysteine and histidine residues in zinc finger proteins to afford different folds and function. *Dalton Trans.* **2011**, *40*, 12619–12632. [CrossRef] [PubMed]
164. Blackshear, P.J. Tristetraprolin and other CCCH tandem zinc-finger proteins in the regulation of mRNA turnover. *Biochem. Soc. Trans.* **2002**, *30 Pt 6*, 945–952. [CrossRef] [PubMed]
165. Andreini, C.; Bertini, I.; Cavallaro, G. Minimal functional sites allow a classification of zinc sites in proteins. *PLoS ONE* **2011**, *6*, e26325. [CrossRef]
166. Decaria, L.; Bertini, I.; Williams, R.J. Zinc proteomes, phylogenetics and evolution. *Met. Integr. Biomet. Sci.* **2010**, *2*, 706–709. [CrossRef]
167. Brandis, J.E.P.; Zalesak, S.M.; Kane, M.A.; Michel, S. Cadmium Exchange with Zinc in the Non-Classical Zinc Finger Protein Tristetraprolin. *Inorg. Chem.* **2021**, *60*, 7697–7707. [CrossRef] [PubMed]
168. Laity, J.H.; Lee, B.M.; Wright, P.E. Zinc finger proteins: New insights into structural and functional diversity. *Curr. Opin. Struct. Biol.* **2001**, *11*, 39–46. [CrossRef]
169. Farmiloe, G.; Lodewijk, G.A.; Robben, S.F.; van Bree, E.J.; Jacobs, F.M.J. Widespread correlation of KRAB zinc finger protein binding with brain-developmental gene expression patterns. *Philos. Trans. R. Soc. Lond. B Biol.* **2020**, *375*, 20190333. [CrossRef] [PubMed]
170. Bu, S.; Lv, Y.; Liu, Y.; Qiao, S.; Wang, H. Zinc Finger Proteins in Neuro-Related Diseases Progression. *Front. Neurosci.* **2021**, *15*, 760567. [CrossRef]
171. Bosch, D.G.; Boonstra, F.N.; Gonzaga-Jauregui, C.; Xu, M.; de Ligt, J.; Jhangiani, S.; Wiszniewski, W.; Muzny, D.M.; Yntema, H.G.; Pfundt, R.; et al. NR2F1 mutations cause optic atrophy with intellectual disability. *Am. J. Hum. Genet.* **2014**, *94*, 303–309. [CrossRef]
172. Mitsui, K.; Matsumoto, A.; Ohtsuka, S.; Ohtsubo, M.; Yoshimura, A. Cloning and characterization of a novel p21(Cip1/Waf1)-interacting zinc finger protein, ciz1. *Biochem. Biophys. Res. Commun.* **1999**, *264*, 457–464. [CrossRef]
173. Coverley, D.; Marr, J.; Ainscough, J. Ciz1 promotes mammalian DNA replication. *J. Cell Sci.* **2005**, *118*, 101–112. [CrossRef] [PubMed]
174. Khan, M.M.; Xiao, J.; Patel, D.; LeDoux, M.S. DNA damage and neurodegenerative phenotypes in aged Ciz1 null mice. *Neurobiol. Aging.* **2018**, *62*, 180–190. [CrossRef] [PubMed]
175. Chen, J.; Cohen, M.L.; Lerner, A.J.; Yang, Y.; Herrup, K. DNA damage and cell cycle events implicate cerebellar dentate nucleus neurons as targets of Alzheimer's disease. *Mol. Neurodegener.* **2010**, *5*, 60. [CrossRef] [PubMed]
176. Silva, A.R.; Santos, A.C.; Farfel, J.M.; Grinberg, L.T.; Ferretti, R.E.; Campos, A.H.; Cunha, I.W.; Begnami, M.D.; Rocha, R.M.; Carraro, D.M.; et al. Repair of oxidative DNA damage, cell-cycle regulation and neuronal death may influence the clinical manifestation of Alzheimer's disease. *PLoS ONE* **2014**, *9*, e99897. [CrossRef]

177. Dahmcke, C.M.; Büchmann-Møller, S.; Jensen, N.A.; Mitchelmore, C. Altered splicing in exon 8 of the DNA replication factor CIZ1 affects subnuclear distribution and is associated with Alzheimer's disease. *Mol. Cell. Neurosci.* **2008**, *38*, 589–594. [CrossRef] [PubMed]
178. Avet-Rochex, A.; Carvajal, N.; Christoforou, C.P.; Yeung, K.; Maierbrugger, K.T.; Hobbs, C.; Lalli, G.; Cagin, U.; Plachot, C.; McNeill, H.; et al. Unkempt is negatively regulated by mTOR and uncouples neuronal differentiation from growth control. *PLoS Genet.* **2014**, *10*, e1004624. [CrossRef]
179. Bateman, J.M. Mechanistic insights into the role of mTOR signaling in neuronal differentiation. *Neurogenesis* **2015**, *2*, e1058684. [CrossRef]
180. Li, N.; Liu, Q.; Xiong, Y.; Yu, J. Headcase and Unkempt Regulate Tissue Growth and Cell Cycle Progression in Response to Nutrient Restriction. *Cell Rep.* **2019**, *26*, 733–747.e3. [CrossRef]
181. Maierbrugger, K.T.; Sousa-Nunes, R.; Bateman, J.M. The mTOR pathway component Unkempt regulates neural stem cell and neural progenitor cell cycle in the Drosophila central nervous system. *Dev. Biol.* **2020**, *461*, 55–65. [CrossRef]
182. Vinsland, E.; Baskaran, P.; Mihaylov, S.R.; Baskaran, P.; Mihaylov, S.R.; Hobbs, C.; Wood, H.; Bouybayoune, I.; Shah, K.; Houart, C.; et al. The zinc finger/RING domain protein Unkempt regulates cognitive flexibility. *Sci. Rep.* **2021**, *11*, 16299. [CrossRef]
183. Huber, K.M.; Klann, E.; Costa-Mattioli, M.; Zukin, R.S. Dysregulation of Mammalian Target of Rapamycin Signaling in Mouse Models of Autism. *J. Neurosci.* **2015**, *35*, 13836–13842. [CrossRef] [PubMed]
184. Murn, J.; Teplova, M.; Zarnack, K.; Shi, Y.; Patel, D.J. Recognition of distinct RNA motifs by the clustered CCCH zinc fingers of neuronal protein Unkempt. *Nat. Struct. Mol. Biol.* **2016**, *23*, 16–23. [CrossRef] [PubMed]
185. Halloran, J.; Hussong, S.A.; Burbank, R.; Podlutskaya, N.; Fischer, K.E.; Sloane, L.B.; Austad, S.N.; Strong, R.; Richardson, A.; Hart, M.J.; et al. Chronic inhibition of mammalian target of rapamycin by rapamycin modulates cognitive and non-cognitive components of behavior throughout lifespan in mice. *Neuroscience* **2012**, *223*, 102–113. [CrossRef] [PubMed]
186. Teng, L.L.; Lu, G.L.; Chiou, L.C.; Lin, W.S.; Cheng, Y.Y.; Hsueh, T.E.; Huang, Y.C.; Hwang, N.H.; Yeh, J.W.; Liao, R.M.; et al. Serotonin receptor HTR6-mediated mTORC1 signaling regulates dietary restriction-induced memory enhancement. *PLoS Biol.* **2019**, *17*, e2007097. [CrossRef] [PubMed]
187. Glatter, T.; Schittenhelm, R.B.; Rinner, O.; Roguska, K.; Wepf, A.; Jünger, M.A.; Köhler, K.; Jevtov, I.; Choi, H.; Schmidt, A.; et al. Modularity and hormone sensitivity of the Drosophila melanogaster insulin receptor/target of rapamycin interaction proteome. *Mol. Syst. Biol.* **2011**, *7*, 547. [CrossRef] [PubMed]
188. Baranzini, S.E.; Srinivasan, R.; Khankhanian, P.; Okuda, D.T.; Nelson, S.J.; Matthews, P.M.; Hauser, S.L.; Oksenberg, J.R.; Pelletier, D. Genetic variation influences glutamate concentrations in brains of patients with multiple sclerosis. *Brain* **2010**, *133*, 2603–2611. [CrossRef]
189. Deans, P.J.M.; Raval, P.; Sellers, K.J.; Gatford, N.; Halai, S.; Duarte, R.; Shum, C.; Warre-Cornish, K.; Kaplun, V.E.; Cocks, G.; et al. Psychosis Risk Candidate ZNF804A Localizes to Synapses and Regulates Neurite Formation and Dendritic Spine Structure. *Biol. Psychiatry* **2017**, *82*, 49–61. [CrossRef]
190. Hess, J.L.; Quinn, T.P.; Akbarian, S.; Glatt, S.J. Bioinformatic analyses and conceptual synthesis of evidence linking ZNF804A to risk for schizophrenia and bipolar disorder. *Am. J. Med. Genetics. Part B Neuropsychiatr. Genet. Off. Publ. Int. Soc. Psychiatr. Genet.* **2015**, *168B*, 14–35. [CrossRef]
191. Tao, R.; Cousijn, H.; Jaffe, A.E.; Burnet, P.W.; Edwards, F.; Eastwood, S.L.; Shin, J.H.; Lane, T.A.; Walker, M.A.; Maher, B.J.; et al. Expression of ZNF804A in human brain and alterations in schizophrenia, bipolar disorder, and major depressive disorder: A novel transcript fetally regulated by the psychosis risk variant rs1344706. *JAMA Psychiatry* **2014**, *71*, 1112–1120. [CrossRef]
192. Chang, E.H.; Kirtley, A.; Chandon, T.S.; Borger, P.; Husain-Krautter, S.; Vingtdeux, V.; Malhotra, A.K. Postnatal neurodevelopmental expression and glutamate-dependent regulation of the ZNF804A rodent homologue. *Schizophr. Res.* **2015**, *168*, 402–410. [CrossRef]
193. Zhou, Y.; Dong, F.; Lanz, T.A.; Reinhart, V.; Li, M.; Liu, L.; Zou, J.; Xi, H.S.; Mao, Y. Interactome analysis reveals ZNF804A, a schizophrenia risk gene, as a novel component of protein translational machinery critical for embryonic neurodevelopment. *Mol. Psychiatry* **2018**, *23*, 952–962. [CrossRef] [PubMed]
194. Kang, H.J.; Kawasawa, Y.I.; Cheng, F.; Zhu, Y.; Xu, X.; Li, M.; Sousa, A.M.; Pletikos, M.; Meyer, K.A.; Sedmak, G.; et al. Spatio-temporal transcriptome of the human brain. *Nature* **2011**, *478*, 483–489. [CrossRef] [PubMed]
195. Hill, M.J.; Jeffries, A.R.; Dobson, R.J.; Price, J.; Bray, N.J. Knockdown of the psychosis susceptibility gene ZNF804A alters expression of genes involved in cell adhesion. *Hum. Mol. Genet.* **2012**, *21*, 1018–1024. [CrossRef]
196. Zhou, Y.; Kaiser, T.; Monteiro, P.; Zhang, X.; Van der Goes, M.S.; Wang, D.; Barak, B.; Zeng, M.; Li, C.; Lu, C.; et al. Mice with Shank3 Mutations Associated with ASD and Schizophrenia Display Both Shared and Distinct Defects. *Neuron* **2016**, *89*, 147–162. [CrossRef] [PubMed]
197. Konopaske, G.T.; Lange, N.; Coyle, J.T.; Benes, F.M. Prefrontal cortical dendritic spine pathology in schizophrenia and bipolar disorder. *JAMA Psychiatry* **2014**, *71*, 1323–1331. [CrossRef]
198. Paternoster, V.; vanborg, M.; Edhager, A.V.; Rajkumar, A.P.; Eickhardt, E.A.; Pallesen, J.; Grove, J.; Qvist, P.; Fryland, T.; Wegener, G.; et al. Brain proteome changes in female Brd1+/- mice unmask dendritic spine pathology and show enrichment for schizophrenia risk. *Neurobiol. Dis.* **2019**, *124*, 479–488. [CrossRef]
199. van Spronsen, M.; Hoogenraad, C.C. Synapse pathology in psychiatric and neurologic disease. *Curr. Neurol. Neurosci. Rep.* **2010**, *10*, 207–214. [CrossRef]

200. Dong, F.; Mao, J.; Chen, M.; Yoon, J.; Mao, Y. Schizophrenia risk ZNF804A interacts with its associated proteins to modulate dendritic morphology and synaptic development. *Mol. Brain* **2021**, *14*, 12. [CrossRef]
201. Huang, Y.; Huang, J.; Zhou, Q.X.; Yang, C.X.; Yang, C.P.; Mei, W.Y.; Zhang, L.; Zhang, Q.; Hu, L.; Hu, Y.Q.; et al. ZFP804A mutant mice display sex-dependent schizophrenia-like behaviors. *Mol. Psychiatry* **2021**, *26*, 2514–2532. [CrossRef]
202. Becker, J.; Czamara, D.; Hoffmann, P.; Landerl, K.; Blomert, L.; Brandeis, D.; Vaessen, A.; Maurer, U.; Moll, K.; Ludwig, K.U.; et al. Evidence for the involvement of ZNF804A in cognitive processes of relevance to reading and spelling. *Transl. Psychiatry* **2012**, *2*, e136. [CrossRef]
203. Cocchella, A.; Malacarne, M.; Forzano, F.; Marciano, C.; Pierluigi, M.; Perroni, L.; Faravelli, F.; Di Maria, E. The refinement of the critical region for the 2q31.2q32.3 deletion syndrome indicates candidate genes for mental retardation and speech impairment. *Am. J. Med. Genet. Part B Neuropsychiatr. Genet. Off. Publ. Int. Soc. Psychiatr. Genet.* **2010**, *153B*, 1342–1346. [CrossRef] [PubMed]
204. Vellutino, F.R.; Fletcher, J.M.; Snowling, M.J.; Scanlon, D.M. Specific reading disability (dyslexia): What have we learned in the past four decades? *J. Child Psychol. Psychiatry Allied Discip.* **2004**, *45*, 2–40. [CrossRef] [PubMed]
205. Liuzzi, J.P.; Pazos, R. Interplay Between Autophagy and Zinc. *J. Trace Elem. Med. Biol.* **2020**, *62*, 126636. [CrossRef] [PubMed]
206. Ravikumar, B.; Sarkar, S.; Davies, J.E.; Futter, M.; Garcia-Arencibia, M.; Green-Thompson, Z.W.; Jimenez-Sanchez, M.; Korolchuk, V.I.; Lichtenberg, M.; Luo, S.; et al. Rubinsztein Regulation of mammalian autophagy in physiology and pathophysiology. *Physiol. Rev.* **2010**, *90*, 1383–1435. [CrossRef]
207. Truong-Tran, A.Q.; Carter, J.; Ruffin, R.E.; Zalewski, P.D. The role of zinc in caspase activation and apoptotic cell death. *Biometals* **2001**, *14*, 315–330. [CrossRef]
208. Wong, S.H.; Zhao, Y.; Schoene, N.W.; Han, C.T.; Shih, R.S.; Lei, K.Y. Zinc deficiency depresses p21 gene expression: Inhibition of cell cycle progression is independent of the decrease in p21 protein level in HepG2 cells. *Am. J. Physiol. Cell Physiol.* **2007**, *292*, C2175–C2184. [CrossRef]
209. Lee, S.J.; Koh, J.Y. Roles of zinc and metallothionein-3 in oxidative stress-induced lysosomal dysfunction, cell death, and autophagy in neurons and astrocytes. *Mol. Brain* **2010**, *3*, 30. [CrossRef]
210. Liuzzi, J.P.; Yoo, C. Role of zinc in the regulation of autophagy during ethanol exposure in human hepatoma cells. *Biol. Trace Elem. Res.* **2013**, *156*, 350–356. [CrossRef]
211. Hwang, J.J.; Kim, H.N.; Kim, J.; Cho, D.H.; Kim, M.J.; Kim, Y.S.; Kim, Y.; Park, S.J.; Koh, J.Y. Zinc(II) ion mediates tamoxifen-induced autophagy and cell death in MCF-7 breast cancer cell line. *Biometals* **2010**, *23*, 997–1013. [CrossRef]
212. Hung, H.H.; Huang, W.P.; Pan, C.Y. Dopamine- and zinc-induced autophagosome formation facilitates PC12 cell survival. *Cell Biol. Toxicol.* **2013**, *29*, 415–429. [CrossRef]
213. Sánchez, C.; Sánchez, M.; López-Jurado, E.; Planells, J.; Llopis, P. Aranda Assessment of iron and Zn intake and related biochemical parameters in an adult Mediterranean population from southern Spain: Influence of lifestyle factors. *J. Nutr. Biochem.* **2009**, *20*, 125–131. [CrossRef] [PubMed]
214. Salminen, A.; Kaarniranta, K. Regulation of the aging process by autophagy. *Trends Mol. Med.* **2009**, *15*, 217–224. [CrossRef] [PubMed]
215. Moloudizargari, M.; Asghari, M.H.; Ghobadi, E.; Fallah, M.; Rasouli, S.; Abdollahi, M. Autophagy, its mechanisms and regulation: Implications in neurodegenerative diseases. *Ageing Res. Rev.* **2017**, *40*, 64–74. [CrossRef]
216. Menzies, F.M.; Fleming, A.; Caricasole, A.; Bento, C.F.; Andrews, S.P.; Ashkenazi, A.; Füllgrabe, J.; Jackson, A.; Karabiyik, C.; Licitra, F.; et al. Rubinsztein Autophagy and neurodegeneration: Pathogenic mechanisms and therapeutic opportunities. *Neuron* **2017**, *93*, 1015–1034. [CrossRef] [PubMed]
217. Schafe, G.E.; Atkins, C.M.; Swank, M.W.; Bauer, E.P.; Sweatt, J.D.; LeDoux, J.E. Activation of ERK/MAP kinase in the amygdala is required for memory consolidation of pavlovian fear conditioning. *J. Neurosci. Off. J. Soc. Neurosci.* **2000**, *20*, 8177–8187. [CrossRef]
218. Perry, G.; Roder, H.; Nunomura, A.; Takeda, A.; Friedlich, A.L.; Zhu, X.; Raina, A.K.; Holbrook, N.; Siedlak, S.L.; Harris, P.L.; et al. Activation of neuronal extracellular receptor kinase (ERK) in Alzheimer disease links oxidative stress to abnormal phosphorylation. *Neuroreport* **1999**, *10*, 2411–2415. [CrossRef]
219. Pei, J.J.; Braak, H.; An, W.L.; Winblad, B.; Cowburn, R.F.; Iqbal, K.; Grundke-Iqbal, I. Up-regulation of mitogen-activated protein kinases ERK1/2 and MEK1/2 is associated with the progression of neurofibrillary degeneration in Alzheimer's disease. Brain research. *Mol. Brain Res.* **2002**, *109*, 45–55. [CrossRef]
220. Pei, J.J.; Gong, C.X.; An, W.L.; Winblad, B.; Cowburn, R.F.; Grundke-Iqbal, I.; Iqbal, K. Okadaic-acid-induced inhibition of protein phosphatase 2A produces activation of mitogen-activated protein kinases ERK1/2, MEK1/2, and p70 S6, similar to that in Alzheimer's disease. *Am. J. Pathol.* **2003**, *163*, 845–858. [CrossRef]
221. Bian, X.; Teng, T.; Zhao, H.; Qin, J.; Qiao, Z.; Sun, Y.; Liun, Z.; Xu, Z. Zinc prevents mitochondrial superoxide generation by inducing mitophagy in the setting of hypoxia/reoxygenation in cardiac cells. *Free Radic. Res.* **2018**, *52*, 80–91. [CrossRef]
222. Zhu, B.; Zhou, Y.; Xu, F.; Shuai, J.; Li, X.; Fang, W. Porcine circovirus type 2 induces autophagy via the AMPK/ERK/TSC2/mTOR signaling pathway in PK-15 cells. *J. Virol.* **2012**, *86*, 12003–12012. [CrossRef]
223. Wang, J.; Whiteman, M.W.; Lian, H.; Wang, G.; Singh, A.; Huang, D.; Denmark, T. A non-canonical MEK/ERK signaling pathway regulates autophagy via regulating Beclin 1. *J. Biol. Chem.* **2009**, *284*, 21412–21424. [CrossRef] [PubMed]
224. Nuttall, J.R.; Oteiza, P.I. Zinc and the ERK kinases in the developing brain. *Neurotox. Res.* **2012**, *21*, 128–141. [CrossRef] [PubMed]
225. Botti, J.; Djavaheri-Mergny, M.; Pilatte, Y.; Codogno, P. Autophagy signaling and the cogwheels of cancer. *Autophagy* **2006**, *2*, 67–73. [CrossRef] [PubMed]

226. Gonnella, R.; Guttieri, L.; Gilardini Montani, M.S.; Santarelli, R.; Bassetti, E.; D'Orazi, G.; Cirone, M. Zinc Supplementation Enhances the Pro-Death Function of UPR in Lymphoma Cells Exposed to Radiation. *Biology* **2022**, *11*, 132. [CrossRef]
227. Lichtlen, P.; Schaffner, W. The metal transcription factor MTF-1: Biological facts and medical implications. *Swiss Med. Wkly.* **2001**, *131*, 647–652.
228. Xiao, W.; Wang, J.; Wang, X.; Cai, S.; Guo, Y.; Ye, L.; Li, D.; Hu, A.; Jin, S.; Yuan, B.; et al. Therapeutic targeting of the USP2-E2F4 axis inhibits autophagic machinery essential for zinc homeostasis in cancer progression. *Autophagy* **2022**, 1–21. [CrossRef]
229. Ramnarine, T.J.S.; Grath, S.; Parsch, J. Natural variation in the transcriptional response of Drosophila melanogaster to oxidative stress. *G3* **2022**, *12*, jkab366. [CrossRef]
230. Actis, C.; Muzio, G.; Autelli, R. Autophagy Triggers Tamoxifen Resistance in Human Breast Cancer Cells by Preventing Drug-Induced Lysosomal Damage. *Cancers* **2021**, *13*, 1252. [CrossRef]
231. Cho, Y.H.; Lee, S.H.; Lee, S.J.; Kim, H.N.; Koh, J.Y. A role of metallothionein-3 in radiation-induced autophagy in glioma cells. *Sci. Rep.* **2020**, *10*, 2015. [CrossRef]
232. Sharif, R.; Thomas, P.; Zalewski, P.; Fenech, M. The role of zinc in genomic stability. *Mutat. Res.* **2012**, *733*, 111–112. [CrossRef]
233. Meiser, N.; Mench, N.; Hengesbach, M. RNA secondary structure dependence in METTL3-METTL14 mRNA methylation is modulated by the N-terminal domain of METTL3. *Biol. Chem.* **2020**, *402*, 89–98. [CrossRef] [PubMed]
234. Shi, H.; Wang, X.; Lu, Z.; Zhao, B.S.; Ma, H.; Hsu, P.J.; Liu, C.; He, C. YTHDF3 facilitates translation and decay of N6-methyladenosine-modified RNA. *Cell Res.* **2017**, *27*, 315–328. [CrossRef] [PubMed]
235. Li, A.; Chen, Y.S.; Ping, X.L.; Yang, X.; Xiao, W.; Yang, Y.; Sun, H.Y.; Zhu, Q.; Baidya, P.; Wang, X.; et al. Cytoplasmic m6A reader YTHDF3 promotes mRNA translation. *Cell Res.* **2017**, *27*, 444–447. [CrossRef] [PubMed]
236. Lin, Y.; Jin, X.; Nie, Q.; Chen, M.; Guo, W.; Chen, L.; Li, Y.; Chen, X.; Zhang, W.; Chen, H.; et al. YTHDF3 facilitates triple-negative breast cancer progression and metastasis by stabilizing ZEB1 mRNA in an m6A-dependent manner. *Ann. Transl. Med.* **2022**, *10*, 83. [CrossRef] [PubMed]
237. Zhang, M.; Bodi, Z.; Mackinnon, K.; Zhong, S.; Archer, N.; Mongan, N.P.; Simpson, G.G.; Fray, R.G. Two zinc finger proteins with functions in m6A writing interact with HAKAI. *Nat. Commun.* **2022**, *13*, 1127. [CrossRef]
238. Bodi, Z.; Zhong, S.; Mehra, S.; Song, J.; Graham, N.; Li, H.; May, S.; Fray, R.G. Adenosine Methylation in Arabidopsis mRNA is Associated with the 3' End and Reduced Levels Cause Developmental Defects. *Front. Plant Sci.* **2012**, *3*, 48. [CrossRef]
239. Zhong, S.; Li, H.; Bodi, Z.; Button, J.; Vespa, L.; Herzog, M.; Fray, R.G. MTA is an Arabidopsis messenger RNA adenosine methylase and interacts with a homolog of a sex-specific splicing factor. *Plant Cell* **2008**, *20*, 1278–1288. [CrossRef]
240. Růžička, K.; Zhang, M.; Campilho, A.; Bodi, Z.; Kashif, M.; Saleh, M.; Eeckhout, D.; El-Showk, S.; Li, H.; Zhong, S.; et al. Identification of factors required for m6A mRNA methylation in Arabidopsis reveals a role for the conserved E3 ubiquitin ligase HAKAI. *New Phytol.* **2017**, *215*, 157–172. [CrossRef]
241. Cuervo, A.M. Autophagy and aging: Keeping that old broom working. *Trends Genet. TIG* **2008**, *24*, 604–612. [CrossRef]
242. Zhou, X.; Li, Y.; Li, C. Autophagy plays a positive role in zinc-induced apoptosis in intestinal porcine epithelial cells. *Toxicol. Vitro* **2017**, *44*, 392–402. [CrossRef]
243. Park, M.H.; Lee, S.J.; Byun, H.R.; Kim, Y.; Oh, Y.J.; Koh, J.J. Hwang Clioquinol induces autophagy in cultured astrocytes and neurons by acting as a zinc ionophore. *Neurobiol. Dis.* **2011**, *42*, 242–251. [CrossRef] [PubMed]
244. Kim, K.W.; Speirs, C.K.; Jung, D.K.; Lu, B. The zinc ionophore PCI-5002 radiosensitizes non-small cell lung cancer cells by enhancing autophagic cell death. *J. Thorac. Oncol.* **2011**, *6*, 1542–1552. [CrossRef] [PubMed]
245. Cho, K.S.; Yoon, Y.H.; Choi, J.A.; Lee, S.J.; Koh, J.Y. Induction of autophagy and cell death by tamoxifen in cultured retinal pigment epithelial and photoreceptor cells. *Investig. Ophthalmol. Vis. Sci.* **2012**, *53*, 5344–5353. [CrossRef]
246. Lee, S.J.; Cho, K.S.; Koh, J.Y. Oxidative injury triggers autophagy in astrocytes: The role of endogenous zinc. *Glia* **2009**, *57*, 1351–1361. [CrossRef] [PubMed]
247. Wen, S.; Li, Y.; Shen, X.; Wang, Z.; Zhang, K.; Zhang, J.; Mei, X. Protective Effects of Zinc on Spinal Cord Injury. *J. Mol. Neurosci. MN* **2021**, *71*, 2433–2440. [CrossRef] [PubMed]
248. Juan, S.M.A.; Adlard, P.A. Ageing and Cognition. *Subcell. Biochem.* **2019**, *91*, 107–122.
249. Paoletti, P.; Vergnano, A.; Barbour, B.; Casado, M. Zinc at glutamatergic synapses. *Neuroscience* **2009**, *158*, 126–136. [CrossRef]
250. Lee, J.Y.; Cole, T.B.; Palmiter, R.D.; Suh, S.W.; Koh, J.Y. Contribution by synaptic zinc to the gender-disparate plaque formation in human Swedish mutant APP transgenic mice. *Proc. Natl. Acad. Sci. USA* **2002**, *99*, 7705–7710. [CrossRef]
251. Lee, J.Y.; Friedman, J.E.; Angel, I.; Kozak, A.; Koh, J.Y. The lipophilic metal chelator DP-109 reduces amyloid pathology in brains of human beta-amyloid precursor protein transgenic mice. *Neurobiol. Aging* **2004**, *25*, 1315–1321. [CrossRef]
252. Ding, W.Q.; Yu, H.J.; Lind, S.E. Zinc-binding compounds induce cancer cell death via distinct modes of action. *Cancer Lett.* **2008**, *271*, 251–259. [CrossRef]
253. Colvin, R.A.; Bush, A.I.; Volitakis, I.; Fontaine, C.P.; Thomas, D.; Kikuchi, K.; Holmes, W.R. Insights into Zn2+ homeostasis in neurons from experimental and modeling studies. *Am. J. Physiol. Cell Physiol.* **2008**, *294*, C726–C742. [CrossRef] [PubMed]
254. Cherny, R.A.; Atwood, C.S.; Xilinas, M.E.; Gray, D.N.; Jones, W.D.; McLean, C.A.; Barnham, K.J.; Volitakis, I.; Fraser, F.W.; Kim, Y.; et al. Treatment with a copper-zinc chelator markedly and rapidly inhibits beta-amyloid accumulation in Alzheimer's disease transgenic mice. *Neuron* **2001**, *30*, 665–676. [CrossRef]
255. Palmiter, R.D.; Cole, T.B.; Quaife, C.J.; Findley, S.D. ZnT-3, a putative transporter of zinc into synaptic vesicles. *Proc. Natl. Acad. Sci. USA* **1996**, *93*, 14934–14939. [CrossRef]

256. Linkous, D.H.; Flinn, J.M.; Koh, J.Y.; Lanzirotti, A.; Bertsch, P.M.; Jones, B.F.; Giblin, L.J.; Frederickson, C.J. Evidence that the ZNT3 protein controls the total amount of elemental zinc in synaptic vesicles. *J. Histochem. Cytochem.* **2008**, *56*, 3–6. [CrossRef]
257. Cole, T.B.; Wenzel, H.J.; Kafer, K.E.; Schwartzkroin, P.A.; Palmiter, R.D. Elimination of zinc from synaptic vesicles in the intact mouse brain by disruption of the *ZnT3* gene. *Proc. Natl. Acad. Sci. USA* **1999**, *96*, 1716–1721. [CrossRef] [PubMed]
258. Adlard, P.A.; Sedjahtera, A.; Gunawan, L.; Bray, L.; Hare, D.; Lear, J.; Doble, P.; Bush, A.I.; Finkelstein, D.I.; Cherny, R.A. A novel approach to rapidly prevent age-related cognitive decline. *Aging Cell* **2014**, *13*, 351–359. [CrossRef] [PubMed]
259. Adlard, P.A.; Parncutt, J.; Lal, V.; James, S.; Hare, D.; Doble, P.; Finkelstein, D.I.; Bush, A.I. Metal chaperones prevent zinc-mediated cognitive decline. *Neurobiol. Dis.* **2015**, *81*, 196–202. [CrossRef]
260. Faux, N.G.; Ritchie, C.W.; Gunn, A.; Rembach, A.; Tsatsanis, A.; Bedo, J.; Harrison, J.; Lannfelt, L.; Blennow, K.; Zetterberg, H.; et al. PBT2 rapidly improves cognition in Alzheimer's disease: Additional phase II analyses. *J. Alzheimers Dis.* **2010**, *20*, 509–516. [CrossRef]
261. Lannfelt, L.; Blennow, K.; Zetterberg, H.; Batsman, S.; Ames, D.; Harrison, J.; Masters, C.L.; Targum, S.; Bush, A.I.; Murdoch, R.; et al. Safety, efficacy, and biomarker findings of PBT2 in targeting Abeta as a modifying therapy for Alzheimer's disease: A phase IIa, double-blind, randomised, placebo-controlled trial. *Lancet Neurol.* **2008**, *7*, 779–786. [CrossRef]
262. Saxton, R.A.; Sabatini, D.A. mTOR Signaling in Growth, Metabolism, and Disease. *Cell* **2017**, *168*, 960–976. [CrossRef]
263. Tramutola, A.; Lanzillotta, C.; Barone, E.; Arena, A.; Zuliani, I.; Mosca, L.; Blarzino, C.; Butterfield, D.A.; Perluigi, M.; Di, D.F. Intranasal rapamycin ameliorates Alzheimer-like cognitive decline in a mouse model of Down syndrome. *Transl. Neurodegener.* **2018**, *7*, 28. [CrossRef]
264. Oddo, S. The role of mTOR signaling in Alzheimer disease. *Front. Biosci.* **2012**, *4*, 941–952. [CrossRef] [PubMed]
265. Pei, J.-J.; Jacques, H. mTOR-dependent signalling in Alzheimer's disease. *J. Cell. Mol. Med.* **2008**, *12*, 2525–2532. [CrossRef] [PubMed]
266. Perluigi, M.; Di Domenico, F.; Barone, E.; Butterfield, D.A. mTOR in Alzheimer disease and its earlier stages: Links to oxidative damage in the progression of this dementing disorder. *Free Radic. Biol. Med.* **2021**, *169*, 382–396. [CrossRef] [PubMed]
267. An, W.-L.; Cowburn, R.F.; Li, L.; Braak, H.; Alafuzoff, I.; Iqbal, K.; Iqbal, I.G.; Winblad, B.; Pei, J.J. Up-regulation of phosphorylated/activated p70 S6 kinase and its relationship to neurofibrillary pathology in Alzheimer's disease. *Am. J. Pathol.* **2003**, *163*, 591–607. [CrossRef]
268. Tang, Z.; Bereczki, E.; Zhang, H.; Wang, S.; Li, C.; Ji, X.; Branca, R.M.; Lehtiö, J.; Guan, Z.; Filipcik, P.; et al. Mammalian target of rapamycin (mTor) mediates tau protein dyshomeostasis: Implication for Alzheimer disease. *J. Biol. Chem.* **2013**, *288*, 15556–15570. [CrossRef]
269. Lai, C.; Chen, Z.; Ding, Y.; Chen, Q.; Su, S.; Liu, H.; Ni, R.; Tang, Z. Rapamycin Attenuated Zinc-Induced Tau Phosphorylation and Oxidative Stress in Rats: Involvement of Dual mTOR/p70S6K and Nrf2/HO-1 Pathways. *Front. Immunol.* **2022**, *13*, 782434. [CrossRef]
270. Tordjman, S.; Chokron, S.; Delorme, R.; Charrier, A.; Bellissant, E.; Jaafari, N.; Fougerou, C. Melatonin: Pharmacology, Functions and Therapeutic Benefits. *Curr. Neuropharmacol.* **2017**, *15*, 434–443. [CrossRef]
271. Grossman, E. Should melatonin be used to lower blood pressure? *Hypertens. Res. Off. J. Jpn. Soc. Hypertens.* **2013**, *36*, 682–683. [CrossRef]
272. Li, Y.; Zhang, J.; Wan, J.; Liu, A.; Sun, J. Melatonin regulates Aβ production/clearance balance and Aβ neurotoxicity: A potential therapeutic molecule for Alzheimer's disease. *Biomed. Pharmacother. = Biomed. Pharmacother.* **2020**, *132*, 110887. [CrossRef]
273. Golpich, M.; Amini, E.; Hemmati, F.; Ibrahim, N.M.; Rahmani, B.; Mohamed, Z.; Raymond, A.A.; Dargahi, L.; Ghasemi, R.; Ahmadiani, A. Glycogen synthase kinase-3 beta (GSK-3β) signaling: Implications for Parkinson's disease. *Pharmacol. Res.* **2015**, *97*, 16–26. [CrossRef] [PubMed]
274. Palomer, E.; Buechler, J.; Salinas, P.C. Wnt Signaling Deregulation in the Aging and Alzheimer's Brain. *Front. Cell. Neurosci.* **2019**, *13*, 227. [CrossRef] [PubMed]
275. Zhao, J.; Han, J.; Jiang, J.; Shi, S.; Ma, X.; Liu, X.; Wang, C.; Nie, X.; He, Y.; Jiang, S.; et al. The downregulation of Wnt/β-catenin signaling pathway is associated with zinc deficiency-induced proliferative deficit of C17. *2 neural stem cells. Brain Res.* **2015**, *1615*, 61–70. [PubMed]

biomolecules

MDPI

Review

The Impact of Zinc and Zinc Homeostasis on the Intestinal Mucosal Barrier and Intestinal Diseases

Yan Wan and Bingkun Zhang *

State Key Laboratory of Animal Nutrition, College of Animal Science and Technology, China Agricultural University, Haidian District, Beijing 100193, China; wy1208@cau.edu.cn
* Correspondence: zhangbk@cau.edu.cn

Abstract: Zinc is an essential trace element for living organisms, and zinc homeostasis is essential for the maintenance of the normal physiological functions of cells and organisms. The intestine is the main location for zinc absorption and excretion, while zinc and zinc homeostasis is also of great significance to the structure and function of the intestinal mucosal barrier. Zinc excess or deficiency and zinc homeostatic imbalance are all associated with many intestinal diseases, such as IBD (inflammatory bowel disease), IBS (irritable bowel syndrome), and CRC (colorectal cancer). In this review, we describe the role of zinc and zinc homeostasis in the intestinal mucosal barrier and the relevance of zinc homeostasis to gastrointestinal diseases.

Keywords: zinc; zinc homeostasis; intestinal mucosal barrier; inflammatory bowel disease; irritable bowel syndrome; colorectal cancer

1. Introduction

Zinc is the second most abundant trace element in the body and is a catalytic cofactor for many enzymes or a structural cofactor for proteins. Indeed, more than 300 enzymes and 1000 transcription factors rely on zinc to perform their biological functions [1,2]. Zinc is essential for various physiological functions, such as growth, development, antioxidant activity, and immunity [3,4]. Additionally, Zn^{2+} is involved in the regulation of cellular signal pathways as a signal molecule [5,6].

Plasma or serum zinc levels in healthy individuals vary from 12 to 16 μM, which corresponds to less than 1% of the whole-body zinc [7]. Both insufficient and excessive zinc can disrupt zinc homeostasis; zinc deficiency dysregulates cellular function, and excess zinc is also toxic to cells. Thus, the levels of zinc are necessarily tightly fine-tuned to maintain zinc homeostasis in organisms.

On the one hand, zinc homeostasis refers to the stabilization of zinc in body fluids, tissues, and organs, which mainly depends on the uptake and excretion of zinc in the body. On the other hand, it refers to the stability of intracellular zinc, which depends on various zinc homeostasis-related proteins, such as zinc transporters and zinc binding proteins. Zinc transporters consist of two families: the ZIP (Zrt- and Irt-like proteins, also known as SLC39A) family and the ZnT (Zn transporter, also known as SLC30A) family. At present, the ZIP family contains 14 members in mammals, and its members transport zinc from the extracellular space and intracellular organelles to the cytoplasm. The ZnT family contains 10 members, and they transport zinc from the cytoplasm to extracellular space or organelles [8–10]. The coordination of the two families' members maintains the homeostasis of intracellular zinc. Metallothionein (MT) is a cysteine-rich metal-binding protein in cells and is the most important zinc-binding protein [11]. It is both a zinc receptor and a zinc donor, which can regulate the absorption, distribution, storage, and release of zinc in the cells.

The intestinal mucosal barrier is composed of an extracellular mucus layer and intestinal epithelial cells (IECs). Functionally, the intestinal mucosal barrier can be divided into a mechanical barrier, chemical barrier, biological barrier, and immune barrier. It not only electively absorbs nutrients from the chyme and transfers them from the intestinal lumen to the metabolic cycle, but it also plays an indispensable role in defending against microorganisms and exogenous antigens. Studies have demonstrated that a variety of intestinal diseases are associated with intestinal mucosal barrier damage, including inflammatory bowel disease (IBD), irritable bowel syndrome (IBS), and colorectal cancer (CRC) [12–14].

Many lines of evidence indicate that zinc homeostasis is essential for the maintenance of intestinal mucosal barrier functions, and they have efficacy in some intestinal disorders. Zinc deficiency causes mucosal barrier dysfunction [15], while zinc supplementation improves these symptoms [16]. In this review, we focus on the roles of zinc and zinc homeostasis in the intestinal mucosal barrier, as well as their functions and mechanisms in the occurrence and development of intestinal diseases.

2. Zinc Homeostasis in the Intestinal Mucosal Barrier

The adult human body contains approximately 2.6 g of zinc, and the fraction in the intestine is about 1.2% [7]. The small intestine is the main place for the absorption and excretion of zinc [7,17]; thus, an injury to the intestinal mucosa can affect the absorption and excretion of zinc, thereby affecting the zinc homeostasis in cells and organisms. Correspondingly, the imbalance of intestinal zinc homeostasis can also negatively affect the structure and function of intestinal mucosal barrier.

2.1. *Zinc Absorption and Excretion*

Exogenous zinc is taken up at the intestinal brush border membrane, where it is transported from the lumen to enterocytes; subsequently, the cation is excreted at the basolateral side of the enterocytes to release the Zn^{2+} into the portal blood, which distributes Zn^{2+} in the body [17,18]. After circulating in the body, zinc is excreted into the small intestinal lumen along with pancreatic secretions and bile; most of the zinc is reabsorbed by the small intestine, and the remaining part is excreted in feces [19]. In cells, zinc exists in three forms: zinc bonded to proteins, zinc stored in vesicles, and cytoplasmic free zinc [17]. The absorption of zinc in the small intestine occurs in two ways: passive transport through ion-channels and active transport dependent on zinc transporters.

Zinc transporters consist of two families: the ZIP family and the ZnT family. Corresponding to the wide distribution of zinc, the distribution of zinc transports in tissues and organs or the localization in cells show specificity and diversity. Intestinal zinc absorption is mainly mediated by ZnT1, ZnT2, ZnT5, ZnT6, ZnT7, and ZnT5B (a splicing variant of ZnT5) of the ZnT family and ZIP1, ZIP4, ZIP5, ZIP10, and ZIP14 of the ZIP family [20,21]. In addition, MT is also involved in the regulation of enterocyte zinc homeostasis through binding and releasing cytoplasmic free Zn^{2+} in cells [11] (as shown in Figure 1).

ZIP4 is localized on the apical membrane of enterocytes and is the main transporter of Zn^{2+} from the lumen to enterocytes. Studies have found that the ZIP4 gene mutations cause zinc absorption disorder and acrodermatitis enteropathica [22]. ZIP4 translocates to the apical membrane during zinc deficiency, thereby increasing Zn^{2+} uptake by enterocytes [23–25]. Under excess zinc conditions, the ZIP4 on the cell surface is rapidly endocytosed, ubiquitinated, and degraded [24,26]. When zinc is sufficient, ZIP5 is localized on the basolateral membrane of enterocytes and participates in cellular zinc excretion. When zinc is deficient, the ZIP5 on the basolateral membrane of enterocytes is endocytosed and degraded to reduce the zinc from the blood circulation to enterocytes [24,25]. ZIP14 is also localized on the basolateral membrane of enterocytes and participates in the excretion of cellular zinc [27], but its protein expression is generally insensitive to the zinc status of the cells.

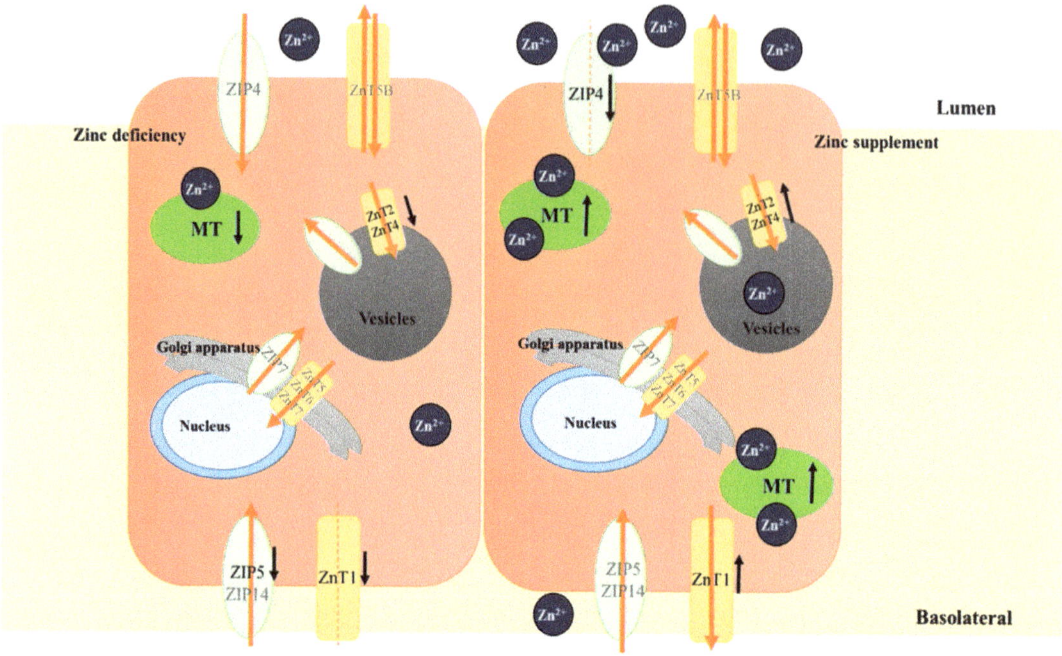

Figure 1. Zinc transporters (ZnT and ZIP) and zinc-binding proteins (MT) are involved in the regulation of intestinal zinc absorption, a potential regulatory mechanism of zinc absorption into enterocytes during zinc excess and zinc supplement. The red arrows indicate the direction of zinc flow, and the black arrows indicate changes in zinc transporters' RNA and/or protein upon changes in the zinc state.

Different from the ZIP family, there are only ZnT1 and ZnT5B localized on the cell membrane of the ZnT family, and most of the others are located in the organelle and nucleus. ZnT1 is located on the basolateral membrane of enterocytes, and it is responsible for the transport of Zn^{2+} from the enterocyte to the extracellular space [28]. The expression of ZnT1 in the intestine at the mRNA and protein levels increases when zinc is sufficient [29] and decreases during zinc deficiency [29,30]. ZnT5B is a splice variant of ZnT5, which is located on the apical membrane of enterocytes. It is capable of the bidirectional transport of zinc [31]. In the IECs, ZnT2 and ZnT4 mainly accumulate in vesicles [28], their gene expression increases under the condition of sufficient zinc, which mediates zinc transport into vesicles and isolates it from the cytoplasm, while under zinc depletion, their gene expression decreases [29]. ZnT6 is found in the ER and Golgi apparatus, but it is incapable of transporting zinc alone; instead, it is capable of transporting zinc by forming heterodimers with ZnT5 and its orthologs [32]. ZnT7 accumulates zinc in the Golgi of cells; thus, it is an important transporter for zinc transport in the Golgi [33]. Studies have shown that zinc depletion in HeLa epithelial cells results in increasing mRNA expression of ZnT5 and ZnT7 [34]; zinc supplementation reduces the mRNA levels of ZnT6 and induces ZnT7 in THP-1 mononuclear cells [35], and in zinc-deficient Raji (B cell line), the expression of ZnT5, 6, and 7 is improved [36]. Nevertheless, in intestinal cells, no study has shown whether zinc homeostasis affects the expression of ZnT5, 6, 7. In addition to zinc transporters, the expression of MT is also closely related to zinc homeostasis. Data have indicated that a high zinc supply can induce an increase in cellular MT in vivo and in vitro [37]. MT is able to bind and release Zn^{2+} which is absorbed into the cells; consequently, changes in MT expression may affect zinc transport kinetics and luminal zinc absorption [17].

In conclusion, specific zinc transporters and zinc-binding proteins work together to maintain the dynamic balance of zinc metabolism in IECs and maintain body zinc homeostasis by regulating zinc absorption and excretion.

2.2. Interactions between the Zinc Homeostasis Imbalance and Intestinal Mucosal Barrier Injury

A zinc deficiency can injure the intestinal mucosal barrier. The study of Zhong et al. [15] showed that treating Caco2 cells with TPEN for zinc deprivation resulted in decreased expression of tight junction proteins, increased intestinal epithelial permeability, and impaired intestinal mucosal barrier function. An in vitro study by Miyoshi et al. [38] also found that the tight junction permeability in Caco2 cells increased in the presence of zinc deficiency, and cell viability was significantly reduced. Furthermore, the repair and regeneration of the intestinal mucosa requires the participation of zinc; so, exogenous supplementation can maintain the integrity of the intestinal mucosal barrier morphology and function [39]. These results suggest that zinc plays an important role in maintaining the intestinal barrier function.

Individuals with a compromised intestinal mucosal barrier develop an imbalance in zinc homeostasis. A recent meta-analysis study demonstrated a significant relationship between serum zinc levels and Crohn's disease (CD), which is characterized by injured intestinal mucosa [40]. The imbalance of zinc homeostasis is due to the absorption of zinc in the small intestine being reduced and the endogenous zinc excretion being increased when the intestinal mucosal barrier is damaged, and the intestinal permeability is decreased. Griffin's study [41] found that plasma zinc levels were reduced in CD patients, and the absorption of zinc in the small intestine was also reduced, while endogenous zinc excretion did not change significantly. However, Main et al. [42] found CD patients' zinc excretion decreased; this difference may be related to the course of CD. In addition to peripheral blood zinc levels, a study also found that ZIP8 mutation in CD patients was associated with changes in intestinal flora [43], suggesting that zinc homeostasis in enterocytes is also important for maintaining the intestinal mucosal barrier.

3. Mechanism of Zinc Homeostasis Affecting the Intestinal Mucosal Barrier

3.1. Intestinal Mucosal Physical and Chemical Barriers

The physical barrier is composed of intestinal epithelial cells, intercellular junctions and the submucosal lamina propria, which can prevent the macromolecular substances and pathogens in the intestinal lumen crossing the intestinal wall. The intestinal epithelium is composed of crypt and villus, which are made up of several different types of cells, including intestinal epithelial stem cells, IECs, and enteroendocrine cells. The chemical barrier is formed by mucin, which is secreted by intestinal goblet cells and various antimicrobial proteins, which are secreted by IECs, which can prevent macromolecular substances from directly contacting the epithelial cell layer and defend against foreign antigen. Zinc and zinc homeostasis is crucial in the maintenance of the physical and chemical barriers. Studies have suggested that zinc supplementation increases the villus height and reduces the crypt depth in the mammalian and avian small intestine [44–48] and increases the number of intestinal goblet cells [49] and the expression of MUC2 [46–48].

Zinc transporters are involved in the continuous self-renewal of the intestinal epithelial cells and the maintenance of the overall structure of the crypt–villus axis. Ohashi et al. [50] illustrated that ZIP7 is essential for intestinal epithelial homeostasis by the regulation of the endoplasmic reticulum function in proliferative progenitor cells and the maintenance of stem cells in the crypt; Zip7 deficiency in the intestinal epithelium leads to the loss of $Olfm4^+$ intestinal stem cells and the degeneration of post-mitotic Paneth cells. The findings provided important evidence for the role of ZIP7 in maintaining intestinal epithelial homeostasis. Furthermore, Geiser et al. [22] showed all Zip4 intestine knockout mice died within two to three weeks. In ZIP4 knockout mice, the expression of Sox9, a marker for Paneth cells decreased, and the lysozyme activity also decreased in the small intestine, accompanied by the disorganization of the crypt and villus morphology and significantly diminished cell division, and the epithelial cells had lost much of their columnar morphology.

In addition to the regeneration and composition of IECs, zinc and zinc homeostasis also affects the tight junctions (TJ) between epithelial cells. Studies have shown that zinc deficiency results in the destruction of intercellular connections and increases intestinal permeability [38,51], while zinc supplementation can repair the injury of TJ. As the cellular pathways underlying this effect (Figure 2), the diosmectite–zinc oxide composite enhanced intestinal barrier restoration by modulating the TGF-β1, ERK1/2, and Akt pathways in acetic acid challenged piglets and weaned pigs [52,53]. Additionally, zinc was shown to have the known insulin-mimetic effects in activating the PI3K/protein kinase B (AKT) signaling cascade via the regulation of gene expression [54]. The activation of PI3K/AKT/mTOR signaling by zinc sulfate is involved in improving the intestinal barrier function by enhancing the cell differentiation and expression of TJ protein ZO-1 in Caco-2 cells [55]. In addition, extracellular Zn^{2+}, via the G protein-coupled receptor (GPR39), directly regulates the epithelial barrier of the colon [54,56,57]. Shao et al. [58] found that zinc via the GPR39 molecule can activate PKCζ to alleviate the intestinal barrier dysfunction induced by *S. typhimurium*. Zinc transporters also contribute to maintain TJ. Gregory et al. [27] indicated that the intestinal permeability of ZIP14 knockout (ZIP14KO) mice was assessed to be higher than wildtype mice, and ZIP14KO mice showed a significant reduction in the amount of threonine-phosphorylated occludin and the expression of claudin1 and claudin2.

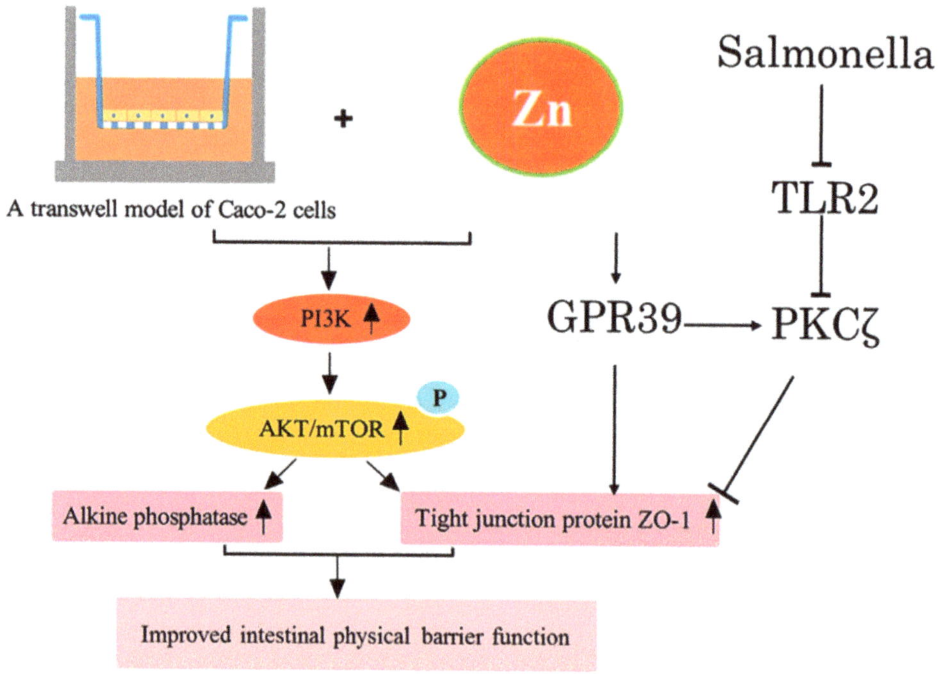

Figure 2. Cellular pathways in which zinc upregulates the function of the mucosal physical barrier. Zinc improves the expression of the tight junction protein by activating PI3K/AKT/mTOR. Zinc directly regulates the epithelial barrier via the G protein-coupled receptor (GPR39). Then, through the GPR39 molecule, it can activate PKCζ to alleviate the intestinal barrier dysfunction induced by *S. typhimurium*.

Zinc deficiency and supplementation affect the number of intestinal goblet cells and the number and composition of mucins [48,49,59]. An in vitro study showed that zinc deficiency affected goblet cell synthesis and the secretion of mucin, along with impairing O-glycan elongation, increasing the production of short sugar chains, and altering the core structure of O-glycans; this altered O-glycan pattern can affect the stability of the intestinal mucus layer [49]. Paneth cells are important sources of lysozyme and defenses in the intestine. Zinc is an essential trace metal for Paneth cells. Zinc coordination affords protection of human defensins against proteolytic degradation by the intestinal protease trypsin and other proteases [60], and ZnO nanoparticles has been shown to maintain the stability of lysozyme [61]. Similarly, the loss of function of some intestinal zinc transporters such as ZIP4 and ZIP7 can lead to a rapid loss of zinc in Paneth cells, which will stop the morphology and cell renewal of Paneth cells [22,50]. ZnT2 is the zinc transporter expressed on the secretory granules in Paneth cells; it can transport Zinc into Paneth cell granules to maintain granulin stability and secretory function [62]. Podany et al. [63] demonstrated in ZnT2Ko mice that the secretory granules were hypodense with less active lysozyme, and there was evidence of autophagosome accumulation and granule degradation in the Paneth cells from ZnT2ko mice.

To summarize, zinc and zinc transports maintain the physical barrier of intestinal mucosa by participating in the renewal of IECs, and the maintenance of the junction between intestinal epithelial cells. Moreover, zinc can also maintain the stability of the mucous layer and the production of antibacterial proteins, which contribute to the stability of the intestinal mucosal chemical barrier.

3.2. Intestinal Mucosal Immune and Biological Barriers

The intestinal mucosal immune barrier mainly consists of gut-associated lymphatic tissue (GALT), its secreted immunoglobulin A (sIgA), antimicrobial peptides, cytokines, and other immune active substances. It can identify abnormal antigens and induce an immune response to protect the intestine from being invaded and prevent excessive immune responses.

A study on poultry found that Zn–Gly chelate supplementation could balance the Th1 and Th2 response in the GALT through the activation of a cascade of cytokines released by stimulated T and B lymphocytes [64]. sIgA is secreted by plasma cells and serves as the first line of the immune barrier, which limits the access of antigens and bacteria in the intestinal tract to the circulation system [65]. Several studies have implied that zinc supplementation can upregulate the expression of sIgA [46,64,66]. Intestinal macrophages are the most abundant immune cell type in the lamina propria, which can clear both pathogens and apoptotic cells. MT can enhance the ability of autophagy and intracellular bacterial clearance through mediating the zinc accumulation of macrophages [67], while the synthesis and degradation of MT in macrophages are regulated by the concentration of cytoplasmic free zinc [68]. Hence, the zinc homeostasis maintained by MT plays a role in regulating the immune function of intestinal macrophages. Defensins are an important class of antimicrobial peptides with antimicrobial activity and are found widely in animals, plants, and fungi. They are an important part of host innate immunity. As previously mentioned, zinc homeostasis affects the morphology and function of Paneth cells [22,50,62,63], which are the main source of intestinal defensins. Some studies have indicated that exogenous addition of zinc can increase the gene expression of intestinal β-defensins in pigs and ducks [46,48]. A study has found the same in humans, the decrease in Paneth cell granules and reduced human defensin 5 (HD5) immunoreactivity are significantly associated with a decrease in the plasma zinc levels [69].

The intestinal mucosal biological barrier consists of the intestinal microbiota that restrict mucosal colonization by pathogens and resists penetration by pathogens.

Zinc deficiency caused by insufficient dietary zinc leads to poor zinc physiological status and induces a decrease in gut microbial diversity and an outgrowth of bacteria particularly suited to low zinc conditions, which eventually lead to an intestinal microecological disorder [70]. Experimental evidence has also shown that high dietary zinc supplementation alters the composition of the gut microbiota [71,72]. Zinc homeostasis affects intestinal microbial balance in two ways. First, as the above results show, zinc homeostasis has an impact on the function of immune cells and immune active substances, which play a crucial role in maintaining the balance of intestinal microorganisms; for example, macrophages have the ability to clear bacteria. Second, the growth and reproduction of gut microbes requires the participation of zinc. A study in mice colonized with *C. difficile* indicated that excess dietary zinc severely exacerbated C. *difficile*-associated disease by increasing its toxicity and altering the host immune response [71]. Based on this, the competition for zinc between the gut microbes and the host can be used to inhibit the infection of some pathogenic bacteria. The host can express some zinc-sequestering proteins, such as calprotectin, to the infection site to reduce the zinc concentration in a single location to limit zinc absorption by bacteria, thereby limiting the growth and virulence of pathogenic bacteria and inhibiting the proliferation of pathogenic bacteria against bacterial infection [73,74].

In conclusion, zinc can help maintain a functional mucosal immunological barrier by regulating the function of the intestinal immune cells and the expression of immune active substances. Simultaneously, zinc also can maintain the composition of gut microbes through immune mechanisms. In addition, zinc homeostasis of both host and microbes can affect the colonization and proliferation of gut microbes, as zinc is an essential component of microbial proteins.

4. Research Advances in the Relationship between Zinc Homeostasis and Gut Diseases

4.1. *Inflammatory Bowel Disease*

IBD is a nonspecific chronic inflammatory disease of the intestine, mainly including ulcerative colitis (UC) and CD. UC is usually confined to colonic tissue, with the occurrence of crypt destruction, erosions, and ulcers with diffuse mucosal inflammation and neutrophils [75]. Compared with UC, CD is an intestinal inflammation that can affect any part of the gastrointestinal tract, but the small intestine is the most common area for CD occurrence, especially the terminal ileum and colon [76]. At present, the pathogenesis of CD and UC are still inconclusive, and it is generally believed that they are related to genetic factors, environmental factors, infectious factors, and immune factors [77,78]. Studies have shown that CD and UC patients have a lower zinc intake and serum zinc levels [41,79,80], and both CD and UC patients with serum zinc deficiency are more likely to have adverse disease-specific outcomes, as these outcomes improve with normalization of zinc levels [80]. However, it is noteworthy that a meta-analysis study found that low serum zinc levels were significantly associated with CD but not with UC [40]. Similarly, a study by Ananthakrishnan et al. [81] also found that dietary zinc supplements were negatively correlated with CD risk but not with UC. Therefore, the relationship between CD or UC progression and zinc level concentration remains controversial, and more research needs to be conducted in the future.

Recently, there has been increasing attention to the mechanism of zinc's effect on the occurrence and development of IBD. Yasuki [82] found that zinc deficiency activated Th17 cells to aggravate colitis, and the activation resulted from the induction of IL-23 expression in macrophages. In addition, zinc deficiency can also lead to lymphoid tissue hypoplasia, decreased NK cell activity, and increased apoptosis of B and T cells, thus affecting the production of antibodies and cell-mediated immunity [83], which can reduce the ability of antivirus and antibacterial activity and the body's immunity and increase the risk of inflammation. Zinc deficiency can also lead to impaired intestinal barrier integrity and leaky gut, enabling pathogenic bacteria and other antigenic substances in the intestine to directly contact IECs, which can stimulate continuous submucosal inflammation [38,51]. Zinc supplementation can effectively reduce the penetration of bacteria through the intestinal mucosal barrier to avoid direct contact with IECs and resist the infection by enteric pathogens, thus reducing intestinal inflammation and improving diarrhea symptoms [53,57,84]. In short, the imbalance of zinc homeostasis increases the risk of impaired intestinal barriers, causes foreign antigens to be more likely to contact IECs directly, increases the risk of infection, and induces intestinal inflammation. It can also lead to the dysbiosis of intestinal flora and the imbalance of organism immunity, which ultimately leads to intestinal inflammation and promotes the onset and progression of CD or UC.

4.2. Irritable Bowel Syndrome

IBS is a common gastrointestinal disorder, which is characterized by chronic or recurrent abdominal pain, in conjunction with relief or aggravation with defecation or altered bowel habits [85]. The etiology and pathogenesis of IBS are not completely clear; it is generally recognized that IBS is a disease closely related to the central nervous system or psychosocial factors, food allergies, gastrointestinal changes, and genetic susceptibility [86]. The central nervous system or psychosocial factors are mainly related to visceral hypersensitivity, abnormal brain–gut axis interaction, and psychosocial dysfunction. Gastrointestinal changes mainly include an imbalance in intestinal flora, a motility disorder in the gastrointestinal tract, low-grade intestinal inflammation, and abnormal immune activation of intestinal mucosa.

A few studies are currently available on the role of zinc in IBS. Studies have found that the dietary zinc intake in IBS patients is significantly lower than those of healthy individuals, but there is no difference in the zinc levels in the plasma or serum [87,88]. A study of diarrhea-predominant irritable bowel syndrome (IBS-D) patients suggested that the serum zinc levels and odds of IBS-D were inversely associated [89]. Abnormal copper–zinc ratios have been found in many mental health disorders that overlap with IBS [90,91]. A North American study also found that the serum copper–zinc ratio was increased in IBS patients, but there was no significant difference in serum copper or zinc levels [92]; the possible reason is that blood zinc levels are less sensitive to zinc deficiency, and it may be difficult to identify a marginal deficiency of zinc. Nevertheless, it still suggests that zinc supplementation may be a way to improve the symptoms of IBS, as it has been found to normalize the copper–zinc ratio and improve symptoms in anxiety and in autism spectrum disorders with concurrent gastrointestinal symptoms [91,93].

There is no obvious intestinal inflammation in IBS patients, but the number and activation of mast cells (MCs) increase as the intestine is stressed [94]. The main structural feature of mast cells is a large number of basophilic granules, which are rich in a variety of bioactive substances; these basophilic particles release various inflammatory mediators, such as histamine and 5-HT, after their activation [95]. One study found that histamine sensitizes the nociceptor transient reporter potential channel V1 (TRPV1) and contributes to IBS symptoms of visceral hypersensitivity in animals [96]. The alterations to 5-HT signaling can result in IBS symptoms such as gastrointestinal dysmotility, visceral hypersensitivity, and abnormal intestinal secretion motility [97,98]. The expression of 5-HT and its receptor in the intestinal mucosa of IBS patients were higher than the normal, and 5-HT receptor antagonist therapy can relieve IBS-D symptoms effectively [99,100]. The microstructure of MCs showed that its granules are rich in zinc [101]. MCs receptor FcεRI stimulation induced an increase in intracellular free zinc [102]. The zinc chelator, N,N,N,N-tetrakis (2-pyridylmethyl) ethylenediamine can significantly inhibit the degranulation of FcεRI and the production of cytokine production such as TNF-α and IL-β [103]. These subjects suggest that zinc figures prominently in the activation of MCs, and it may regulate the degranulation of mast cells to affect the release of bioactive substances from mast cells and mediate the occurrence of IBS symptoms.

As a functional intestinal disease, IBS is closely related to the psychological and mental health factors of patients; hence, the dysfunction of the brain–gut axis is increasingly believed to be one of the main mechanisms underlying IBS. Stress and emotion trigger neuroimmune and neuroendocrine responses through the brain–gut axis, thus affecting the gastrointestinal tract and inducing IBS symptoms. Zinc is related to the secretion of neurotransmitters. Zinc deficiency can critically impair the GABAergic neurotransmitter system [104], while it is generally recognized that GABA plays a regulatory role in gastrointestinal motility and visceral sensation [105]. In addition, zinc homeostasis also affects the transmission of neurotransmitters, including 5-HT, norepinephrine, dopamine, etc. [106,107], and these neurotransmitters contribute to the pathogenesis of IBS by activating or inhibiting the central or intestinal nervous system [108,109]. As stated in Section 3.2, zinc homeostasis affects the structural and functional integrity of the intestinal mucosa, and intestinal mucosal permeability and intestinal microbial composition are also hypothetical etiological mechanisms of IBS [85]. In summary, zinc may regulate the central nervous system and gastrointestinal system by affecting the intestinal mucosal barrier function, mast cell activation, and neurotransmitter secretion, relative to the initiation and development of IBS.

4.3. Colorectal Cancer

CRC is a common malignant tumor of the digestive tract with high morbidity and mortality, and it is the third deadliest cancer in the world. The etiology of CRC is relatively clear; dietary factors, genetic factors, and pathological factors are the main causes of CRC, but the mechanism of its occurrence and development is not yet clear.

Whether there is a correlation between CRC and zinc homeostasis is controversial. It has been found that there is no substantial association between serum zinc levels and rectal cancer in the representative general population in the USA [110]. However, in Iran's CRC population, a survey found that serum zinc levels were significantly lower than in healthy individuals [111]. Christudoss et al. [112] found that in rats with colon cancer induced by dimethylhydrazine, the decrease in plasma zinc, tissue zinc, and the activity of zinc-related enzymes was associated with the development of precancerous lesions, and these biochemical parameters further decreased with the progression to carcinoma in the colon. In addition, there is evidence that zinc intake has an effect on the risk of CRC among a certain demographic. A meta-analysis found a significant correlation between zinc intake and CRC risk in female patients but not in male. In addition, it was found that zinc intake was significantly associated with the risk of colorectal cancer in Asia, but not in the USA and Europe [113]. Further, Li et al. [113] found a U-shaped relationship (the cut-point dose was 22 mg/d) between the dose of zinc intake and colorectal cancer risk. In vitro studies also found that excessive Zn^{2+} induced epithelial transformation through the GSK3/mTOR signaling pathway to promote the migration and invasion capabilities of CRC cells [114]. This indicates that both too much and too little zinc intake can increase the risk of CRC.

Zinc deficiency increases the production of reactive oxygen species (ROS) and the sensitivity to oxidative damage [115–117], causing antioxidant dysfunction and resulting in an increased possibility of tumorigenesis. Zinc supplementation can effectively alleviate oxidative stress [115] and prevent oxidative DNA damage in cells to maintain the integrity of DNA in cells [118–120]. In addition, zinc participates in DNA transcription and replication through zinc finger protein and the p53 tumor suppressor protein, and so on, and zinc deficiency can affect the active DNA-binding expression of numerous transcription factors that are involved in cancer promotion and progression, such as AP-1, NF-κB, and p53 [118,121], which can lead to increased DNA damage, mutation, genomic instability, and increased inflammatory levels in the body, resulting in an increased risk of cancer.

Although it is still controversial whether the level of zinc can change in patients with CRC, improper intake of zinc does increase the risk of cancer, such as CRC, which indicates that the change in zinc homeostasis is indeed related to the occurrence and development of CRC. Hence, it is not surprising that researchers have focused on the role of zinc transporters. Zinc is involved in various life processes such as the proliferation and migration of tumor cells, and the maintenance of intracellular zinc levels is of great significance to the life progress of tumor cells. Sheng et al. [122] found that the expression of ZIP7 in the colon tissue of CRC patients was significantly increased, and ZIP7 could mediate the extracellular influx of Zn^{2+}. It was also shown that knockdown of ZIP7 interfered with cell cycle progression and induced G2/M cell cycle arrest, as well as boosting apoptosis in colorectal cancer cells [122]. Similarly, Barresi et al. [123] found differential gene expression of ZnT and ZIP transporters in CRC cells to maintain high zinc turnover predominantly required in the ER vesicles of tumoral cells where appropriate folding, assembly, and modification of proteins occur.

To summarize, zinc is a necessary trace element for the resistance to CRC; zinc deficiency can lead to DNA damage, impair antioxidant function, and increase the risk of cancer. Zinc transporters are also involved in the development of CRC through the maintenance of intracellular zinc levels in the CRC cells. This also suggests the importance of maintaining zinc homeostasis in the prevention and treatment of CRC.

4.4. Celiac Disease

Celiac disease (CeD) is a kind of chronic intestinal malabsorption syndrome caused by gluten-intake intolerance in genetically susceptible individuals [124]. It is an autoimmune disease, typically characterized by chronic diarrhea, abdominal pain, and other gastrointestinal symptoms [125] and often leads to growth retardation in children [126]. At present, the pathogenesis of CeD is not completely clear. It is mainly believed that CeD is the result of the interaction of genetic, immune, and environmental factors [127].

A recent study of a U.S. population found that serum zinc levels were decreased significantly in children aged 6–19 years old with celiac disease, but there was no significant change in patients over 20 years old [128]. In fact, zinc deficiency is common in patients with celiac disease [129]. The possible reason for zinc deficiency in patients with CeD is that the atrophy of the proximal intestinal villiin in CeD patients results in malabsorption of zinc [130]; in addition, the chelation of zinc by fatty acids, excessive loss due to protein-losing enteropathy, or excessive utilization due to enhanced enterocyte turnover can also led to zinc deficiency in CeD patients [129]. The symptoms of slower growth, later sexual maturity, and slower wound healing, as well as intellectual disabilities in patients with CeD may be related to zinc deficiency. Although zinc deficiency is common in patients with CeD, there is no evidence that the imbalance in zinc homeostasis is the cause of CeD. However, because zinc and zinc homeostasis are conducive to the maintenance of intestinal mucosal barrier function, they could be a therapeutic modality in CeD by way of reducing gluten and other luminal antigen permeation across the small bowel epithelium of CeD patients.

5. Conclusions

Zinc and zinc homeostasis strongly contribute to the maintenance of the intestinal mucosal barrier structure and function. An imbalance in zinc homeostasis can cause impairment of the epithelial barrier function, altered immune responses, and dysbiosis of the gut microbiota, and it is related to the occurrence and development of intestinal diseases such as IBD, IBS, and CRC. However, whether the improvement of intestinal barrier function with zinc can alleviate and treat these intestinal diseases still needs further clinical verification, and the molecular mechanism by which zinc regulates these intestinal diseases also remains poorly understood; Moreover, the function of zinc transporters has not been clearly discussed. More experimental and clinical studies are needed to further specify the precise role of zinc and zinc homeostasis in intestinal diseases, so as to provide new methods of intestinal disease diagnosis and therapy.

Author Contributions: Conceptualization, Y.W. and B.Z.; writing—original draft preparation, Y.W.; literature search, Y.W. and B.Z. All authors have read and agreed to the published version of the manuscript.

Funding: This study was supported by the National Natural Science Foundation of China (No. 32072750) and the 2115 Talent Development Program of China Agricultural University.

Conflicts of Interest: The authors declare no conflict of interest.

References

1. Maret, W. Molecular aspects of human cellular zinc homeostasis: Redox control of zinc potentials and zinc signals. *Biometals* **2009**, *22*, 149–157. [CrossRef] [PubMed]
2. Vallee, B.L.; Auld, D.S. Zinc Coordination, Function, and Structure of Zinc Enzymes and Other Proteins. *Biochemistry* **1990**, *29*, 5647–5659. [CrossRef] [PubMed]
3. Shankar, A.H.; Prasad, A.S. Zinc and immune function: The biological basis of altered resistance to infection. *Am. J. Clin. Nutr.* **1998**, *68* (Suppl. S2), 447S–463S. [CrossRef] [PubMed]
4. Prasad, A.S. Zinc: An antioxidant and anti-inflammatory agent: Role of zinc in degenerative disorders of aging. *J. Trace Elem. Med. Biol.* **2014**, *28*, 364–371. [CrossRef] [PubMed]
5. Murakami, M.; Hirano, T. Intracellular zinc homeostasis and zinc signaling. *Cancer Sci.* **2008**, *99*, 1515–1522. [CrossRef] [PubMed]
6. Hirano, T.; Murakami, M.; Fukada, T.; Nishida, K.; Yamasaki, S.; Suzuki, T. Roles of zinc and zinc signaling in immunity: Zinc as an intracellular signaling molecule. *Adv. Immunol.* **2008**, *97*, 149–176. [PubMed]
7. Maares, M.; Haase, H. A Guide to Human Zinc Absorption: General Overview and Recent Advances of In Vitro Intestinal Models. *Nutrients* **2020**, *12*, 762. [CrossRef]
8. Kambe, T.; Taylor, K.M.; Fu, D. Zinc transporters and their functional integration in mammalian cells. *J. Biol. Chem.* **2021**, *296*, 100320. [CrossRef]
9. Kambe, T.; Hashimoto, A.; Fujimoto, S. Current understanding of ZIP and ZnT zinc transporters in human health and diseases. *Cell Mol. Life Sci.* **2014**, *71*, 3281–3295. [CrossRef]
10. Lichten, L.A.; Cousins, R.J. Mammalian Zinc Transporters: Nutritional and Physiologic Regulation. *Annu. Rev. Nutr.* **2009**, *29*, 153–176. [CrossRef]
11. Kimura, T.; Kambe, T. The Functions of Metallothionein and ZIP and ZnT Transporters: An Overview and Perspective. *Int. J. Mol. Sci.* **2016**, *17*, 336. [CrossRef]
12. Oshima, T.; Miwa, H. Gastrointestinal mucosal barrier function and diseases. *J. Gastroenterol.* **2016**, *51*, 768–778. [CrossRef]
13. Dorofeyev, A.E.; Vasilenko, I.V.; Rassokhina, O.A.; Kondratiuk, R.B. Mucosal Barrier in Ulcerative Colitis and Crohn's Disease. *Gastroenterol. Res. Pract.* **2013**, *2013*, 431231. [CrossRef]
14. Gonzalez-Castro, A.M.; Martinez, C.; Salvo-Romero, E.; Fortea, M.; Pardo-Camacho, C.; Perez-Berezo, T.; Alonso-Cotoner, C.; Santos, J.; Vicario, M. Mucosal pathobiology and molecular signature of epithelial barrier dysfunction in the small intestine in irritable bowel syndrome. *J. Gastroenterol. Hepatol.* **2017**, *32*, 53–63. [CrossRef]
15. Zhong, W.; McClain, C.J.; Cave, M.; Kang, Y.J.; Zhou, Z.X. The role of zinc deficiency in alcohol-induced intestinal barrier dysfunction. *Am. J. Physiol. Gastrointest. Liver Physiol.* **2010**, *298*, G625–G633. [CrossRef]
16. Sturniolo, G.C.; Fries, W.; Mazzon, E.; Di Leo, V.; Barollo, M.; D'Inca, R. Effect of zinc supplementation on intestinal permeability in experimental colitis. *J. Lab. Clin. Med.* **2002**, *139*, 311–315. [CrossRef]
17. Krebs, N.F. Overview of zinc absorption and excretion in the human gastrointestinal tract. *J. Nutr.* **2000**, *130* (Suppl. S5), 1374S–1377S. [CrossRef]
18. Menard, M.P.; Cousins, R.J. Zinc Transport by Brush-Border Membrane-Vesicles from Rat Intestine. *J. Nutr.* **1983**, *113*, 1434–1442. [CrossRef] [PubMed]
19. King, J.C.; Shames, D.M.; Woodhouse, L.R. Zinc homeostasis in humans. *J. Nutr.* **2000**, *130*, 1360S–1366S. [CrossRef] [PubMed]
20. Baltaci, A.K.; Yuce, K. Zinc Transporter Proteins. *Neurochem. Res.* **2018**, *43*, 517–530. [CrossRef] [PubMed]
21. Hennigar, S.R.; McClung, J.P. Zinc Transport in the Mammalian Intestine. *Compr. Physiol.* **2018**, *9*, 59–74.
22. Geiser, J.; Venken, K.J.; De Lisle, R.C.; Andrews, G.K. A mouse model of acrodermatitis enteropathica: Loss of intestine zinc transporter ZIP4 (Slc39a4) disrupts the stem cell niche and intestine integrity. *PLoS Genet.* **2012**, *8*, e1002766. [CrossRef]
23. Hashimoto, A.; Nakagawa, M.; Tsujimura, N.; Miyazaki, S.; Kizu, K.; Goto, T.; Komatsu, Y.; Matsunaga, A.; Shirakawa, H.; Narita, H.; et al. Properties of Zip4 accumulation during zinc deficiency and its usefulness to evaluate zinc status: A study of the effects of zinc deficiency during lactation. *Am. J. Physiol. Regul. Integr. Comp. Physiol.* **2016**, *310*, R459–R468. [CrossRef]
24. Dufner-Beattie, J.; Kuo, Y.M.; Gitschier, J.; Andrews, G.K. The adaptive response to dietary zinc in mice involves the differential cellular localization and zinc regulation of the zinc transporters ZIP4 and ZIP5. *J. Biol. Chem.* **2004**, *279*, 49082–49090. [CrossRef]
25. Weaver, B.P.; Dufner-Beattie, J.; Kambe, T.; Andrews, G.K. Novel zinc-responsive post-transcriptional mechanisms reciprocally regulate expression of the mouse Slc39a4 and Slc39a5 zinc transporters (Zip4 and Zip5). *Biol. Chem.* **2007**, *388*, 1301–1312. [CrossRef]
26. Mao, X.Q.; Kim, B.E.; Wang, F.; Eide, D.J.; Pettis, M.J. A histidine-rich cluster mediates the ubiquitination and degradation of the human zinc transporter, hZIP4, and protects against zinc cytotoxicity. *Biophys. J.* **2007**, *282*, 6992–7000. [CrossRef]
27. Guthrie, G.J.; Aydemir, T.B.; Troche, C.; Martin, A.B.; Chang, S.M.; Cousins, R.J. Influence of ZIP14 (slc39A14) on intestinal zinc processing and barrier function. *Am. J. Physiol. Gastrointest. Liver Physiol.* **2015**, *308*, G171–G178. [CrossRef]
28. Liuzzi, J.P.; Bobo, J.A.; Cui, L.; McMahon, R.J.; Cousins, R.J. Zinc transporters 1, 2 and 4 are differentially expressed and localized in rats during pregnancy and lactation. *J. Nutr.* **2003**, *133*, 342–351. [CrossRef]
29. Liuzzi, J.P.; Blanchard, R.K.; Cousins, R.J. Differential regulation of zinc transporter 1, 2, and 4 mRNA expression by dietary zinc in rats. *J. Nutr.* **2001**, *131*, 46–52. [CrossRef]

30. Jou, M.Y.; Hall, A.G.; Philipps, A.F.; Kelleher, S.L.; Lonnerdal, B. Tissue-specific alterations in zinc transporter expression in intestine and liver reflect a threshold for homeostatic compensation during dietary zinc deficiency in weanling rats. *J. Nutr.* **2009**, *139*, 835–841. [CrossRef]
31. Valentine, R.A.; Jackson, K.A.; Christie, G.R.; Mathers, J.C.; Taylor, P.M.; Ford, D. ZnT5 variant B is a bidirectional zinc transporter and mediates zinc uptake in human intestinal Caco-2 cells. *J. Biol. Chem.* **2007**, *282*, 14389–14393. [CrossRef]
32. Fukunaka, A.; Suzuki, T.; Kurokawa, Y.; Yamazaki, T.; Fujiwara, N.; Ishihara, K.; Migaki, H.; Okumura, K.; Masuda, S.; Yamaguchi-Iwai, Y.; et al. Demonstration and characterization of the heterodimerization of ZnT5 and ZnT6 in the early secretory pathway. *J. Biol. Chem.* **2009**, *284*, 30798–30806. [CrossRef]
33. Kirschke, C.P.; Huang, L.P. ZnT7, a novel mammalian zinc transporter, accumulates zinc in the Golgi apparatus. *J. Biol. Chem.* **2003**, *278*, 4096–4102. [CrossRef]
34. Devergnas, S.; Chimienti, F.; Naud, N.; Pennequin, A.; Coquerel, Y.; Chantegrel, J.; Favier, A.; Seve, M. Differential regulation of zinc efflux transporters ZnT-1, ZnT-5 and ZnT-7 gene expression by zinc levels: A real-time RT-PCR study. *Biochem. Pharmacol.* **2004**, *68*, 699–709. [CrossRef]
35. Cousins, R.J.; Blanchard, R.K.; Popp, M.P.; Liu, L.; Cao, J.; Moore, J.B.; Green, C.L. A global view of the selectivity of zinc deprivation and excess on genes expressed in human THP-1 mononuclear cells. *Proc. Natl. Acad. Sci. USA* **2003**, *100*, 6952–6957. [CrossRef]
36. Overbeck, S.; Uciechowski, P.; Ackland, M.L.; Ford, D.; Rink, L. Intracellular zinc homeostasis in leukocyte subsets is regulated by different expression of zinc exporters ZnT-1 to ZnT-9. *J. Leukoc. Biol.* **2008**, *83*, 368–380. [CrossRef]
37. Martin, L.; Lodemann, U.; Bondzio, A.; Gefeller, E.M.; Vahjen, W.; Aschenbach, J.R.; Zentek, J.; Pieper, R. A High Amount of Dietary Zinc Changes the Expression of Zinc Transporters and Metallothionein in Jejunal Epithelial Cells in Vitro and in Vivo but Does Not Prevent Zinc Accumulation in Jejunal Tissue of Piglets. *J. Nutr.* **2013**, *143*, 1205–1210. [CrossRef]
38. Miyoshi, Y.; Tanabe, S.; Suzuki, T. Cellular zinc is required for intestinal epithelial barrier maintenance via the regulation of claudin-3 and occludin expression. *Am. J. Physiol. Gastrointest. Liver Physiol.* **2016**, *311*, G105–G116. [CrossRef]
39. Camilleri, M. What is the leaky gut? Clinical considerations in humans. *Curr. Opin. Clin. Nutr. Metab. Care* **2021**, *24*, 473–482. [CrossRef]
40. Zhang, M.Q.; Yi, Z.Y.; Ma, L.Y.; Li, S.S.; Han, L.; Cao, Q. Zinc Level in Pathogenesis of Inflammatory Bowel Disease A Systematic Review and Meta-analysis. *Top. Clin. Nutr.* **2021**, *36*, 319–330. [CrossRef]
41. Griffin, I.J.; Kim, S.C.; Hicks, P.D.; Liang, L.K.; Abrams, S.A. Zinc metabolism in adolescents with Crohn's disease. *Pediatr. Res.* **2004**, *56*, 235–239. [CrossRef] [PubMed]
42. Main, A.N.; Hall, M.J.; Russell, R.I.; Fell, G.S.; Mills, P.R.; Shenkin, A. Clinical experience of zinc supplementation during intravenous nutrition in Crohn's disease: Value of serum and urine zinc measurements. *Gut* **1982**, *23*, 984–991. [CrossRef] [PubMed]
43. Li, D.L.; Achkar, J.P.; Haritunians, T.; Jacobs, J.P.; Hui, K.Y.; D'Amato, M.; Brand, S.; Radford-Smith, G.; Halfvarson, J.; Niess, J.H.; et al. A Pleiotropic Missense Variant in SLC39A8 Is Associated with Crohn's Disease and Human Gut Microbiome Composition. *Gastroenterology* **2016**, *151*, 724–732. [CrossRef] [PubMed]
44. Zhu, C.; Lv, H.; Chen, Z.; Wang, L.; Wu, X.; Chen, Z.; Zhang, W.; Liang, R.; Jiang, Z. Dietary Zinc Oxide Modulates Antioxidant Capacity, Small Intestine Development, and Jejunal Gene Expression in Weaned Piglets. *Biol. Trace Elem. Res.* **2017**, *175*, 331–338. [CrossRef]
45. Hu, C.H.; Qian, Z.C.; Song, J.; Luan, Z.S.; Zuo, A.Y. Effects of zinc oxide-montmorillonite hybrid on growth performance, intestinal structure, and function of broiler chicken. *Poult. Sci.* **2013**, *92*, 143–150. [CrossRef]
46. Wen, M.; Zhao, H.; Liu, G.M.; Chen, X.L.; Wu, B.; Tian, G.; Cai, J.Y.; Jia, G. Effect of Zinc Supplementation on Growth Performance, Intestinal Development, and Intestinal Barrier-Related Gene Expression in Pekin Ducks. *Biol. Trace Elem. Res.* **2018**, *183*, 351–360. [CrossRef]
47. Pearce, S.C.; Fernandez, M.V.S.; Torrison, J.; Wilson, M.E.; Baumgard, L.H.; Gabler, N.K. Dietary organic zinc attenuates heat stress-induced changes in pig intestinal integrity and metabolism. *J. Anim. Sci.* **2015**, *93*, 4702–4713. [CrossRef]
48. Liu, P.; Pieper, R.; Tedin, L.; Martin, L.; Meyer, W.; Rieger, J.; Plendl, J.; Vahjen, W.; Zentek, J. Effect of dietary zinc oxide on jejunal morphological and immunological characteristics in weaned piglets. *J. Anim. Sci.* **2014**, *92*, 5009–5018. [CrossRef]
49. Maares, M.; Keil, C.; Straubing, S.; Robbe-Masselot, C.; Haase, H. Zinc Deficiency Disturbs Mucin Expression, O-Glycosylation and Secretion by Intestinal Goblet Cells. *Int. J. Mol. Sci.* **2020**, *21*, 6149. [CrossRef]
50. Ohashi, W.; Kimura, S.; Iwanaga, T.; Furusawa, Y.; Irie, T.; Izumi, H.; Watanabe, T.; Hijikata, A.; Hara, T.; Ohara, O.; et al. Zinc Transporter SLC39A7/ZIP7 Promotes Intestinal Epithelial Self-Renewal by Resolving ER Stress. *PLoS Genet.* **2016**, *12*, e1006349. [CrossRef]
51. Finamore, A.; Massimi, M.; Conti Devirgiliis, L.; Mengheri, E. Zinc deficiency induces membrane barrier damage and increases neutrophil transmigration in Caco-2 cells. *J. Nutr.* **2008**, *138*, 1664–1670. [CrossRef]
52. Song, Z.H.; Xiao, K.; Ke, Y.L.; Jiao, L.F.; Hu, C.H. Zinc oxide influences mitogen-activated protein kinase and TGF-beta 1 signaling pathways, and enhances intestinal barrier integrity in weaned pigs. *Innate Immun.* **2015**, *21*, 341–348. [CrossRef]
53. Song, Z.H.; Ke, Y.L.; Xiao, K.; Jiao, L.F.; Hong, Q.H.; Hu, C.H. Diosmectite-zinc oxide composite improves intestinal barrier restoration and modulates TGF-beta 1, ERK1/2, and Akt in piglets after acetic acid challenge. *J. Anim. Sci.* **2015**, *93*, 1599–1607. [CrossRef]

54. Cohen, L.; Sekler, I.; Hershfinkel, M. The zinc sensing receptor, ZnR/GPR39, controls proliferation and differentiation of colonocytes and thereby tight junction formation in the colon. *Cell Death Dis.* **2014**, *5*, e1307. [CrossRef]
55. Shao, Y.X.; Wolf, P.G.; Guo, S.S.; Guo, Y.M.; Gaskins, H.R.; Zhang, B.K. Zinc enhances intestinal epithelial barrier function through the PI3K/AKT/mTOR signaling pathway in Caco-2 cells. *J. Nutr. Biochem.* **2017**, *43*, 18–26. [CrossRef]
56. Sharir, H.; Zinger, A.; Nevo, A.; Sekler, I.; Hershfinkel, M. Zinc Released from Injured Cells Is Acting via the Zn^{2+}-sensing Receptor, ZnR, to Trigger Signaling Leading to Epithelial Repair. *J. Biol. Chem.* **2010**, *285*, 26097–26106. [CrossRef]
57. Sunuwar, L.; Medini, M.; Cohen, L.; Sekler, I.; Hershfinkel, M. The zinc sensing receptor, ZnR/GPR39, triggers metabotropic calcium signalling in colonocytes and regulates occludin recovery in experimental colitis. *Philos. Trans. R. Soc. Lond. B Biol. Sci.* **2016**, *371*, 20150420.
58. Shao, Y.X.; Lei, Z.; Wolf, P.G.; Gao, Y.; Guo, Y.M.; Zhang, B.K. Zinc Supplementation, via GPR39, Upregulates PKC zeta to Protect Intestinal Barrier Integrity in Caco-2 Cells Challenged by Salmonella enterica Serovar Typhimurium. *J. Nutr.* **2017**, *147*, 1282–1289. [CrossRef]
59. Quarterman, J.; Jackson, F.A.; Morrison, J.N. The effect of zinc deficiency on sheep intestinal mucin. *Life Sci.* **1976**, *19*, 979–986. [CrossRef]
60. Zhang, Y.F.; Cougnon, F.B.L.; Wanniarachchi, Y.A.; Hayden, J.A.; Nolan, E.M. Reduction of Human Defensin 5 Affords a High-Affinity Zinc-Chelating Peptide. *ACS Chem. Biol.* **2013**, *8*, 1907–1911. [CrossRef]
61. Chakraborti, S.; Chatterjee, T.; Joshi, P.; Poddar, A.; Bhattacharyya, B.; Singh, S.P.; Gupta, V.; Chakrabarti, P. Structure and Activity of Lysozyme on Binding to ZnO Nanoparticles. *Langmuir* **2010**, *26*, 3506–3513. [CrossRef]
62. Podany, A.; Geletzke, A.; Lee, S.; Soybel, D.; Kelleher, S. The Zinc Transporter ZnT2 is Necessary for Stabilization of Granule Contents in Intestinal Paneth Cell Granules. *FASEB J.* **2015**, *29*, 1011.1. [CrossRef]
63. Podany, A.B.; Wright, J.; Lamendella, R.; Soybel, D.I.; Kelleher, S.L. ZnT2-Mediated Zinc Import Into Paneth Cell Granules Is Necessary for Coordinated Secretion and Paneth Cell Function in Mice. *Cell. Mol. Gastroenterol. Hepatol.* **2016**, *2*, 369–383. [CrossRef]
64. Jarosz, L.; Marek, A.; Gradzki, Z.; Kwiecien, M.; Zylinska, B.; Kaczmarek, B. Effect of feed supplementation with zinc glycine chelate and zinc sulfate on cytokine and immunoglobulin gene expression profiles in chicken intestinal tissue. *Poult. Sci.* **2017**, *96*, 4224–4235. [CrossRef]
65. Pabst, O. New concepts in the generation and functions of IgA. *Nat. Rev. Immunol.* **2012**, *12*, 821–832. [CrossRef]
66. Han, X.Y.; Ma, Y.F.; Lv, M.Y.; Wu, Z.P.; Qian, L.C. Chitosan-zinc chelate improves intestinal structure and mucosal function and decreases apoptosis in ileal mucosal epithelial cells in weaned pigs. *Br. J. Nutr.* **2014**, *111*, 1405–1411. [CrossRef]
67. Lahiri, A.; Abraham, C. Activation of pattern recognition receptors up-regulates metallothioneins, thereby increasing intracellular accumulation of zinc, autophagy, and bacterial clearance by macrophages. *Gastroenterology* **2014**, *147*, 835–846. [CrossRef]
68. Laurin, D.E.; Barnes, D.M.; Klasing, K.C. Rates of metallothionein synthesis, degradation and accretion in a chicken macrophage cell line. *Proc. Soc. Exp. Biol. Med.* **1990**, *194*, 157–164. [CrossRef]
69. Kelly, P.; Feakins, R.; Domizio, P.; Murphy, J.; Bevins, C.; Wilson, J.; Mcphail, G.; Poulsom, R.; Dhaliwal, W. Paneth cell granule depletion in the human small intestine under infective and nutritional stress. *Clin. Exp. Immunol.* **2004**, *135*, 303–309. [CrossRef]
70. Glahn, R.P.; Tako, E.; Reed, S.; Neuman, H.; Moscovich, S.; Koren, O. Chronic Zinc Deficiency Alters Chick Gut Microbiota Composition and Function. *Nutrients* **2015**, *7*, 9768–9784.
71. Zackular, J.P.; Moore, J.L.; Jordan, A.T.; Juttukonda, L.J.; Noto, M.J.; Nicholson, M.R.; Crews, J.D.; Semler, M.W.; Zhang, Y.; Ware, L.B.; et al. Dietary zinc alters the microbiota and decreases resistance to Clostridium difficile infection. *Nat. Med.* **2016**, *22*, 1330–1334. [CrossRef] [PubMed]
72. Vahjen, W.; Pieper, R.; Zentek, J. Bar-Coded Pyrosequencing of 16S rRNA Gene Amplicons Reveals Changes in Ileal Porcine Bacterial Communities Due to High Dietary Zinc Intake. *Appl. Environ. Microb.* **2010**, *76*, 6689–6691. [CrossRef] [PubMed]
73. Watly, J.; Potocki, S.; Rowinska-Zyrek, M. Zinc Homeostasis at the Bacteria/Host Interface-From Coordination Chemistry to Nutritional Immunity. *Chem.-Eur. J.* **2016**, *22*, 15992–16010. [CrossRef] [PubMed]
74. Kehl-Fie, T.E.; Skaar, E.P. Nutritional immunity beyond iron: A role for manganese and zinc. *Curr. Opin. Chem. Biol.* **2010**, *14*, 218–224. [CrossRef]
75. Geboes, K.; Riddell, R.; Ost, A.; Jensfelt, B.; Persson, T.; Lofberg, R. A reproducible grading scale for histological assessment of inflammation in ulcerative colitis. *Gut* **2000**, *47*, 404–409. [CrossRef]
76. Naftali, T.; Reshef, L.; Kovacs, A.; Porat, R.; Amir, I.; Konikoff, F.M.; Gophna, U. Distinct Microbiotas are Associated with Ileum-Restricted and Colon-Involving Crohn's Disease. *Inflamm. Bowel Dis.* **2016**, *22*, 293–302. [CrossRef]
77. Lakatos, P.L. Environmental Factors Affecting Inflammatory Bowel Disease: Have We Made Progress? *Dig. Dis.* **2009**, *27*, 215–225. [CrossRef]
78. Karlinger, K.; Gyorke, T.; Mako, E.; Mester, A.; Tarjan, Z. The epidemiology and the pathogenesis of inflammatory bowel disease. *Eur. J. Radiol.* **2000**, *35*, 154–167. [CrossRef]
79. Wasser, T.E.; Reed, J.F.; Moser, K.; Robson, P.; Faust, L.; Fink, L.L.; Wunderler, D. Nutritional Assessment and Disease-Activity for Patients with Inflammatory Bowel-Disease. *Can. J. Gastroenterol.* **1995**, *9*, 131–136. [CrossRef]
80. Siva, S.; Rubin, D.T.; Gulotta, G.; Wroblewski, K.; Pekow, J. Zinc Deficiency is Associated with Poor Clinical Outcomes in Patients with Inflammatory Bowel Disease. *Inflamm. Bowel Dis.* **2017**, *23*, 152–157. [CrossRef]

81. Ananthakrishnan, A.N.; Khalili, H.; Song, M.; Higuchi, L.M.; Richter, J.M.; Chan, A.T. Zinc intake and risk of Crohn's disease and ulcerative colitis: A prospective cohort study. *Int. J. Epidemiol.* **2015**, *44*, 1995–2005. [CrossRef]
82. Higashimura, Y.; Takagi, T.; Naito, Y.; Uchiyama, K.; Mizushima, K.; Tanaka, M.; Hamaguchi, M.; Itoh, Y. Zinc Deficiency Activates the IL-23/Th17 Axis to Aggravate Experimental Colitis in Mice. *J. Crohns Colitis.* **2020**, *14*, 856–866. [CrossRef]
83. Maywald, M.; Wessels, I.; Rink, L. Zinc Signals and Immunity. *Int. J. Mol. Sci.* **2017**, *18*, 2222. [CrossRef]
84. Wiegand, S.; Zakrzewski, S.S.; Eichner, M.; Schulz, E.; Gunzel, D.; Pieper, R.; Rosenthal, R.; Barmeyer, C.; Bleich, A.; Dobrindt, U.; et al. Zinc treatment is efficient against Escherichia coli alpha-haemolysin-induced intestinal leakage in mice. *Sci. Rep.* **2017**, *7*, 45649. [CrossRef]
85. El-Salhy, M. Recent developments in the pathophysiology of irritable bowel syndrome. *World J. Gastroenterol.* **2015**, *21*, 7621–7636. [CrossRef]
86. Ohman, L.; Simren, M. New insights into the pathogenesis and pathophysiology of irritable bowel syndrome. *Dig. Liver. Dis.* **2007**, *39*, 201–215. [CrossRef]
87. Bahrami, A.; Gonoodi, K.; Khayyatzadeh, S.S.; Tayefi, M.; Darroudi, S.; Bahrami-Taghanaki, H.; Eslami, S.; Jaberi, N.; Ferns, G.A.; Farahmand, K.; et al. The association of trace elements with premenstrual syndrome, dysmenorrhea and irritable bowel syndrome in adolescents. *Eur. J. Obstet. Gynecol. Reprod. Biol.* **2019**, *233*, 114–119. [CrossRef]
88. Torres, M.J.; Sabate, J.M.; Bouchoucha, M.; Buscail, C.; Hercberg, S.; Julia, C. Food consumption and dietary intakes in 36,448 adults and their association with irritable bowel syndrome: Nutrinet-Sante study. *Ther. Adv. Gastroenterol.* **2018**, *11*, 1756283X17746625. [CrossRef]
89. Rezazadegan, M.; Soheilipour, M.; Tarrahi, M.J.; Amani, R. Correlation between Zinc Nutritional Status with Serum Zonulin and Gastrointestinal Symptoms in Diarrhea-Predominant Irritable Bowel Syndrome: A Case-Control Study. *Dig. Dis. Sci.* **2022**. [CrossRef]
90. Vela, G.; Stark, P.; Socha, M.; Sauer, A.K.; Hagmeyer, S.; Grabrucker, A.M. Zinc in Gut-Brain Interaction in Autism and Neurological Disorders. *Neural Plast.* **2015**, *2015*, 972791. [CrossRef]
91. Russo, A.J. Decreased zinc and increased copper in individuals with anxiety. *Nutr. Metab. Insights* **2011**, *4*, 1–5. [CrossRef]
92. Hujoel, I.A. Nutritional status in irritable bowel syndrome: A North American population-based study. *JGH Open Access J. Gastroenterol. Hepatol.* **2020**, *4*, 656–662. [CrossRef]
93. Russo, A.J. Increased Copper in Individuals with Autism Normalizes Post Zinc Therapy More Efficiently in Individuals with Concurrent GI Disease. *Nutr. Metab. Insights* **2011**, *4*, 49–54. [CrossRef]
94. Zhang, L.; Song, J.; Hou, X.H. Mast Cells and Irritable Bowel Syndrome: From the Bench to the Bedside. *J. Neurogastroenterol.* **2016**, *22*, 181–192. [CrossRef]
95. Ding, N.; Jiang, J.; Qin, P.P.; Wang, Q.X.; Hu, J.T.; Li, Z.G. Mast cells are important regulator of acupoint sensitization via the secretion of tryptase, 5-hydroxytryptamine, and histamine. *PLoS ONE* **2018**, *13*, e0194022. [CrossRef]
96. Wouters, M.M.; Balemans, D.; Van Wanrooy, S.; Dooley, J.; Cibert-Goton, V.; Alpizar, Y.A.; Valdez-Morales, E.E.; Nasser, Y.; Van Veldhoven, P.P.; Vanbrabant, W.; et al. Histamine Receptor H1-Mediated Sensitization of TRPV1 Mediates Visceral Hypersensitivity and Symptoms in Patients With Irritable Bowel Syndrome. *Gastroenterology* **2016**, *150*, 875–887.e9. [CrossRef]
97. Crowell, M.D.; Shetzline, M.A.; Moses, P.L.; Mawe, G.M.; Talley, N.J. Enterochromaffin cells and 5-HT signaling in the pathophysiology of disorders of gastrointestinal function. *Curr. Opin. Investig. Drugs* **2004**, *5*, 55–60.
98. Keating, D.J.; Spencer, N.J. Release of 5-Hydroxytryptamine from the Mucosa Is Not Required for the Generation or Propagation of Colonic Migrating Motor Complexes. *Gastroenterology* **2010**, *138*, 659–670.E2. [CrossRef]
99. Fu, R.; Chen, M.X.; Chen, Y.; Mao, G.Q.; Liu, S.Y. Expression and clinical significance of 5-HT and 5-HT3R in the intestinal mucosa of patient with diarrhea-type irritable bowel syndrome. *Exp. Ther. Med.* **2019**, *17*, 3077–3082. [CrossRef]
100. Gunn, D.; Garsed, K.; Lam, C.; Singh, G.; Lingaya, M.; Wahl, V.; Niesler, B.; Henry, A.; Hall, I.P.; Whorwell, P.; et al. Abnormalities of mucosal serotonin metabolism and 5-HT3 receptor subunit 3C polymorphism in irritable bowel syndrome with diarrhoea predict responsiveness to ondansetron. *Aliment. Pharm. Ther.* **2019**, *50*, 538–546. [CrossRef]
101. Pihl, E.; Gustafso, G.T. Heavy Metals in Rat Mast Cell Granules. *Lab. Investig.* **1967**, *17*, 588–598.
102. Yamasaki, S.; Sakata-Sogawa, K.; Hasegawa, A.; Suzuki, T.; Kabu, K.; Sato, E.; Kurosaki, T.; Yamashita, S.; Tokunaga, M.; Nishida, K.; et al. Zinc is a novel intracellular second messenger. *J. Cell Biol.* **2007**, *177*, 637–645. [CrossRef] [PubMed]
103. Kabu, K.; Yamasaki, S.; Kamimura, D.; Ito, Y.; Hasegawa, A.; Sato, E.; Kitamura, H.; Nishida, K.; Hirano, T. Zinc is required for Fc epsilon RI-mediated mast cell activation. *J. Immunol.* **2006**, *177*, 1296–1305. [CrossRef] [PubMed]
104. Takeda, A.; Itoh, H.; Imano, S.; Oku, N. Impairment of GABAergic neurotransmitter system in the amygdala of young rats after 4-week zinc deprivation. *Neurochem. Int.* **2006**, *49*, 746–750. [CrossRef]
105. Moloney, R.D.; O'Mahony, S.M.; Dinan, T.G.; Cryan, J.F. Stress-induced visceral pain: Toward animal models of irritable-bowel syndrome and associated comorbidities. *Front. Psychiatry* **2015**, *6*, 15. [CrossRef]
106. Lepping, P.; Huber, M. Role of Zinc in the Pathogenesis of Attention-Deficit Hyperactivity Disorder Implications for Research and Treatment. *CNS Drugs* **2010**, *24*, 721–728.
107. Toth, K. Zinc in Neurotransmission. *Annu. Rev. Nutr.* **2011**, *31*, 139–153. [CrossRef]
108. Posserud, I.; Agerforz, P.; Ekman, R.; Bjornsson, E.S.; Abrahamsson, H.; Simren, M. Altered visceral perceptual and neuroendocrine response in patients with irritable bowel syndrome during mental stress. *Gut* **2004**, *53*, 1102–1108. [CrossRef]

109. Galligan, J.J. 5-hydroxytryptamine, ulcerative colitis, and irritable bowel syndrome: Molecular connections. *Gastroenterology* **2004**, *126*, 1897–1899. [CrossRef]
110. Zhang, C.F.; Cheng, R.Q.; Ding, J.; Li, X.J.; Niu, H.W.; Li, X. Serum Copper and Zinc Levels and Colorectal Cancer in Adults: Findings from the National Health and Nutrition Examination 2011–2016. *Biol. Trace Elem. Res.* **2021**, *200*, 2033–2039. [CrossRef]
111. Khoshdel, Z.; Naghibalhossaini, F.; Abdollahi, K.; Shojaei, S.; Moradi, M.; Malekzadeh, M. Serum Copper and Zinc Levels Among Iranian Colorectal Cancer Patients. *Biol. Trace Elem. Res.* **2016**, *170*, 294–299. [CrossRef]
112. Christudoss, P.; Selvakumar, R.; Pulimood, A.B.; Fleming, J.J.; Mathew, G. Zinc and Zinc Related Enzymes in Precancerous and Cancerous Tissue in the Colon of Dimethyl Hydrazine Treated Rats. *Asian Pac. J. Cancer Prev.* **2012**, *13*, 487–492. [CrossRef]
113. Li, P.W.; Xu, J.M.; Shi, Y.; Ye, Y.; Chen, K.; Yang, J.; Wu, Y.H. Association between zinc intake and risk of digestive tract cancers: A systematic review and meta-analysis. *Clin. Nutr.* **2014**, *33*, 415–420. [CrossRef]
114. He, G.Y.; Zhu, H.F.; Yao, Y.K.; Chai, H.R.; Wang, Y.Q.; Zhao, W.L.; Fu, S.Z.; Wang, Y.X. Cysteine-rich intestinal protein 1 silencing alleviates the migration and invasive capability enhancement induced by excessive zinc supplementation in colorectal cancer cells. *Am. J. Transl. Res.* **2019**, *11*, 3578–3588.
115. Prasad, A.S.; Beck, F.W.J.; Bao, B.; Fitzgerald, J.T.; Snell, D.C.; Steinberg, J.D.; Cardozo, L.J. Zinc supplementation decreases incidence of infections in the elderly: Effect of zinc on generation of cytokines and oxidative stress. *Am. J. Clin. Nutr.* **2007**, *85*, 837–844. [CrossRef]
116. Shaheen, A.A.; Elfattah, A.A.A. Effect of Dietary Zinc on Lipid-Peroxidation, Glutathione, Protein Thiols Levels and Superoxide-Dismutase Activity in Rat-Tissues. *Int. J. Biochem. Cell Biol.* **1995**, *27*, 89–95. [CrossRef]
117. Zhang, Q.R.; Xue, Y.; Fu, Y.X.; Bao, B.W.; Guo, M.Y. Zinc Deficiency Aggravates Oxidative Stress Leading to Inflammation and Fibrosis in Lung of Mice. *Biol. Trace Elem. Res.* **2021**. [CrossRef]
118. Al-saran, N.; Subash-Babu, P.; Al-Nouri, D.M.; Alfawaz, H.A.; Alshatwi, A.A. Zinc enhances CDKN2A, pRb1 expression and regulates functional apoptosis via upregulation of p53 and p21 expression in human breast cancer MCF-7 cell. *Environ. Toxicol. Pharmacol.* **2016**, *47*, 19–27. [CrossRef]
119. Ho, E. Zinc deficiency, DNA damage and cancer risk. *J. Nutr. Biochem.* **2004**, *15*, 572–578. [CrossRef]
120. Song, Y.; Leonard, S.W.; Traber, M.G.; Ho, E. Zinc Deficiency Affects DNA Damage, Oxidative Stress, Antioxidant Defenses, and DNA Repair in Rats. *J. Nutr.* **2009**, *139*, 1626–1631. [CrossRef]
121. Ho, E.; Ames, B.N. Low intracellular zinc induces oxidative DNA damage, disrupts p53, NF kappa B, and AP1 DNA binding, and affects DNA repair in a rat glioma cell line. *Proc. Natl. Acad. Sci. USA* **2002**, *99*, 16770–16775. [CrossRef] [PubMed]
122. Sheng, N.Q.; Yan, L.; You, W.Q.; Tan, G.W.; Gong, J.F.; Chen, H.Q.; Yang, Y.; Hu, L.D.; Wang, Z.G. Knockdown of SLC39A7 inhibits cell growth and induces apoptosis in human colorectal cancer cells. *Acta Biochim. Biophys. Sin.* **2017**, *49*, 926–934. [CrossRef] [PubMed]
123. Barresi, V.; Valenti, G.; Spampinato, G.; Musso, N.; Castorina, S.; Rizzarelli, E.; Condorelli, D.F. Transcriptome analysis reveals an altered expression profile of zinc transporters in colorectal cancer. *J. Cell. Biochem.* **2018**, *119*, 9707–9719. [CrossRef] [PubMed]
124. Cichewicz, A.B.; Mearns, E.S.; Taylor, A.; Boulanger, T.; Gerber, M.; Leffler, D.A.; Drahos, J.; Sanders, D.S.; Thomas Craig, K.J.; Lebwohl, B. Diagnosis and Treatment Patterns in Celiac Disease. *Dig. Dis. Sci.* **2019**, *64*, 2095–2106. [CrossRef]
125. Ryan, M.; Grossman, S. Celiac disease: Implications for patient management. *Gastroenterol. Nurs.* **2011**, *34*, 225–228. [CrossRef]
126. Fasano, A. Clinical presentation of celiac disease in the pediatric population. *Gastroenterology* **2005**, *128* (Suppl. S1), S68–S73. [CrossRef]
127. Ciclitira, P.J.; Johnson, M.W.; Dewar, D.H.; Ellis, H.J. The pathogenesis of coeliac disease. *Mol. Asp. Med.* **2005**, *26*, 421–458. [CrossRef]
128. Li, Z.; Zhang, L.; Li, L.; Du, Z. Evaluation of Serum Levels of Copper and Zinc in Patients with Celiac Disease Seropositivity: Findings from the National Health and Nutrition Examination Survey. *Biol. Trace Elem. Res.* **2022**. [CrossRef]
129. Caruso, R.; Pallone, F.; Stasi, E.; Romeo, S.; Monteleone, G. Appropriate nutrient supplementation in celiac disease. *Ann. Med.* **2013**, *45*, 522–531. [CrossRef]
130. Skrovanek, S.; DiGuilio, K.; Bailey, R.; Huntington, W.; Urbas, R.; Mayilvaganan, B.; Mercogliano, G.; Mullin, J.M. Zinc and gastrointestinal disease. *World J. Gastrointest. Pathophysiol.* **2014**, *5*, 496–513. [CrossRef]

biomolecules

MDPI

Review

Zinc in Regulating Protein Kinases and Phosphatases in Neurodegenerative Diseases

Hui-Liang Zhang, Xiao-Chuan Wang and Rong Liu *

Department of Pathophysiology, Key Laboratory of Ministry of Education for Neurological Disorders, School of Basic Medicine, Tongji Medical College, Huazhong University of Science and Technology, Wuhan 430030, China; d202181578@hust.edu.cn (H.-L.Z.); wxch@mails.tjmu.edu.cn (X.-C.W.)

* Correspondence: rong.liu@hust.edu.cn

Abstract: Zinc is essential for human growth and development. As a trace nutrient, zinc plays important roles in numerous signal transduction pathways involved in distinct physiologic or pathologic processes. Protein phosphorylation is a posttranslational modification which regulates protein activity, degradation, and interaction with other molecules. Protein kinases (PKs) and phosphatases (PPs), with their effects of adding phosphate to or removing phosphate from certain substrates, are master regulators in controlling the phosphorylation of proteins. In this review, we summarize the disturbance of zinc homeostasis and role of zinc disturbance in regulating protein kinases and protein phosphatases in neurodegenerative diseases, with the focus of that in Alzheimer's disease, providing a new perspective for understanding the mechanisms of these neurologic diseases.

Keywords: zinc; protein kinases; protein phosphatases; neurodegenerative diseases; Alzheimer's disease

1. Introduction

1.1. Zinc Balance in the Human Body

Minerals in the human body that are less than 0.01% of body weight are called micronutrients. Zinc is an indispensable trace nutrient for life. Initially, zinc ions were only found to be essential for the normal growth and development of plants and animals [1,2], and it was not until 1961 that zinc was identified as an essential micronutrient for human beings. It is currently ranked first among the 14 essential micronutrients identified by the World Health Organization. Zinc is also the second most abundant micronutrient in the human body, following only iron. The adult human body contains about 2–3 g of zinc, of which 60% is found in skeletal muscle, 30% in bone, 5% in liver and skin, 1.5% in brain, and the kidneys, heart, and other tissues in total contribute less than 2% [3]. Serum zinc occupies only 0.1% of total body zinc, which is exactly the daily level of zinc intake. Zinc homeostasis is tightly controlled through the coordinated regulation of zinc absorption, efflux, distribution, and storage (Figure 1) [4,5]. Zinc performs unique and broad functions in numerous biological processes, including cell division, growth, differentiation, and signal transduction. In these processes, zinc ions serve as structural, catalytic, or regulatory components in proteins, such as transcription factors, enzymes, transporter proteins, and receptors [5–7]. Hence, zinc is extremely valuable to the organism. Disruption in zinc homeostasis is found to be associated with a variety of diseases, including asthma, diabetes, and Alzheimer's disease [8–10].

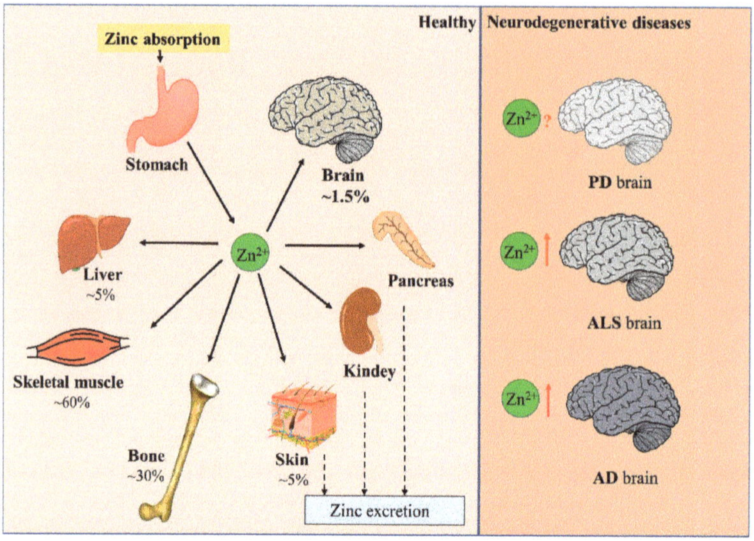

Figure 1. Distribution of zinc in the body. Dietary zinc is absorbed by the digestive system and then distributed to peripheral tissues. Approximately 60% of zinc is stored in skeletal muscle, 30% in bone, and 5% in liver and skin. Excess zinc is excreted through gastrointestinal secretions and mucosal cells. Brain zinc levels change in patients with neurodegenerative diseases such as PD, ALS, and AD.

1.2. Distribution of Zinc and the Maintenance of Zinc Homeostasis in the Brain

The adult human brain contains 0.04 g of zinc, accounting for only 1.5% of the total body zinc [3]. There are three zinc pools in the brain: vesicular zinc, protein-bound zinc, and free zinc ions. Ultrastructural observation shows that abundant free zinc ions are present in presynaptic glutamatergic vesicles called "zincergic" synapses. Vesicular zinc is found in the cerebral cortex, amygdala, hippocampus, and olfactory bulb, which accounts for 5~15% of the brain's zinc content and is especially abundant in the hippocampus [11,12]. Endogenous zinc homeostasis is mainly regulated by zinc-binding proteins, metallothioneins (MTs), and zinc transporter proteins. There are four MTs in human and mammalian cells, of which MT-3 is dominated in the brain. MT-3 serves to bind and store zinc and thus protect cells from the toxic effects of excessive free zinc ions [13]. Zinc transport proteins are divided into two families: the ZnTs (zinc transporter proteins) and the ZIPs (Zrt and Trt-like proteins). ZnTs are responsible for removing zinc out of the cells or transporting zinc into cellular organelles, thus reducing the concentration of zinc in the cytoplasm, while ZIPs play the opposite role [14,15]. There is a growing body of evidence showing that ZnTs and ZIPs play an instrumental role in maintaining zinc homeostasis, which profoundly affects human health, directly or indirectly [16].

Zinc is required for normal brain development, and it also plays important roles in the physiological function of the central nervous system (CNS) [17]. The average intracellular zinc ion concentration in brain cells is about 150 μM [18,19]. Zinc ion concentration in the synaptic cleft is 15–30 μM under the resting state and can be up to about 300 μM during neural activity [20,21]. Sub-toxic doses of zinc can cause the activation of microglia (>30 μM) or astrocytes (>100 μM) via the NF-κB or Stat3 pathways, respectively [22,23]; higher concentrations of exogenously applied zinc (>300 μM) can cause neuronal death in vitro [24]. The increase in intracellular zinc ions caused by noxious stimuli can also lead to extensive neuronal death [25]. Owing to the multifaceted roles of zinc in biological processes, zinc dyshomeostasis in the brain participates in the development of many neurodegenerative diseases [26].

1.3. Limited Effect of Peripheral Zinc Supplementation on Brain Zinc Level

Zinc homeostasis in the brain is tightly controlled and is not easily altered by a peripheral zinc load. It was shown that a slight increase in ZnT2 was observed after treatment of the BBB model with "moderate zinc excess" (50 μM), with no significant difference in the expression of ZIP1. Our previous investigation also showed that high-dose zinc supplementation significantly increased serum zinc levels in mice but did not increase brain zinc levels. On the contrary, decreased zinc levels were observed in the hippocampal brain region, which may be a result of the loss of hippocampal neurons (abundant with zinc) induced by the toxicity of excessive zinc supplementation [27].

2. Disturbance of Zinc Homeostasis in Neurodegenerative Diseases

Due to the multiple roles of zinc in lots of physiological processes, the maintenance of zinc homeostasis is of particular importance for normal brain function. Either zinc deficiency or overload may be harmful and may exacerbate brain damage in neurological disorders [26]. In addition, even though the total zinc in the brain is normal, the mislocalization of zinc ions, for example, abnormal zinc accumulation with pathologic proteins, or excessive zinc release from intracellular zinc-binding proteins may also result in dysfunction of the nervous system. The important role of zinc in the brain has led to the exploration of the effects of zinc disturbance in many neurodegenerative diseases, particularly in Parkinson's disease (PD), amyotrophic lateral sclerosis (ALS), and Alzheimer's disease (AD) [28–31].

2.1. Parkinson's Disease (PD)

Parkinson's disease (PD) is a severe motor disorder characterized by tonicity, tremor, and bradykinesia. In the PD brain, α-synuclein aggregates, and neurofibrillary tangles are formed in the substantia nigra pars compacta, with a selective loss of dopaminergic neurons [32]. The deficit of dopaminergic neurons in substantia nigra causes a striatal dopamine deficiency and subsequent alteration in basal ganglia physiology, which is thought to be the key mechanism leading to the development of clinical symptoms [33,34]. Dopamine neurons in the ventral tegmental area (VTA), which is adjacent to the substantia nigra and regulates emotional behavior, are less damaged [35,36].

The relationship between zinc dyshomeostasis and PD is ambiguous, with some studies showing reduced serum zinc levels [37,38], with others showing increased serum zinc levels in PD patients. No significant change in cerebrospinal fluid zinc levels was observed [39,40]. Evidence supporting the relationship between low zinc and PD is that methamphetamine causes dopaminergic cell death through the production of reactive oxygen species and increasing the total amount of α-synuclein (α-syn), a major component of Lewy bodies, and zinc pretreatment reverses the above phenomena by increasing metallothionein expression in vitro [41,42]. In the Drosophila PD model, zinc supplementation also increased the lifespan and motility [43]. On the other hand, there is numerous evidence showing the toxicity of excess zinc ions in PD-related degenerated dopaminergic neurons. Zinc exposure is an environmental risk factor for PD [44]. The autopsy results showed zinc deposition in the substantia nigra and striatum in PD patients [45]. Consistent with these findings, zinc treatment significantly enhanced MPTP-induced dopaminergic neuronal loss in mice [34,46]. Taken together, these studies suggest that the imbalance of zinc may play a dual role in the development of PD, which depends on the different signaling pathways influenced by zinc in different disease stages.

2.2. Amyotrophic Lateral Sclerosis (ALS)

Amyotrophic lateral sclerosis (ALS) is an incurable neurodegenerative disease in which motor neurons in the spinal cord and brain progressively deteriorate, leading to progressive muscle atrophy, paralysis, and ultimately death. The most commonly studied risk factors for ALS include genetics, environment, and lifestyle. Metals with a potential relationship with ALS progress include copper, aluminum, arsenic, cadmium, cobalt, zinc, vanadium, and uranium, all of which are significantly elevated in the cerebrospinal fluid of ALS patients [47], while the precise roles of these metals in the brain in disease progression need further exploration.

It is suspected that aggregates of a protein called copper–zinc superoxide dismutase 1 (SOD1) in nerve cells may play a crucial role in disease development. More than 140 SOD1 gene mutations have been found associated with ALS, with varying degrees of severity. TDP-43 gene mutation is also found in 5–10% of familial ALS cases. Significantly increased levels of Mn, Cu, and Zn were found in the spinal cord of TDP-43 A315T mice [48]. The immunoreactive activity of metallothionein was found to be elevated in the brain and liver in ALS patients [49]. Similarly, in SOD1-Gly93Ala transgenic mice, increased MT1, MT2, and MT3 expression in astrocytes and increased MT3 in neurons were observed. The elevated expression of metallothionein may be compensatory. Taking together, it seems that zinc exerts deleterious effects during ALS development.

2.3. Alzheimer's Disease (AD)

Alzheimer's disease (AD) is a progressive neurodegeneration that is currently receiving the most attention in aging-related disorders. It is expected that by 2050, there will be 130 million AD patients worldwide [50,51]. AD patients exhibit progressive cognitive decline manifested by memory loss, learning and orientation difficulties, behavioral changes, and dementia [51]. Amyloid plaques aggregated by β-amyloid (Aβ) peptides and neuronal fibrillary tangles (NFTs) formed by hyperphosphorylated tau protein are two typical pathological features in AD brain [52,53]. Zinc dyshomeostasis was one of the dominated pathological characters in AD. The levels of zinc ions in the amygdala, hippocampus, cerebellum, and superior temporal gyrus, which are all susceptible brain regions in AD, were found elevated [54–56]. These changes may be a result of the capture of zinc by Aβ. Zinc was highly enriched in senile plaques in the brains of AD mice [57]. Amyloid precursor protein (APP) and Aβ have zinc-binding sites [58], and studies show that zinc promotes Aβ aggregation directly in vitro [59,60]. On the other side, zinc exacerbated Aβ toxicity, and zinc chelation into amyloid deposits led to the loss of functional zinc at the synapse [61]. Besides zinc ions, zinc transporters also showed significant changes in AD brain. The levels of ZIPs, ZnTs, and MTs, which regulate brain zinc levels, were increased or decreased in the early or late stages of AD [62–64]. Postmortem studies of AD brains showed a significant decrease in ZnT-1, along with an increase in ZnT-6 in the hippocampus/parahippocampal gyrus compared to normal control subjects. These processes correlate with disease progression and the severity of cognitive impairment [64,65]. The expression of ZnTs (ZnT1, ZnT3, ZnT4, ZnT6, and ZnT7) were extensively increased in the senile plaque of APP/PS1 mouse models [16,58]. Till now, what is the initial cause of zinc disturbance, and whether the changes of zinc transporters are causes or results of zinc disturbance in AD, remain unclarified.

The pathologic role of zinc in AD is also manifested by the effects of zinc on promoting tauopathy [61,66]. Zinc can cause tau hyperphosphorylation through inhibiting the protein-phosphatase 2A (PP-2A) [67] and activating tau kinases [68,69]. All of these mechanisms lead to increased Aβ deposition and P-tau, ultimately exacerbating the pathological progression of AD.

3. Zinc-Regulated Protein Kinases and Protein Phosphatases

Due to the multiple functions of zinc in cell growth and differentiation, metabolism, apoptosis, and neurogenesis, zinc disturbance may participate in the development of neurodegenerative diseases through distinct molecular mechanisms in different cell types and disease stages. Especially in the development of AD, these mechanisms have been thoroughly discussed in several outstanding reviews [70–72]. Protein phosphorylation is a posttranslational modification that regulates the protein activity, degradation, and interaction with other molecules. Protein kinases (PKs) and phosphatases (PPs), with their effects of adding phosphate to or removing phosphate from the substrates, are key regulators in controlling the phosphorylation of proteins. Abnormal protein phosphorylation is a common pathologic event in neurodegenerative diseases. For example, APP and tau were hyperphosphorylated in the AD brain, and hyperphosphorylated α-synuclein was found in the PD brain, while for ALS, it was TDP-43 that was hyperphosphorylated [70,72]. In this review, we discuss the role of zinc in regulating protein phosphorylation by influencing the structures or activities of protein kinases and protein phosphatases, with the focus on tau kinases and phosphatases (Figure 2).

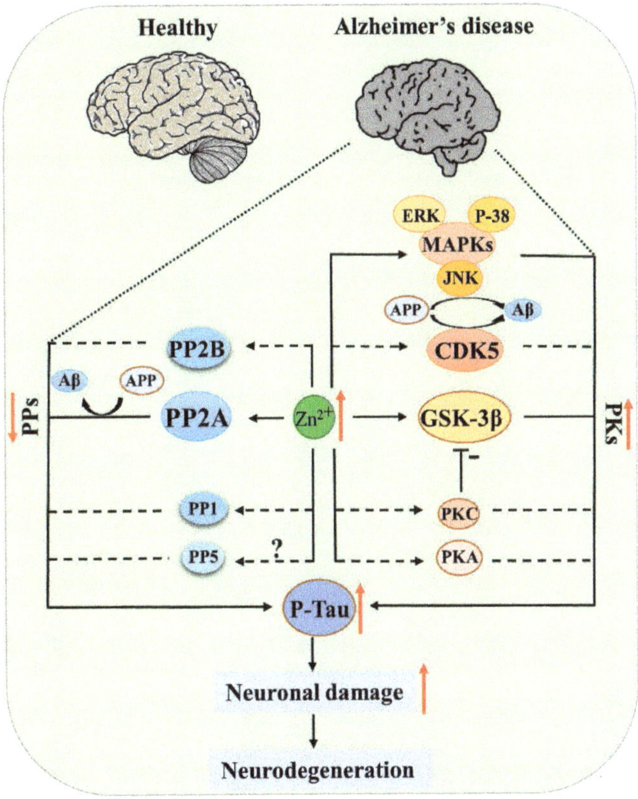

Figure 2. Protein kinases (PKs) and protein phosphatases (PPs) of tau regulated by zinc in AD. Elevated zinc causes activation of kinases (GSK-3β, CDK5, MAPK, PKA, PKC) and simultaneous inhibition of phosphatases (PP-2A, PP-2B), leading to tau pathogenesis; the inhibition of PP-2A and activation of JNK, CDK5 also promote APP hyperphosphorylation and Aβ deposition, ultimately resulting in neuronal damage and disease development. Solid lines indicate there is known evidence from literatures, and dashed lines indicate that direct evidence is still absent, but current data support the hypothesis.

3.1. Zinc and Protein Kinases

Kinases that phosphorylate tau are classified into two types: proline-directed kinases, such as glycogen synthase kinase 3β (GSK-3β), cyclin-dependent kinase 5 (CDK5), MAP kinase (p38, ERK, c-JNK), and other stress kinases; and non-proline-directed kinases, such as protein kinase A (PKA), protein kinase C (PKC), calmodulin (CaM) kinase II, and MARK kinase [73–77].

3.1.1. Proline-Directed Kinases

GSK-3β

Glycogen synthase kinase-3beta (GSK-3β) is an evolutionarily conserved serine/threonine kinase that was first discovered for its role in glycogen synthesis [78]. GSK-3β exerts its function in numerous cellular processes including cell proliferation, DNA repair, cell cycle, and metabolic signaling pathways. So far, GSK-3β is considered one of the most favorable tau kinases [79,80]. Zinc ions can activate GSK-3β. Treatment of SY5Y neuroblastoma cells with 100 μM zinc sulfate induced GSK-3β phosphorylation at Ser9 and Tyr216, which were reversed by PI3K and MEK1/2 inhibitors respectively, together with increased phosphorylation of tau [81]. It is known that GSK-3β phosphorylation at Tyr216 activates GSK-3β, whereas Ser9 phosphorylation results in GSK-3β inhibition [82]. Although the activity of GSK-3β was not measured in this study, another experiment indicated that in cultured hippocampal progenitor H19-7 cells, upon exposure to zinc (10 μM) and zinc-specific ionophore pyrithione, GSK-3β was phosphorylated at Ser9 and tyrosine sites (not identified), and enzyme activity assay indicated the activation of GSK-3β. At the same time, tau was hyperphosphorylated. Inhibition of PI3K/Akt or overexpression of kinase-deficient Akt (KD-Akt) rescued these changes induced by zinc; tau phosphorylation was reversed by the overexpression of kinase-deficient GSK-3β or its dominant-negative mutant [83]. These data collectively indicate that zinc can activate GSK-3β, thus promoting tau phosphorylation. However, the precise molecular mechanism of GSK-3β activation by zinc needs further clarification. For example, tyrosine kinases/phosphatases contributing to GSK-3β Tyr216 phosphorylation in zinc treatment have not been identified.

In addition to regulating tau phosphorylation, GSK-3β increased BACE1 gene expression through its effect on the BACE1 promoter, which in turn promoted Aβ production, whereas the administration of ARA (a GSK inhibitor) decreased BACE1-mediated APP processing and the production of C99 and Aβ in vitro [84]. The ability of GSK-3β to contribute to the phosphorylation of tau and Aβ production provides a link between the two pathologies and suggests a key role for GSK-3β in the pathogenesis of Alzheimer's disease, where zinc ions may be involved through activating GSK-3β.

CDK5

Cyclin-dependent kinase 5 (CDK5) is a member of the cyclin-dependent kinases (CDKs), and its full activity in cells requires the binding of p35 or p25 regulatory subunits, which are highly enriched in neurons. Different groups have identified the phosphorylation sites of CDK5 on tau using amino acid sequencing or mass spectrometry, with Ser202, Thr205, Ser235, and Ser404 being the most favorable [85,86]. Zinc can induce CDK5 activation through phosphorylation of CDK5 at Tyr15 in rat brain with ischemic injury [87]. In addition, it is known that iron ions can induce the phosphorylation of tau via CDK5 in the mouse hippocampus, while it remains to be explored whether zinc ions can exert a similar effect [88]. APP can also be phosphorylated at Thr668 by CDK5 in neuronal cultures in vitro [89], suggesting an important role of CDK5 in promoting amyloidosis in AD. Based on these findings, it is reasonable to speculate that tau and APP phosphorylation by CDK5 may be induced by zinc, which needs further identification.

MAPKs (p38, ERK, JNK)

MAPKs are serine/threonine protein kinases that regulate the cellular signals in response to a wide range of extracellular stimuli and play important roles in the regulation of cell proliferation, differentiation, survival, and apoptosis. The mitogen-activated protein kinase (MAPK) cascade reaction consists of three core kinases: MKKK, MKK and MAPK. The top kinase of the triple kinase module is MAPK kinase kinase (MKKK). MKKK is a serine/threonine kinase that, when activated, phosphorylates the next kinase in the module, MAPK kinase (MKK). In turn, these kinases activate MAPK by directly phosphorylating the Tyr and Thr residues in their activation loop motifs. MAPK is the final kinase in the triple kinase module, phosphorylating substrates on serine and threonine residues [90]. In mammalian cells, several distinct MAPKs have been identified, mainly including p38

MAPK, c-jun N-terminal kinase (JNK), and extracellular signal-regulated kinase (ERK 1/2). Multiple studies have shown the activation of MAPKs in AD brains and their important roles in mediating tau phosphorylation in cell and animal models [91].

Zinc may regulate the activity of MAPKs directly or indirectly through regulating the upstream MKKKs or MKKs. Most evidence is from cells treated by extracellular zinc incubation or inducible intracellular zinc release, and evidence for the direct regulation of zinc on enzyme activities of MAPKs is lacking. DTDP-induced intracellular zinc release was found to activate p38 and cause neuronal death in vitro [92]. Elevated zinc ions can activate ERK through the MAPK cascade of Ras/Raf/MEK and eventually lead to differentiated PC12 cells death. The researchers also observed the activation of JNK, but the signal pathway downstream of zinc leading to JNK activation was not further explored since JNK activation seemed to not participate in cell death by zinc in this experiment [93]. In the presence of 100 μM zinc ions, astrocytes can be activated through the ERK/Stat3 pathway in vitro [22].

In relating to tau pathology, zinc ions can activate ERK via the Ras/Raf/MEK pathway, ultimately causing tau S214 phosphorylation in human wild-type SH-SY5Y cells expressing $tau_{1\text{-}441}$ [69]. A study on APP phosphorylation showed that JNK also plays an important role in APP metabolism, as the JNK inhibitory peptide (D-JNKI1) prevents APP phosphorylation at Thr668, leading to a significant decrease in βAPP and Aβ levels in H4-APP cell line [94]. Administration of SP600125, a JNK-specific inhibitor, significantly reversed amyloid plaque burden and synaptic loss in APP/PS1 transgenic mice [95]. Given the role of MAPKs in APP and tau phosphorylation, several inhibitors targeting ERK, JNK or p38 have been developed as potential drugs for AD therapy [95–97].

3.1.2. Non-Proline-Directed Kinases

PKA

Protein kinase A (PKA) is a tetramer composed of two regulatory protein subunits, R, and two catalytic subunits, C. When the second messenger cAMP binds to the R subunit, it causes PKA activation. The phosphorylation of tau at Ser214 and PKA activity are increased simultaneously in the dorsolateral prefrontal association cortex of macaques, suggesting that tau phosphorylation may result from PKA activation. Moreover, the activation of PKA by forskolin (Fsk) produced a dose-dependent increase in the phosphorylation of tau at Ser214 in mouse primary cortical neurons [98,99]. In a study on sperm motility, extracellular zinc was identified to bind and activate the Zn^{2+} sensing receptor, which further activates adenylyl-cyclase (AC) to increase [cAMP]i and activate PKA [100]. Another experiment also showed that zinc exposure caused the activation of the cAMP/PKA pathway in human bone marrow-derived mesenchymal stem cells [101]. However, till now there is no direct evidence supporting that zinc induces the phosphorylation of tau via the cAMP/PKA pathway.

PKC

Protein kinase C (PKC), consisting of 12 isozymes, is divided into three subfamilies (classic PKC, novel PKC, and atypical PKC). The second messenger DAG and Ca^{2+} play a master role in PKC activation [102–104]. PKC not only induces the phosphorylation of tau at the Thr231 and Ser313 sites, but also the phosphorylation of GSK3β at Ser-9, thereby inhibiting GSK3β activity as well as GSK3β-induced phosphorylation of tau at Ser202, Thr205, and Thr181 in vitro [105–107]. These results suggest a complex effect of PKC on GSK3β and tau. In neuronal cultures, zinc induces the activation of PKC, which activates NADPH oxidase, accompanied by the production of nitric oxide synthase (NOS) [108]. Short-term exposure to 300 μM zinc caused increased PKC activity and cortical neuronal death in vitro, which were eliminated by Ca-EDTA or PKC inhibitors. This suggests that PKC also plays an important role in zinc-induced neuronal death [109]. Insulin-induced increases in $A\beta_{40/42}$ and phosphorylation of tau at Thr231 were blocked by a PKC inhibitor ACPD in both cultured neuronal cells and mouse hippocampal slices [110].

The above results suggest that PKC is also a potential kinase mediating zinc-induced tau/APP phosphorylation in AD.

3.2. Zinc and Protein Phosphatases

Proteomic analysis of 6600 phosphorylation sites on thousands of human proteins showed that phosphoserine (pSer), phosphothreonine (pThr) and phosphotyrosine (pTyr) accounted for 86.4%, 11.8% and 1.8% of phosphorylated amino acids, respectively [111]. Protein serine/threonine phosphatases (PSPs), which are responsible for the removal of phosphate from target proteins on serine and threonine sites, participate in many cellular functions, such as cell cycle, growth, metabolism, transformation, and apoptosis [6]. PSPs are classified as three families: phosphoprotein phosphatases (PPP), metal-dependent protein phosphatases (PPM), and aspartic acid-based phosphatases (FCP/SCP, TFIIF-associating component of RNA polymerase II CTD phosphatase/small CTD phosphatase). The general classification of such phosphatases is based on the primary structure and catalytic mechanism of the enzyme. The PPM family are metal-dependent protein phosphatases that act as single-subunit enzymes bound to Mn^{2+} or Mg^{2+}. This family is also known as the PP2C family as members of PPP. It has been reported that the dysregulation of PPM phosphatases may be associated with the development of tumorigenesis, metabolic disorders, and cancers. Currently, its role in neurodegenerative diseases is under determination [112]; FCP/SCP uses an aspartic acid-based catalytic mechanism, and its only known substrate is the C-terminal structural domain (CTD) of RNA polymerase II, which contains a tandem repeat sequence of a serine-rich heptapeptide [113]. Among three PSPs, zinc ions are rarely reported with PPM or FCP/SCP; here, we mainly focus on PPP, which can be further classified into three types: PP-1, PP-2 (PP-2A, PP-2B), and PP-5.

3.2.1. PP-1

Protein phosphatase 1 is a ubiquitously expressed serine/threonine phosphatase. In AD brain, the activity of PP-1 is decreased by ~20% [114]. In a study comparing the dephosphorylation ability toward tau of different tau phosphatases, PP-1 was found to account for ~11% of total brain tau phosphatase activity [115]. However, autopsy of AD brain revealed that phosphorylated tau at Thr212 could only be dephosphorylated by purified PP-1 but not by PP-2A/PP-2B [116]. PP-1 can be inhibited by many toxins, such as OA and tautomycin, as well as endogenous protein inhibitors [113,117]. For the recombinant catalytic subunit of PP-1, its activation requires the addition of Mn^{2+} [118]. Chu et al. further investigated the effects of other divalent cations on PP-1 activity and found that treatment of the recombinant PP-1 catalytic subunit with a combination of Fe^{2+} and Zn^{2+} (but not the individual metal ions) significantly activated the phosphatase. These results suggest that PP-1 may be an iron/zinc metalloprotein in vivo [119]. Till now, there are still no studies that investigated the effect of elevated extracellular zinc on PP-1 activity in cells. In a study exploring the toxicity of aluminum in rats, Al treatment increased the protein level of GSK-3, whereas it decreased the protein level of PP-1α in the brain, which were reversed upon zinc supplementation in drinking water. However, the underlying mechanisms were not further explored [120].

3.2.2. PP-2A

Protein phosphatase 2A (PP-2A) is the major serine/threonine phosphatase in eukaryotic cells, which contributes to ~71% of total brain tau phosphatase activity [115]. In the AD brain, the activity of PP-1, PP-2A, and PP-5 are decreased [114,121]. PP-2A inhibition can cause the hyperphosphorylation of tau at multiple AD-related phosphorylation sites [122], as well as APP phosphorylation at Thr668 [123]. Both tau hyperphosphorylation and Aβ deposition are typical pathological features of AD [124,125]. Just as PP-1, the activity of PP-2A catalytic subunit is also regulated by ions. A study compared the activity of the core enzyme of PP-2A, which is composed of a catalytic C subunit and a scaffolding A subunit, under different ion concentrations in vitro [126]. In the mammalian brain, the major core enzyme of PP-2A is CA [127,128]. The researchers reported that the Mn^{2+}-dependent C′A′ and -independent CA PP2A can be purified from human erythrocytes, and CA is a Zn^{2+}- and Fe^{2+}-metalloenzyme. For the Mn^{2+}-independent PP-2A (C′A′), $ZnCl_2$ at a concentration lower than 10 μM stimulated the phosphatase activity, whereas at higher concentrations, this effect was attenuated. The PP and PTP activities of CA were unaffected by the same zinc and/or iron treatment [126]. We explored the direct effect of zinc on PP-2A activity through incubating neuronal cell lysates or purified recombinant PP-2A catalytic subunit with zinc in vitro, and found that zinc at a concentration of 10 μM inhibited PP-2A directly, and the binding site located in segment PP2Ac (51–270) [129]. Thus, whether excessive zinc inhibits PP-2A in the brain, depends on the subunit composition of PP-2A and the exact zinc concentration. While zinc concentration may exceed 10 μM under lots of pathologic conditions, such as during ischemia, or in zinc-accumulated senile plaques in AD, it is reasonable to speculate a direct inhibitory effect of zinc on PP-2A exists in the development of brain disorders, such as stroke and AD.

In the brain, excessive zinc may also regulate the PP-2A activity indirectly through cell signaling transduction. Y307 phosphorylation of PP-2A catalytic subunit was found to potently inhibit PP-2A activity [130]. In cultured rat hippocampal slices, synaptic released zinc promotes tau hyperphosphorylation through inhibiting PP-2A [67]. In cells and the rat brain, zinc induced PP-2A inactivation and tau hyperphosphorylation through Src-dependent PP2A-Y307 phosphorylation [67,131]. These findings suggest that zinc disturbance is an important upstream factor contributing to PP-2A inhibition in AD.

3.2.3. Calcineurin/PP-2B

Calcineurin is a calcium- and calmodulin-dependent enzyme found in many cell types, especially enriched in neuronal soma and processes. Calcineurin comprises a catalytic "A" subunit, which interacts with calmodulin in a calcium-dependent manner, and a regulatory "B" subunit [132]. Calcineurin can dephosphorylate tau at a few sites, but it accounts for only ~7% of total brain tau phosphatase activity [115,133,134]. Decreased expression of calcineurin contributes to abnormal tau hyperphosphorylation in mouse models of Huntington's disease, an autosomal dominant neurodegenerative disorder [135]. Calcineurin contains Zn^{2+} in its catalytic domain and its activity can also be directly regulated by exogenous zinc. It was reported that Zn^{2+} at a concentration of 10 nM–10 μM inhibited Ni^{2+}-stimulated-calcineurin activity in a dose-dependent manner in vitro [136]. The inhibitory effect of zinc on calcineurin was also observed in cell experiments. Zinc at 30–100 μM inhibited the osteoclastogesis induced by receptor activator of NF-κB ligand (RANKL) in mouse bone marrow-derived monocytes. During this process, the calcineurin activity decreased in response to zinc but its protein level was unchanged. A similar result was observed in doxorubicin (DOX)-treated H9c2 rat cardiac muscle cells: pretreatment of H9c2 cells with exogenous zinc attenuated the DOX-activated calcineurin signaling in a dose-dependent manner [137,138]. These findings collectively suggest an inhibitory effect of zinc on calcineurin. In neurological-related diseases, this inhibition may promote the hyperphosphorylation of calcineurin substrates, such as tau protein.

3.2.4. PP-5

Protein phosphatase 5 is ubiquitously expressed in all mammalian tissues, with a high level in the brain. In the presence of arachidonic acid, PP-5 dephosphorylates tau at many of the sites that are hyperphosphorylated in AD in vitro [139]. The activity, but not the protein level of PP5, was found to be decreased by approximately 20% in AD neocortex [114,121]. Since PP-5 is a late-identified phosphatase, its interaction with zinc remains to be explored.

A summary of the changes of zinc-regulated kinases/phosphatases in neurodegenerative diseases, and the possible regulating mechanisms of zinc, was list in Table 1.

Table 1. Zinc disruption in regulating protein kinases and protein phosphatases in neurodegenerative diseases.

Protein Kinase or Phosphatase	Expression Level or Activity in Disease Progression	Effect and Regulating Pathways of Zinc in Disease	Reference
GSK-3β	AD↑	Activation zinc-MEK/ERK/GSK-3β	[81,140]
CDK5	AD↑; ALS↑; PD ↑	Activation, To be identified	[141–143]
ERK	AD↑; PD↑	Activation zinc-Ras/Raf/MEK/ERK	[144–147]
JNK	AD↑; ALS↑	Activation To be identified	[148–150]
p38	AD↑	Activation To be identified	[147,148]
PKA	AD↑; PD↑	Not explored	[151,152]
PKC	AD↓	Not explored	[153,154]
PP1	AD↓	Not explored	[114]
PP2A	AD↓	Inhibition Direct effect, or through zinc-Src-PP2A(Y307)	[131,155]
PP2B	HD↓	Inhibition To be identified	[135]
PP5	AD↓	Not explored	[156]

AD, Alzheimer's disease; ALS, Amyotrophic lateral sclerosis; CDK5, Cyclin-dependent kinase 5; ERK, Extracellular signal-regulated kinase; HD, Huntington's disease; GSK-3β, Glycogen synthase kinase 3β; JNK, c-jun N-terminal kinase; PD, Parkinson's disease; PKA, Protein kinase A; PKC, Protein kinase C; PP1, Protein phosphatase; PP2A, Protein phosphatase 2A; PP2B, Protein phosphatase 2B; PP5, Protein phosphatase 5.

4. Conclusions

Zinc is one of the most important micronutrients. Zinc homeostasis is of great importance for the central nervous system. Disturbed zinc metabolism is prominent in neurodegenerative disorders, such as PD, ALS, and AD, which may participate in the disease progress in multiple pathways. The abnormal hyperphosphorylation of proteins, such as tau and APP, is a hallmark in these diseases. Zinc can regulate the activity of related protein kinases and phosphatases, accelerating this pathologic process. Solid data support that elevated zinc may activate kinase GSK-3β, MAPKs and inhibit phosphatase PP-2A, thus promoting the hyperphosphorylation of tau and APP. Although the causal relationship between zinc disturbance and tau/Aβ pathology remains to be further clarified, targeting the zinc dyshomeostasis, thus preventing the disturbance of related kinases and phosphatases, may be a new approach for the intervention of these neurodegenerative diseases, which is worthy of further extensive exploration.

Author Contributions: Writing—original draft preparation, H.-L.Z.; writing—review and editing, X.-C.W. and R.L.; funding acquisition, R.L. All authors have read and agreed to the published version of the manuscript.

Funding: This research was funded by the National Natural Science Foundation of China (grant number 31970964, 82171426).

Informed Consent Statement: Not applicable.

Data Availability Statement: Not applicable.

Conflicts of Interest: The authors declare no conflict of interest.

References

1. O'Dell, B.L.; Newberne, P.M.; Savage, J.E. Significance of dietary zinc for the growing chicken. *J. Nutr.* **1958**, *65*, 503–518. [CrossRef] [PubMed]
2. Sommer, A.L.; Lipman, C.B. Evidence on the indispensable nature of zinc and boron for higher green plants. *Plant Physiol.* **1926**, *1*, 231–249. [CrossRef] [PubMed]
3. Jackson, M. Physiology of zinc: General aspects. In *Zinc in Human Biology*; Springer: Berlin/Heidelberg, Germany, 1989; pp. 1–14.
4. Liuzzi, J.P.; Bobo, J.A.; Lichten, L.A.; Samuelson, D.A.; Cousins, R.J. Responsive transporter genes within the murine intestinal-pancreatic axis form a basis of zinc homeostasis. *Proc. Natl. Acad. Sci. USA* **2004**, *101*, 14355–14360. [CrossRef] [PubMed]
5. Vallee, B.L.; Falchuk, K.H. The biochemical basis of zinc physiology. *Physiol. Rev.* **1993**, *73*, 79–118. [CrossRef]
6. Fukada, T.; Kambe, T. Molecular and genetic features of zinc transporters in physiology and pathogenesis. *Metallomics* **2011**, *3*, 662–674. [CrossRef]
7. Maret, W.; Li, Y. Coordination dynamics of zinc in proteins. *Chem. Rev.* **2009**, *109*, 4682–4707. [CrossRef]
8. Devirgiliis, C.; Zalewski, P.D.; Perozzi, G.; Murgia, C. Zinc fluxes and zinc transporter genes in chronic diseases. *Mutat. Res.* **2007**, *622*, 84–93. [CrossRef]
9. Fraker, P.J.; King, L.E. Reprogramming of the immune system during zinc deficiency. *Annu. Rev. Nutr.* **2004**, *24*, 277–298. [CrossRef]
10. Prasad, A.S. Discovery of human zinc deficiency and studies in an experimental human model. *Am. J. Clin. Nutr.* **1991**, *53*, 403–412. [CrossRef]
11. Frederickson, C.J. Neurobiology of zinc and zinc-containing neurons. *Int. Rev. Neurobiol.* **1989**, *31*, 145–238. [CrossRef]
12. Ibata, Y.; Otsuka, N. Electron microscopic demonstration of zinc in the hippocampal formation using Timm's sulfide silver technique. *J. Histochem. Cytochem.* **1969**, *17*, 171–175. [CrossRef]
13. Coyle, P.; Philcox, J.C.; Carey, L.C.; Rofe, A.M. Metallothionein: The multipurpose protein. *Cell. Mol. Life Sci.* **2002**, *59*, 627–647. [CrossRef] [PubMed]
14. Kambe, T.; Hashimoto, A.; Fujimoto, S. Current understanding of ZIP and ZnT zinc transporters in human health and diseases. *Cell. Mol. Life Sci.* **2014**, *71*, 3281–3295. [CrossRef] [PubMed]
15. Lichten, L.A.; Cousins, R.J. Mammalian zinc transporters: Nutritional and physiologic regulation. *Annu. Rev. Nutr.* **2009**, *29*, 153–176. [CrossRef] [PubMed]
16. Xu, Y.; Xiao, G.; Liu, L.; Lang, M. Zinc transporters in Alzheimer's disease. *Mol. Brain* **2019**, *12*, 106. [CrossRef]
17. Bitanihirwe, B.K.; Cunningham, M.G. Zinc: The brain's dark horse. *Synapse* **2009**, *63*, 1029–1049. [CrossRef]
18. Choi, D.W.; Koh, J.Y. Zinc and brain injury. *Annu. Rev. Neurosci.* **1998**, *21*, 347–375. [CrossRef]
19. Minami, A.; Takeda, A.; Yamaide, R.; Oku, N. Relationship between zinc and neurotransmitters released into the amygdalar extracellular space. *Brain Res.* **2002**, *936*, 91–94. [CrossRef]
20. Assaf, S.Y.; Chung, S.H. Release of endogenous Zn^{2+} from brain tissue during activity. *Nature* **1984**, *308*, 734–736. [CrossRef]
21. Vogt, K.; Mellor, J.; Tong, G.; Nicoll, R. The actions of synaptically released zinc at hippocampal mossy fiber synapses. *Neuron* **2000**, *26*, 187–196. [CrossRef]
22. Huiliang, Z.; Mengzhe, Y.; Xiaochuan, W.; Hui, W.; Min, D.; Mengqi, W.; Jianzhi, W.; Zhongshan, C.; Caixia, P.; Rong, L. Zinc induces reactive astrogliosis through ERK-dependent activation of Stat3 and promotes synaptic degeneration. *J. Neurochem.* **2021**, *159*, 1016–1027. [CrossRef] [PubMed]
23. Kauppinen, T.M.; Higashi, Y.; Suh, S.W.; Escartin, C.; Nagasawa, K.; Swanson, R.A. Zinc triggers microglial activation. *J. Neurosci.* **2008**, *28*, 5827–5835. [CrossRef] [PubMed]
24. Choi, D.W.; Yokoyama, M.; Koh, J. Zinc neurotoxicity in cortical cell culture. *Neuroscience* **1988**, *24*, 67–79. [CrossRef]
25. Sensi, S.L.; Yin, H.Z.; Carriedo, S.G.; Rao, S.S.; Weiss, J.H. Preferential Zn^{2+} influx through Ca^{2+}-permeable AMPA/kainate channels triggers prolonged mitochondrial superoxide production. *Proc. Natl. Acad. Sci. USA* **1999**, *96*, 2414–2419. [CrossRef]
26. Plum, L.M.; Rink, L.; Haase, H. The essential toxin: Impact of zinc on human health. *Int. J. Environ. Res. Public Health* **2010**, *7*, 1342–1365. [CrossRef]
27. Yang, Y.; Jing, X.P.; Zhang, S.P.; Gu, R.X.; Tang, F.X.; Wang, X.L.; Xiong, Y.; Qiu, M.; Sun, X.Y.; Ke, D.; et al. High dose zinc supplementation induces hippocampal zinc deficiency and memory impairment with inhibition of BDNF signaling. *PLoS ONE* **2013**, *8*, e55384. [CrossRef]
28. Barnham, K.J.; Bush, A.I. Biological metals and metal-targeting compounds in major neurodegenerative diseases. *Chem. Soc. Rev.* **2014**, *43*, 6727–6749. [CrossRef]
29. Calap-Quintana, P.; González-Fernández, J.; Sebastiá-Ortega, N.; Llorens, J.V.; Moltó, M.D. Drosophila melanogaster Models of Metal-Related Human Diseases and Metal Toxicity. *Int. J. Mol. Sci.* **2017**, *18*, 1456. [CrossRef]
30. Shaw, B.F.; Valentine, J.S. How do ALS-associated mutations in superoxide dismutase 1 promote aggregation of the protein? *Trends Biochem. Sci.* **2007**, *32*, 78–85. [CrossRef]
31. Xiao, G.; Zhou, B. What can flies tell us about zinc homeostasis? *Arch. Biochem. Biophys.* **2016**, *611*, 134–141. [CrossRef]
32. Samii, A.; Nutt, J.G.; Ransom, B.R. Parkinson's disease. *Lancet* **2004**, *363*, 1783–1793. [CrossRef]
33. Calabresi, P.; Di Filippo, M.; Gallina, A.; Wang, Y.; Stankowski, J.N.; Picconi, B.; Dawson, V.L.; Dawson, T.M. New synaptic and molecular targets for neuroprotection in Parkinson's disease. *Mov. Disord.* **2013**, *28*, 51–60. [CrossRef] [PubMed]

34. Granzotto, A.; Sensi, S.L. Intracellular zinc is a critical intermediate in the excitotoxic cascade. *Neurobiol. Dis.* **2015**, *81*, 25–37. [CrossRef]
35. Surmeier, D.J.; Obeso, J.A.; Halliday, G.M. Selective neuronal vulnerability in Parkinson disease. *Nat. Rev. Neurosci.* **2017**, *18*, 101–113. [CrossRef] [PubMed]
36. Tang, J.; Lu, L.; Wang, Q.; Liu, H.; Xue, W.; Zhou, T.; Xu, L.; Wang, K.; Wu, D.; Wei, F.; et al. Crocin Reverses Depression-Like Behavior in Parkinson Disease Mice via VTA-mPFC Pathway. *Mol. Neurobiol.* **2020**, *57*, 3158–3170. [CrossRef]
37. Ahmed, S.S.; Santosh, W. Metallomic profiling and linkage map analysis of early Parkinson's disease: A new insight to aluminum marker for the possible diagnosis. *PLoS ONE* **2010**, *5*, e11252. [CrossRef]
38. Hegde, M.L.; Shanmugavelu, P.; Vengamma, B.; Rao, T.S.; Menon, R.B.; Rao, R.V.; Rao, K.S. Serum trace element levels and the complexity of inter-element relations in patients with Parkinson's disease. *J. Trace Elem. Med. Biol.* **2004**, *18*, 163–171. [CrossRef]
39. Alimonti, A.; Bocca, B.; Pino, A.; Ruggieri, F.; Forte, G.; Sancesario, G. Elemental profile of cerebrospinal fluid in patients with Parkinson's disease. *J. Trace Elem. Med. Biol.* **2007**, *21*, 234–241. [CrossRef]
40. Forte, G.; Bocca, B.; Senofonte, O.; Petrucci, F.; Brusa, L.; Stanzione, P.; Zannino, S.; Violante, N.; Alimonti, A.; Sancesario, G. Trace and major elements in whole blood, serum, cerebrospinal fluid and urine of patients with Parkinson's disease. *J. Neural Transm.* **2004**, *111*, 1031–1040. [CrossRef]
41. Ajjimaporn, A.; Phansuwan-Pujito, P.; Ebadi, M.; Govitrapong, P. Zinc protects SK-N-SH cells from methamphetamine-induced alpha-synuclein expression. *Neurosci. Lett.* **2007**, *419*, 59–63. [CrossRef]
42. Ajjimaporn, A.; Shavali, S.; Ebadi, M.; Govitrapong, P. Zinc rescues dopaminergic SK-N-SH cell lines from methamphetamine-induced toxicity. *Brain Res. Bull.* **2008**, *77*, 361–366. [CrossRef] [PubMed]
43. Saini, N.; Schaffner, W. Zinc supplement greatly improves the condition of parkin mutant Drosophila. *Biol. Chem.* **2010**, *391*, 513–518. [CrossRef] [PubMed]
44. Pals, P.; Van Everbroeck, B.; Grubben, B.; Viaene, M.K.; Dom, R.; van der Linden, C.; Santens, P.; Martin, J.J.; Cras, P. Case-control study of environmental risk factors for Parkinson's disease in Belgium. *Eur. J. Epidemiol.* **2003**, *18*, 1133–1142. [CrossRef]
45. Dexter, D.T.; Carayon, A.; Javoy-Agid, F.; Agid, Y.; Wells, F.R.; Daniel, S.E.; Lees, A.J.; Jenner, P.; Marsden, C.D. Alterations in the levels of iron, ferritin and other trace metals in Parkinson's disease and other neurodegenerative diseases affecting the basal ganglia. *Brain* **1991**, *114 Pt 4*, 1953–1975. [CrossRef] [PubMed]
46. Hussain, S.; Ali, S.F. Zinc potentiates 1-methyl-4-phenyl-1,2,3,6-tetrahydropyridine induced dopamine depletion in caudate nucleus of mice brain. *Neurosci. Lett.* **2002**, *335*, 25–28. [CrossRef]
47. Bruijn, L.I.; Miller, T.M.; Cleveland, D.W. Unraveling the mechanisms involved in motor neuron degeneration in ALS. *Annu. Rev. Neurosci.* **2004**, *27*, 723–749. [CrossRef] [PubMed]
48. Dang, T.N.; Lim, N.K.; Grubman, A.; Li, Q.X.; Volitakis, I.; White, A.R.; Crouch, P.J. Increased metal content in the TDP-43(A315T) transgenic mouse model of frontotemporal lobar degeneration and amyotrophic lateral sclerosis. *Front. Aging Neurosci.* **2014**, *6*, 15. [CrossRef]
49. Sillevis Smitt, P.A.; Mulder, T.P.; Verspaget, H.W.; Blaauwgeers, H.G.; Troost, D.; de Jong, J.M. Metallothionein in amyotrophic lateral sclerosis. *Biol. Signals* **1994**, *3*, 193–197. [CrossRef]
50. Winblad, B.; Amouyel, P.; Andrieu, S.; Ballard, C.; Brayne, C.; Brodaty, H.; Cedazo-Minguez, A.; Dubois, B.; Edvardsson, D.; Feldman, H.; et al. Defeating Alzheimer's disease and other dementias: A priority for European science and society. *Lancet. Neurology* **2016**, *15*, 455–532. [CrossRef]
51. Wortmann, M. Dementia: A global health priority—Highlights from an ADI and World Health Organization report. *Alzheimer's Res. Ther.* **2012**, *4*, 40. [CrossRef]
52. Lashley, T.; Schott, J.M.; Weston, P.; Murray, C.E.; Wellington, H.; Keshavan, A.; Foti, S.C.; Foiani, M.; Toombs, J.; Rohrer, J.D.; et al. Molecular biomarkers of Alzheimer's disease: Progress and prospects. *Dis. Models Mech.* **2018**, *11*, dmm031781. [CrossRef] [PubMed]
53. Overk, C.R.; Masliah, E. Pathogenesis of synaptic degeneration in Alzheimer's disease and Lewy body disease. *Biochem. Pharmacol.* **2014**, *88*, 508–516. [CrossRef] [PubMed]
54. Danscher, G.; Jensen, K.B.; Frederickson, C.J.; Kemp, K.; Andreasen, A.; Juhl, S.; Stoltenberg, M.; Ravid, R. Increased amount of zinc in the hippocampus and amygdala of Alzheimer's diseased brains: A proton-induced X-ray emission spectroscopic analysis of cryostat sections from autopsy material. *J. Neurosci. Methods* **1997**, *76*, 53–59. [CrossRef]
55. Deibel, M.A.; Ehmann, W.D.; Markesbery, W.R. Copper, iron, and zinc imbalances in severely degenerated brain regions in Alzheimer's disease: Possible relation to oxidative stress. *J. Neurol. Sci.* **1996**, *143*, 137–142. [CrossRef]
56. Religa, D.; Strozyk, D.; Cherny, R.A.; Volitakis, I.; Haroutunian, V.; Winblad, B.; Naslund, J.; Bush, A.I. Elevated cortical zinc in Alzheimer disease. *Neurology* **2006**, *67*, 69–75. [CrossRef]
57. Lovell, M.A.; Robertson, J.D.; Teesdale, W.J.; Campbell, J.L.; Markesbery, W.R. Copper, iron and zinc in Alzheimer's disease senile plaques. *J. Neurol. Sci.* **1998**, *158*, 47–52. [CrossRef]
58. Lovell, M.A. A potential role for alterations of zinc and zinc transport proteins in the progression of Alzheimer's disease. *J. Alzheimer's Dis.* **2009**, *16*, 471–483. [CrossRef]
59. Lee, M.C.; Yu, W.C.; Shih, Y.H.; Chen, C.Y.; Guo, Z.H.; Huang, S.J.; Chan, J.C.C.; Chen, Y.R. Zinc ion rapidly induces toxic, off-pathway amyloid-β oligomers distinct from amyloid-β derived diffusible ligands in Alzheimer's disease. *Sci. Rep.* **2018**, *8*, 4772. [CrossRef]

60. Sharma, A.K.; Pavlova, S.T.; Kim, J.; Kim, J.; Mirica, L.M. The effect of Cu(2+) and Zn(2+) on the Aβ42 peptide aggregation and cellular toxicity. *Metallomics* **2013**, *5*, 1529–1536. [CrossRef]
61. Bush, A.I.; Pettingell, W.H.; Multhaup, G.; d Paradis, M.; Vonsattel, J.P.; Gusella, J.F.; Beyreuther, K.; Masters, C.L.; Tanzi, R.E. Rapid induction of Alzheimer A beta amyloid formation by zinc. *Science* **1994**, *265*, 1464–1467. [CrossRef]
62. Adlard, P.A.; West, A.K.; Vickers, J.C. Increased density of metallothionein I/II-immunopositive cortical glial cells in the early stages of Alzheimer's disease. *Neurobiol. Dis.* **1998**, *5*, 349–356. [CrossRef] [PubMed]
63. Lovell, M.A.; Smith, J.L.; Xiong, S.; Markesbery, W.R. Alterations in zinc transporter protein-1 (ZnT-1) in the brain of subjects with mild cognitive impairment, early, and late-stage Alzheimer's disease. *Neurotox. Res.* **2005**, *7*, 265–271. [CrossRef] [PubMed]
64. Lyubartseva, G.; Smith, J.L.; Markesbery, W.R.; Lovell, M.A. Alterations of zinc transporter proteins ZnT-1, ZnT-4 and ZnT-6 in preclinical Alzheimer's disease brain. *Brain Pathol.* **2010**, *20*, 343–350. [CrossRef]
65. Whitfield, D.R.; Vallortigara, J.; Alghamdi, A.; Howlett, D.; Hortobágyi, T.; Johnson, M.; Attems, J.; Newhouse, S.; Ballard, C.; Thomas, A.J.; et al. Assessment of ZnT3 and PSD95 protein levels in Lewy body dementias and Alzheimer's disease: Association with cognitive impairment. *Neurobiol. Aging* **2014**, *35*, 2836–2844. [CrossRef] [PubMed]
66. Bush, A.I.; Pettingell, W.H., Jr.; Paradis, M.D.; Tanzi, R.E. Modulation of A beta adhesiveness and secretase site cleavage by zinc. *J. Biol. Chem.* **1994**, *269*, 12152–12158. [CrossRef]
67. Sun, X.Y.; Wei, Y.P.; Xiong, Y.; Wang, X.C.; Xie, A.J.; Wang, X.L.; Yang, Y.; Wang, Q.; Lu, Y.M.; Liu, R.; et al. Synaptic released zinc promotes tau hyperphosphorylation by inhibition of protein phosphatase 2A (PP2A). *J. Biol. Chem.* **2012**, *287*, 11174–11182. [CrossRef]
68. Boom, A.; Authelet, M.; Dedecker, R.; Frédérick, C.; Van Heurck, R.; Daubie, V.; Leroy, K.; Pochet, R.; Brion, J.P. Bimodal modulation of tau protein phosphorylation and conformation by extracellular Zn^{2+} in human-tau transfected cells. *Biochim. Biophys. Acta* **2009**, *1793*, 1058–1067. [CrossRef]
69. Kim, I.; Park, E.J.; Seo, J.; Ko, S.J.; Lee, J.; Kim, C.H. Zinc stimulates tau S214 phosphorylation by the activation of Raf/mitogen-activated protein kinase-kinase/extracellular signal-regulated kinase pathway. *Neuroreport* **2011**, *22*, 839–844. [CrossRef] [PubMed]
70. Frederickson, C.J.; Koh, J.Y.; Bush, A.I. The neurobiology of zinc in health and disease. *Nat. Rev. Neurosci.* **2005**, *6*, 449–462. [CrossRef]
71. Sensi, S.L.; Granzotto, A.; Siotto, M.; Squitti, R. Copper and Zinc Dysregulation in Alzheimer's Disease. *Trends Pharmacol. Sci.* **2018**, *39*, 1049–1063. [CrossRef]
72. Sensi, S.L.; Paoletti, P.; Bush, A.I.; Sekler, I. Zinc in the physiology and pathology of the CNS. *Nat. Rev. Neurosci.* **2009**, *10*, 780–791. [CrossRef] [PubMed]
73. Baudier, J.; Lee, S.H.; Cole, R.D. Separation of the different microtubule-associated tau protein species from bovine brain and their mode II phosphorylation by Ca^{2+}/phospholipid-dependent protein kinase C. *J. Biol. Chem.* **1987**, *262*, 17584–17590. [CrossRef]
74. Drewes, G.; Lichtenberg-Kraag, B.; Döring, F.; Mandelkow, E.M.; Biernat, J.; Goris, J.; Dorée, M.; Mandelkow, E. Mitogen activated protein (MAP) kinase transforms tau protein into an Alzheimer-like state. *EMBO J.* **1992**, *11*, 2131–2138. [CrossRef] [PubMed]
75. Goedert, M.; Hasegawa, M.; Jakes, R.; Lawler, S.; Cuenda, A.; Cohen, P. Phosphorylation of microtubule-associated protein tau by stress-activated protein kinases. *FEBS Lett.* **1997**, *409*, 57–62. [CrossRef]
76. Imahori, K.; Uchida, T. Physiology and pathology of tau protein kinases in relation to Alzheimer's disease. *J. Biochem.* **1997**, *121*, 179–188.
77. Scott, C.W.; Spreen, R.C.; Herman, J.L.; Chow, F.P.; Davison, M.D.; Young, J.; Caputo, C.B. Phosphorylation of recombinant tau by cAMP-dependent protein kinase. Identification of phosphorylation sites and effect on microtubule assembly. *J. Biol. Chem.* **1993**, *268*, 1166–1173. [CrossRef]
78. Embi, N.; Rylatt, D.B.; Cohen, P. Glycogen synthase kinase-3 from rabbit skeletal muscle. Separation from cyclic-AMP-dependent protein kinase and phosphorylase kinase. *Eur. J. Biochem.* **1980**, *107*, 519–527. [CrossRef]
79. Pei, J.J.; Braak, E.; Braak, H.; Grundke-Iqbal, I.; Iqbal, K.; Winblad, B.; Cowburn, R.F. Distribution of active glycogen synthase kinase 3beta (GSK-3beta) in brains staged for Alzheimer disease neurofibrillary changes. *J. Neuropathol. Exp. Neurol.* **1999**, *58*, 1010–1019. [CrossRef]
80. Pei, J.J.; Tanaka, T.; Tung, Y.C.; Braak, E.; Iqbal, K.; Grundke-Iqbal, I. Distribution, levels, and activity of glycogen synthase kinase-3 in the Alzheimer disease brain. *J. Neuropathol. Exp. Neurol.* **1997**, *56*, 70–78. [CrossRef]
81. An, W.L.; Bjorkdahl, C.; Liu, R.; Cowburn, R.F.; Winblad, B.; Pei, J.J. Mechanism of zinc-induced phosphorylation of p70 S6 kinase and glycogen synthase kinase 3beta in SH-SY5Y neuroblastoma cells. *J. Neurochem.* **2005**, *92*, 1104–1115. [CrossRef]
82. Frame, S.; Cohen, P.; Biondi, R.M. A common phosphate binding site explains the unique substrate specificity of GSK3 and its inactivation by phosphorylation. *Mol. Cell* **2001**, *7*, 1321–1327. [CrossRef]
83. Min, Y.K.; Lee, J.E.; Chung, K.C. Zinc induces cell death in immortalized embryonic hippocampal cells via activation of Akt-GSK-3beta signaling. *Exp. Cell Res.* **2007**, *313*, 312–321. [CrossRef] [PubMed]
84. Ly, P.T.; Wu, Y.; Zou, H.; Wang, R.; Zhou, W.; Kinoshita, A.; Zhang, M.; Yang, Y.; Cai, F.; Woodgett, J.; et al. Inhibition of GSK3β-mediated BACE1 expression reduces Alzheimer-associated phenotypes. *J. Clin. Investig.* **2013**, *123*, 224–235. [CrossRef] [PubMed]
85. Arioka, M.; Tsukamoto, M.; Ishiguro, K.; Kato, R.; Sato, K.; Imahori, K.; Uchida, T. Tau protein kinase II is involved in the regulation of the normal phosphorylation state of tau protein. *J. Neurochem.* **1993**, *60*, 461–468. [CrossRef] [PubMed]

86. Lund, E.T.; McKenna, R.; Evans, D.B.; Sharma, S.K.; Mathews, W.R. Characterization of the in vitro phosphorylation of human tau by tau protein kinase II (cdk5/p20) using mass spectrometry. *J. Neurochem.* **2001**, *76*, 1221–1232. [CrossRef]
87. Tuo, Q.Z.; Liuyang, Z.Y.; Lei, P.; Yan, X.; Shentu, Y.P.; Liang, J.W.; Zhou, H.; Pei, L.; Xiong, Y.; Hou, T.Y.; et al. Zinc induces CDK5 activation and neuronal death through CDK5-Tyr15 phosphorylation in ischemic stroke. *Cell Death Dis.* **2018**, *9*, 870. [CrossRef]
88. Guo, C.; Wang, P.; Zhong, M.L.; Wang, T.; Huang, X.S.; Li, J.Y.; Wang, Z.Y. Deferoxamine inhibits iron induced hippocampal tau phosphorylation in the Alzheimer transgenic mouse brain. *Neurochem. Int.* **2013**, *62*, 165–172. [CrossRef]
89. Iijima, K.; Ando, K.; Takeda, S.; Satoh, Y.; Seki, T.; Itohara, S.; Greengard, P.; Kirino, Y.; Nairn, A.C.; Suzuki, T. Neuron-specific phosphorylation of Alzheimer's beta-amyloid precursor protein by cyclin-dependent kinase 5. *J. Neurochem.* **2000**, *75*, 1085–1091. [CrossRef]
90. Johnson, G.L.; Lapadat, R. Mitogen-activated protein kinase pathways mediated by ERK, JNK, and p38 protein kinases. *Science* **2002**, *298*, 1911–1912. [CrossRef]
91. Kim, E.K.; Choi, E.J. Compromised MAPK signaling in human diseases: An update. *Arch. Toxicol.* **2015**, *89*, 867–882. [CrossRef]
92. McLaughlin, B.; Pal, S.; Tran, M.P.; Parsons, A.A.; Barone, F.C.; Erhardt, J.A.; Aizenman, E. p38 activation is required upstream of potassium current enhancement and caspase cleavage in thiol oxidant-induced neuronal apoptosis. *J. Neurosci.* **2001**, *21*, 3303–3311. [CrossRef] [PubMed]
93. Seo, S.R.; Chong, S.A.; Lee, S.I.; Sung, J.Y.; Ahn, Y.S.; Chung, K.C.; Seo, J.T. Zn^{2+}-induced ERK activation mediated by reactive oxygen species causes cell death in differentiated PC12 cells. *J. Neurochem.* **2001**, *78*, 600–610. [CrossRef] [PubMed]
94. Colombo, A.; Bastone, A.; Ploia, C.; Sclip, A.; Salmona, M.; Forloni, G.; Borsello, T. JNK regulates APP cleavage and degradation in a model of Alzheimer's disease. *Neurobiol. Dis.* **2009**, *33*, 518–525. [CrossRef]
95. Zhou, Q.; Wang, M.; Du, Y.; Zhang, W.; Bai, M.; Zhang, Z.; Li, Z.; Miao, J. Inhibition of c-Jun N-terminal kinase activation reverses Alzheimer disease phenotypes in APPswe/PS1dE9 mice. *Ann. Neurol.* **2015**, *77*, 637–654. [CrossRef] [PubMed]
96. Lee, J.K.; Kim, N.J. Recent Advances in the Inhibition of p38 MAPK as a Potential Strategy for the Treatment of Alzheimer's Disease. *Molecules* **2017**, *22*, 1287. [CrossRef] [PubMed]
97. Siano, G.; Caiazza, M.C.; Ollà, I.; Varisco, M.; Madaro, G.; Quercioli, V.; Calvello, M.; Cattaneo, A.; Di Primio, C. Identification of an ERK Inhibitor as a Therapeutic Drug Against Tau Aggregation in a New Cell-Based Assay. *Front. Cell. Neurosci.* **2019**, *13*, 386. [CrossRef]
98. Carlyle, B.C.; Nairn, A.C.; Wang, M.; Yang, Y.; Jin, L.E.; Simen, A.A.; Ramos, B.P.; Bordner, K.A.; Craft, G.E.; Davies, P.; et al. cAMP-PKA phosphorylation of tau confers risk for degeneration in aging association cortex. *Proc. Natl. Acad. Sci. USA* **2014**, *111*, 5036–5041. [CrossRef] [PubMed]
99. Paspalas, C.D.; Carlyle, B.C.; Leslie, S.; Preuss, T.M.; Crimins, J.L.; Huttner, A.J.; van Dyck, C.H.; Rosene, D.L.; Nairn, A.C.; Arnsten, A.F.T. The aged rhesus macaque manifests Braak stage III/IV Alzheimer's-like pathology. *Alzheimer's Dement.* **2018**, *14*, 680–691. [CrossRef]
100. Allouche-Fitoussi, D.; Breitbart, H. The Role of Zinc in Male Fertility. *Int. J. Mol. Sci.* **2020**, *21*, 7796. [CrossRef]
101. Park, K.H.; Choi, Y.; Yoon, D.S.; Lee, K.M.; Kim, D.; Lee, J.W. Zinc Promotes Osteoblast Differentiation in Human Mesenchymal Stem Cells Via Activation of the cAMP-PKA-CREB Signaling Pathway. *Stem Cells Dev.* **2018**, *27*, 1125–1135. [CrossRef]
102. Kang, J.H.; Toita, R.; Kim, C.W.; Katayama, Y. Protein kinase C (PKC) isozyme-specific substrates and their design. *Biotechnol. Adv.* **2012**, *30*, 1662–1672. [CrossRef] [PubMed]
103. Mochly-Rosen, D.; Das, K.; Grimes, K.V. Protein kinase C, an elusive therapeutic target? *Nat. Rev. Drug Discov.* **2012**, *11*, 937–957. [CrossRef]
104. Steinberg, S.F. Structural basis of protein kinase C isoform function. *Physiol. Rev.* **2008**, *88*, 1341–1378. [CrossRef] [PubMed]
105. Isagawa, T.; Mukai, H.; Oishi, K.; Taniguchi, T.; Hasegawa, H.; Kawamata, T.; Tanaka, C.; Ono, Y. Dual effects of PKNalpha and protein kinase C on phosphorylation of tau protein by glycogen synthase kinase-3beta. *Biochem. Biophys. Res. Commun.* **2000**, *273*, 209–212. [CrossRef] [PubMed]
106. Takashima, A. GSK-3 is essential in the pathogenesis of Alzheimer's disease. *J. Alzheimer's Dis.* **2006**, *9*, 309–317. [CrossRef] [PubMed]
107. Correas, I.; Díaz-Nido, J.; Avila, J. Microtubule-associated protein tau is phosphorylated by protein kinase C on its tubulin binding domain. *J. Biol. Chem.* **1992**, *267*, 15721–15728. [CrossRef]
108. Noh, K.M.; Koh, J.Y. Induction and activation by zinc of NADPH oxidase in cultured cortical neurons and astrocytes. *J. Neurosci.* **2000**, *20*, Rc111. [CrossRef]
109. Noh, K.M.; Kim, Y.H.; Koh, J.Y. Mediation by membrane protein kinase C of zinc-induced oxidative neuronal injury in mouse cortical cultures. *J. Neurochem.* **1999**, *72*, 1609–1616. [CrossRef]
110. Sajan, M.P.; Hansen, B.C.; Higgs, M.G.; Kahn, C.R.; Braun, U.; Leitges, M.; Park, C.R.; Diamond, D.M.; Farese, R.V. Atypical PKC, PKCλ/ι, activates β-secretase and increases Aβ(1-40/42) and phospho-tau in mouse brain and isolated neuronal cells, and may link hyperinsulinemia and other aPKC activators to development of pathological and memory abnormalities in Alzheimer's disease. *Neurobiol. Aging* **2018**, *61*, 225–237. [CrossRef]
111. Olsen, J.V.; Blagoev, B.; Gnad, F.; Macek, B.; Kumar, C.; Mortensen, P.; Mann, M. Global, in vivo, and site-specific phosphorylation dynamics in signaling networks. *Cell* **2006**, *127*, 635–648. [CrossRef]
112. Kamada, R.; Kudoh, F.; Ito, S.; Tani, I.; Janairo, J.I.B.; Omichinski, J.G.; Sakaguchi, K. Metal-dependent Ser/Thr protein phosphatase PPM family: Evolution, structures, diseases and inhibitors. *Pharmacol. Ther.* **2020**, *215*, 107622. [CrossRef] [PubMed]

113. Shi, Y. Serine/threonine phosphatases: Mechanism through structure. *Cell* **2009**, *139*, 468–484. [CrossRef] [PubMed]
114. Gong, C.X.; Singh, T.J.; Grundke-Iqbal, I.; Iqbal, K. Phosphoprotein phosphatase activities in Alzheimer disease brain. *J. Neurochem.* **1993**, *61*, 921–927. [CrossRef] [PubMed]
115. Liu, F.; Grundke-Iqbal, I.; Iqbal, K.; Gong, C.X. Contributions of protein phosphatases PP1, PP2A, PP2B and PP5 to the regulation of tau phosphorylation. *Eur. J. Neurosci.* **2005**, *22*, 1942–1950. [CrossRef]
116. Rahman, A.; Grundke-Iqbal, I.; Iqbal, K. Phosphothreonine-212 of Alzheimer abnormally hyperphosphorylated tau is a preferred substrate of protein phosphatase-1. *Neurochem. Res.* **2005**, *30*, 277–287. [CrossRef] [PubMed]
117. Kelker, M.S.; Page, R.; Peti, W. Crystal structures of protein phosphatase-1 bound to nodularin-R and tautomycin: A novel scaffold for structure-based drug design of serine/threonine phosphatase inhibitors. *J. Mol. Biol.* **2009**, *385*, 11–21. [CrossRef]
118. Zhang, A.J.; Bai, G.; Deans-Zirattu, S.; Browner, M.F.; Lee, E.Y. Expression of the catalytic subunit of phosphorylase phosphatase (protein phosphatase-1) in *Escherichia coli*. *J. Biol. Chem.* **1992**, *267*, 1484–1490. [CrossRef]
119. Chu, Y.; Lee, E.Y.; Schlender, K.K. Activation of protein phosphatase 1. Formation of a metalloenzyme. *J. Biol. Chem.* **1996**, *271*, 2574–2577. [CrossRef]
120. Singla, N.; Dhawan, D.K. Regulatory role of zinc during aluminium-induced altered carbohydrate metabolism in rat brain. *J. Neurosci. Res.* **2012**, *90*, 698–705. [CrossRef]
121. Gong, C.X.; Shaikh, S.; Wang, J.Z.; Zaidi, T.; Grundke-Iqbal, I.; Iqbal, K. Phosphatase activity toward abnormally phosphorylated tau: Decrease in Alzheimer disease brain. *J. Neurochem.* **1995**, *65*, 732–738. [CrossRef]
122. Sun, L.; Liu, S.Y.; Zhou, X.W.; Wang, X.C.; Liu, R.; Wang, Q.; Wang, J.Z. Inhibition of protein phosphatase 2A- and protein phosphatase 1-induced tau hyperphosphorylation and impairment of spatial memory retention in rats. *Neuroscience* **2003**, *118*, 1175–1182. [CrossRef]
123. Sontag, E.; Nunbhakdi-Craig, V.; Sontag, J.M.; Diaz-Arrastia, R.; Ogris, E.; Dayal, S.; Lentz, S.R.; Arning, E.; Bottiglieri, T. Protein phosphatase 2A methyltransferase links homocysteine metabolism with tau and amyloid precursor protein regulation. *J. Neurosci.* **2007**, *27*, 2751–2759. [CrossRef]
124. Ingebritsen, T.S.; Cohen, P. The protein phosphatases involved in cellular regulation. 1. Classification and substrate specificities. *Eur. J. Biochem.* **1983**, *132*, 255–261. [CrossRef]
125. Sangodkar, J.; Farrington, C.C.; McClinch, K.; Galsky, M.D.; Kastrinsky, D.B.; Narla, G. All roads lead to PP2A: Exploiting the therapeutic potential of this phosphatase. *FEBS J.* **2016**, *283*, 1004–1024. [CrossRef]
126. Nishito, Y.; Usui, H.; Tanabe, O.; Shimizu, M.; Takeda, M. Interconversion of Mn(2+)-dependent and -independent protein phosphatase 2A from human erythrocytes: Role of Zn(2+) and Fe(2+) in protein phosphatase 2A. *J. Biochem.* **1999**, *126*, 632–638. [CrossRef] [PubMed]
127. Cohen, P. The structure and regulation of protein phosphatases. *Annu. Rev. Biochem.* **1989**, *58*, 453–508. [CrossRef]
128. Janssens, V.; Goris, J. Protein phosphatase 2A: A highly regulated family of serine/threonine phosphatases implicated in cell growth and signalling. *Biochem. J.* **2001**, *353*, 417–439. [CrossRef]
129. Xiong, Y.; Luo, D.J.; Wang, X.L.; Qiu, M.; Yang, Y.; Yan, X.; Wang, J.Z.; Ye, Q.F.; Liu, R. Zinc binds to and directly inhibits protein phosphatase 2A in vitro. *Neurosci. Bull.* **2015**, *31*, 331–337. [CrossRef] [PubMed]
130. Chen, J.; Martin, B.L.; Brautigan, D.L. Regulation of protein serine-threonine phosphatase type-2A by tyrosine phosphorylation. *Science* **1992**, *257*, 1261–1264. [CrossRef]
131. Xiong, Y.; Jing, X.P.; Zhou, X.W.; Wang, X.L.; Yang, Y.; Sun, X.Y.; Qiu, M.; Cao, F.Y.; Lu, Y.M.; Liu, R.; et al. Zinc induces protein phosphatase 2A inactivation and tau hyperphosphorylation through Src dependent PP2A (tyrosine 307) phosphorylation. *Neurobiol. Aging* **2013**, *34*, 745–756. [CrossRef]
132. Ferreira, A.; Kincaid, R.; Kosik, K.S. Calcineurin is associated with the cytoskeleton of cultured neurons and has a role in the acquisition of polarity. *Mol. Biol. Cell* **1993**, *4*, 1225–1238. [CrossRef]
133. Gong, C.X.; Singh, T.J.; Grundke-Iqbal, I.; Iqbal, K. Alzheimer's disease abnormally phosphorylated tau is dephosphorylated by protein phosphatase-2B (calcineurin). *J. Neurochem.* **1994**, *62*, 803–806. [CrossRef]
134. Wei, Q.; Holzer, M.; Brueckner, M.K.; Liu, Y.; Arendt, T. Dephosphorylation of tau protein by calcineurin triturated into neural living cells. *Cell. Mol. Neurobiol.* **2002**, *22*, 13–24. [CrossRef]
135. Gratuze, M.; Noël, A.; Julien, C.; Cisbani, G.; Milot-Rousseau, P.; Morin, F.; Dickler, M.; Goupil, C.; Bezeau, F.; Poitras, I.; et al. Tau hyperphosphorylation and deregulation of calcineurin in mouse models of Huntington's disease. *Hum. Mol. Genet.* **2015**, *24*, 86–99. [CrossRef] [PubMed]
136. Takahashi, K.; Akaishi, E.; Abe, Y.; Ishikawa, R.; Tanaka, S.; Hosaka, K.; Kubohara, Y. Zinc inhibits calcineurin activity in vitro by competing with nickel. *Biochem. Biophys. Res. Commun.* **2003**, *307*, 64–68. [CrossRef]
137. Merten, K.E.; Jiang, Y.; Kang, Y.J. Zinc inhibits doxorubicin-activated calcineurin signal transduction pathway in H9c2 embryonic rat cardiac cells. *Exp. Biol. Med.* **2007**, *232*, 682–689.
138. Park, K.H.; Park, B.; Yoon, D.S.; Kwon, S.H.; Shin, D.M.; Lee, J.W.; Lee, H.G.; Shim, J.H.; Park, J.H.; Lee, J.M. Zinc inhibits osteoclast differentiation by suppression of Ca2+-Calcineurin-NFATc1 signaling pathway. *Cell Commun. Signal.* **2013**, *11*, 74. [CrossRef]
139. Gong, C.X.; Liu, F.; Wu, G.; Rossie, S.; Wegiel, J.; Li, L.; Grundke-Iqbal, I.; Iqbal, K. Dephosphorylation of microtubule-associated protein tau by protein phosphatase 5. *J. Neurochem.* **2004**, *88*, 298–310. [CrossRef] [PubMed]

140. Leroy, K.; Yilmaz, Z.; Brion, J.P. Increased level of active GSK-3beta in Alzheimer's disease and accumulation in argyrophilic grains and in neurones at different stages of neurofibrillary degeneration. *Neuropathol. Appl. Neurobiol.* **2007**, *33*, 43–55. [CrossRef]
141. Lee, K.Y.; Clark, A.W.; Rosales, J.L.; Chapman, K.; Fung, T.; Johnston, R.N. Elevated neuronal Cdc2-like kinase activity in the Alzheimer disease brain. *Neurosci. Res.* **1999**, *34*, 21–29. [CrossRef]
142. Nguyen, M.D.; Larivière, R.C.; Julien, J.P. Deregulation of Cdk5 in a mouse model of ALS: Toxicity alleviated by perikaryal neurofilament inclusions. *Neuron* **2001**, *30*, 135–147. [CrossRef]
143. Smith, P.D.; Crocker, S.J.; Jackson-Lewis, V.; Jordan-Sciutto, K.L.; Hayley, S.; Mount, M.P.; O'Hare, M.J.; Callaghan, S.; Slack, R.S.; Przedborski, S.; et al. Cyclin-dependent kinase 5 is a mediator of dopaminergic neuron loss in a mouse model of Parkinson's disease. *Proc. Natl. Acad. Sci. USA* **2003**, *100*, 13650–13655. [CrossRef] [PubMed]
144. Ferrer, I.; Blanco, R.; Carmona, M.; Ribera, R.; Goutan, E.; Puig, B.; Rey, M.J.; Cardozo, A.; Viñals, F.; Ribalta, T. Phosphorylated map kinase (ERK1, ERK2) expression is associated with early tau deposition in neurones and glial cells, but not with increased nuclear DNA vulnerability and cell death, in Alzheimer disease, Pick's disease, progressive supranuclear palsy and corticobasal degeneration. *Brain Pathol.* **2001**, *11*, 144–158. [CrossRef]
145. Kulich, S.M.; Chu, C.T. Sustained extracellular signal-regulated kinase activation by 6-hydroxydopamine: Implications for Parkinson's disease. *J. Neurochem.* **2001**, *77*, 1058–1066. [CrossRef] [PubMed]
146. Perry, G.; Roder, H.; Nunomura, A.; Takeda, A.; Friedlich, A.L.; Zhu, X.; Raina, A.K.; Holbrook, N.; Siedlak, S.L.; Harris, P.L.; et al. Activation of neuronal extracellular receptor kinase (ERK) in Alzheimer disease links oxidative stress to abnormal phosphorylation. *Neuroreport* **1999**, *10*, 2411–2415. [CrossRef]
147. Zhu, X.; Castellani, R.J.; Takeda, A.; Nunomura, A.; Atwood, C.S.; Perry, G.; Smith, M.A. Differential activation of neuronal ERK, JNK/SAPK and p38 in Alzheimer disease: The 'two hit' hypothesis. *Mech. Ageing Dev.* **2001**, *123*, 39–46. [CrossRef]
148. Atzori, C.; Ghetti, B.; Piva, R.; Srinivasan, A.N.; Zolo, P.; Delisle, M.B.; Mirra, S.S.; Migheli, A. Activation of the JNK/p38 pathway occurs in diseases characterized by tau protein pathology and is related to tau phosphorylation but not to apoptosis. *J. Neuropathol. Exp. Neurol.* **2001**, *60*, 1190–1197. [CrossRef]
149. Migheli, A.; Piva, R.; Atzori, C.; Troost, D.; Schiffer, D. c-Jun, JNK/SAPK kinases and transcription factor NF-kappa B are selectively activated in astrocytes, but not motor neurons, in amyotrophic lateral sclerosis. *J. Neuropathol. Exp. Neurol.* **1997**, *56*, 1314–1322. [CrossRef]
150. Zhu, X.; Raina, A.K.; Rottkamp, C.A.; Aliev, G.; Perry, G.; Boux, H.; Smith, M.A. Activation and redistribution of c-jun N-terminal kinase/stress activated protein kinase in degenerating neurons in Alzheimer's disease. *J. Neurochem.* **2001**, *76*, 435–441. [CrossRef]
151. Dagda, R.K.; Das Banerjee, T. Role of protein kinase A in regulating mitochondrial function and neuronal development: Implications to neurodegenerative diseases. *Rev. Neurosci.* **2015**, *26*, 359–370. [CrossRef]
152. Howells, D.W.; Porritt, M.J.; Wong, J.Y.; Batchelor, P.E.; Kalnins, R.; Hughes, A.J.; Donnan, G.A. Reduced BDNF mRNA expression in the Parkinson's disease substantia nigra. *Exp. Neurol.* **2000**, *166*, 127–135. [CrossRef] [PubMed]
153. Govoni, S.; Bergamaschi, S.; Racchi, M.; Battaini, F.; Binetti, G.; Bianchetti, A.; Trabucchi, M. Cytosol protein kinase C downregulation in fibroblasts from Alzheimer's disease patients. *Neurology* **1993**, *43*, 2581–2586. [CrossRef]
154. Wang, H.Y.; Pisano, M.R.; Friedman, E. Attenuated protein kinase C activity and translocation in Alzheimer's disease brain. *Neurobiol. Aging* **1994**, *15*, 293–298. [CrossRef]
155. Pei, J.J.; Gong, C.X.; Iqbal, K.; Grundke-Iqbal, I.; Wu, Q.L.; Winblad, B.; Cowburn, R.F. Subcellular distribution of protein phosphatases and abnormally phosphorylated tau in the temporal cortex from Alzheimer's disease and control brains. *J. Neural Transm.* **1998**, *105*, 69–83. [CrossRef]
156. Liu, F.; Iqbal, K.; Grundke-Iqbal, I.; Rossie, S.; Gong, C.X. Dephosphorylation of tau by protein phosphatase 5: Impairment in Alzheimer's disease. *J. Biol. Chem.* **2005**, *280*, 1790–1796. [CrossRef] [PubMed]

MDPI AG
Grosspeteranlage 5
4052 Basel
Switzerland
Tel.: +41 61 683 77 34

Biomolecules Editorial Office
E-mail: biomolecules@mdpi.com
www.mdpi.com/journal/biomolecules

www.ingramcontent.com/pod-product-compliance
Lightning Source LLC
Chambersburg PA
CBHW061443210926
51850CB00008B/1592